# Experimental Testing and Constitutive Modelling of Pavement Materials

# Experimental Testing and Constitutive Modelling of Pavement Materials

Editors

Xueyan Liu
Linbing Wang
Zhanping You
Yuqing Zhang
Changhong Zhou

Basel • Beijing • Wuhan • Barcelona • Belgrade • Novi Sad • Cluj • Manchester

*Editors*

Xueyan Liu
Delft University of Technology
Delft
The Netherlands

Linbing Wang
University of Georgia
Athens
GA
USA

Zhanping You
Michigan Technological University
Houghton
MI
USA

Yuqing Zhang
Aston University
Birmingham
UK

Changhong Zhou
Guilin University of Electronic Technology
Guilin
China

*Editorial Office*
MDPI
St. Alban-Anlage 66
4052 Basel, Switzerland

This is a reprint of articles from the Special Issue published online in the open access journal *Materials* (ISSN 1996-1944) (available at: https://www.mdpi.com/journal/materials/special_issues/pavement_testing_modelling).

For citation purposes, cite each article independently as indicated on the article page online and as indicated below:

Lastname, A.A.; Lastname, B.B. Article Title. *Journal Name* **Year**, *Volume Number*, Page Range.

**ISBN 978-3-0365-8474-4 (Hbk)**
**ISBN 978-3-0365-8475-1 (PDF)**
doi.org/10.3390/books978-3-0365-8475-1

© 2023 by the authors. Articles in this book are Open Access and distributed under the Creative Commons Attribution (CC BY) license. The book as a whole is distributed by MDPI under the terms and conditions of the Creative Commons Attribution-NonCommercial-NoDerivs (CC BY-NC-ND) license.

# Contents

Xueyan Liu, Yuqing Zhang, Zhanping You, Linbing Wang and Changhong Zhou
Experimental Testing and Constitutive Modelling of Pavement Materials
Reprinted from: *Materials* 2023, 16, 4186, doi:10.3390/ma16114186 . . . . . . . . . . . . . . . . . . 1

Jie Yuan, Wenhao Li, Yuening Li, Lukuan Ma and Jiake Zhang
Fatigue Models for Airfield Concrete Pavement: Literature Review and Discussion
Reprinted from: *Materials* 2021, 14, 6579, doi:10.3390/ma14216579 . . . . . . . . . . . . . . . . . . 5

Bangwei Wu, Chufan Luo, Zhaohui Pei, Chuangchuang Chen, Ji Xia and Peng Xiao
Evaluation of the Aging of Styrene-Butadiene-Styrene Modified Asphalt Binder with Different
Polymer Additives
Reprinted from: *Materials* 2021, 14, 5715, doi:10.3390/ma14195715 . . . . . . . . . . . . . . . . . . 25

Chao Peng, Hanneng Yang, Zhanping You, Hongchao Ma, Fang Xu, Lingyun You, et al.
Investigation of Adhesion Performance of Wax Based Warm Mix Asphalt with Molecular
Dynamics Simulation
Reprinted from: *Materials* 2022, 15, 5930, doi:10.3390/ma15175930 . . . . . . . . . . . . . . . . . . 39

Jingyao Yang, Gang Xu, Peipei Kong and Xianhua Chen
Characterization of Desulfurized Crumb Rubber/Styrene–Butadiene–Styrene Composite
Modified Asphalt Based on Rheological Properties
Reprinted from: *Materials* 2021, 14, 3780, doi:10.3390/ma14143780 . . . . . . . . . . . . . . . . . . 57

Hui Xu, Yiren Sun, Jingyun Chen, Jiyang Li, Bowen Yu, Guoqing Qiu, et al.
Investigation into Rheological Behavior of Warm-Mix Recycled Asphalt Binders with High
Percentages of RAP Binder
Reprinted from: *Materials* 2023, 16, 1599, doi:10.3390/ma16041599 . . . . . . . . . . . . . . . . . . 71

Tingting Xie, Kang Zhao and Linbing Wang
Reinforcement Effect of Different Fibers on Asphalt Mastic
Reprinted from: *Materials* 2022, 15, 8304, doi:10.3390/ma15238304 . . . . . . . . . . . . . . . . . . 99

Jingsheng Pan, Hua Zhao, Yong Wang and Gang Liu
The Influence of Aeolian Sand on the Anti-Skid Characteristics of Asphalt Pavement
Reprinted from: *Materials* 2021, 14, 5523, doi:10.3390/ma14195523 . . . . . . . . . . . . . . . . . . 119

Miao Yu, Yao Kong, Zhanping You, Jue Li, Liming Yang and Lingyun Kong
Anti-Skid Characteristics of Asphalt Pavement Based on Partial Tire Aquaplane Conditions
Reprinted from: *Materials* 2022, 15, 4976, doi:10.3390/ma15144976 . . . . . . . . . . . . . . . . . . 133

Kang Zhao, Hailu Yang, Wentao Wang and Linbing Wang
Characterization of Rutting Damage by the Change of Air-Void Characteristics in the Asphalt
Mixture Based on Two-Dimensional Image Analysis
Reprinted from: *Materials* 2022, 15, 7190, doi:10.3390/ma15207190 . . . . . . . . . . . . . . . . . . 155

Lei Zhang, Guiping Zheng, Kai Zhang, Yongfeng Wang, Changming Chen, Liting Zhao, et al.
Study on the Extraction of CT Images with Non-Uniform Illumination for the Microstructure of
Asphalt Mixture
Reprinted from: *Materials* 2022, 15, 7364, doi:10.3390/ma15207364 . . . . . . . . . . . . . . . . . . 175

Chao Wang, Hui Xu, Yan Zhang, Yiren Sun, Weiying Wang and Jingyun Chen
Improved Procedure for the 3D Reconstruction of Asphalt Concrete Mesostructures
Considering the Similarity of Aggregate Phase Geometry between Adjacent CT Slices
Reprinted from: *Materials* 2023, 16, 234, doi:10.3390/ma16010234 . . . . . . . . . . . . . . . . . . 191

**Xiaodong Zhou, Dongzhao Jin, Dongdong Ge, Siyu Chen and Zhanping You**
Identify the Micro-Parameters for Optimized Discrete Element Models of Granular Materials in Two Dimensions Using Hexagonal Close-Packed Structures
Reprinted from: *Materials* **2023**, *16*, 3073, doi:10.3390/ma16083073 . . . . . . . . . . . . . . . . . . **213**

**Zhoujing Ye, Yanxia Cai, Chang Liu, Kaiji Lu, Dylan G. Ildefonzo and Linbing Wang**
Optimization of Embedded Sensor Packaging Used in Rollpave Pavement Based on Test and Simulation
Reprinted from: *Materials* **2022**, *15*, 2283, doi:10.3390/ma15062283 . . . . . . . . . . . . . . . . . . **231**

**Bangwei Wu, Weijie Meng, Ji Xia and Peng Xiao**
Influence of Basalt Fibers on the Crack Resistance of Asphalt Mixtures and Mechanism Analysis
Reprinted from: *Materials* **2022**, *15*, 744, doi:10.3390/ma15030744 . . . . . . . . . . . . . . . . . . . . **251**

**Bo Tan, Tao Yang, Heying Qin and Qi Liu**
Laboratory Study on the Stability of Large-Size Graded Crushed Stone under Cyclic Rotating Axial Compression
Reprinted from: *Materials* **2021**, *14*, 1584, doi:10.3390/ma14071584 . . . . . . . . . . . . . . . . . . **265**

**Weidong Chen, Bing Hui and Ali Rahman**
Interlayer Shear Characteristics of Bridge Deck Pavement through Experimental and Numerical Analysis
Reprinted from: *Materials* **2022**, *15*, 7001, doi:10.3390/ma15197001 . . . . . . . . . . . . . . . . . . **287**

**Shi Xu, Xueyan Liu, Amir Tabaković and Erik Schlangen**
Experimental Investigation of the Performance of a Hybrid Self-Healing System in Porous Asphalt under Fatigue Loadings
Reprinted from: *Materials* **2021**, *14*, 3415, doi:10.3390/ma14123415 . . . . . . . . . . . . . . . . . . **303**

**Jun Li, Mingliang Li and Hao Wu**
Key Performance Analysis of Emulsified Asphalt Cold Recycling Mixtures of the Middle Layer of Pavement Structure
Reprinted from: *Materials* **2023**, *16*, 1613, doi:10.3390/ma16041613 . . . . . . . . . . . . . . . . . . **323**

**Yanxia Cai, Zhi Lin, Jingrui Zhang, Kaiji Lu, Linbing Wang, Yue Zhao, et al.**
Dosage Effect of Wet-Process Tuff Silt Powder as an Alternative Material of Sand on the Performance of Reactive Powder Concrete
Reprinted from: *Materials* **2022**, *15*, 3930, doi:10.3390/ma15113930 . . . . . . . . . . . . . . . . . . **339**

**Yan Zhang and Yiren Sun**
Fast-Acquiring High-Quality Prony Series Parameters of Asphalt Concrete through Viscoelastic Continuous Spectral Models
Reprinted from: *Materials* **2022**, *15*, 716, doi:10.3390/ma15030716 . . . . . . . . . . . . . . . . . . . . **357**

**Qipeng Zhang, Xingyu Gu, Zilu Yu, Jia Liang and Qiao Dong**
Viscoelastic Damage Characteristics of Asphalt Mixtures Using Fractional Rheology
Reprinted from: *Materials* **2021**, *14*, 5892, doi:10.3390/ma14195892 . . . . . . . . . . . . . . . . . . **381**

**Xunli Jiang, Zhiyi Huang and Xue Luo**
An Improved Mechanistic-Empirical Creep Model for Unsaturated Soft and Stabilized Soils
Reprinted from: *Materials* **2021**, *14*, 4146, doi:10.3390/ma14154146 . . . . . . . . . . . . . . . . . . **401**

*Editorial*

# Experimental Testing and Constitutive Modelling of Pavement Materials

Xueyan Liu [1,*], Yuqing Zhang [2], Zhanping You [3], Linbing Wang [4] and Changhong Zhou [5]

1. Section of Pavement Engineering, Department of Engineering Structures, Delft University of Technology, 2628 CN Delft, The Netherlands
2. Aston Institute of Materials Research (AIMR), Aston University, Birmingham B4 7ET, UK; y.zhang10@aston.ac.uk
3. Department of Civil and Environmental Engineering, Michigan Technological University, 1400 Townsend Drive, Houghton, MI 49931, USA; zyou@mtu.edu
4. School of Environmental, Civil, Agricultural and Mechanical Engineering, 1254 STEM Research Building II, University of Georgia, Athens, GA 30602, USA; linbing.wang@uga.edu
5. School of Architecture and Transportation Engineering, Guilin University of Electronic Technology, Guilin 541004, China; czhou@guet.edu.cn
* Correspondence: x.liu@tudelft.nl

Pavement materials such as asphalt mixtures, granular aggregates and soils exhibit complex material properties and engineering performance under external loading and environmental conditions. For instance, the asphalt mixture shows highly nonlinear viscoelastic and viscoplastic properties at high temperatures, and it presents fatigue cracking damage and fracture properties at intermediate or low temperatures. The granular aggregate materials show an obvious anisotropic and stress-dependent resilient modulus. Their permanent deformation is fundamentally determined by stress levels, moisture and the number of load cycles. Constitutive models based on mechanics theories have been the kernel of performance prediction of pavement infrastructures and materials. They lay down a solid foundation for material selection, design and pavement structural evaluation, and maintenance decisions. Advances in mechanics modeling and the associated experimental testing for pavement infrastructures and construction materials are emerging constantly, such as nonlinear viscoelasticity, viscoplasticity, fracture and damage mechanics models. Meanwhile, various numerical modeling technologies are being developed and implemented to solve the multiscale and multi-physical equations and models for the pavement structures and materials. Examples include finite element, discrete element and micromechanics or molecular dynamics simulations at different dimensions and scales. These are being applied to both existing traditional pavement materials and novel or emerging materials such as recycled, modified or alternative materials. All the aforementioned advances have been leading to a large number of new studies and discoveries in the relevant areas.

This Special Issue provides a unique platform to collect and present these novel studies and new discoveries in the areas of mechanics, numerical modeling and the experimental testing of pavement infrastructures and materials. It includes the studies of various pavement materials such as asphalt concretes, granular materials, soils, recycled materials and additives. In addition, different testing and modeling technologies including discrete element modelling (DEM), computed tomography (CT) and molecular dynamics (MD) simulation are included.

A review paper summarizes the fatigue models of cement concrete pavements based on different testing scales [1]. Recommendations in terms of the data source, stress calculation method and regression analysis process were proposed for the improvement of current fatigue models for the cement concrete pavements.

Four papers focus on the characterization of different asphalt binders (e.g., polymer-modified, warm mix recycled and wax-modified binders) via experiments and molecular

Citation: Liu, X.; Zhang, Y.; You, Z.; Wang, L.; Zhou, C. Experimental Testing and Constitutive Modelling of Pavement Materials. *Materials* 2023, *16*, 4186. https://doi.org/10.3390/ma16114186

Received: 28 April 2023
Accepted: 3 June 2023
Published: 5 June 2023

Copyright: © 2023 by the authors. Licensee MDPI, Basel, Switzerland. This article is an open access article distributed under the terms and conditions of the Creative Commons Attribution (CC BY) license (https://creativecommons.org/licenses/by/4.0/).

dynamics simulations [2–5]. The Dynamic Shear Rheometer (DSR)- and Bending Beam Rheometer (BBR)-based rheological tests are the mainstream methods to evaluate the high-, intermediate- and low-temperature performance of the asphalt binders. Fourier Transform Infrared (FTIR) Spectroscopy is widely used in terms of the chemical characterization of the functional groups in the asphalt binders. The Multiple Steep Creep and Recovery (MSCR) test and Linear Amplitude Sweep (LAS) test are used to investigate the effects of lignin and carbon fiber on the physical and mechanical properties' changes in asphalt mastics [6].

Two papers focus on the evaluation of the asphalt pavement skid resistance. One investigated the effect of sand accumulation on the skid resistance of asphalt pavements using the British Pendulum Number (BPN) test on two types of asphalt mixtures [7]. Another one presented a finite element model of radial tire–asphalt pavement interaction to investigate the pavement skid resistance under partial tire aquaplane conditions [8]. The results showed that the vertical contact force between the tire and pavement is greatly reduced because of the partial aquaplane state.

Three papers utilize digital image processing (DIP) technology for either the performance test or meso-structure reconstruction of asphalt mixtures. The relationship between the rutting damage and the air void change was investigated via a 2D image technology [9]. An adaptive image processing method for CT images of asphalt mixtures was proposed to improve the accuracy of the meso-structure reconstruction of asphalt mixtures [10]. An improved procedure of the meso-structure reconstruction of asphalt mixtures considering the similarity of aggregate phase geometry was proposed, and the results indicated that the proposed approach can maintain the 3D spatial distribution features and contour characteristics of asphalt mixtures' mesostructured [11]. One paper used the hexagonal close-packed (HCP) structure to establish the discrete model of asphalt mixtures for better simulating the shear failure [12]. The embedded sensor packaging of the rollpave pavements was optimized via experimental and numerical investigations [13]. This paper improved the compatibility of the embedded sensors and road materials in a prefabricated pavement structure, so the real-time in situ monitoring of the pavement response will be more accurate.

Six papers used laboratory tests and numerical simulations to assess the performance of different road materials and structures, including emulsified cold recycling asphalt mixtures, self-healing asphalt binder, reactive powder concrete and bridge deck pavement. The findings provide in-depth understandings in terms of various road materials key performance [14–19].

An efficient approach to obtain the parameters of the Prony series was proposed for the asphalt mixtures [20]. This method can simultaneously determine the retardation and relaxation spectra, which is more effective than the current approach. A fractional viscoelastic and damage constitutive relation of asphalt mixtures was proposed to characterize the three-stage creep process [21]. The model prediction results agreed well with the laboratory uniaxial compressive creep tests with different stress levels and temperatures. An improved mechanistic–empirical creep model considering the stress dependence and moisture sensitivity was proposed for the unsaturated soft and stabilized soils [22]. This developed model can predict the soil creep deformation under arbitrary water content and stress levels.

**Conflicts of Interest:** The authors declare no conflict of interest.

# References

1. Yuan, J.; Li, W.; Li, Y.; Ma, L.; Zhang, J. Fatigue Models for Airfield Concrete Pavement: Literature Review and Discussion. *Materials* **2021**, *14*, 6579. [CrossRef] [PubMed]
2. Wu, B.; Luo, C.; Pei, Z.; Chen, C.; Xia, J.; Xiao, P. Evaluation of the Aging of Styrene-Butadiene-Styrene Modified Asphalt Binder with Different Polymer Additives. *Materials* **2021**, *14*, 5715. [CrossRef] [PubMed]
3. Peng, C.; Yang, H.; You, Z.; Ma, H.; Xu, F.; You, L.; Diab, A.; Lu, L.; Hu, Y.; Liu, Y.; et al. Investigation of Adhesion Performance of Wax Based Warm Mix Asphalt with Molecular Dynamics Simulation. *Materials* **2022**, *15*, 5930. [CrossRef] [PubMed]

4. Yang, J.; Xu, G.; Kong, P.; Chen, X. Characterization of Desulfurized Crumb Rubber/Styrene–Butadiene–Styrene Composite Modified Asphalt Based on Rheological Properties. *Materials* **2021**, *14*, 3780. [CrossRef] [PubMed]
5. Xu, H.; Sun, Y.; Chen, J.; Li, J.; Yu, B.; Qiu, G.; Zhang, Y.; Xu, B. Investigation into Rheological Behavior of Warm-Mix Recycled Asphalt Binders with High Percentages of RAP Binder. *Materials* **2023**, *16*, 1599. [CrossRef]
6. Xie, T.; Zhao, K.; Wang, L. Reinforcement Effect of Different Fibers on Asphalt Mastic. *Materials* **2022**, *15*, 8304. [CrossRef]
7. Pan, J.; Zhao, H.; Wang, Y.; Liu, G. The Influence of Aeolian Sand on the Anti-Skid Characteristics of Asphalt Pavement. *Materials* **2021**, *14*, 5523. [CrossRef]
8. Yu, M.; Kong, Y.; You, Z.; Li, J.; Yang, L.; Kong, L. Anti-Skid Characteristics of Asphalt Pavement Based on Partial Tire Aquaplane Conditions. *Materials* **2022**, *15*, 4976. [CrossRef]
9. Zhao, K.; Yang, H.; Wang, W.; Wang, L. Characterization of Rutting Damage by the Change of Air-Void Characteristics in the Asphalt Mixture Based on Two-Dimensional Image Analysis. *Materials* **2022**, *15*, 7190. [CrossRef]
10. Zhang, L.; Zheng, G.; Zhang, K.; Wang, Y.; Chen, C.; Zhao, L.; Xu, J.; Liu, X.; Wang, L.; Tan, Y.; et al. Study on the Extraction of CT Images with Non-Uniform Illumination for the Microstructure of Asphalt Mixture. *Materials* **2022**, *15*, 7364. [CrossRef]
11. Wang, C.; Xu, H.; Zhang, Y.; Sun, Y.; Wang, W.; Chen, J. Improved Procedure for the 3D Reconstruction of Asphalt Concrete Mesostructures Considering the Similarity of Aggregate Phase Geometry between Adjacent CT Slices. *Materials* **2023**, *16*, 234. [CrossRef] [PubMed]
12. Zhou, X.; Jin, D.; Ge, D.; Chen, S.; You, Z. Identify the Micro-Parameters for Optimized Discrete Element Models of Granular Materials in Two Dimensions Using Hexagonal Close-Packed Structures. *Materials* **2023**, *16*, 3073. [CrossRef] [PubMed]
13. Ye, Z.; Cai, Y.; Liu, C.; Lu, K.; Ildefonzo, D.G.; Wang, L. Optimization of Embedded Sensor Packaging Used in Rollpave Pavement Based on Test and Simulation. *Materials* **2022**, *15*, 2283. [CrossRef] [PubMed]
14. Wu, B.; Meng, W.; Xia, J.; Xiao, P. Influence of Basalt Fibers on the Crack Resistance of Asphalt Mixtures and Mechanism Analysis. *Materials* **2022**, *15*, 744. [CrossRef]
15. Tan, B.; Yang, T.; Qin, H.; Liu, Q. Laboratory Study on the Stability of Large-size Graded Crushed Stone under Cyclic Rotating Axial Compression. *Materials* **2021**, *14*, 1584. [CrossRef]
16. Chen, W.; Hui, B.; Rahman, A. Interlayer Shear Characteristics of Bridge Deck Pavement through Experimental and Numerical Analysis. *Materials* **2022**, *15*, 7001. [CrossRef]
17. Xu, S.; Liu, X.; Tabaković, A.; Schlangen, E. Experimental Investigation of the Performance of a Hybrid Self-Healing System in Porous Asphalt under Fatigue Loadings. *Materials* **2021**, *14*, 3415. [CrossRef]
18. Li, J.; Li, M.; Wu, H. Key Performance Analysis of Emulsified Asphalt Cold Recycling Mixtures of the Middle Layer of Pavement Structure. *Materials* **2023**, *16*, 1613. [CrossRef]
19. Cai, Y.; Lin, Z.; Zhang, J.; Lu, K.; Wang, L.; Zhao, Y.; Huang, Q. Dosage Effect of Wet-Process Tuff Silt Powder as an Alternative Material of Sand on the Performance of Reactive Powder Concrete. *Materials* **2022**, *15*, 3930. [CrossRef]
20. Zhang, Y.; Sun, Y. Fast-Acquiring High-Quality Prony Series Parameters of Asphalt Concrete through Viscoelastic Continuous Spectral Models. *Materials* **2022**, *15*, 716. [CrossRef]
21. Zhang, Q.; Gu, X.; Yu, Z.; Liang, J.; Dong, Q. Viscoelastic Damage Characteristics of Asphalt Mixtures Using Fractional Rheology. *Materials* **2021**, *14*, 5892. [CrossRef] [PubMed]
22. Jiang, X.; Huang, Z.; Luo, X. An Improved Mechanistic-Empirical Creep Model for Unsaturated Soft and Stabilized Soils. *Materials* **2021**, *14*, 4146. [CrossRef] [PubMed]

**Disclaimer/Publisher's Note:** The statements, opinions and data contained in all publications are solely those of the individual author(s) and contributor(s) and not of MDPI and/or the editor(s). MDPI and/or the editor(s) disclaim responsibility for any injury to people or property resulting from any ideas, methods, instructions or products referred to in the content.

*Review*

# Fatigue Models for Airfield Concrete Pavement: Literature Review and Discussion

Jie Yuan [1,2], Wenhao Li [1], Yuening Li [3], Lukuan Ma [1,2,*] and Jiake Zhang [1,2,*]

1. Key Laboratory of Road and Traffic Engineering of the Ministry of Education, Tongji University, Shanghai 201804, China; yuanjie@tongji.edu.cn (J.Y.); li_wenhao@tongji.edu.cn (W.L.)
2. Key Laboratory of Infrastructure Durability and Operation Safety in Airfield of CAAC, Tongji University, Shanghai 201804, China
3. School of Civil Engineering, University of Sydney, Sydney, NSW 2006, Australia; winnieli321@163.com
* Correspondence: malukuan5071@163.com (L.M.); zhjiake@tongji.edu.cn (J.Z.)

**Abstract:** The fatigue model plays an important role in the mechanistic–empirical design procedure of airfield pavement. As for cement concrete pavement, the fatigue model represents the relationship between the stress and the number of load repetitions. To further understand the fatigue model, a literature review was performed in this paper along with the discussion. In this paper, the developed fatigue models available now were classified as the full-scale testing-based fatigue model and the concrete beam testing-based fatigue model, according to the data source. Then, the regression analysis process and stress calculation method of each fatigue model were summarized. Besides, the fatigue model proposed by the Federal Aviation Administration (FAA) was compared with the fatigue model of the Civil Aviation Administration of China (CAAC). The design thicknesses using the two models were obtained based on the finite element analysis. The results show that the designed slab using the fatigue model of FAA is thicker than that of CAAC, meaning that the fatigue model of FAA is comparatively conservative. Moreover, it can be concluded that the differences in the slab thickness become more significant with the increase in the wheel load and the foundation strength. Finally, the recommendation was proposed to refine the fatigue model in the future study from three aspects: data source, stress calculation method, and regression analysis process.

**Keywords:** airfield; concrete pavement; fatigue model; slab thickness; improvement method

## 1. Introduction

Cement concrete pavement is a common structural type in the airfield. Under mechanical loading, the concrete slab often experiences structural damage while the stress is far below the ultimate strength of the concrete slab. This kind of damage is the fatigue cracking caused by repeated loading. The development of fatigue cracking deteriorates the pavement performance, which has a detrimental influence on the service life of the pavement structure. Therefore, it is necessary to ensure that the pavement structure has sufficient thickness to resist fatigue cracking. In the mechanistic–empirical design procedure of airfield pavement, it is essential to calculate the ultimate load repetitions to the fatigue failure of concrete slabs. The basis of the calculation is the fatigue model that describes the relationship between the stress of the concrete slab induced by load and the number of load repetitions [1–4]. In practice, the fatigue model for airfield pavements plays an important role, not only in the design of pavement structure, but also in the evaluation of the remaining service life for in-service pavements [5].

The fatigue mechanism of concrete pavement is complicated, because the pavement performances are affected by the pavement structure, the surrounding environment, the wheel loading, etc. [6,7]. The fatigue mechanism reflected by the mechanistic model of the concrete structure is different from that of the pavement structure in service. Thus, the fatigue models for airfield pavements are commonly developed based on the regression analysis of the fatigue test data. Accordingly, the fatigue models are generally divided into two types: one is proposed based on the on-site full-scale test data, and the other is developed based on the laboratory concrete beam test data. Due to the difference in test data and regression analysis processes, the parameters of the fatigue models are usually different.

As for airfield concrete pavement, the fatigue model is associated with the stress calculation theory and critical stress location [8]. In the design of airfield concrete pavement, it is common that the designed thicknesses are obviously different due to different fatigue models. It is difficult for the designers to make a trade-off between the reasonable structural design and the economy [9]. Therefore, the development process and critical mechanism of the fatigue model for airfield concrete pavements are reviewed in this paper. Subsequently, the thickness differences based on typical fatigue models in the current airfield concrete pavement design method are analyzed. Moreover, this paper proposes how to improve the fatigue model of airfield concrete pavements in the future.

## 2. Literature Review of Fatigue Models of Airfield Concrete Pavements

### 2.1. Full-Scale Testing-Based Fatigue Models

The early fatigue model was proposed by the USA Army Corps of Engineers (COE). In the 1940s, COE proposed the fatigue model for airfield concrete pavement based on the full-scale test data at Lockbourne and the Westergaard edge stress theory [10,11]. The pavement structure information and traffic loading for Lockbourne No.1 test sections are shown in Figure 1 [12]. COE assumed that the concrete slab can withstand 5000 coverages to satisfy the service life [13]. The fatigue failure of concrete slab was defined as 50 percent of the slabs cracking. The fatigue model is shown as follows [14]:

$$DF = \frac{MR}{0.75\sigma_e} = 1.3 \tag{1}$$

where:

$DF$ is the design factor;
$MR$ is the concrete modulus of rupture, which is equal to the flexural strength of concrete;
$\sigma_e$ is the edge stress calculated by the Westergaard stress theory;
0.75 is a stress reduction coefficient.

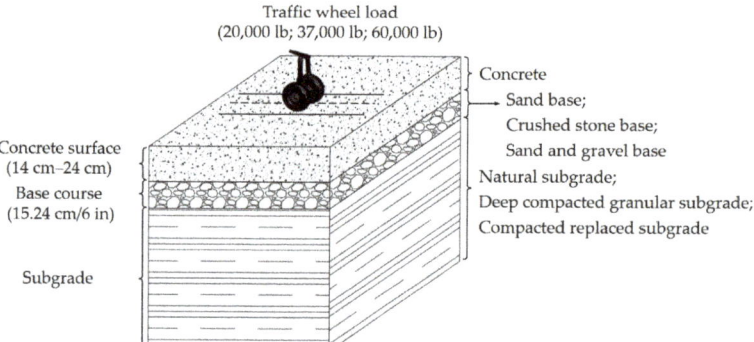

**Figure 1.** The structure and loading of the full-scale test at Lockbourne.

In the fatigue model of COE, the design factor (DF) was defined as the ratio of the concrete modulus of rupture to the stress at the edge of the slab. The DF of 1.3 was determined to consider factors affecting fatigue failure, such as the effect of loading, thermal curling, and the support of the base layer and the foundation [15]. The service life of the concrete slab can be obtained when the design factor is greater than 1.3.

The fatigue model of COE indicated that the calculated stress at the bottom of the concrete slab was the maximum stress multiplied by 0.75, which considers the effect of the joint load transfer. The effect of joint load transfer on stress reduction proposed by COE had been adopted by the subsequent researches. Although the formula and parameters of the fatigue model proposed by COE provided guidance for the follow-up research, it was oversimplified for the complex pavement condition.

Then, COE made efforts to improve the initial fatigue model. The pavement thickness designed for 5000 coverages was regarded as the criterion. The segmented fatigue models for the pavement structure with different service life were proposed, as shown in Equations (2) and (3) [16]:

$$RH = 1 + 0.07058 \times (\log C - 3.69897) \text{ for } C < 5000 \tag{2}$$

$$RH = 1 + 0.15603 \times (\log C - 3.69897) \text{ for } C > 5000 \tag{3}$$

where:
$RH$ is the relative thickness factor;
$C$ is the coverage to failure.

Compared with the initial fatigue model, this fatigue model established the relationship between the coverage and the relative design thickness of the concrete slab. The slope of the fatigue curve of the pavement structure with more than 5000 coverages had been adjusted, instead of simplifying the influence of various factors to 1.3 [17,18]. The adjusted fatigue model had a broader scope of application and fitted better with the real pavement structure.

In 1979, COE systematically reanalyzed the full-scale test data from 1943 to 1973, adopting the layered elastic analysis approach used in pavements. The modified fatigue model is shown as follows [6,19]:

$$DF_{LEA} = \frac{MR}{\sigma_{LEA}} = 0.58901 + 0.35486 \times \log_{10} C \tag{4}$$

where:
$DF_{LEA}$ is the design factor based on layered elastic analysis;
$MR$ is the modulus of rupture;
$\sigma_{LEA}$ is the maximum principal tensile stress at the bottom of the concrete slab based on layered elastic analysis;
$C$ is the coverage to failure.

In Equation (4), the empirical relationship between the critical stress at the bottom of the concrete slab and the number of the coverages to failure was developed for the first time. The results showed that the design factor is linear with the logarithmic value of coverage. With the increase in the service life of the airport pavement, the critical coverages to fatigue increased. When the number of coverages to fatigue failure was increased, the design factor calculated by the fatigue model was also increased. Then, a thicker slab is needed to satisfy the design requirements.

The subsequent fatigue models were mainly obtained by improving the fatigue model proposed by COE in 1979. In 1988, Rollings proposed the Structural Condition Index (SCI) to characterize the damage of pavement structure. Compared to the Pavement Condition Index (PCI), SCI was deducted by structural distresses induced by load, and the distresses resulted from the non-load case were ignored. It was found that the SCI deteriorates as a linear function of the logarithm of coverages, as shown in Figure 2. The modified fatigue model based on the definition of SCI is shown as follows [20]:

$$SCI = \frac{DF - 0.2967 - (0.3881 + 0.000039 \times SCI) \times \log_{10} C}{0.002269} \tag{5}$$

where:

$SCI$ is the structural condition index.

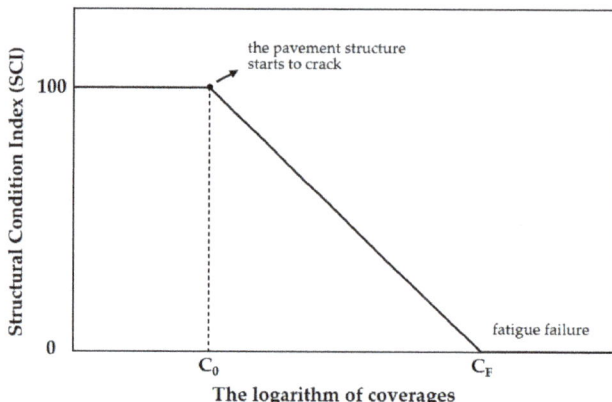

**Figure 2.** The relationship between SCI and the logarithm of coverages.

Using Equation (5), the fatigue model based on the full-scale test data of COE from 1943 to 1973 was developed by the regression analysis of SCI, design factors, and coverages. When the value of SCI was different, the design factor and coverage had a different regression relationship. In fact, the SCI of 80 was equivalent to the fatigue failure of pavement structure assumed by COE. The form of the fatigue model with 80 of SCI was the same as that proposed by COE in 1979, which is shown as follows:

$$DF = 0.4782 + 0.3912 \times \log_{10}(C_{80}) \tag{6}$$

where:
$DF$ is the design factor ($MR/\sigma_{LEA}$);
$C_{80}$ is the coverage to an SCI of 80.

Furthermore, Darter recalculated the stresses in the full-scale test data of COE from 1943 to 1979 by the H-51 computer program instead of the Westergaard edge stress theory. The H-51 program can calculate the stresses quickly by implementing the Pickett and Ray influence charts [21,22]. The exponential fatigue model was proposed by Darter, as shown in Equation (7) [23]. The fatigue model of Darter was limited by the assumption of the infinite slab in the H-51 computer program. It was also a lack of the consideration of the temperature curling influence.

$$\log_{10} N = 2.13 \left( \frac{MR}{\sigma_e} \right)^{1.2} \tag{7}$$

where:
$N$ is the number of coverages for 50 percent of the slabs cracking;
$\sigma_e$ is the critical edge stress calculated by H-51.

Another important fatigue model was developed by Foxworthy according to the full-scale test data conducted by COE, as expressed in Equation (8). To obtain the stresses closer to the real pavement condition, Foxworthy reanalyzed the COE test data by the ILLI-SLAB finite element program [24]. Compared with other fatigue models, the fatigue model proposed by Foxworthy was more conservative, especially in a high level of coverages. Although Foxworthy conducted a comprehensive analysis of the slab stresses to develop the pavement evaluation method, the influence of temperature curling on the slab stresses was also not considered.

$$\log_{10} N = 1.323 \left( \frac{MR}{\sigma} \right) + 0.588 \tag{8}$$

where:
$N$ is the number of coverages for 50 percent of the slabs cracking;
$\sigma$ is the critical edge stress calculated by ILLI-SLAB.

For all the improved fatigue models, the stress calculation method varied from the Westergaard edge stress theory method to computer calculation procedures, while the essence of the fatigue model remained almost the same. However, the fatigue models were affected by the limitations of the full-scale test conducted by COE, such as the location of the test site and the loading conditions. Moreover, the effects of the temperature and environmental factors, which contributed to the fatigue characteristics of the real pavement structure, were ignored. The limitations above resulted in the difference between the theoretical calculation and the real structure. Considering the limitation, in 1992, Thompson and Barenberg proposed the NCHRP 1-26 fatigue model by recalculating the stresses in the full-scale test data of COE and the road test data of AASHO [25–27]. The NCHRP 1-26 fatigue model is shown as follows:

$$\log_{10} N = -1.7136 \left( \frac{\sigma}{MR} \right) + 4.284 \quad \text{for} \quad \frac{\sigma}{MR} > 1.25 \tag{9}$$

$$\log_{10} N = 2.8127 \left( \frac{\sigma}{MR} \right)^{-1.2214} \quad \text{for} \quad \frac{\sigma}{MR} < 1.25 \tag{10}$$

where:
$N$ is the number of coverages for 50 percent of the slabs cracking;
$\sigma$ is the critical edge stress calculated by ILLI-SLAB.

The fatigue model comprehensively considered the effects of aircraft landing gear loading and wheel loading on the fatigue deterioration of concrete slabs. It also considered the influence of thermal curling on the slab stresses in the process of developing the fatigue model for the first time. Therefore, it was widely used in the airfield pavement design and highway pavement design due to the comprehensive analysis of the influencing factors of the fatigue characteristics of concrete slabs.

In recent years, the Federal Aviation Administration (FAA) has improved the fatigue model based on the research of COE. In the Federal Aviation Administration Rigid and Flexible Iterative Elastic Layer Design (FAARFIELD) released by FAA, the fatigue model is as follows [28]:

$$DF = \frac{R}{0.75\sigma_e} = \left[ \frac{F'_S bd}{(1-\alpha)(b-d) + F'_S b} \right] \times \log_{10} C + \left[ \frac{(1-\alpha)(ad-bc) + F'_S bc}{(1-\alpha)(b-d) + F'_S b} \right] \tag{11}$$

$$\alpha = \frac{SCI}{100} \tag{12}$$

where:
$DF$ is the design factor;
$\sigma_e$ is the critical edge stress calculated by the FAARFIELD procedure;
$R$ is the concrete flexural strength;
$F'_S$ is the stabilization factor;
$a, b, c, d$ are the regression coefficients;
$a$ is 0.5878, $b$ is 0.2523, $c$ is 0.7409, $d$ is 0.2465.

FAA used the three-dimensional finite element software to recalculate the stresses in the historical full-scale test data conducted by COE [29,30]. The concept of SCI was adopted based on the study of Rollings. In addition, the new data points were supplemented from the National Airport Pavement Test Facility (NAPTF) Construction Cycle 2 (CC2) test. The CC2 test items, including MRC, MRG, MRS, were designed with different pavement structures and loading methods [31]. The information on the pavement structure and loading method for the test items as well as the test strip is shown in Figure 3 [31–33].

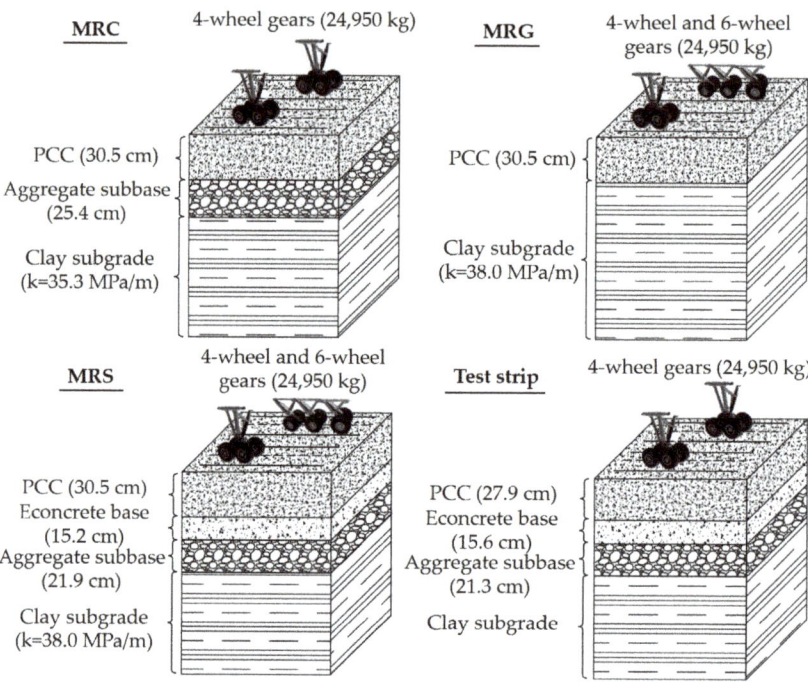

**Figure 3.** The pavement structure and loading method for the NAPTF CC2 test items.

The test data from the test items and the test strip were supplemented to the historical full-scale database conducted by COE. The dataset for regression analysis in the development of this fatigue model consisted of the 30 data points derived from the historical full-scale test and 7 data points from NAPTF. The details of the data sources are shown in Table 1. Based on the research of Rollings [20], the relationship of SCI, DF, and coverages was reestablished, with consideration of the support of the base layer and foundation. Since the stresses calculated by the three-dimensional finite element software were closer to the real pavement structure condition, the fatigue model of FAA more accurately represented the fatigue characteristics.

*2.2. Concrete Beam Testing-Based Fatigue Model*

Different from the full-scale testing-based fatigue model, the concrete beam testing-based fatigue model was analyzed by the regression of the data from the laboratory concrete beam fatigue test. The test was performed by applying sinusoidal load at constant magnitude into the concrete beam directly, as shown in Figure 4. It is possible to clearly obtain the interior stresses' change and structural performance deterioration of the concrete beam in a short time. The number of load repetitions to fatigue failure under different loading conditions can be obtained quickly, and it is easy to analyze the fatigue characteristic of cement concrete. It was shown that the relationship between the stress ratio (the ratio of stress to the modulus of rupture) and the number of load repetitions to failure is effective to describe the fatigue characteristics [1,34,35].

Table 1. The full-scale test data used for the regression analysis in the fatigue model of FAA.

| Test Sites | Number of Data Points |
|---|---|
| Lockbourne No. 1 | 15 |
| Lockbourne No. 2 | 3 |
| Sharonville Heavy Load Tests | 1 |
| Multiple Wheel Heavy Gear Load (MWHGL) Tests | 4 |
| Keyed Longitudinal Joint Study (KLJS) | 4 |
| Soil Stabilization Pavement Study (SSPS) | 3 |
| National Airport Pavement Test Facility (NAPTF) | 7 |
| Total | 37 |

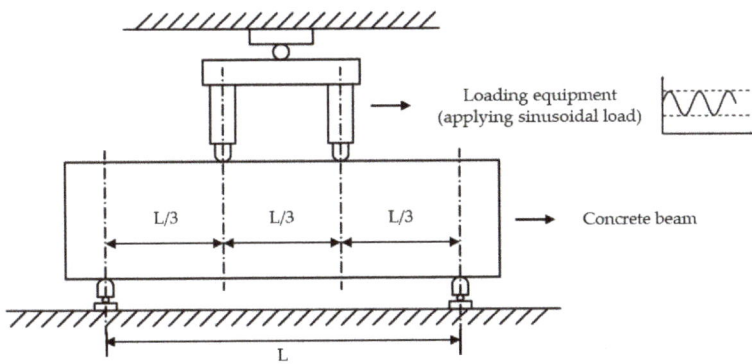

Figure 4. The schematic diagram of concrete beam fatigue test.

From 1922 to 1966, the Portland Cement Association (PCA) initially developed the fatigue model based on the regression analysis of the laboratory concrete beam fatigue test data. The fatigue model is shown in Equation (13) [36–39]. In Equation (13), PCA proposed the interior stress and the cumulative fatigue damage concepts. Furthermore, PCA assumed that a load with less than 50% of the flexural tensile stress had almost no effects on the fatigue deterioration of the concrete beam. This assumption was widely adopted by subsequent research.

$$\log_{10} N = 11.810 - 12.165 \times \left(\frac{\sigma}{MR}\right) \quad \text{for } 0.5 < \frac{\sigma}{MR} < 1.0 \tag{13}$$

where:

$N$ is the number of load repetitions to flexural failure of the concrete beam;
$MR$ is the modulus of rupture;
$\sigma$ is the stress at the bottom of the concrete beam.

Similarly, in the 1970s, the Federal Highway Administration (FHWA) proposed a zero-maintenance fatigue model, as shown in Equation (14) [40–42]. The data used in the fatigue model of FHWA were from the concrete beam tests conducted by Kelser, Ballinger, and Raithby from 1953 to 1974 [43–45]. For the fatigue testing of the concrete beam, the fatigue failure was defined as the fracture of the beam, and the stresses were calculated by the simple bending equation. The researchers recorded the data of the stress-to-strength ratio of the concrete beam under different loading conditions and the number of load repetitions for fatigue cracking. Based on the recorded data, the fatigue model was developed by the least square regression analysis. This fatigue model was employed in the pavement design procedure by the Illinois Department of Transportation in the 1990s.

$$\log_{10} N = 17.61 - 17.61 \times \left(\frac{\sigma}{MR}\right) \tag{14}$$

where:

$N$ is the number of load repetitions to flexural failure of the concrete beam;
$MR$ is the modulus of rupture;
$\sigma$ is the stress at the bottom of the concrete beam.

Based on the fatigue characteristics of concrete beams, researchers found that the range of stresses applied during the test also affected the fatigue strength of concrete [46–48]. It can easily be included that the R-value, which is the ratio of the minimum stress ($\sigma_{min}$) to the maximum stress ($\sigma_{max}$), should be involved in the fatigue model as well as the stress ratio. The fatigue model proposed by Aas-Jakobsen, which includes the R-value, is shown as follows [49]:

$$\frac{\sigma_{max}}{f_c} = 1 - \beta(1-R)\log_{10} N_f \quad (15)$$

where:

$N_f$ is the number of load repetitions;
$\sigma_{max}$ is the maximum applied stress;
$f_c$ is the concrete strength;
$\beta$ is an empirical coefficient with a value of 0.0640;
$R$ is the ratio of the minimum stress to the maximum stress.

After that, Domenichini and Marchionna modified the empirical coefficient proposed by Tepfers to account for the differences between the laboratory test and real pavement structural conditions, including the influence of environment and the concrete slab properties [46,50]. However, it is challenging to introduce the R-value to the pavement design process due to the complexity of the stress fluctuations in the real pavement structure. The conclusions of the effects of stress fluctuations on the fatigue characteristics have not been widely used in airfield pavement design.

In China, the early research on fatigue models of cement concrete were mainly focused on highway pavements. The fatigue model was developed based on the relationship between the stress-to-strength ratio and the number of load repetitions obtained by laboratory concrete beam fatigue tests. The initial fatigue model in 1984 is a semi-logarithmic equation, as shown in Equation (16), and the fatigue test used for regression analysis was performed at an R-value of 0.1. Later, a large number of laboratory concrete beam fatigue tests with different stress ratios and stress range were conducted at Tongji University. The semi-logarithmic and double-logarithmic fatigue equations including stress ranges and stress-to-strength ratio were developed, as shown in Equations (17) and (18).

$$S = \frac{\sigma_{max}}{f_c} = 0.961 - 0.0631 \log_{10} N_f \quad (16)$$

where:

$S$ is the ratio of the maximum stress to the flexural strength;
$N_f$ is the number of load repetitions to fatigue failure;

$$\log_{10} S = \log_{10} A - 0.0422(1-R) \times \log_{10} N \quad (17)$$

$$S = \frac{\sigma_{max}}{f_c} = B - 0.0724(1-R) \times \log_{10} N \quad (18)$$

where:

$R$ is the ratio of the minimum stress to the maximum stress.
$A$ and $B$ are the regression coefficients, and $A$ is 1.0380 and $B$ is 0.9993 when the probability of failure is 50%.

Afterward, the Civil Aviation Administration of China (CAAC) investigated the discrepancy between the fatigue characteristics reflected in the laboratory concrete beam fatigue tests and the airfield pavement structure. The fatigue model for airfield concrete pavement was developed based on the research into fatigue characteristics of highway pavement. Whether the fatigue strength concept proposed in 1995 or the allowable number of aircraft loading proposed in 2010, the essence of the fatigue model represents the relationship between the stress-to-strength ratio of concrete beam and the number of load repetitions [51,52]. The fatigue models from 1995 and 2010 are as follows:

$$\frac{\sigma}{f_{cm}} = 0.885 - 0.0631 \times \log_{10} N_e \quad \text{in 1995} \quad (19)$$

$$\frac{\sigma}{f_{cm}} = 0.9293 - 0.06615 \times \log_{10} N_e \quad \text{in 2010} \quad (20)$$

where:

$N_e$ is the number of load repetitions;
$f_{cm}$ is the flexural strength of concrete;
$\sigma$ is the slab stress.

*2.3. A Brief Summary*

Based on the research above, it can be clearly found that the full-scale testing-based fatigue model is different from the concrete beam testing-based fatigue model in the process of regression analysis, the loading method, the definition of fatigue failure, and the stress calculation method. Table 2 shows the key factors that play an important role in the development of the two types of fatigue models.

Summarily, the full-scale testing-based fatigue models were mainly based on the regression analysis of the full-scale test data. The loading was directly performed by applying the load of landing gear to the concrete slab. The fatigue failure was defined as 50% of the concrete slabs cracking, and the stress calculation methods vary from the Westergaard edge stress theory to the layered elastic analysis approach, and then to the finite element software. Comparably, the concrete beam testing-based fatigue models were mainly based on the laboratory concrete beam fatigue test. The concrete beams were directly applied the sinusoidal load at constant magnitude. The fatigue failure was defined as the bottom cracking of the concrete beams. The stresses were obtained with the bending equation of the simply supported beam.

Different loading methods and definitions of fatigue failure contributed to the number of load repetitions to fatigue failure of concrete slabs or beams. Different stress calculation methods also had an impact on the determination of stresses in pavement structures. The distinctions of these factors led to a large difference in fatigue curve and design results, as shown in Figure 5. Due to the difference in fatigue curves, the designed pavement structures using different fatigue models under the same traffic loading would also have a large difference.

Table 2. Summary of fatigue models for airfield concrete pavements [4].

| Types | Fatigue Model | Regression Data | Stress Type | Failure Definition | Stress Calculation Method |
|---|---|---|---|---|---|
| Full-scale testing-based fatigue models | COE (1946) | COE field data | Load only | 50% of slabs cracking | Westergaard edge stress theory |
|  | Improved COE (1957) | COE field data | Load only | 50% of slabs cracking | Westergaard edge stress theory |
|  | COE-LEA (1979) | COE field data | Load only | 50% of slabs cracking | Layered elastic analysis |
|  | Rollings (1990) | COE field data | Load only | 50% of slabs cracking | Layered elastic analysis |
|  | Foxworthy (1985) | COE field data | Load only | 50% of slabs cracking | Finite element (ILLI-SLAB) |
|  | Darter (1990) | COE field data | Load only | 50% of slabs cracking | H-51 |
|  | NCHRP 1-26 (1992) | COE field data & AASHO road test data | Load and temperature curling | 50% of slabs cracking | Finite element (ILLI-SLAB) |
|  | FAA (2010) | COE field data & NAPTF data | Load only | 50% of slabs cracking | Finite element (3D-FE) |
| Concrete beam testing-based fatigue models | PCA (1963) | Concrete beams | Load only | Beam fracture | Beam bending equation |
|  | Aas-Jakobsen (1970) | Concrete beams | Load only | Beam fracture | Beam bending equation |
|  | FHWA (1977) | Concrete beams | Load only | Beam fracture | Beam bending equation |
|  | CAAC (1995&2010) | Concrete beams | Load only | Beam fracture | Beam bending equation |

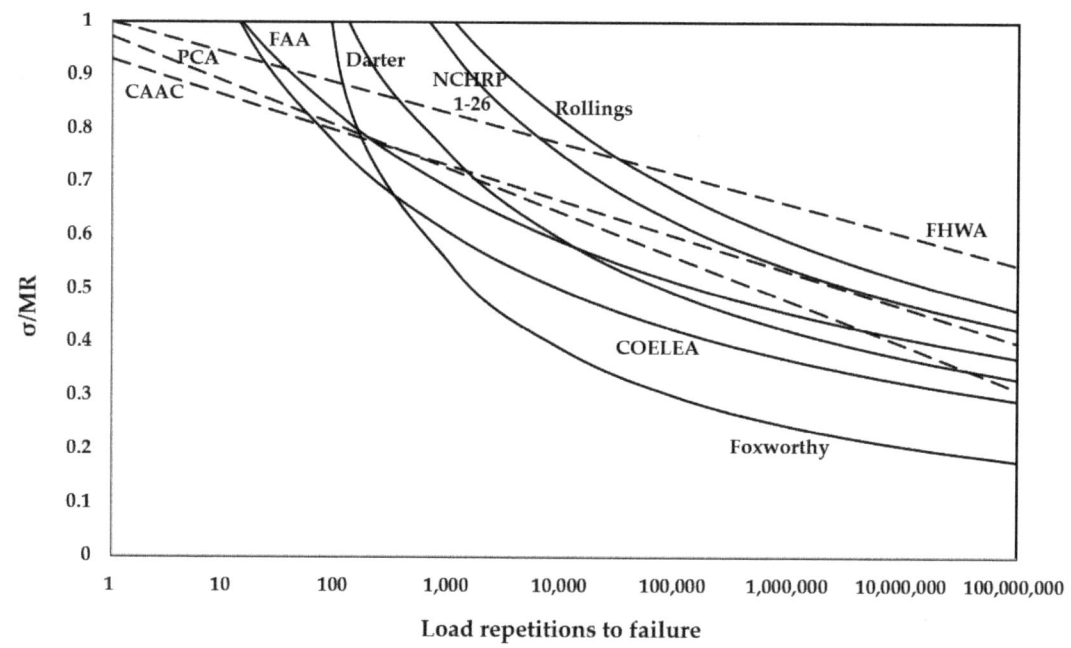

Figure 5. Fatigue curves of the fatigue models involved in this paper.

## 3. Comparison Analysis of Full-Scale Testing-Based and Concrete Beam Testing-Based Fatigue Models

### 3.1. Analysis Method

As mentioned above, it is meaningful to compare the existing fatigue models, especially for the two types of fatigue models: full-scale testing-based fatigue models and concrete beam testing-based fatigue models. In this paper, the full-scale testing-based model proposed by FAA Equation (11) was compared with the concrete beam testing-based fatigue model proposed by CAAC Equation (20). The thickness of the concrete slab was selected as the analysis parameter, due to its contribution to the fatigue resistance of the pavement structure. The procedure of the comparison analysis is shown in Figure 6.

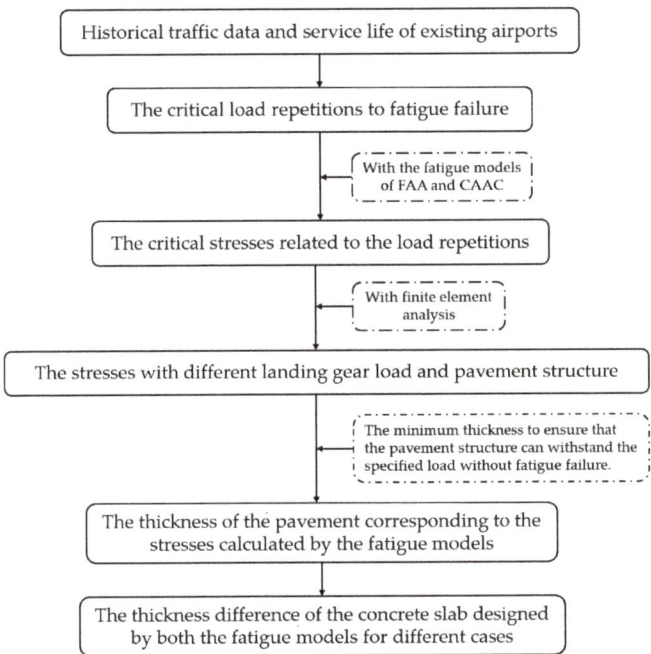

**Figure 6.** The flowchart of comparison analysis.

In this paper, the three-dimensional (3D) finite element models (FEM) were established by the ABAQUS program shown in Figure 7. In the 3D FEM, the concrete slab and base layer were modeled as plates and the subgrade was modeled as Winkler foundation. The size of the concrete slab was set as 10 m by 10 m [53]. The width of the base extension was set to 2.5 m, meaning that the size of the base layer was 15 m by 15 m. The concrete slab and base layer materials were assumed to be isotropic and linearly elastic, whose properties were described by the modulus of elasticity and Poisson's ratio. As for the contact interaction between the upper slab and base layer, the tangential behavior was assumed to be frictionless and the normal behavior was determined as hard contact [54]. For the boundary conditions, the normal displacements of the base layer on four sides were restrained. For the concrete slab, the displacements in both tangential directions on the opposite side of the load-acting side were restrained.

To balance the accuracy and speed of the calculation, the elements of both the upper slab and base layer were determined as eight-node linear brick elements with reduced integration (C3D8R) [55]. Besides, the sizes of the element for the concrete slab and base layer were set as 10 cm and 20 cm, respectively. The aircraft landing gear load, in the form of the static load, was applied at the critical load location of the concrete slab. The maximum principal stresses at the bottom of the slab were obtained for the comparison analysis. The parameters of structure and material properties used in the 3D FEM are presented in Table 3.

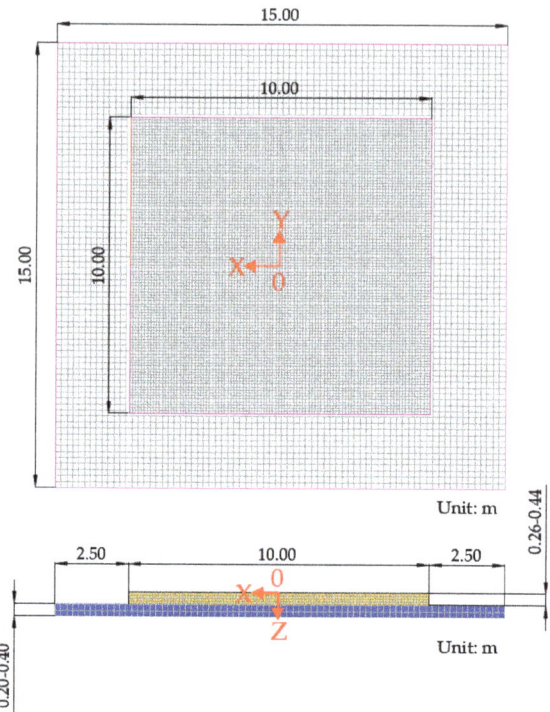

**Figure 7.** The three-dimensional finite element models developed in this paper.

**Table 3.** The parameters of pavement structure and material properties used in the FEM.

| Pavement Structure | Parameters | Values |
|---|---|---|
| Concrete slab | Size of plate | 10 m × 10 m |
| | Elasticity modulus | 29 GPa |
| | Poisson's ratio | 0.15 |
| | Elements | C3D8R |
| | Element size | 10 cm |
| Base layer | Size of plate | 15 m × 15 m |
| | Elasticity modulus | 2000 MPa |
| | Poisson's ratio | 0.20 |
| | Elements | C3D8R |
| | Element size | 20 cm |

To investigate the effect of pavement structure and aircraft loading on the difference in fatigue models, different cases were analyzed. The parameters for different cases were assumed based on the traffic data of the existing airports and the actual pavement structure conditions. The specific values of the parameters are given in Table 4, and the loading parameters of the landing gears for selected typical aircraft are given in Table 5.

Table 4. Variation parameters of analyzed cases in this paper.

| Parameters | Values |
|---|---|
| Load repetitions | 100,000, 200,000, 500,000, 1,000,000 |
| Aircraft types | B737-800, B747-400, B777-300ER |
| Thickness of base layer | 20 cm, 30 cm, 40 cm |
| Subgrade strength | High strength subgrade (k = 150 MPa/m) [1], Medium strength (k = 80 MPa/m), Low strength (k = 40 MPa/m), Ultra-low strength (k = 20 MPa/m) |

[1] The k-value represents the modulus of subgrade reaction.

Table 5. Loading parameters of the landing gears for selected typical aircraft.

| Aircraft | Single Wheel Load (kN) | Tire Pressure (MPa) | Number of Landing Gears | Number of Wheels | Axle Spacing (m) | Wheel Spacing (m) |
|---|---|---|---|---|---|---|
| B737-800 | 187.6 | 1.40 | 2 | 2 | - | 0.86 |
| B747-400 | 236.2 | 1.38 | 4 | 4 | 1.12 | 1.47 |
| B777-300ER | 265.3 | 1.50 | 2 | 6 | 1.40 | 1.45/1.48 |

## 3.2. Results and Analysis

### 3.2.1. The Influence of Load Repetitions

In this analysis, the thickness of the base layer is determined as 30 cm. The subgrade strength category is assumed as low strength (k = 40 MPa/m). The aircraft is chosen as B737-800. The number of load repetitions varies from 100,000 to 1,000,000. Accordingly, the calculated critical stress for fatigue models of CAAC and FAA are given in Table 6, respectively.

Table 6. The critical stresses for different load repetitions.

| Load Repetitions | Critical Stress, MPa | |
|---|---|---|
| | CAAC | FAA |
| 100,000 | 3.990 | 3.420 |
| 200,000 | 3.857 | 3.294 |
| 500,000 | 3.682 | 3.141 |
| 1,000,000 | 3.549 | 3.035 |

According to the results of the critical stress, the thickness for different load repetitions can be determined, as shown in Figure 8. Figure 8 indicates that the design thickness increases with the number of load repetitions and the growth rate gradually slows down. The variation trend is similar to the fatigue curves shown in Figure 5, where the decrease in the stress-to-strength ratio is slowed down when the number of load repetitions keeps increasing. This is due to the design thickness of the slab being determined by the stresses calculated by the fatigue model.

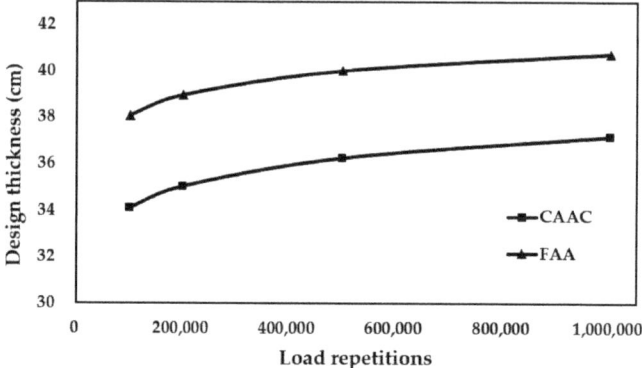

**Figure 8.** Designed thickness variation of concrete slabs with load repetitions.

For the same number of load repetitions, the concrete slab designed with the fatigue model of FAA is thicker than that of CAAC. It can be clearly found that the fatigue model of FAA is more conservative than that of CAAC. In addition, the difference in thickness between the two models is always about 4 cm. With the increase in the number of load repetitions, the percentage value of the difference relative to the design thickness of CAAC continues to decrease due to the increase in slab thickness. Additionally, it can be seen that the absolute value of the difference between the design thickness of the two models is constant with the increase in load repetitions. This indicates that the two fatigue models are similar in their perception of the relationship between the slab thickness and the number of load repetitions.

3.2.2. The Influence of Aircraft Loading

In this analysis, the thickness of the base layer is determined as 30 cm. The subgrade strength category is assumed as low strength (k = 40 MPa/m). The number of load repetitions is assumed as 200,000. The aircraft varies from B737-800 to B777-300ER. The thicknesses of the concrete slab designed by the fatigue model of FAA and CAAC and the relative difference (the percentage value of the difference relative to the design thickness of CAAC) are shown in Figure 9, respectively.

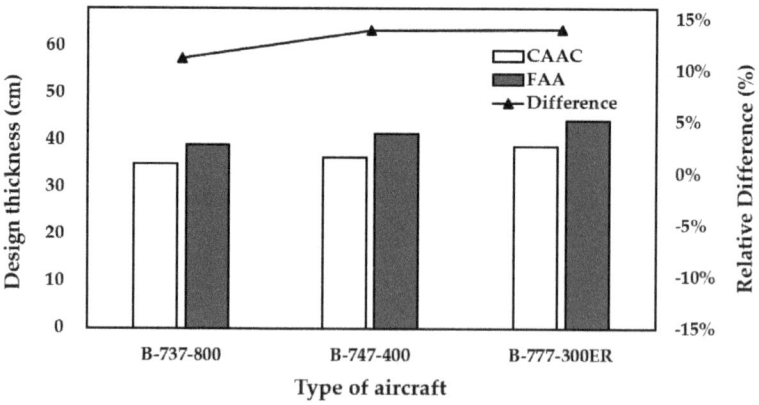

**Figure 9.** Designed thickness of concrete slabs for different aircraft.

From B737-800 to B777-300ER, both the thicknesses designed with the fatigue models of FAA and CAAC increase. The thickness differences also increase with the increase in aircraft loading. The influence of aircraft on the design thickness is mainly due to the different landing gear wheel loads. From B737-800 to B777-300ER, the landing gear wheel load is increasing, which leads to different damage caused by different aircraft in the pavement structure under the same load repetitions. Therefore, both the concrete slabs designed by the two fatigue models become thicker. Moreover, the thickness differences increase with the increase in wheel load. This indicates that the thickness of the concrete slab designed by the fatigue model of FAA is more conservative for larger aircraft.

### 3.2.3. The Influence of the Thickness of Base Layer and Subgrade Strength

In this analysis, the number of load repetitions is assumed as 200,000. The design aircraft is assumed as B737-800. The thickness of the base layer varies from 20 cm to 40 cm. The subgrade strength varies from ultra-low strength to high strength. The thicknesses of the concrete slab designed with the fatigue model of FAA and CAAC for different thicknesses of the base layer are shown in Figure 10. The thicknesses for different subgrade strengths are shown in Figure 11.

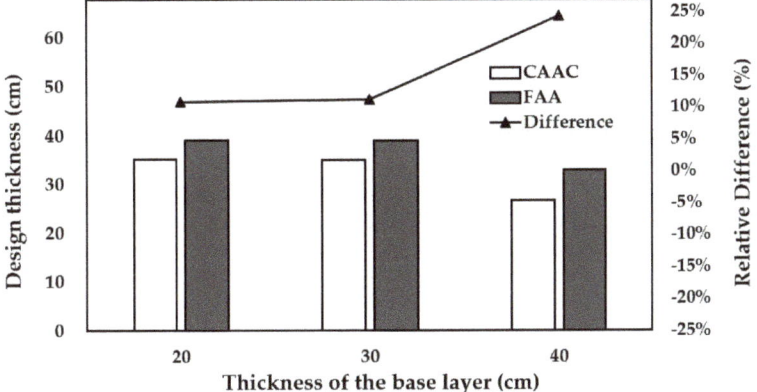

**Figure 10.** Designed thickness of concrete slabs for different thicknesses of the base layer.

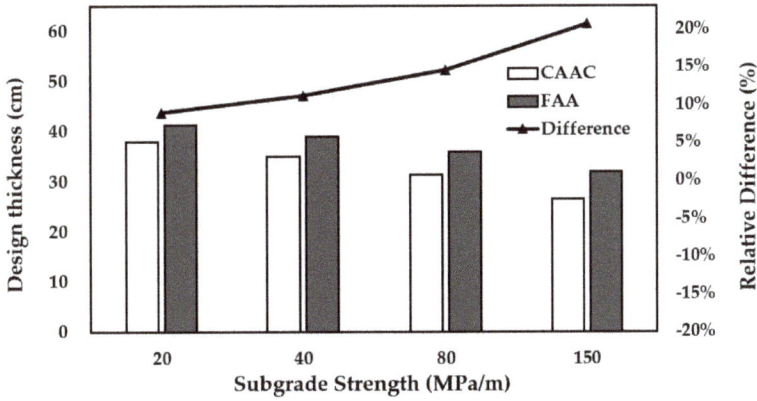

**Figure 11.** Designed thickness of concrete slabs for different subgrade strengths.

With the increase in both the thickness of the base layer and the subgrade strength, the thicknesses of the concrete slab designed by the two fatigue models decrease, while the slab designed by FAA is always thicker than that designed by CAAC. It indicates that increasing the thickness of the base layer and the subgrade strength can be deemed as increasing the strength of the substructure beneath the concrete slab. The substructure can share a partial effect of the load. The increase in the strength of the substructure can improve the resistance to fatigue cracking of the overall pavement structure. Therefore, the thickness of the concrete slab designed by the two fatigue models presents a decreasing trend with the increase in base layer thickness and subgrade strength. Meanwhile, the difference in design thickness between the two fatigue models increases with the substructure strength increasing. This indicates that the fatigue model of FAA in pavement design would give a more conservative upper concrete slab when the strength of the substructure is high.

Moreover, the variation trend in Figure 11 can be mainly contributed to the different test processes of the full-scale testing-based fatigue model and the concrete beam testing-based fatigue model. A full-scale test is carried out by casting the concrete slab of the same size as the pavement structure on site. In the process of the full-scale test, the substructure beneath the concrete slab can share partial effects of the load, while in the concrete beam test, the concrete beam can resist all the fatigue cracking induced by the load repetitions. The difference in the test process results in the difference of the thickness of the concrete slab designed by the two fatigue models. With the increase in substructure strength, the substructure would have a larger effect to resist fatigue cracking in the fatigue model of CAAC model compared to that of FAA, which leads to a larger reduction in the thickness of the concrete slab. Thus, the design thickness of FAA is more conservative than that of CAAC as the strength of the substructure increases.

## 4. Recommendations for Fatigue Models of Airfield Concrete Pavements in the Future

For the fatigue model proposed by FAA, most of the data used in the regression analysis were conducted by COE in Lockbourne, Sharonville, and other sites from 1943 to 1973. Since the historical full-scale tests were conducted a long time ago, the test conditions and pavement structure were limited by the research level and the aircraft load, as well as the pavement structure at that time. Therefore, the fatigue characteristics of the concrete structure reflected by the historical full-scale data could not be well fitted to the current airfield concrete pavement. Further, the fatigue model proposed by CAAC was mainly obtained by the concrete beam fatigue test. Compared to the concrete beams, the stresses of real pavement structures are more complex. In the operation, the factors, such as temperature and humidity, would have a certain impact on the fatigue characteristics of the concrete slab.

Therefore, the current fatigue models cannot be widely and effectively used for the design of airfield concrete pavement structures due to the limitations of their historical test data. Based on the previous research of the development process of the fatigue model, the future fatigue model can be improved from the following three aspects:

- Data source.

The development of fatigue models in the future should be based on historical full-scale test databases supplemented with the local test data in different regions. However, it is costly in terms of manpower and investment to conduct full-scale tests in different regions, considering multiple influencing factors, such as loading method, pavement structure, and environmental conditions. Luckily, many field evaluations on airfield pavements have been performed and a great deal of evaluation data have been collected. Therefore, the Structural Condition Index (SCI) and the related air traffic data from the evaluated pavements can be added to the historical database. It can solve the problem that the fatigue characteristics reflected by the historical database do not match the local airport pavement structure during the development of fatigue models.

- Stress calculation method.

Historical stress calculation methods have developed from Westergaard edge stress theory to layered elastic analysis and then to finite element analysis. It can be seen that the development of stress calculation methods is a process of progressively more accurate calculation of stresses in pavement structures. The process of developing fatigue equations in the future is suggested to be based on numerical simulations by finite element analysis. The accuracy and efficiency of stress calculation can be improved by adjusting the conditions and parameters of the finite element model. In the future, artificial intelligence algorithms, such as artificial neural networks (ANNs), can be introduced in the structural stress calculation of cement concrete [56–58]. A database of stress analysis can be developed through field tests and numerical simulations, after which the artificial intelligence algorithm model can be trained based on the database to achieve stress prediction.

- Regression analysis process

After obtaining data that are sufficient to reflect the fatigue characteristics of the local pavement structure, the SCI data from different periods can be linearly fitted to the logarithm of coverages, for which has been observed by Rollings that SCI deteriorates as a linear function of the logarithm of coverages [29]. After linear regression analysis, the coverages when the pavement structure starts to crack (when SCI deteriorates from 100) and the coverages to fatigue failure (when SCI equals 0) can be obtained. These data and stress-to-strength ratios will be supplemented to the historical database for re-regression analysis in order to obtain the modified coefficients for the improved fatigue model [28,59]. During the process, if the structural conditions of the pavement structures are similar in different testing periods, resulting in obtained data that cannot be regarded as valid being added to the historical database, it is suggested that the fatigue model be developed by conducting full-scale tests and regression analysis locally in different regions.

The improved method of the fatigue model based on the three aspects mentioned above is proposed in this paper. It can help develop fatigue models to be more suitable for the local pavement structures, based on the previous studies, in the future. By applying the improved fatigue model to the design of pavement structure, the fatigue resistance of the pavement structure will be enhanced and the risk of fatigue cracking could be reduced. Meanwhile, it is also useful to estimate more accurately the service situation and remaining life of the concrete structure in the evaluation of the pavement.

## 5. Conclusions

1. According to the regression analysis process, the fatigue models are divided into two types: the full-scale testing-based fatigue model and the concrete beam testing-based fatigue model. This paper reviews the development process of two types of fatigue models for airfield concrete pavement. It can be clearly found that the full-scale testing-based fatigue model is different from the concrete beam testing-based fatigue model in the process of regression analysis, the loading method, the definition of fatigue failure, and the stress calculation method.
2. Compared with the fatigue model of CAAC, the fatigue model of FAA always tends to design a more conservative and safer thickness of concrete slab with the same load repetitions in different cases, thus making the pavement structure have sufficient resistance to fatigue cracking. The difference between the thickness of the two models is less influenced by the number of load repetitions but largely affected by factors such as aircraft type, the thickness of the base layer, and subgrade strength. This is caused by the different test methods and test conditions between the two fatigue models.
3. Moreover, due to the limitations of the historical test data, the current fatigue models cannot be applied to various pavement structures in different regions. Therefore, this paper proposes an improved method to refine the future fatigue model from three aspects: data source, stress calculation method, and regression analysis process.

**Author Contributions:** Conceptualization, J.Y.; methodology, J.Y., L.M.; investigation, W.L.; writing—original draft preparation, W.L., Y.L.; writing—review and editing, L.M., J.Z.; supervision, J.Y.; L.M., J.Z. All authors have read and agreed to the published version of the manuscript.

**Funding:** This research was funded by the National Key R&D Program of China (Project No. 2019YFB1310600) and National Science Foundation of China (Project No. U1933116).

**Institutional Review Board Statement:** Not applicable.

**Informed Consent Statement:** Not applicable.

**Data Availability Statement:** Not applicable.

**Conflicts of Interest:** The authors declare no conflict of interest.

## References

1. Roesler, J.R. Fatigue of Concrete Beams and Slabs. Ph.D. Thesis, University of Illinois at Urbana-Champaign, Ann Arbor, MI, USA, 1998.
2. Titus-Glover, L.; Mallela, J.; Darter, M.I.; Voigt, G.; Waalkes, S. Enhanced Portland cement concrete fatigue model for StreetPave. *Transp. Res. Rec.* **2005**, *1919*, 29–37. [CrossRef]
3. Rao, C.B.; Barenberg, E.J. Fatigue Implications in Rigid Airport Pavements. In *The 2020 Vision of Air Transportation: Emerging Issues and Innovative Solutions, Proceedings of the 26th International Air Transportation Conference, San Francisco, CA, USA, 19–21 June 2000*; ASCE: San Francisco, CA, USA, 2000; pp. 298–312.
4. Smith, K.D.; Roesler, J.R. Review of fatigue models for concrete airfield pavement design. In *Airfield Pavements: Challenges and New Technologies, Proceedings of the Airfield Pavements Specialty Conference, Las Vegas, NV, USA, 21–24 September 2004*; ASCE: Las Vegas, NV, USA, 2004; pp. 231–258.
5. Federal Aviation Administration (FAA). *AC 150/5320-6G, Airport Pavement Design and Evaluation*; FAA: Washington, DC, USA, 2021.
6. Parker, F., Jr.; Barker, W.R.; Gunkel, R.C.; Odom, E.C. *Development of a Structural Design Procedure for Rigid Airport Pavements*; U.S. Army Engineer Waterways Experiment Station Geotechnical Laboratory: Washington, DC, USA, 1979.
7. Hiller, J.E.; Roesler, J.R. Determination of critical concrete pavement fatigue damage locations using influence lines. *J. Transp. Eng.* **2005**, *131*, 599–607. [CrossRef]
8. Foxworthy, P.T. Concepts for the Development of a Nondestructive Testing and Evaluation System for Rigid Airfield Pavements. Ph.D. Thesis, University of Illinois at Urbana-Champaign, Farmington Hills, MI, USA, 1985.
9. Treybig, H.J.; Smith, P.; VonQuintus, H. *Overlay Design and Reflection Cracking Analysis for Rigid Pavements: Development of New Design Criteria*; Federal Highway Administration: Washington, DC, USA, 1977; Volume 1.
10. Philippe, R.R. Structural Behavior of Concrete Airfield Pavements-The Test Program. In Proceedings of the Twenty-Fourth Annual Meeting of the Highway Research Board (Unassembled), Cincinnati, OH, USA, 22–24 November 1944.
11. Westergaard, H. New formulas for stresses in concrete pavements of airfields. *Trans. Am. Soc. Civ. Eng.* **1948**, *113*, 425–439. [CrossRef]
12. COE. *Lockbourne No. 1 Test Track Final Report*; War Department Corps of Engineers: Mariemont, OH, USA, 1946.
13. Sale, J.P.; Hutchinson, R.L. Development of rigid pavement design criteria for military airfields. *J. Air Transp. Div.* **1959**, *85*, 129–151. [CrossRef]
14. USACOE. Airfield Pavement Design, Rigid Pavements. In *Engineering Manual for War Department Construction*; USA Army Corps of Engineers: Washington, DC, USA, 1946.
15. Hutchinson, R.L. *Basis of Rigid Pavement Design for Military Airfields*; Ohio River Division Laboratories: Cincinnati, OH, USA, 1966.
16. Lee, Y.-H. Mechanistic reappraisal of the current design methodology for rigid airfield pavements. *Transp. Res. Rec.* **1999**, *1684*, 90–100. [CrossRef]
17. Mellinger, F.M.; Sale, J.P.; Wathen, T.R. Heavy Wheel Load Traffic on Concrete Airfield Pavements. In Proceedings of the Thirty-Sixth Annual Meeting of the Highway Research Board, Washington, DC, USA, 7–11 January 1957.
18. Sale, J.P. Rigid Pavement Design for Airfields. In Proceedings of the International Conference on Concrete Pavement Design, West Lafayette, IN, USA, 15–17 February 1977.
19. Barker, W. Introduction to a Rigid Pavement Design Procedure. In Proceedings of the 2nd International Conference on Concrete Pavement Design, West Lafayette, IN, USA, 14–16 April 1981.
20. Rollings, R.S.; Witczak, M.W. Structural Deterioration Model for Rigid Airfield Pavements. *J. Transp. Eng.* **1990**, *116*, 479–491. [CrossRef]
21. Pickett, G.; Ray, G.K. Influence charts for concrete pavements. *Trans. Am. Soc. Civ. Eng.* **1951**, *116*, 49–73. [CrossRef]
22. Kreger, W.C. *Computerized Aircraft Ground Flotation Analysis-Edge-Loaded Rigid Pavement*; General Dynamics Corporation: Fort Worth, TX, USA, 1967.
23. Darter, M.I. Concrete slab vs beam fatigue models. In Proceedings of the Second International Workshop on the theoretical design of Concrete Pavements, Siguenza, Spain, 4–5 October 1990.
24. Ioannides, A.M. Analysis of Slabs-on-grade for a Variety of Loading and Support Conditions. Ph.D. Thesis, University of Illinois at Urbana-Champaign, Ann Arbor, MI, USA, 1984.
25. Thompson, M.R. *Calibrated Mechanistic Structural Analysis Procedures for Pavements: Phase 2*; National Cooperative Highway Research Program: Washington, DC, USA, 1992.
26. Meyerhof, G. Load-carrying capacity of concrete pavements. *J. Soil Mech. Found. Div.* **1962**, *88*, 89–116. [CrossRef]

27. Losberg, A. *Structurally Reinforced Concrete Pavements*; Chalmers Tekniska Hogskola: Goteborg, Sweden, 1960.
28. Brill, D.R. *Calibration of FAARFIELD Rigid Pavement Design Procdure*; Federal Aviation Administration William J. Hughes Technical Center: Washington, DC, USA, 2010.
29. Rollings, R.S. *Design of Overlays for Rigid Airport Pavements*; Federal Aviation Administration: Washington, DC, USA, 1988.
30. Maker, B.N.; Ferencz, R.M.; Hallquist, J.O. *NIKE3D: A Nonlinear, Implicit, Three-Dimensional Finite Element Code for Solid and Structural Mechanics—User's Manual*; Lawrence Livermore National Laboratory: Livermore, CA, USA, 1995.
31. Brill, D.; Hayhoe, G.; Ricalde, L. Analysis of CC2 Rigid Pavement Test Data from the FAA's National Airport Pavement Test Facility. In Proceedings of the International Conferences on the Bearing Capacity of Roads, Railways and Airfields, Trondheim, Norway, 25–27 June 2005.
32. Hayhoe, G.F. Traffic testing results from the FAA's national airport pavement test facility. In Proceedings of the 2nd International Conference on Accelerated Pavement Testing, Minneapolis, MN, USA, 26–29 September 2004.
33. Ricalde, L. Analysis of HWD data from CC2 traffic tests at the National Airport Pavement Test Facility. In Proceedings of the 2007 FAA Airport Technology Transfer Conference, Atlantic City, NJ, USA, 16–18 April 2007.
34. Lee, Y.-H.; Carpenter, S.H. PCAWIN Program for jointed concrete pavement design. *J. Appl. Sci. Eng.* **2001**, *4*, 293–300.
35. Ioannides, A. Pavement fatigue concepts: A historical review. In Proceedings of the 6th International Purdue Conference on Concrete Pavement Design and Materials for High Performance, Indianapolis, IN, USA, 18–21 November 1997.
36. Packard, R.G. *Design of Concrete Airport Pavement*; Portland Cement Association: Skokie, IL, USA, 1973.
37. Packard, R.G. Fatigue concepts for concrete airport pavement design. *Transp. Eng. J. ASCE* **1974**, *100*, 567–582. [CrossRef]
38. Hilsdorf, H.K. Fatigue strength of concrete under varying flexural stresses. In *Journal Proceedings*; ACI Materials Journal: Farmington Hills, MI, USA, 1966; pp. 1059–1076.
39. Hilsdorf, H.K.; Kesler, C.E. *The Behavior of Concrete in Flexure under Varying Repeated Loads*; University of Illinois, Department of Theoretical and Applied Mechanics: Farmington Hills, MI, USA, 1960.
40. Darter, M.I.; Barenberg, E.J. Zero-Maintenance Design for Plain Jointed Concrete Pavements. In Proceedings of the International Conference on Pavement Design, West Lafayette, IN, USA, 15–17 February 1977.
41. Darter, M.I.; Barenberg, E.J. *Design of Zero-Maintenance Plain Jointed Concrete Pavement, Vol. II-Design Manual*; Federal Highway Administration: Washington, DC, USA, 1977.
42. Darter, M.I. *Design of a Zero-Maintenance Plain Jointed Concrete Pavement, Volume One-Development of Design Procedures*; Federal Highway Administration: Washington, DC, USA, 1977.
43. Raithby, K.; Galloway, J. Effects of moisture condition age, and rate of loading on fatigue of plain concrete. *Spec. Publ.* **1974**, *41*, 15–35.
44. Ballinger, C.A. Cumulative fatigue damage characteristics of plain concrete. *Highw. Res. Rec.* **1971**, *370*, 48–60.
45. Kesler, C.E. Effect of speed of testing on flexural fatigue strength of plain concrete. In Proceedings of the Highway Research Board Proceedings, Washington, DC, USA, 13–16 January 1953.
46. Tepfers, R. Tensile fatigue strength of plain concrete. In *Journal Proceedings*; ACI Materials Journal: Farmington Hills, MI, USA, 1979; pp. 919–934.
47. Tepfers, R.; Kutti, T. Fatigue strength of plain, ordinary, and lightweight concrete. In *Journal Proceedings*; ACI Materials Journal: Farmington Hills, MI, USA, 1979; pp. 635–652.
48. Murdock, J.W.; Kesler, C.E. Effect of range of stress on fatigue strength of plain concrete beams. In *Journal Proceedings*; ACI Materials Journal: Farmington Hills, MI, USA, 1958; pp. 221–231.
49. Aas-Jakobsen, K. *Fatigue of Concrete Beams and Columns*; Division of Concrete Structures, Norwegian Institute of Technology, University of Trondheim: Trondheim, Norway, 1970.
50. Domenichini, L.; Marchionna, A. Influence of stress range on plain concrete pavement fatigue design. In Proceedings of the 2nd International Conference on Concrete Pavement Design, West Lafayette, IN, USA, 14–16 April 1981.
51. Civil Aviation Administration of China (CAAC). *Specifications for Cement Concrete Pavement Design for Civil Transport Airports*; CAAC: Beijing, China, 1995.
52. Civil Aviation Administration of China (CAAC). *Specifications for Airport Cement Concrete Pavement Design*; CAAC: Beijing, China, 2010.
53. Brill, D.R.; Parsons, I.D. Three-dimensional finite element analysis in airport pavement design. *Int. J. Geomech.* **2001**, *1*, 273–290. [CrossRef]
54. Hammons, M.J. *Advanced Pavement Design: Finite Element Modeling for Rigid Pavement Joints, Report II: Model Development*; Army Engineer Waterways Experiment Station: Washington, DC, USA, 1998.
55. Edward, G.; Lia, R.; Izydor, K. *FAA Finite Element Design Procedure for Rigid Pavements*; US Department of Transportation, Federal Aviation Administration: Washington, DC, USA, 2007.
56. Vishnu, B.S.; Simon, K.M.; Raj, B. Fatigue Life Prediction of Reinforced Concrete Using Artificial Neural Network. In Proceedings of the International Conference on Structural Engineering and Construction Management, Cham, Switzerland, 12–15 May 2021; pp. 265–271.
57. Abambres, M.; Lantsoght, E.O.L. ANN-Based Fatigue Strength of Concrete under Compression. *Materials* **2019**, *12*, 3787. [CrossRef] [PubMed]

58. DeSantis, J.W.; Vandenbossche, J.M.; Alland, K.; Harvey, J. Development of artificial neural networks for predicting the response of bonded concrete Overlays of asphalt for use in a faulting prediction model. *Transp. Res. Rec.* **2018**, *2672*, 360–370. [CrossRef]
59. Gücbilmez, E.; Yüce, R. Mechanistic evaluation of rigid airfield pavements. *J. Transp. Eng.* **1995**, *121*, 468–475. [CrossRef]

Article

# Evaluation of the Aging of Styrene-Butadiene-Styrene Modified Asphalt Binder with Different Polymer Additives

Bangwei Wu [1], Chufan Luo [1], Zhaohui Pei [1], Chuangchuang Chen [1], Ji Xia [1] and Peng Xiao [1,2,*]

[1] College of Civil Science and Engineering, Yangzhou University, Yangzhou 225127, China; wubw@yzu.edu.cn (B.W.); MZ220190387@yzu.edu.cn (C.L.); MZ220200335@yzu.edu.cn (Z.P.); MZ120200967@yzu.edu.cn (C.C.); MZ120200952@yzu.edu.cn (J.X.)
[2] Research Center for Basalt Fiber Composite Construction Materials, Yangzhou University, Yangzhou 225127, China
* Correspondence: pengxiao@yzu.edu.cn; Tel.: +86-0514-8797-9418

**Abstract:** A wide variety of polymer additives have been widely used in recent years. However, the effect of different polymer additives on the durability of asphalt binders has not been investigated thoroughly. To evaluate the aging property of styrene-butadiene-styrene (SBS) asphalt binder with different polymer additives, three polymer modifiers, namely high modulus modifier (HMM), anti-rutting agent (ARA), and high viscosity modifier (HVM), were added to it. First, the Thin Film Over Test (TFOT) and Pressure Aging Vessel (PAV) was performed on the asphalt binders. The rheological properties of the four asphalt binders before and after aging were then checked by the Dynamic Shear Rheometer Test (DSR). The chemical compositions of the asphalt binders were determined by the Fourier Transform Infrared Spectrometer (FTIR) test. Several aging indicators were adopted to reflect the aging degree of the asphalt binders. The results show that when polymer additives are added to the SBS asphalt binder, the complex modulus, storage modulus, loss modulus, and rutting factor substantially increase and the phase angle decreases. All the test parameters become higher after aging. The phase angle of the SBS asphalt binder is the highest at both unaged and aged states, while its other parameters values are the smallest. Moreover, the Carbonyl Aging Indicator (CAI) of SBS with polymer additives becomes lower under both TFOT and PAV conditions, indicating that polymer additives can improve the aging resistance of SBS asphalt, of which HVM modifies the aging resistance best. Complex Modulus Aging Indicator (CMAI) and Storage Modulus Aging Indicator (SMAI) have the best correlation coefficients with CAI, and the two aging indicators can be used to predict the aging degree of polymer modified asphalt binders.

**Keywords:** SBS; polymer additive; aging property; rheological properties; FTIR

## 1. Introduction

With the increasing traffic load in China, many asphalt pavements may suffer from severe early damage within a few years of completion. People have paid more attention to the durability of asphalt pavements [1,2]. Therefore, in the past decades, in addition to SBS, many other types of modifiers have been used for asphalt mixtures to improve the durability of mixtures, such as the high modulus modifier (HMM) [3], anti-rutting agent (ARA) [3,4], and high viscosity modifier (HVM) [5].

The HMM was initially used in France to improve asphalt mixtures' rutting resistance and fatigue resistance [3]. The dynamic modulus of high modulus asphalt concrete (HMAC) is about 14 GPa, far higher than that of ordinary asphalt mixtures. Wu [6] investigated the anti-rutting property of HMAC using the wheel track test, and the results show that HMAC has better high-temperature stability than ordinary asphalt mixtures. Lee [7] found that using HMM in asphalt mixtures can increase the stiffness modulus, resulting in a higher deformation resistance of asphalt mixtures.

ARA is also used to modify the anti-rutting property of the asphalt mixture. Chen [8] compared the pavement performance of ARA modified SMA-13 and SBS modified SMA-13 and found that the rutting resistance of the former is far better than that of the latter. A similar conclusion was also drawn by Ulucayli [9]. Sun [10] argued that the elastic component in ARA helped to reduce the shear deformation of asphalt mixtures, resulting in a minor rutting depth in asphalt pavement. Chen [11] thought that the ARA particle could fill the voids in the aggregates, increasing the denseness of asphalt mixtures and bonding adjacent aggregates to resist deformation.

HVM has been widely used for asphalt mixtures in recent years. It can increase the asphalt viscosity significantly. The zero-shear viscosity of HVM modified asphalt at 60 °C can be higher than 40,000 Pa·s [12]. Yang [13] found that HVM modified the adhesion ability of asphalt with mineral aggregate, resulting in asphalt mixtures with better resistance to cracking, deformation, and moisture damage. Tan [14] compared the properties of HVM asphalt and SBS asphalt. The segregation test showed that the compatibility and stability of HVM asphalt were better and the cohesion of the HVM asphalt before and after aging was more stable. The Performance Grade (PG) test results showed that HVM asphalt had a better high and low temperature performance. Li [15] studied the performances of HVM porous asphalt mixtures. After a serial of tests, which included the multi-stress repeated creep test, accelerated fatigue test, temperature sweep, and other tests, Li argued that HVM modified asphalt helped improve the stability and cracking resistance and prevent the loosening of porous asphalt pavements.

It can be seen that types of modifiers improve the pavement performance of asphalt mixtures. However, current research has been conducted into new mixtures, and studies on the aging property of the asphalt binders are lacking. Many studies have proved that the aging property of polymer modified asphalt is more complicated than that of pure asphalt. For polymer-modified asphalt, the aging process includes the aging of the pure asphalt and the degradation and possible chemical reactions of the polymer modifier [16–18]. Many researchers have studied the aging behavior of SBS asphalt binder and mixtures, while the durability of other polymer additives has not been studied yet. Cortizo [19] found the degradation of SBS after thermal aging. Sugano [20] insisted that such degradation of SBS decreased the durability of asphalt mixtures. Zhao [21] used Fourier transform infrared spectrum (FTIR) to explore the aging mechanism of SBS asphalt. Zhao found that the chemical structure of SBS changed over the aging process. Thus, the objective of this study is to evaluate and compare the effect of different polymer modifiers on the aging property of the asphalt binder. Three polymer modifiers, namely HMM, ARA, and HVM, were added to the SBS asphalt binder. Short-term aging and long-term aging on the asphalt binders were performed, respectively. A serial of microscopic and macroscopic tests was then conducted to check the aging behavior of the asphalt binders.

## 2. Test Materials
### 2.1. Polymer Additives

Three kinds of polymer additives (HVM, HAA, and HMM) were used in this paper. The macro images of the three additives are shown in Figure 1. All three additives were provided by Wanpu Traffic Technology Co., Ltd., Wuxi, China, and they were all black particles at room temperature. The supplier provided the basic properties of the three additives, as presented in Table 1.

**Figure 1.** The appearance of the three modifiers.

**Table 1.** Properties of additives.

| Test Items | HMM | ARA | HVM |
|---|---|---|---|
| Particle size (mm) | 2–4 | 2–3 | 1–3 |
| Melting point (°C) | 120–130 | 120–150 | 120–135 |
| Density (g/cm$^3$) | 0.9–0.98 | 0.92–0.99 | 0.9–0.95 |
| Melt index (g/10 min) | 5–11 | 5–11 | 6–12 |
| Polymer content (%) | | ≥95 | |
| Exterior | | Black solid particles | |

## 2.2. Asphalt Binder

In this paper, the SBS asphalt was chosen as the base asphalt binder. It was provided by Tongsha Asphalt Technology Co., Ltd., Nantong, China. The properties of the SBS asphalt were tested according to Chinese specification (JTG E20-2011) Test Standard Methods of Bitumen and Bituminous Mixtures for Highway Engineering [22]. The results are listed in Table 2. The SBS asphalt properties satisfy the requirements in (JTG F40-2004) Technical Specification for Construction of Highway Asphalt Pavements [23]. The polymer additives were then added into the base binder to fabricate a polymer-modified asphalt binder. Thus, a total of four types of asphalt binders were used in this paper, named SBS, HMM-SBS, ARA-SBS, and HVM-SBS for convenience.

**Table 2.** Properties of SBS asphalt.

| Index | Results | Requirements |
|---|---|---|
| Penetration at 25 °C (0.1 mm) | 68 | 60~80 |
| Penetration Index | 0.4 | ≮−0.4 |
| Ductility at 5 °C (cm) | 45 | ≮30 |
| Softening point (°C) | 62 | ≮55 |
| Viscosity at 135 °C (Pa·s) | 1.4 | ≯3.0 |
| Elastic recovery at 25 °C (%) | 76 | ≮65 |

According to our team's previous research results [24], the polymer-modified asphalt binders were prepared by the following steps: Heating the SBS asphalt to 185 °C; adding polymer additive to the SBS asphalt; blending them for 60 min at a speed of 1500 r/min.

## 3. Research Scope and Test Methods

### 3.1. Research Scope

In this paper, the effect of different polymer additives on the aging behavior of SBS asphalt binder was explored. First, short-term aging and long-term aging were performed on the four kinds of asphalt binders. Then, the rheological properties of the four asphalt binders before and after aging were checked, and the chemical compositions of the asphalt binders were also determined by the FTIR. At last, based on other scholars' research results [25], several indicators calculated by the test results were adopted to reflect the aging degree of the different asphalt binders. These indicators are shown in Table 2.

Many factors affect asphalt aging, and it is the oxidation that mainly causes the aging of the asphalt binder. Therefore, the change in the carbonyl absorption peak can reflect the aging level of the asphalt binder [26]. The Carbonyl Index (CI) calculated by the FTIR test results was used by many researchers to evaluate the aging degree of the asphalt binder [21,27,28]. Thus, this paper used CI as the basic aging indicator, and its correlations with other indicators were analyzed. The calculation method of CI is shown in Table 3.

Table 3. Aging indicators in this paper.

| Indicators | Calculation Methods |
|---|---|
| Phase Angle Aging Indicator (PAAI) (%) | $100 \times |(\delta_{aged} - \delta_{unaged})| / \delta_{unaged}$ |
| Complex Modulus Aging Indicator (CMAI) (%) | $100 \times |(G^*_{aged} - G^*_{unaged})| / G^*_{unaged}$ |
| Storage Modulus Aging Indicator (SMAI) (%) | $100 \times |(G'_{aged} - G'_{unaged})| / G'_{unaged}$ |
| Loss Modulus Aging Indicator (LMAI) (%) | $100 \times |(G''_{aged} - G''_{unaged})| / G''_{unaged}$ |
| Rutting Factor Aging Indicator (RFAI) (%) | $100 \times |(G^*_{aged}/\sin\delta - G^*_{unaged}/\sin\delta)| / (G^*_{aged}/\sin\delta)$ |
| Carbonyl Aging Indicator (CAI) (%) | $CAI = 100 \times |(CI_{aged} - CI_{unaged})| / CI_{unaged}$; |

$G^*$ is the complex shear modulus; $\delta$ is the phase angle; $G'$ is the storage modulus, $G' = G^*\cos\delta$; $G''$ is the loss modulus, $G'' = G^*\cos\delta$; $CI = A1700\ cm^{-1}/A1376\ cm^{-1}$; $A\ Xcm^{-1}$ is area at $X\ cm^{-1}$ peak of the FTIR figure.

### 3.2. Test Methods

#### 3.2.1. Aging Methods

In this paper, the short-term aging and long-term aging of the asphalt binder were performed by the Thin Film Over Test (TFOT) and Pressure Aging Vessel (PAV) test, respectively.

The TFOT was used to simulate the thermal oxygen aging of the asphalt binder during storage, transportation, and paving. The PAV test was used to simulate the oxidative aging of asphalt binders during in-service. They were conducted following the steps in ASTM D1754 [29] and ASTM D6521 [30], respectively.

#### 3.2.2. FTIR Test

FTIR adopted in this paper was made by Perkinelmer Instruments Co., Ltd., Shanghai branch, China. This instrument was used in this paper to analyze the changes of chemical functional groups of the asphalt binder before and after aging, thus the mechanism of the difference in the durability of different modifiers could be better understood. The wavenumber used in the test was 600–4000 $cm^{-1}$.

#### 3.2.3. Dynamic Shear Rheometer (DSR) Test

The DSR adopted in this paper is specified in AASHTO T315 [31]. This test method is suitable for determining the phase angle and dynamic shear complex modulus of the asphalt binder. The diameter of the test piece was 25 mm and the thickness was 1 mm. A sinusoidal vibration load with an angular frequency of 10 rad/s was applied to the test piece. The test temperature ranged from 52 °C to 82 °C with an increment of 6 °C/min. The instrument used in this paper was made by TA Instruments Co., Ltd., New Castle, DE, USA.

## 4. Results and Discussion

### 4.1. FTIR Test Results Analysis

#### 4.1.1. FTIR Characteristics before Aging

The FTIR pictures of three polymer additives and the four kinds of asphalt binders are presented in Figure 2.

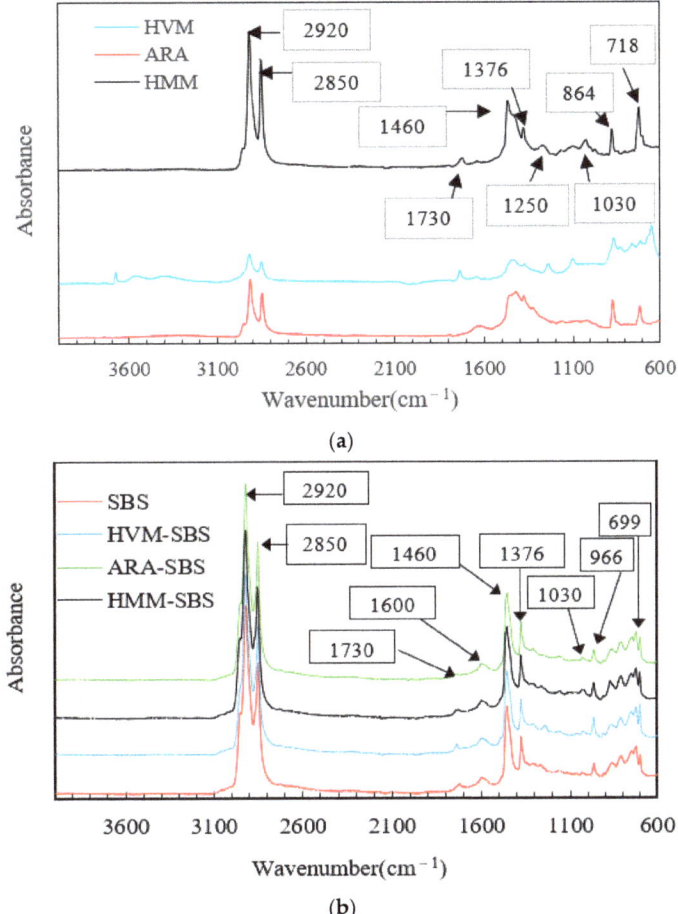

**Figure 2.** FTIR of polymer additives and asphalt binders before aging: (**a**) polymer additives; (**b**) asphalt binders.

It can be seen from Figure 2a that the FTIR characteristic peaks of the three polymer additives were not the same. For HMM, the absorption peaks mainly appeared at 2920 cm$^{-1}$, 2850 cm$^{-1}$, 1730 cm$^{-1}$, 1460 cm$^{-1}$, 1376 cm$^{-1}$, 1250 cm$^{-1}$, 1030 cm$^{-1}$, 864 cm$^{-1}$, and 718 cm$^{-1}$ etc. The peaks at 2920 cm$^{-1}$ and 2850 cm$^{-1}$ were due to the antisymmetric stretching vibration and symmetric stretching vibration of methylene (CH$_2$), separately. The peak at 1460 cm$^{-1}$ was caused by the deformation vibration of the C–H bond. The peak at 1376 cm$^{-1}$ was formed by a specific vibration of methyl (CH$_3$). The peak at 864 cm$^{-1}$ was caused by C–C stretch. The peaks at 1030 cm$^{-1}$ and 718 cm$^{-1}$ were due to the S–O stretch and C–S bend, respectively [32,33]. The peaks at 1730 cm$^{-1}$ and 1250 cm$^{-1}$ were the characteristic peaks of esters, which may have been caused by the plasticizer in HMM [34]. HVM showed a similar FTIR characteristic peak to HMM. However, a very slight observable change in peak position was clear in the FTIR spectra of HMM and HVM, indicating that the plasticizer or crosslinker used in HVM and HMM was different. For the FTIR spectra of ARA, there were no characteristic peaks of esters.

Compared with the infrared spectrum of polymer, the infrared spectrum of polymer-asphalt binders (seen as Figure 2b) had several more characteristic peaks, which mainly appeared at 1600 cm$^{-1}$, 966 cm$^{-1}$, and 699 cm$^{-1}$. The 1600 cm$^{-1}$ peak reflected the C=C bond in the benzene ring and the stretching vibration of the C-H bond. The peaks at 966 cm$^{-1}$ and 699 cm$^{-1}$ were the characteristic peaks of the SBS [31]. The FTIR characteristics of the four asphalt binders were similar, indicating that no new chemical functional groups were generated after other polymer additives were added to the SBS asphalt. Other polymer additives did not change the molecular properties of the SBS asphalt binder, and the interaction between the other polymer additives and the SBS was virginly physical.

4.1.2. FTIR Characteristics after Aging

TFOT and PAV were adopted to age different asphalt binders. The FTIR pictures of asphalt binders after aging are shown in Figure 3.

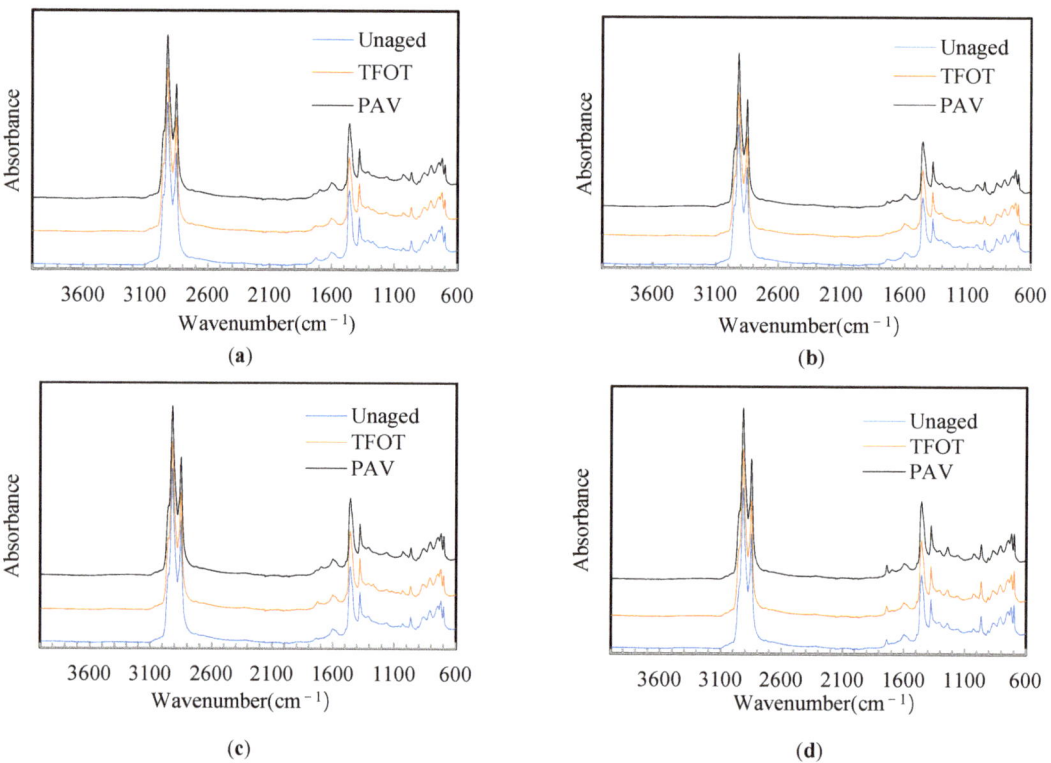

**Figure 3.** FTIR of asphalt binders after aging: (**a**) SBS; (**b**) HMM-SBS; (**c**) ARA-SBS; (**d**) HVM-SBS.

From Table 4, it can be observed that the aging of the asphalt binders showed the same pattern. The peak areas at 1700 cm$^{-1}$ and 1030 cm$^{-1}$ increased, while the peak areas at 966 cm$^{-1}$ and 699 cm$^{-1}$ decreased, and the peak area at 1376 cm$^{-1}$ was stable. This phenomenon implies that the aging process increases the amount of carbonyl and sulfoxide in the asphalt binder and decreases the amount of butadiene and styrene in the SBS. When the polymer-modified asphalt was aged, some chemical bonds, such as C–C, and C=C, dismantled and reacted with oxygen, or sulfur-based compounds in the asphalt reacted with oxygen to form a new C=O bond and S=O bond [35]. Such chemical reactions resulted in the increase of peak areas at 1700 cm$^{-1}$ and 1030 cm$^{-1}$. Thus, many researchers used the carbonyl index to reflect the aging level of asphalt binders. To further compare the aging level of different asphalt binders, CAI (as shown in Table 3) of the asphalt binders were calculated, and the results are presented in Table 4.

**Table 4.** CAI of the asphalt binders.

| Binder Type | SBS | HMM-SBS | ARA-SBS | HVM-SBS |
|---|---|---|---|---|
| TFOT | 467.8 | 385.4 | 438.5 | 330.1 |
| PAV | 3798.6 | 2441.2 | 2772.6 | 1450.5 |

From Table 4, some points can be observed. (1) When the aging pattern changed from TFOT to PAV, the CAI of all the asphalt binders increased. (2) After the SBS asphalt binder was added into other polymer additives, CAI became lower under both TFOT and PAV conditions, indicating that HMM, ARA, and HVM can improve the aging resistance. (3) The ranking of CAI was HVM-SBS < HMM-SBS < ARA-SBS < SBS, indicating that HVM-SBS has the best aging resistance. This phenomenon may be due to the synergistic effect of the esters in the plasticizer and the SBS in the HVM-SBS.

*4.2. DSR Test Results Analysis*

4.2.1. DSR Test Results before Aging

According to the DSR test results, the rheology properties of the asphalt binders before aging are given in Figure 4. The complex modulus, storage modulus, loss modulus, phase angle, and rutting factor are compared here.

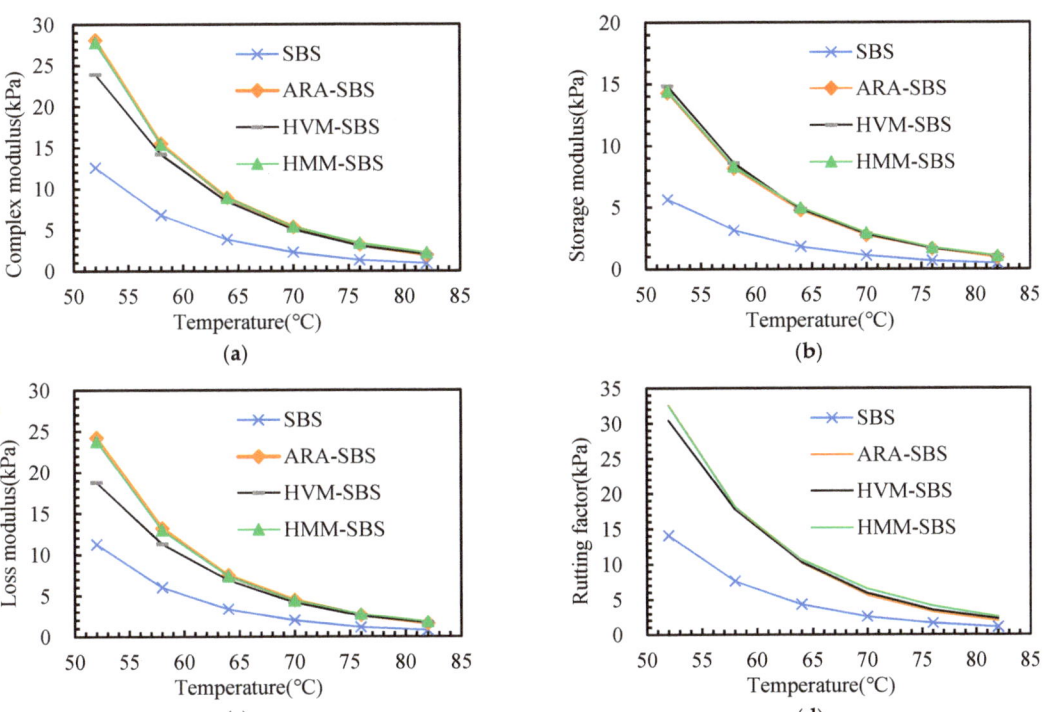

**Figure 4.** *Cont.*

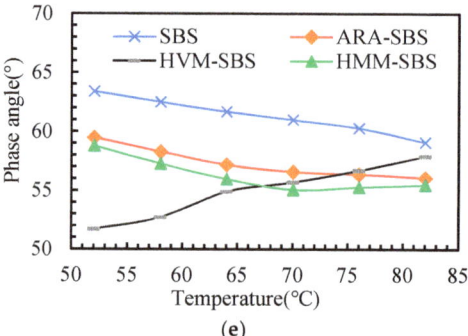

(e)

**Figure 4.** Rheology property of the asphalt binders before aging: (**a**) G*; (**b**) G′; (**c**) G″; (**d**) Rutting factor; (**e**) δ.

According to Figure 4, some points can be observed. (1) When polymer additives were added to the SBS asphalt binder, the complex modulus, storage modulus, and loss modulus were substantially increased, and the improvement caused by ARA and HMM was comparable and higher than that of HVM. Moreover, the rutting factor also increased obviously, indicating that polymer additives enhance the rutting resistance of the SBS asphalt binder. (2) The phase angle decreased after the polymer additives were added. A lower phase angle means that there were more elastic components than viscous components in the material. Thus, Figure 4e shows that the addition of polymer additives to the SBS asphalt binder could improve its elasticity, which enhanced the shear-deformation resistance of the SBS asphalt binder. Additionally, the phase angle of the HVM-SBS increased with the temperature, while the other three asphalt binders showed an opposite rule. (3) By comparing Figure 4b,c, it can be observed that the storage modulus of the three polymer asphalt binders was almost the same, while the loss modulus of HVM-SBS was lower than that of HMM-SBS and ARA-SBS. This phenomenon shows that the three polymers had a similar ability to improve the elastic part of SBS asphalt. In contrast, HMM and ARA had a better ability to improve the viscous part of SBS asphalt than HVM. The phase angle refleceds the proportional relationship between the viscous and elastic parts of the material. The different improvements on the elastic and viscous parts of SBS asphalt by the three polymers caused the different phase angles of the three polymer-modified asphalts.

4.2.2. DSR Test Results after Aging

The rheology properties of the asphalt binders after aging are analyzed in this section. Considering the fact that (1) the storage modulus, loss modulus, and rutting factor can be calculated with complex modulus and phase angle, (2) the complex modulus, storage modulus, loss modulus, and rutting factor showed a similar change pattern with the aging degree of the asphalt binders deepening, thus this paper takes the complex modulus and phase angle, for instance, to analyze the rheology property of the asphalt binders after aging. The complex modulus and phase angle of the asphalt binders after aging are presented in Figures 5 and 6.

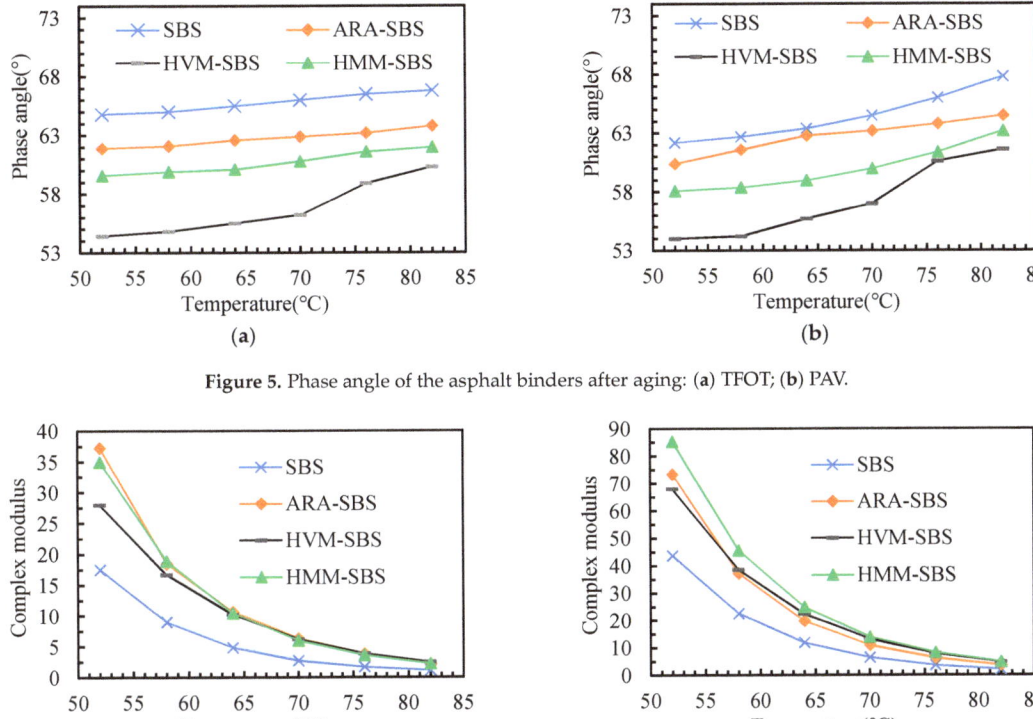

Figure 5. Phase angle of the asphalt binders after aging: (a) TFOT; (b) PAV.

Figure 6. Complex modulus of the asphalt binders after aging: (a) TFOT; (b) PAV.

According to Figure 5, several points can be observed. (1) Comparing the phase angle before and after aging, it can be observed that the phase angle of the four kinds of asphalt binders became higher after aging, indicating that the aging process increases the proportion of viscous components in the asphalt binders. Generally speaking, the higher the molecular weight of the polymer, the higher the viscosity. The aging process made the small molecules in the asphalt change to large molecules, causing the proportion of viscous components in the asphalt binder to rise and the phase angle to become higher. (2) The phase angle of the SBS asphalt binder was the greatest at both the short-term aging state and long-term aging state, followed by ARA-SBS, HMM-SBS, and HVM-SBS. This indicates that the polymer additives help to improve the elasticity of the SBS asphalt binder. Aging increased the viscosity of asphalt. Thus, the modification on the elasticity caused by polymer additives decreased the aging sensitivity of the SBS asphalt binder.

From Figure 6, it can be seen that, at both TFPT and PAV conditions, the complex modulus of ARA-SBS and HMM-SBS was higher, the complex modulus of HVM-SBS was lower, and the complex modulus of SBS was the lowest. This phenomenon was similar to the complex modulus before aging. Moreover, comparing the complex modulus before and after aging, it can be found that the complex modulus of asphalt binders increased after both TFOT and PAV. When the aging pattern changed from TFOT to PAV, the increment in complex modulus was more obvious. The aging process made the asphalt binder stiffer, resulting in a greater complex modulus.

The storage modulus, loss modulus, and rutting factor showed the same change pattern as the complex modulus. That is, the deeper the aging degree of the asphalt binder, the higher the values. The aging process increased the storage and loss modulus, resulting in a more excellent rutting resistance of asphalt binders.

### 4.2.3. Aging Indicators Analysis

This paper adopted several aging indicators (as shown in Table 3) to explore the effect of polymer additives on the aging sensitivity of asphalt binders. According to the DSR test results, these indicators are calculated and presented in Figures 7–11.

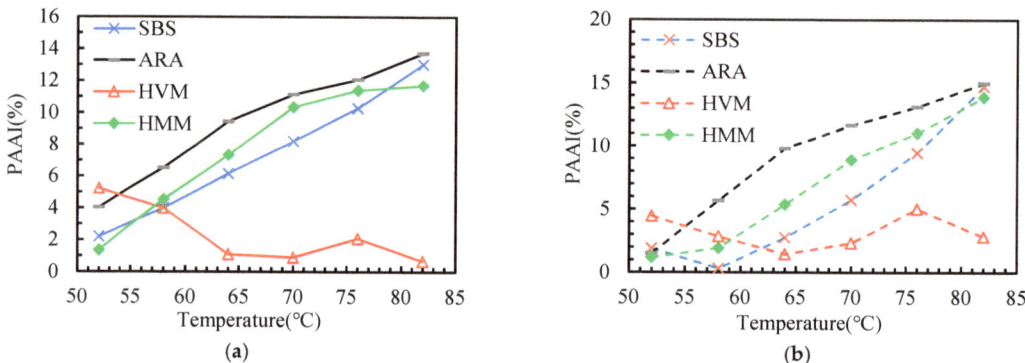

**Figure 7.** Phase Angle Aging Indicator (PAAI) of the asphalt binders after aging: (**a**) TFOT; (**b**) PAV.

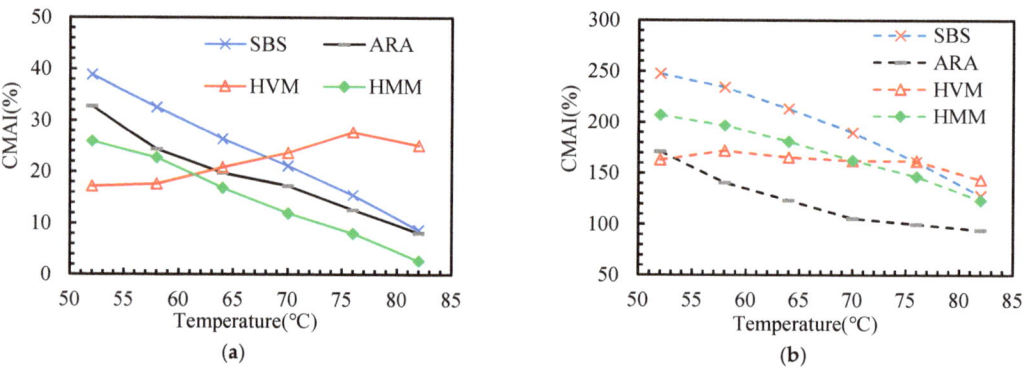

**Figure 8.** Complex Modulus Aging Indicator (CMAI) of the asphalt binders after aging: (**a**) TFOT; (**b**) PAV.

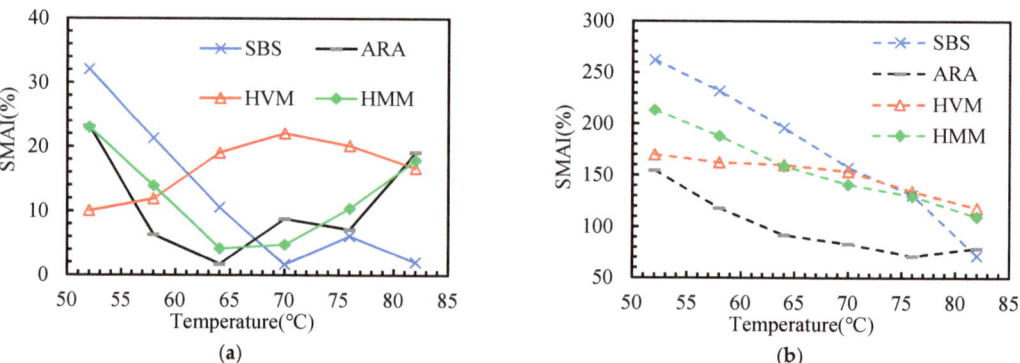

**Figure 9.** Storage Modulus Aging Indicator (SMAI) of the asphalt binders after aging: (**a**) TFOT; (**b**) PAV.

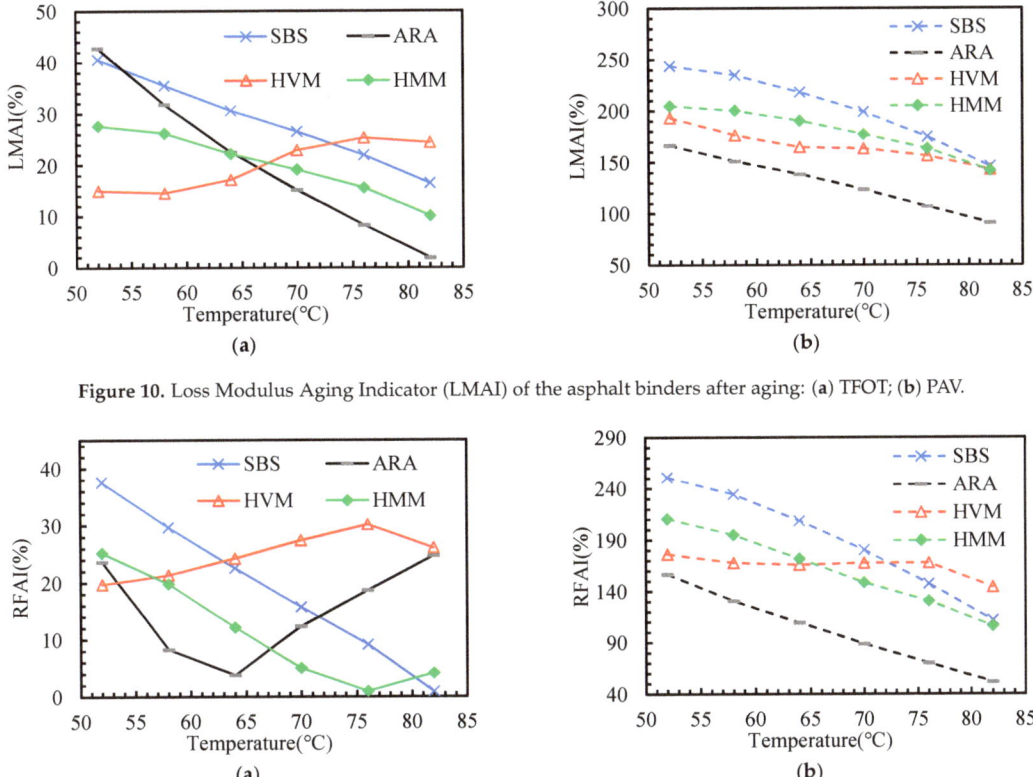

Figure 10. Loss Modulus Aging Indicator (LMAI) of the asphalt binders after aging: (**a**) TFOT; (**b**) PAV.

Figure 11. Rutting Factor Aging Indicator (RFAI) of the asphalt binders after aging: (**a**) TFOT; (**b**) PAV.

In this paper, the PAAI, CMAI, SMAI, LMAI, and RFAI were calculated to reflect the aging degree of asphalt binders. A higher aging indicator implies a greater aging degree of the asphalt binder. From Figures 7–11, some general commonalities can be seen. (1) The result was not the same when the aging degree was evaluated with different aging indicators. For example, the PAAI of ARA-SBS was higher than that of SBS, indicating that ARA-SBS experienced more severe aging. However, the CMAI of ARA-SBS was lower than that of SBS, implying an opposite conclusion. (2) The value of the aging indicators increased significantly when the aging pattern was changed from TFOT to PAV. This phenomenon follows the result of CAI. (3) In general, all the aging indicators, except PAAI of SBS, were higher than that of the other three polymer binders, indicating that polymer additives modify the aging resistance of SBS. However, it is hard to judge which polymer additive modified the aging resistance most. When the improvement on the aging resistance was evaluated with different aging indicators, the evaluation results were not the same. For example, the PAAI of ARA-SBS was higher than that of HMM-SBS, implying that HMM is more beneficial to improve the aging resistance of SBS. However, when this situation was evaluated with SMAI, an opposite conclusion could be drawn. (4) In most cases, the aging indicators of HVM-SBS increased with the temperature, while that of ARA-SBS and HMM-SBS showed the opposite rule.

The above analysis indicates that the aging degree of asphalt binders evaluated by the adopted aging indicators may result in different evaluation results. This phenomenon may be related to a variety of factors. For example, the phase angle and complex modulus of asphalt binders were both temperature dependent. Their temperature sensitivity was strongly related to the molecular weight distribution of various substances within the asphalt [36]. Therefore, when data at different temperatures were used to calculate the aging indicator, it was highly likely that different results would be obtained. Another possible reason is that aging causes changes in the molecular structure of certain substances within the asphalt [37]. These changes have different effects on the rheological indexes, such as phase angle and complex modulus of the asphalt. Thus, different results were obtained when using different aging indicators calculated by different rheological indices to evaluate the aging degree of asphalt binder.

There were six aging indicators in this paper, of which CAI was calculated by FTIR results, and the other five indicators were calculated by DSR results. Because the carbonyl index has proven to be an important indicator of asphalt aging degree [26,27], the aging indicator closely related to CAI can be a helpful indicator to characterize the aging degree of asphalt binders. Therefore, the correlation between CAI and other aging indicators was analyzed. The linear correlation coefficients (denoted by $R^2$) calculated by EXCEL software are shown in Table 5.

Table 5. The correlation coefficients between CAI and other aging indicators.

| Indicators | $R^2$ at the Following Temperatures (°C) | | | | | |
|---|---|---|---|---|---|---|
| | 52 | 58 | 64 | 70 | 76 | 82 |
| PAAI | 0.237 | 0.374 | 0.006 | 0.006 | 0.067 | 0.189 |
| CMAI | 0.912 | 0.851 | 0.815 | 0.759 | 0.692 | 0.661 |
| SMAI | 0.886 | 0.828 | 0.747 | 0.677 | 0.552 | 0.459 |
| LMAI | 0.843 | 0.847 | 0.843 | 0.801 | 0.748 | 0.699 |
| RFAI | 0.871 | 0.836 | 0.779 | 0.686 | 0.645 | 0.469 |

From Table 5, it can be seen that the correlation coefficients between CAI and other aging indicators declined with temperature in general. The rheological properties of asphalt were temperature dependent. Changes in temperature caused phase changes of certain substances within the asphalt, affecting the rheological properties, resulting in a wide variation in aging indexes at different temperatures. The FTIR of asphalt binders were measured at room temperature, which can lead to a better correlation of rheological property-based aging indexes with CAI at lower temperatures. Moreover, PAAI showed the worst correlation with CAI in all conditions. This may be because PAAI was affected by asphalt aging and was related to other properties of asphalt (such as glass transition temperature). This phenomenon also indicates that PAAI is not applicable to evaluate the aging sensitivity of asphalt, which agrees with the findings of other researchers [25]. When the temperature was lower than 64 °C, CMAI showed the best correlation with CAI, and the highest $R^2$ was 0.912. On the contrary, when the temperature was over 64 °C, the correlation coefficient between CAI and LMAI was the greatest. Therefore, the aging degree of polymer-modified asphalt binders can be predicted by CMAI and LMAI.

## 5. Conclusions

In this paper, the aging property of the SBS asphalt binder with different polymer additives were evaluated. The following conclusions can be drawn.

1. The SBS asphalt binder with polymer additives showed a lower CAI under both TFOT and PAV conditions, indicating that polymer additives can improve the aging resistance of the SBS asphalt binder. The ranking of CAI was HVM-SBS < HMM-SBS < ARA-SBS < SBS.
2. When polymer additives were added to the SBS asphalt binder, the complex modulus, storage modulus, loss modulus, and rutting factor were substantially increased, and the improvement caused by ARA and HMM was comparable and higher than that of HVM. The phase angle decreased after the polymer additives were added.

3. The phase angle, complex modulus, storage modulus, loss modulus, and rutting factor of asphalt binders became higher after aging. The phase angle of the SBS asphalt binder was the highest at both unaged and aged states, while its other indicator values were the smallest.
4. CMAI and LMAI had the best correlation coefficients with CAI. The two aging indicators can predict the aging degree of polymer-modified asphalt binders.

In future work, the effect of different polymer additives on the aging behavior of asphalt mixtures should be explored, and the aging mechanism of different polymer-modified asphalt should be studied more thoroughly.

**Author Contributions:** Conceptualization, B.W. and P.X.; methodology, C.L.; software, Z.P.; validation, C.C. and J.X.; formal analysis, C.L.; investigation, C.C.; resources, J.X.; data curation, C.L.; writing—original draft preparation, Z.P.; writing—review and editing, B.W., C.C and J.X.; visualization, P.X.; supervision, P.X.; project administration, P.X.; funding acquisition, B.W. All authors have read and agreed to the published version of the manuscript.

**Funding:** This research was funded by National Natural Science Foundation of China, grant number 52008365.

**Institutional Review Board Statement:** Not applicable.

**Informed Consent Statement:** Not applicable.

**Data Availability Statement:** The data presented in this study are available on request from the corresponding author.

**Conflicts of Interest:** The authors declare no conflict of interest.

# References

1. Zheng, J.L.; Lv, S.T.; Liu, C.C. Technical system, key scientific problems and technical frontier of long-life pavement. *Sci. Bull.* **2020**, *65*, 3219–3227.
2. Zhang, C.G. Factors of affecting the durability of asphalt pavement and improvement measure. *Pet. Sci. Technol.* **2007**, *21*, 66–69.
3. Ma, T.; Ding, X.; Zhang, D.; Huang, X.; Chen, J. Experimental study of recycled asphalt concrete modified by high-modulus agent. *Constr. Build. Mater.* **2016**, *128*, 128–135. [CrossRef]
4. Xiao, F.; Ma, D.; Wang, J.; Cai, D.; Lou, L.; Yuan, J. Impacts of high modulus agent and anti-rutting agent on performances of airfield asphalt pavement. *Constr. Build. Mater.* **2019**, *204*, 1–9. [CrossRef]
5. Hu, J.; Ma, T.; Zhu, Y.; Huang, X.; Xu, J.; Chen, L. High-viscosity modified asphalt mixtures for double-layer porous asphalt pavement: Design optimization and evaluation metrics. *Constr. Build. Mater.* **2021**, *271*, 121893. [CrossRef]
6. Wu, C.Y.; Jing, B.; Li, X.Y. Performance Evaluation of High-Modulus Asphalt Mixture. *Adv. Mat. Res.* **2011**, *311*, 2138–2141. [CrossRef]
7. Lee, H.J.; Lee, J.H.; Park, H.M. Performance evaluation of high modulus asphalt mixtures for long-life asphalt pavements. *Constr. Build. Mater.* **2007**, *21*, 1079–1087. [CrossRef]
8. Chen, B.; Guo, M.; Li, P.; Xia, J.; Chen, C.; Kang, A. The influence of anti-rutting agent and SBS modifier on the road performance of SMA-13 asphalt mixture. *Highway* **2016**, *61*, 196–200.
9. Ulucayli, M. Utilization of Anti-rutting additives and experience gained in Turkey. In Proceedings of the 3rd International Conference on Bituminous Mixtures and Pavements, Thessaloniki, Greece, 5–6 November 2002.
10. Sun, J.; Huang, L.; Li, C. Research on the technology applying anti-rutting additive to asphalt mixture. In Proceedings of the International Conference on Mechanical Engineering and Control Systems, Wuhan, China, 26–27 July 2015.
11. Chen, Z.; Zhang, X.; Cong, L.; Lu, H.; Yang, J. Rutting resistance performance evaluation of superpave mixes with antirutting additives. In Proceedings of the Transportation Research Board 89th Annual Meeting, Washington, DC, USA, 10–14 January 2010.
12. Li, L.H.; Geng, H.; Sun, Y.N. Viscosity evaluation indicators and criteria of high-viscosity modified asphalt. *J. Tongji Univ. Nat. Sci.* **2010**, *38*, 1155–1160.
13. Yang, Y.S.; Dong, Q. Study on durability of granulated crumb rubber asphalt pavement based on TPS high-viscosity asphalt. *Appl. Mech. Mater.* **2014**, *587*, 985–989. [CrossRef]
14. Tan, Y.; Zhang, H.; Cao, D.; Xia, L.; Du, R.; Shi, Z.; Dong, R.; Wang, X. Study on cohesion and adhesion of high-viscosity modified asphalt. *Int. J. Transp. Sci. Technol.* **2019**, *8*, 394–402.
15. Li, M.; Zeng, F.; Xu, R.; Cao, D.; Li, J. Study on compatibility and rheological properties of high-viscosity modified asphalt prepared from low-grade asphalt. *Materials* **2019**, *12*, 3776. [CrossRef]
16. Wang, S.; Huang, W. Investigation of aging behavior of terminal blend rubberized asphalt with SBS polymer. *Constr. Build. Mater.* **2021**, *267*, 120870. [CrossRef]

17. Ruan, Y.; Davison, R.R.; Glover, C.J. The effect of long-term oxidation on the rheological properties of polymer modified asphalts. *Fuel* **2003**, *82*, 1763–1773. [CrossRef]
18. Lin, P.; Yan, C.; Huang, W.; Li, Y.; Zhou, L.; Tang, N.; Xiao, F.; Zhang, Y.; Lv, Q. Rheological, chemical and aging characteristics of high content polymer modified asphalt. *Constr. Build. Mater.* **2019**, *207*, 616–629. [CrossRef]
19. Cortizo, M.S.; Larsen, D.O.; Bianchetto, H.; Alessandrini, J.L. Effect of the thermal degradation of SBS copolymers during the ageing of modified asphalts. *Polym. Degrad. Stab.* **2004**, *86*, 275–282. [CrossRef]
20. Sugano, M.; Iwabuchi, Y.; Watanabe, T.; Kajita, J.; Iwata, K.; Hirano, K. Relations between thermal degradations of SBS copolymer and asphalt substrate in polymer modified asphalt. *Clean Technol. Environ. Policy* **2010**, *12*, 653–659. [CrossRef]
21. Zhao, Y.; Gu, F.; Xu, J.; Jin, J. Analysis of aging mechanism of SBS polymer modified asphalt based on Fourier transform infrared spectrum. *J. Wuhan Univ. Technol. Mater. Sci. Ed.* **2010**, *25*, 1047–1052. [CrossRef]
22. JTG E20-2011. *Test Standard Methods of Bitumen and Bituminous Mixtures for Highway Engineering*; China Communications Press: Beijing, China, 2011.
23. JTG F40-2004. *Technical Specification for Construction of Highway Asphalt Pavements*; China Communications Press: Beijing, China, 2004.
24. Xiao, P.; Chen, Y. A study on comparative experiment of performance of physical and chemical crumb rubber modified asphalt. *Highway* **2010**, *30*, 130–133.
25. Wang, S.; Huang, W.; Lv, Q.; Yan, C.; Lin, P.; Zheng, M. Influence of different high viscosity modifiers on the aging behaviors of SBSMA. *Constr. Build. Mater.* **2020**, *253*, 119214. [CrossRef]
26. Hofko, B.; Porot, L.; Falchetto Cannone, A. FTIR spectral analysis of bituminous binders: Reproducibility and impact of ageing temperature. *Mater.Struct.* **2018**, *52*, 44–59. [CrossRef]
27. Liu, M.; Ferry, M.A.; Davison, R.R. Oxygen uptake as correlated to carbonyl growth in aged asphalts and asphalt Corbett fractions. *Ind. Eng. Chem. Res.* **1998**, *37*, 4669–4674. [CrossRef]
28. Morian, N.; Zhu, C.; Hajj, E.Y. Rheological indexes: Phenomenological aspects of asphalt binder aging evaluations. *Transp.Res. Rec.* **2015**, *2505*, 32–40. [CrossRef]
29. ASTM D1754/1754M. *Standard Test Method for Effect of Heat and Air on Asphaltic Materials (Thin-Film Oven Test)*; ASTM International: West Conshohocken, PA, USA, 2009.
30. ASTM D6521. *Standard Practice for Accelerated Aging of Asphalt Binder Using a Pressurized Aging Vessel (PAV)*; ASTM International: West Conshohocken, PA, USA, 2019.
31. AASHTO T315. *Standard Method of Test for Determining the Rheological Properties of Asphalt Binder Using a Dynamic Shear Rheometer (DSR)*; AASHTO: Washington, DC, USA, 2016.
32. Xu, Z.; Chen, Z.; Chang, Y. Application of infrared spectroscopy to detect the dosage of SBS in modified asphalt. *J. Chang. Univ. (Nat. Sci. Ed.)* **2015**, *35*, 7–12.
33. Diego, O.; Larsen, J.L.A.; Alejandra, B. Micro-structural and rheological characteristics of SBS-asphalt blends during their manufacturing. *Constr. Build. Mater.* **2009**, *23*, 2769–2774.
34. Zhang, F.; Hu, C. Preparation and properties of high viscosity modified asphalt. *Polym. J.* **2017**, *38*, 936–946. [CrossRef]
35. Qian, G.; Yu, H.; Jin, D. Different water environment coupled with ultraviolet radiation on ageing of asphalt binder. *Road Mater. Pavement Des.* **2020**, 1–14. [CrossRef]
36. Kumar, A.; Choudhary, R.; Kumar, A. Aging characteristics of asphalt binders modified with waste tire and plastic pyrolytic chars. *PLoS ONE* **2021**, *16*, e0256030.
37. Alghrafy, Y.M.; Abd Alla El-Sayed, M.; El-Badawy, S.M. Rheological properties and aging performance of sulfur extended asphalt modified with recycled polyethylene waste. *Constr. Build. Mater.* **2021**, *273*, 121771. [CrossRef]

Article

# Investigation of Adhesion Performance of Wax Based Warm Mix Asphalt with Molecular Dynamics Simulation

Chao Peng [1,2,*], Hanneng Yang [1], Zhanping You [2,*], Hongchao Ma [1], Fang Xu [1], Lingyun You [3], Aboelkasim Diab [4], Li Lu [1], Yudong Hu [1], Yafeng Liu [1], Jing Dai [5] and Zhibo Li [1]

[1] Faculty of Engineering, China University of Geosciences, Wuhan 430074, China
[2] Department of Civil and Environmental Engineering, Michigan Technological University, Houghton, MI 49931-1295, USA
[3] School of Civil and Hydraulic Engineering, Huazhong University of Science and Technology, Wuhan 430074, China
[4] Department of Civil Engineering, Aswan University, Aswan 81542, Egypt
[5] Key Laboratory of Advanced Technology for Specially Functional Materials, School of Materials Science and Engineering, Wuhan University of Technology, Wuhan 430070, China
* Correspondence: pengchao@cug.edu.cn (C.P.); zyou@mtu.edu (Z.Y.)

**Citation:** Peng, C.; Yang, H.; You, Z.; Ma, H.; Xu, F.; You, L.; Diab, A.; Lu, L.; Hu, Y.; Liu, Y.; et al. Investigation of Adhesion Performance of Wax Based Warm Mix Asphalt with Molecular Dynamics Simulation. *Materials* 2022, 15, 5930. https://doi.org/10.3390/ma15175930

Academic Editor: Simon Hesp

Received: 26 July 2022
Accepted: 24 August 2022
Published: 27 August 2022

**Publisher's Note:** MDPI stays neutral with regard to jurisdictional claims in published maps and institutional affiliations.

**Copyright:** © 2022 by the authors. Licensee MDPI, Basel, Switzerland. This article is an open access article distributed under the terms and conditions of the Creative Commons Attribution (CC BY) license (https://creativecommons.org/licenses/by/4.0/).

**Abstract:** Compared with traditional hot mix asphalt (HMA), wax based warm mix asphalt (WWMA) can be mixed with the aggregate at a lower temperature and achieve the desired compaction. However, the adhesion performance of WWMA on aggregate is uncertain. To evaluate the adhesion performance of asphalt and aggregate, researchers used contact angle test, pull-off test, and ultrasonic washing experiments. However, these tests cannot adequately explain the microscopic mechanism of the interface between asphalt and aggregate. Molecular dynamics (MD) can better explain the adhesion mechanism of asphalt aggregates because they can be simulated at the molecular scale. So, the purpose of this research is to use the MD method to study the adhesion performance between WWMA and aggregate. Two aggregate oxides ($CaCO_3$ and $SiO_2$) models, the matrix asphalt model and WWMA models, were built in Materials Studio (MS) software. The adhesion work of asphalt and aggregate oxides was calculated. With the increase of wax modifier content, the adhesion work of asphalt and aggregate oxides ($CaCO_3$ and $SiO_2$) first increases and then decreases. When the wax modifier is increased to 3 wt%, the adhesion works of the WWMA-$SiO_2$ and WWMA-$CaCO_3$ increase by 31.2% and 14.0%, compared with that of matrix asphalt. In this study, the accuracy of the MD calculation result was verified by the pull-off experiments and the contact angle experiments. WWMA was prepared by a high-shear mixer emulsifier. In the pull-off experiments and the contact angle experiments, the tensile strength and the adhesion work between the aggregate and the asphalt containing 3% wax modifier reaches peak values. These values are 140.7% and 124.9%, compared with those between the aggregate and the matrix asphalt. In addition, the results of the pull-off experiments and the contact angle experiments are in good agreement with that of the MD simulation. Finally, Fourier transform infrared spectroscopy (FTIR) shows that the carbonyl content of WWMA is greater than that of matrix asphalt. It explains well that the wax modifier promotes the adhesion between asphalt and aggregate. This paper provides an important theoretical basis to understand the adhesion performance of WWMA and aggregate.

**Keywords:** molecular dynamics; wax warm mix asphalt; contact angle test; pull-off test; Fourier transform infrared spectroscopy; adhesion work

## 1. Introduction

Compared with hot mix asphalt (HMA), warm mix asphalt (WMA) can be mixed with aggregate at lower temperatures and reduces energy consumption. The mixing temperature of WMA is usually 20 to 40 °C lower than that of HMA [1–4]. WMA can improve the workability of asphalt binders and reduce the emission of harmful gases. It is a green environmental protection technology with low energy consumption and emission.

There are three main methods for producing WMA by use of synthetic or organic additives, which are zeolite [5,6], surfactant [7,8] and wax [9,10]. The zeolite and wax reduce the mixing temperature of asphalt and aggregate by reducing the viscosity of the asphalt. The surfactant reduces the mixing temperature of asphalt aggregate by reducing the surface tension between asphalt and aggregate. Compared with surfactant and zeolite, wax can improve the moisture susceptibility and anti-rutting performance of asphalt mixtures [11,12]. Therefore, wax modifiers are more widely used to reduce the mixing temperature of asphalt and aggregate.

The adhesion performance of asphalt and aggregate is directly related to the water damage resistance of asphalt pavement [13]. Enhancing the adhesion of wax based warm mix asphalt (WWMA) and aggregate is of great significance for prolonging the service life of WWMA pavement. In recent years, many researchers have studied the adhesion performance of WWMA and aggregate. Akhtar et al. [14] evaluate the adhesion performance of Sasobit (a type of wax modifier) modified asphalt and aggregate by atomic force microscope (AFM). They found that Sasobit improves the adhesion between asphalt and aggregate. Wen [15] calculated the adhesion work of Sasobit modified asphalt and matrix asphalt on aggregate using surface free energy to evaluate the adhesion strength. The result shows that the adhesion strength between Sasobit modified asphalt and aggregate is better than that of matrix asphalt and aggregate. Yang et al. [16] measured the contact angle of different liquids on the surfaces of penetration grade 60–70 asphalt (Pen 60–70 asphalt) and WWMA to calculate the surface energy and adhesion work of asphalt and aggregate. They found that the adhesion work of WWMA on aggregate is increased by 6% compared with Pen 60–70 asphalt on aggregate. This shows that WWMA has better adhesion strength than Pen 60–70 asphalt. To sum up, most of the above researchers used macroscopic experiments to study the adhesion performance between WWMA and aggregate. However, the microscopic mechanism related to adhesion performance between WWMA and aggregate is still unclear.

In recent years, MD simulation has been an effective method to study the microscopic mechanism of material components from the perspective of molecular motion. Many researchers have evaluated the adhesion performance between asphalt and aggregate by calculating the adhesion work of the asphalt-aggregate model in MD simulation. Tarefder et al. [17] used the MD method to determine the thermodynamic properties of asphaltene before and after oxidative aging. The results showed that the glass transition temperature is related to the viscosity, hardness, and rigidity of asphalt, and decreases with the increased oxidative aging degree. Qu et al. [18] studied paraffin's effect on the mechanical properties of asphalt binder through MD simulation. The results showed that paraffin could reduce the high-temperature stability and self-healing rate of asphalt binder. Chu et al. [19] established asphalt-aggregates (quartz and calcite) models in MS software. They discovered that the adhesion strength of asphalt and calcite is greater than that of asphalt and quartz. Gao et al. [20] constructed the interface models of asphalt and four aggregates (quartz, calcite, albite, and microcline) in MS software. They revealed that the adhesion work of asphalt and microcline is larger than asphalt and other aggregates. Xu et al. [21] adopted the ReaxFF force field to simulate the interaction between the main molecules in asphalt and aggregate ($SiO_2$). They found that the adhesion work of phenol molecules and $SiO_2$ is much larger than that of other molecules and $SiO_2$. The above researches show that MD has been widely used to study the adhesion performance of asphalt-aggregate. However, there are few studies on the adhesion performance of WWMA and aggregate by using the MD method.

In this study, we studied the adhesion between WWMA and aggregates using the MD method. Twelve different molecules were first selected to represent the asphalt component models. Then, wax molecules were chosen as the wax modifier to construct WWMA models. The adhesion work between asphalt and two aggregate oxides ($CaCO_3$ and $SiO_2$)

was calculated by MD simulation. Then, the accuracy of the MD simulation calculation results was verified by the pull-off experiments and the contact angle experiments. The Fourier transform infrared spectroscopy (FTIR) was conducted to reveal the chemical mechanism of adhesion performance between WWMA and aggregate.

## 2. Experiments and Methods

### 2.1. MD Simulation Models

#### 2.1.1. Force Field Selection

MD simulation is to analyze the basic properties of materials by simulating the motion and interaction of material atoms. The principles of MD are based on statistical mechanics and Newton's laws of motion. Condensed-phase Optimized Molecular Potential for Atomistic Simulation Studies (COMPASS) can accurately simulate the interaction of organic and inorganic molecules. The COMPASS force field has been successfully applied to simulate the adhesion performance of asphalt and aggregate in previous study [22]. The COMPASS force field was selected in this work and can be described by Equations (1)–(3) [23].

$$E_{total} = E_{val} + E_{non-bond} \tag{1}$$

$$E_{val} = E_b + E_\theta + E_\varphi + E_\chi + E_{bb'} + E_{b\theta} + E_{b\varphi} + E_{\theta\varphi} + E_{\theta\theta'} + E_{\theta\theta'\varphi} \tag{2}$$

$$E_{non-bond} = E_{elec} + E_{LJ} \tag{3}$$

where $E_{total}$ is total potential energy, $E_{val}$ is valence state energy, $E_{non-bond}$ is non-bond energy, $E_b$ is bond stretching energy, $E_\theta$ is angular bending energy, $E_\varphi$ is internal torsion energy, $E_\chi$ is out-of-plane bending energy, $E_{bb'}$, $E_{b\theta}$, $E_{b\varphi}$, $E_{\theta\varphi}$, $E_{\theta\theta'}$, and $E_{\theta\theta'\varphi}$ are the interactions of crosscoupling terms, $E_{elec}$ is Coulomb electrostatic energy, and $E_{LJ}$ is the van der Waals energy.

#### 2.1.2. Matrix Asphalt Model

The American Society for Testing and Materials (ASTM, West Conshohocken, PA, USA) D4124-09 proposed the SARA (saturated, aromatic, resin, and asphaltene) classification scheme. Li and Greenfield [24] developed three 12-component asphalt models, namely AAA-1, AAK-1 and AAM-1. Compared with other model systems, the density and thermal expansion coefficient of the AAA-1 model system are closer to the experimental data of Jones et al. [25]. In this paper, the AAA-1 asphalt model is selected as the model of matrix asphalt. The specific parameters of matrix asphalt are listed in Table 1. The constructed four-component molecular models of matrix asphalt are shown in Figure 1.

Table 1. Matrix asphalt model parameters.

| Molecules in Model | | Molar Mass | Molecular Formula | Number of Molecules |
|---|---|---|---|---|
| Group | Label | | | |
| Saturate | A1 | 422.9 | $C_{30}H_{62}$ | 4 |
| | B1 | 483.0 | $C_{35}H_{62}$ | 4 |
| Aromatic | A2 | 464.8 | $C_{35}H_{44}$ | 11 |
| | B2 | 406.8 | $C_{30}H_{46}$ | 13 |
| | A3 | 554.0 | $C_{40}H_{59}N$ | 4 |
| | B3 | 573.1 | $C_{40}H_{60}S$ | 4 |
| Resin | C3 | 414.8 | $C_{29}H_{50}O$ | 5 |
| | D3 | 530.9 | $C_{36}H_{57}N$ | 4 |
| | E3 | 290.4 | $C_{18}H_{10}S_2$ | 15 |
| | AS1 | 575.0 | $C_{42}H_{54}O$ | 3 |
| Asphaltene | AS2 | 888.5 | $C_{66}H_{81}N$ | 2 |
| | AS3 | 707.2 | $C_{51}H_{63}S$ | 3 |

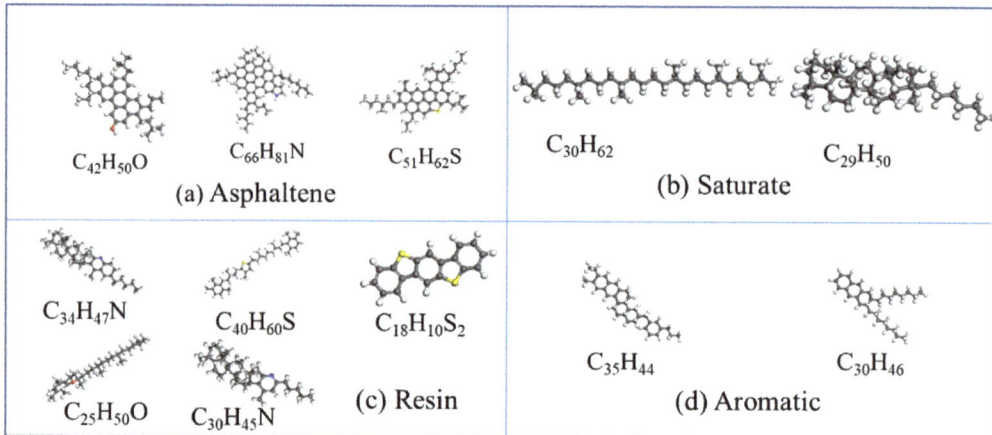

**Figure 1.** (a) Asphaltene molecular models; (b) Saturate molecular models; (c) Resin molecular models; (d) Aromatic molecular models.

First, the Amorphous Cell Calculation module in the MS software was used to construct an amorphous asphalt model. The initial density of asphalt was set to 0.1 g/cm³. Then, the asphalt was placed in a periodic cube box. The constructed matrix asphalt model is shown in Figure 2.

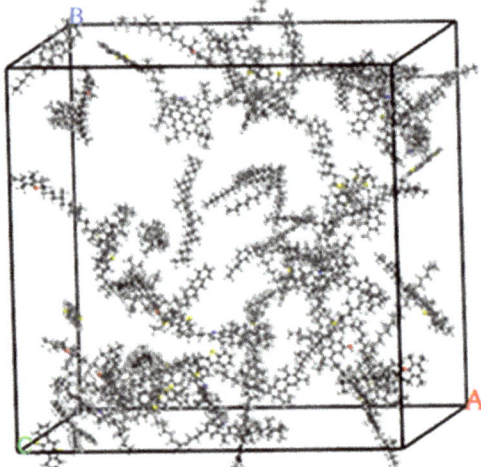

**Figure 2.** Matrix asphalt model.

2.1.3. WWMA Models

The wax molecular model constructed by Samieadel et al. [26] was used in this study. The molecular formula of wax is $C_{11}H_{24}$. The molecular model of wax is shown in Figure 3.

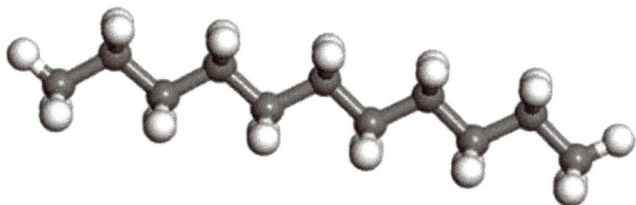

**Figure 3.** Wax molecular model.

The molar mass of the matrix asphalt model is 32,710.4 g/mol. The molar mass of wax is 156.4 g/mol. A total of 2, 4, 6, and 8, mol of wax molecules were added to the matrix asphalt to construct WWMA models. The WWMA models are shown in Figure 4.

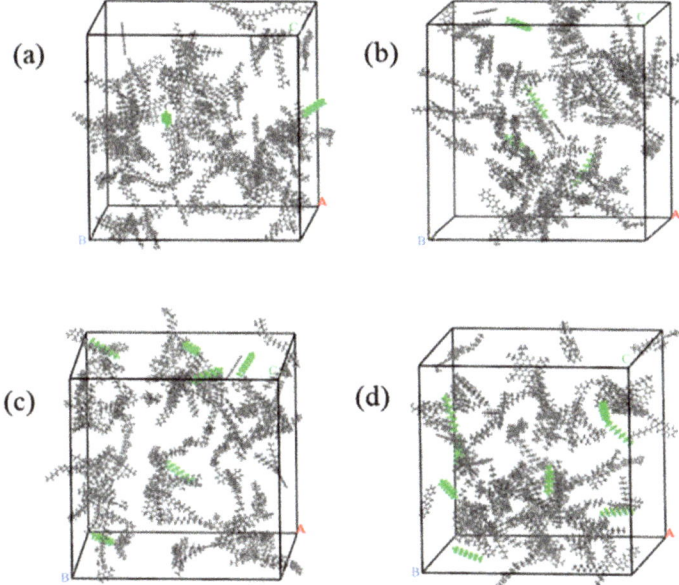

**Figure 4.** (a) WWMA-1 model; (b) WWMA-2 model; (c) WWMA-3 model; (d) WWMA-4 model.

2.1.4. Model Optimization

The matrix asphalt and WWMA models were geometrically optimized in the Forcite module. In order to minimize the energy of the asphalt system, the asphalt models were subjected to 5000 iterations under the COMPASS force field to achieve geometric optimization. Then, the asphalt models were used for dynamics simulation in the NPT (constant pressure and constant temperature) ensemble and the NVT (constant temperature and constant volume) ensemble for 100 ps to be stable. Finally, asphalt models with stable volume and energy fluctuations were obtained. The asphalt models after geometric optimization and dynamics simulation are shown in Figure 5. Compared with Figures 2 and 4, the structure of each asphalt model is more compact after geometric optimization and dynamics simulation.

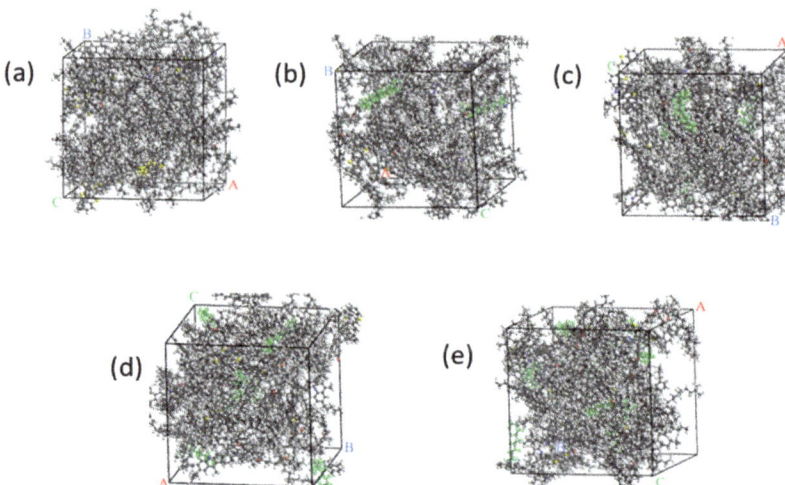

**Figure 5.** (a) Optimized matrix asphalt model; (b) Optimized WWMA-1 model; (c) Optimized WWMA-2 model; (d) Optimized WWMA-3 model; (e) Optimized WWMA-4 model.

## 2.2. Adhesion Work

### 2.2.1. Single Cell Oxide Model

Limestone, basalt, granite, diabase, and amphibolite, are mostly used as aggregates. The aggregates mainly contain $SiO_2$, $CaCO_3$, $Al_2O_3$, MgO, and other components. In this paper, two oxide molecules ($SiO_2$ and $CaCO_3$) were selected in MD simulation. The $SiO_2$ and $CaCO_3$ models are shown in Figure 6. The single-cell model parameters of $SiO_2$ and $CaCO_3$ are listed in Table 2.

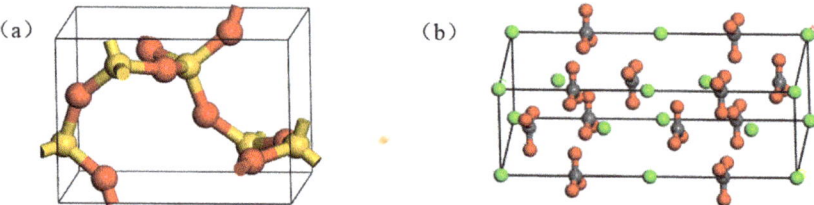

**Figure 6.** (a) The single-cell model of $SiO_2$; (b) The single-cell model of $CaCO_3$.

**Table 2.** Single-cell model parameters of $SiO_2$ and $CaCO_3$.

| Oxide Type | Edge Length (Å) | | | Cross Angle (°) | | |
|---|---|---|---|---|---|---|
| | a | b | c | α | β | γ |
| $SiO_2$ | 4.98 | 4.98 | 6.95 | 90 | 90 | 90 |
| $CaCO_3$ | 4.99 | 4.99 | 17.06 | 90 | 90 | 120 |

### 2.2.2. Oxide Supercell Model

In MS, the aggregate model is represented by a supercell. First, the single-cell oxide was cut along the (0,0,1) direction to expose the surface. Then, the Supercell function in the Build toolbar was used to build the supercell model. A 10 Å vacuum layer was inserted in the Z direction of the constructed supercell model. Finally, geometric optimization was used to optimize its structure. The $SiO_2$ supercell model and the $CaCO_3$ supercell model are shown in Figure 7.

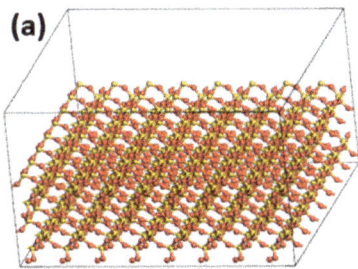 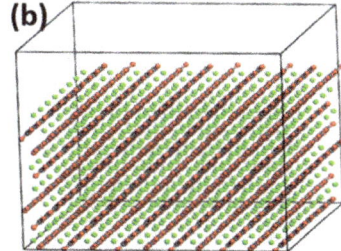

**Figure 7.** (a) The supercell model of SiO$_2$; (b) The supercell model of CaCO$_3$.

2.2.3. Calculation of Adhesion Work

The asphalt-aggregate interface system was constructed by attaching an asphalt layer to the surface of the aggregate. Then, a 30 Å vacuum layer was placed on top of the asphalt layer. In the MD simulation, the asphalt-aggregate models were first geometrically optimized. Then, the dynamic balance of the NVT ensemble of 100 ps was performed to optimize the structure of the asphalt-aggregate models further. In this study, all the MD simulated temperatures were performed at 298 K, and the specific parameters refer to the literature [27].

In order to quantitatively evaluate the adhesion strength of asphalt and aggregate, this study calculated the adhesion work of the asphalt-aggregate model. Adhesion work is defined as the energy required to separate an interface per unit area into two free surfaces in a vacuum [28]. The adhesion work corresponds to the adhesion strength of the asphalt-aggregate system. The adhesion work of asphalt-aggregate per unit area ($W_{MB}$) and the adhesion work of the asphalt-aggregate ($\nabla E_{MB}$) can be calculated by Equations (4) and (5) [29].

$$W_{MB} = \frac{\nabla E_{MB}}{A} \quad (4)$$

$$\nabla E_{MB} = E_M + E_B - E_{MB} \quad (5)$$

In Equation (4), $A$ is the contact area between asphalt and aggregate. In Equation (5), $E_M$ and $E_B$ are the potential energy of aggregate and asphalt at thermodynamic equilibrium, respectively. $E_{MB}$ is the asphalt-aggregate system in thermodynamics potential energy at equilibrium. Since the $A$ of the asphalt-aggregate system is constant, $\nabla E_{MB}$ can directly reflect the adhesion work of the asphalt-aggregate system.

*2.3. Macroscopic Adhesion Performance Experiment*

2.3.1. Raw Materials

In this study, the used 90# matrix asphalt is produced by Maoming Petrochemical Company in Guangdong Province, China. Its physical properties are listed in Table 3. The wax modifier used to prepare WWMA comes from AkzoNobel Co., Ltd., Jiaxing, China. Wax modifier is light white solid particles. The physical properties of the wax modifier are listed in Table 4.

**Table 3.** Physical properties of 90# matrix asphalt.

| Physical Properties | Unit | Value | Technical Requirements | Test Method (JTG F40-2004) |
|---|---|---|---|---|
| Softening Point | °C | 49.6 | >42 | T0606 |
| Penetration (25°) | 0.1 mm | 86.1 | 80–100 | T0604 |
| Ductility (15°) | mm | 153 | >100 | T0605 |
| Flashing point | °C | 345 | >245 | T0611 |

Table 4. The physical properties of wax modifier.

| Physical Properties | Unit | Value |
| --- | --- | --- |
| Density | g/cm$^3$ | 0.9 |
| Melting point | °C | 99 |
| Flash point | °C | 285 |

### 2.3.2. Preparation of WWMA

First, 90# matrix asphalt was heated. Then, 1 wt%, 2 wt%, 3 wt%, and 4 wt%, of wax modifier were mixed with the hot 90# matrix asphalt, respectively. The WWMA were prepared by a high-shear mixer and stirred at a speed of 5000 r/min for 15 min. The WWMA samples containing 1 wt%, 2 wt%, 3 wt%, and 4 wt%, of wax modifier are referred to as WWMA-1, WWMA-2, WWMA-3, and WWMA-4, respectively.

### 2.3.3. Pull-Off Test

In this experiment, the ZQS6 pull-off instrument was used to measure the tensile strength. The pull-off instrument is shown in Figure 8. First, limestone was cut into 1.5 cm × 1.5 cm × 1 cm cubes (the main component is CaCO$_3$). Then, the limestone cubes and asphalt were placed in an oven at 170 °C. When the asphalt was heated into a fluid state, asphalt was dropped on to the surface of the bottom limestone cube. After the asphalt spreads evenly to the surface of the bottom limestone, the upper limestone was placed on the bottom limestone. When the asphalt sample was cooled to 20 °C, the upper and bottom limestone cubes adhered to the pull head and the marble slab with epoxy glue, respectively. Afterward, the height of the pull head was adjusted until the pull force reading of the pull-off instrument was 0. Pull force was applied to the pull head through the handle until the asphalt breaks. In the breaking process of asphalt, the tensile strength can be recorded by a digital display sensor. The tensile strength of each asphalt sample and limestone was tested in three replicates to obtain an average value.

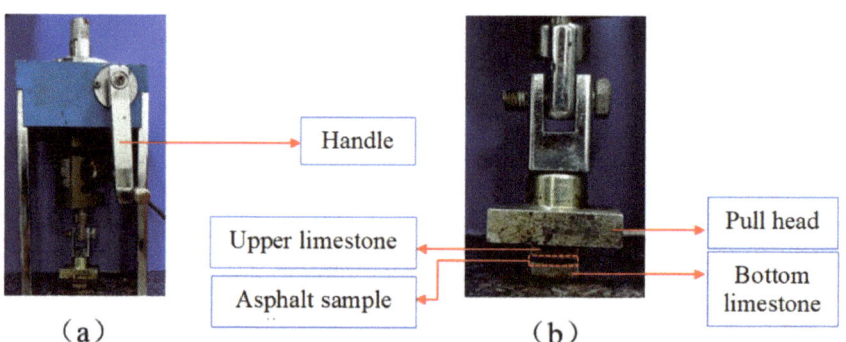

**Figure 8.** (a) The photo of the pull-off instrument; (b) The photo of asphalt sample and the pull head.

The results of the pull-off experiment can directly evaluate the adhesion strength of each asphalt sample and limestone. The tensile strength ($\sigma$) of each asphalt sample and limestone can be calculated by Equation (6).

$$\sigma = \frac{F_d}{A} \tag{6}$$

where $F_d$ is the maximum tensile force when the asphalt is broken. $A$ is the contact area of asphalt and limestone cube.

### 2.3.4. Contact Angle Experiment

The SDC-100 contact angle instrument was used in this experiment. The photo of the contact angle instrument and the liquid drop on the sample surface is shown in Figure 9. The sessile drop method was used to measure the contact angles of water and glycerin on the asphalt surface. First, the asphalt sample and glass slide were heated in an oven at 170 °C. When the asphalt became liquid, the asphalt was dropped on to the glass slide. When the asphalt spreads evenly over the entire surface of the glass slide, the glass slide was cooled at room temperature until the asphalt became solid. Water and glycerin with known surface energy parameters were used in the contact angle experiment to calculate the surface free energy of each asphalt sample and limestone. The surface free energy parameters of water and glycerin are listed in Table 5 [30]. The contact angle of water and glycerin on each sample's surface was measured three times to obtain an average value.

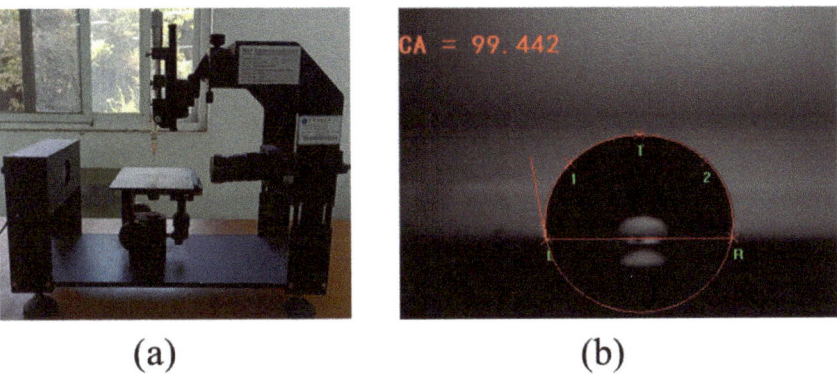

**Figure 9.** (a) The photo of contact angle instrument; (b) The photo of a liquid drop on the sample surface.

**Table 5.** Surface free energy parameters of water and glycerin (25 °C, unit: mJ/m$^2$).

| Surface Free Energy Parameters | Water | Glycerol |
|---|---|---|
| $\gamma_f^d$ | 21.8 | 37.0 |
| $\gamma_f^p$ | 51.0 | 26.4 |
| $\gamma$ | 72.8 | 63.4 |

In Table 5, $\gamma_f^d$ is the dispersion component of the water or glycerol; $\gamma_f^p$ is polar component of the water or glycerol; $\gamma$ is the surface free energy.

### 2.3.5. FTIR Experiment

The FTIR of 90# matrix asphalt, WWMA-3, and WWMA-4, were tested in this study. First, each sample was dissolved in carbon disulfide. Then, the prepared solution was taken out and dropped on to the potassium bromide wafer. After waiting for the carbon disulfide to volatilize completely, the FTIR of the samples was measured. The ordinate and abscissa of the FTIR are transmittance and wave number, respectively. The infrared spectrum scanned from 4000 to 500 cm$^{-1}$ wavenumbers. The number of scans was 64 times and the resolution was 4 cm$^{-1}$.

## 3. Results and Discussion

### 3.1. The Rationality Verification of the Asphalt Model

The density and solubility parameters of the optimized asphalt models are listed in Table 6. The actual asphalt density is 1.00–1.04 g/cm$^3$ [31]. After geometric optimization and dynamics simulation, their density is in the range of 0.981–0.987 g/cm$^3$. Compared with the actual asphalt, the difference value in density between the asphalt models and the actual asphalt is within 5%. The actual reference value of asphalt solubility is 15.3–23.0 (J/cm$^3$)$^{0.5}$. The solubility range of the asphalt models constructed is 17.398–17.697 (J/cm$^3$)$^{0.5}$ within the reference value range. It shows that the constructed asphalt models can reflect the properties of the natural asphalt from the perspective of density and solubility parameters.

**Table 6.** Density of matrix asphalt and WWMA.

| Sample | Density (g/cm$^3$) | Reference Value (g/cm$^3$) | Solubility Parameter (J/cm$^3$)$^{0.5}$ | Reference Value (J/cm$^3$)$^{0.5}$ |
|---|---|---|---|---|
| Matrix asphalt | 0.987 | | 17.398 | |
| WWMA-1 | 0.986 | | 17.586 | |
| WWMA-2 | 0.983 | 1.00–1.04 | 17.544 | 15.3–23.0 |
| WWMA-3 | 0.981 | | 17.514 | |
| WWMA-4 | 0.983 | | 17.697 | |

### 3.2. Adhesion Work between Asphalt and Aggregate Oxide

The adhesion work of each asphalt and two aggregate oxides is shown in Figure 10. It can be concluded that the WWMA and the two aggregate oxides have all increased compared with the matrix asphalt. With the increase of wax modifier, the adhesion works of asphalt and two aggregate oxides first increase and then decrease. When the content of wax modifier in asphalt increases to 3 wt%, the adhesion work of WWMA with $SiO_2$ and $CaCO_3$ increased to 248.3 kcal/mol and 271.5 kcal/mol, respectively. Besides, the adhesion works of WWMA-3 with $SiO_2$ and $CaCO_3$ are higher than those of others. This shows that 3 wt% wax modifier improves the adhesion strength of asphalt with $SiO_2$ and $CaCO_3$ most significantly. In addition, it can also be seen that the adhesion work between asphalt and $CaCO_3$ is greater than that between asphalt and $SiO_2$. This indicates that the adhesion strength of asphalt and $CaCO_3$ is higher than that of asphalt and $SiO_2$. This is consistent with the results of previous laboratory measurements [32,33]. The optimized interface models between each asphalt and two aggregate oxides are shown in Figure 11. It can be seen that the interface distance between asphalt and $SiO_2$ is greater than that between asphalt and $CaCO_3$. This indicates that the adhesion strength of asphalt and $SiO_2$ is less than that of asphalt and $CaCO_3$. This is because the adhesion strength between aggregate and asphalt is determined by their atomic interaction force. The atomic interaction force mainly includes Coulomb electrostatic force and van der Waals force [34]. The increase of the interface distance reduces the Coulomb electrostatic force and van der Waals force, resulting in that the adhesion strength of asphalt and $CaCO_3$ is better than that of $SiO_2$. The follow-up of this article will systematically study the adhesion mechanism of asphalt and $CaCO_3$.

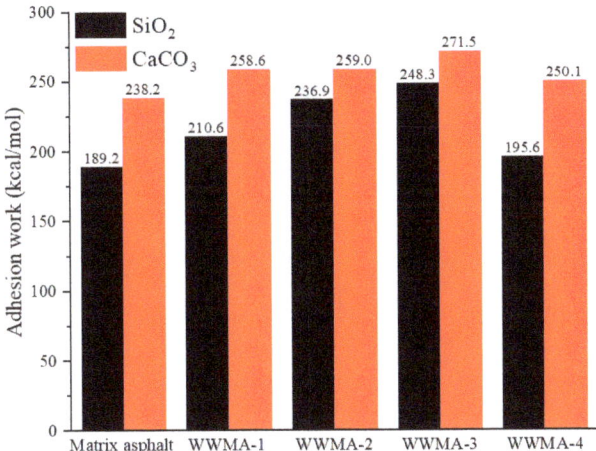

**Figure 10.** The adhesion work of each asphalt with $SiO_2$ and $CaCO_3$.

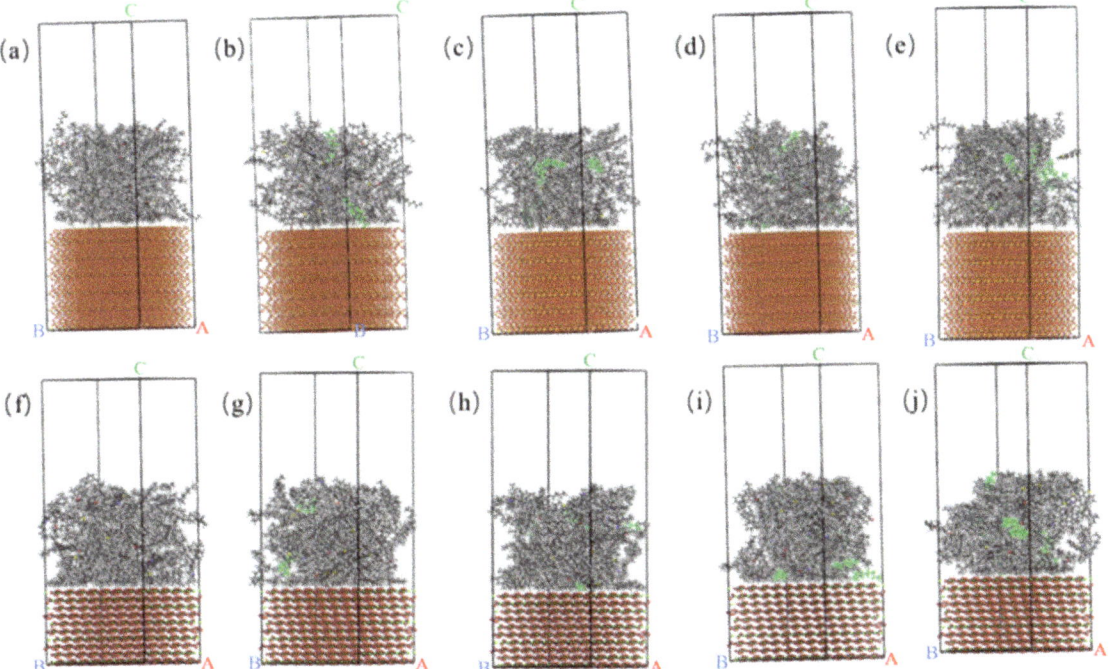

**Figure 11.** (**a**) Matrix asphalt and $SiO_2$ interface model; (**b**) WWMA-1 and $SiO_2$ interface model; (**c**) WWMA-2 and $SiO_2$ interface model; (**d**) WWMA-3 and $SiO_2$ interface model; (**e**) WWMA-4 and $SiO_2$ interface model; (**f**) MA and $CaCO_3$ interface model; (**g**) WWMA-1 and $CaCO_3$ interface model; (**h**) WWMA-2 and $CaCO_3$ interface model; (**i**) WWMA-3 and $CaCO_3$ interface model; (**j**) WWMA-4 and $CaCO_3$ interface model.

## 3.3. Relative Concentration Analysis

In order to explore the adhesion mechanism of WWMA and aggregate, the relative concentration curves of matrix asphalt, WWMA-3, WWMA-4, and CaCO$_3$ in the Z direction, derived in the MS software, are shown in Figure 12. The relative concentration at the distance of 32 angstroms correspond to asphalt molecules. The relative concentration of matrix asphalt, WWMA-3, and WWMA-4, is 1.9, 2.2, and 2.4, respectively. It can be seen that the relative concentration of WWMA-3 is larger than that of matrix asphalt and WWMA-4. This phenomenon indicates that the interaction force between WWMA-3 and aggregate is greater than that of matrix asphalt and WWMA-4. In other words, the interaction force between WWMA-3 and aggregate attracts more asphalt molecules at the asphalt-aggregate interface. This also explains that the adhesion strength between WWMA-3 and CaCO$_3$ in Figure 10 is higher than those between other WWMA samples and CaCO$_3$.

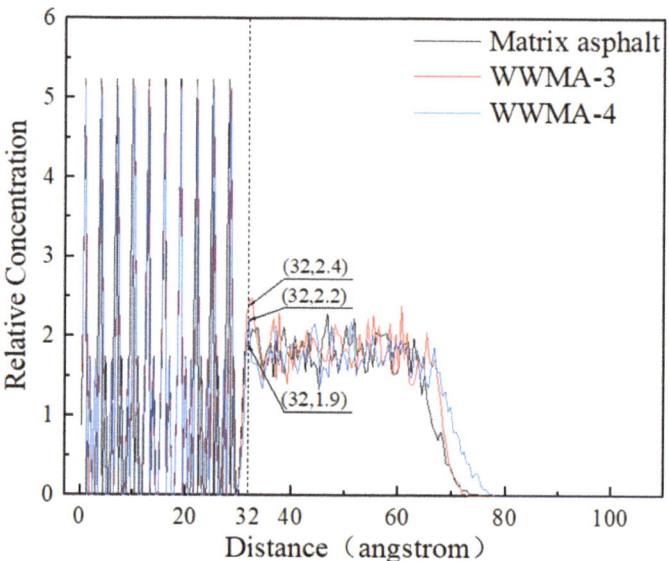

**Figure 12.** The relative concentration distribution of matrix asphalt, WWMA-3, WWMA-4, and CaCO$_3$.

## 3.4. Pull-Off Test Results

The tensile strength of asphalt and limestone cube are shown in Figure 13. It can be seen that the tensile strength of the matrix asphalt is 631 kPa. The tensile strength of WWMA and limestone cube is higher than that of matrix asphalt and limestone cube. In particular, the tensile strength of WWMA-3 and limestone cube is increased to 888 kPa, which is 40.7% higher than that of matrix asphalt and limestone cube. In addition, it can be seen that the tensile strength of asphalt and limestone cube first increases and then decreases with the increase of wax modifier. WWMA-3 has a higher tensile strength with limestone cube than other asphalts, indicating that WWMA-3 has the best adhesion strength with limestone cube. Pull-off test results are in good agreement with the MD simulation results. This proves that MD can simulate the adhesion performance between asphalt and aggregate well.

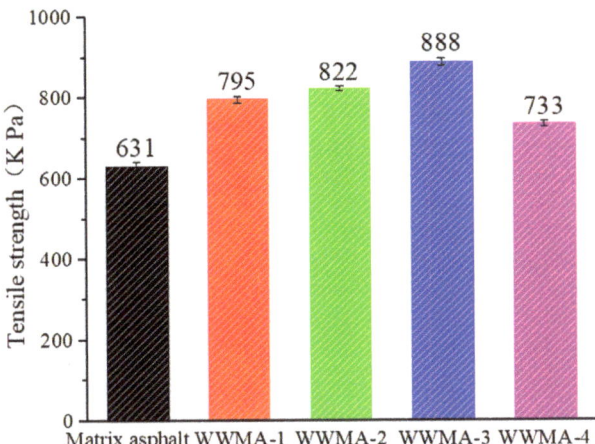

**Figure 13.** Tensile strength of asphalt and limestone cube.

*3.5. Contact Angle Experiment*

The surface free energy theory can be used to study the adhesion of asphalt and aggregate [35,36]. The surface tension of the solid or liquid is composed of the dispersion component caused by the Newtonian force, and the polar component caused by the non-Newtonian force [37,38]. Contact angles of two liquids (water and glycerin) on the surface of all asphalt samples can be measured by a contact angle instrument to calculate the surface free energy of the asphalt. Contact angles of the two liquids on the surface of all asphalt samples are listed in Table 7.

**Table 7.** Contact angles of the liquid on the surface of all asphalt samples (°).

| Sample | limestone | Matrix Asphalt | WWMA-1 | WWMA-2 | WWMA-3 | WWMA-4 |
| --- | --- | --- | --- | --- | --- | --- |
| Water | 61.71 | 99.44 | 97.09 | 96.01 | 94.31 | 97.67 |
| Glycerin | 73.02 | 97.33 | 96.95 | 96.85 | 95.79 | 96.39 |

In Table 7, it can be seen that contact angles of water and glycerin on the WWMA surface have become smaller, compared with those on the matrix asphalt surface. The contact angles of water and glycerin on the WWMA-3 surface are 94.31° and 95.79°. These contact angles are lower than those on the other asphalt surfaces. It indicates that the wax modifier improves the hydrophilicity of the asphalt.

According to the measured contact angles of water and glycerin on all asphalt surfaces in Table 7, the surface free energy parameters of asphalt and limestone cube can be calculated by Equations (7) and (8) [39].

$$\frac{\gamma_f}{2\sqrt{\gamma_f^d}} \cdot (1+cos\theta) = \sqrt{\frac{\gamma_f^p}{\gamma_f^d}} \cdot \sqrt{\gamma_a^p} + \sqrt{\gamma_a^d} \qquad (7)$$

$$\gamma = \gamma^d + \gamma^p \qquad (8)$$

In Equation (7): $\gamma_f$ is the surface energy of the liquid; $\gamma_a^d$ is the dispersion component of the solid; $\gamma_a^p$ is polar component of the solid; $\gamma_f^d$ is dispersion component of liquid; $\gamma_f^p$ is polar component of the liquid; $\theta$ is contact angle of the liquid on the solid surface. In Equation (8): $\gamma^p$ is the polar component of the asphalt or limestone, and it has the same meaning as $\gamma_a^p$; $\gamma^d$ is the dispersion component of the asphalt or limestone, and it has the same meaning as $\gamma_a^d$; $\gamma$ is the indicated surface free energy of the asphalt or limestone.

The surface free energy parameters of all asphalt samples and limestone are shown in Figure 14. It can be seen that the surface free energy of the WWMA is significantly higher than that of the matrix asphalt. It is worth noting that the surface free energy of the WWMA-3 is 16.1 mJ/m², which is higher than those of other asphalts.

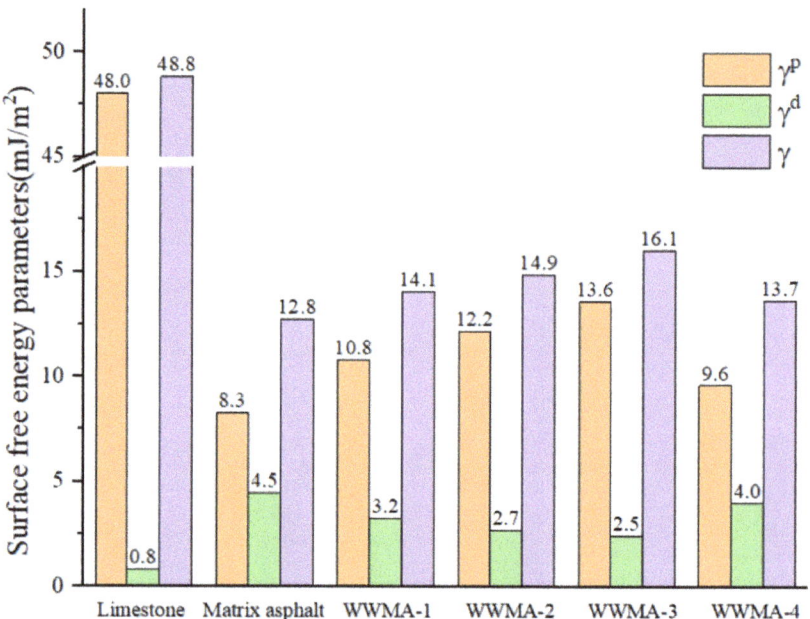

**Figure 14.** The surface free parameters of asphalt samples and limestone.

From a microscopic point of view, the adhesion work is expressed as the energy required to separate the asphalt from the aggregate. According to the dispersion and polar components obtained in Figure 14, the adhesion work ($W_{adhesion}$) of asphalt and limestone can be calculated by Equation (9) [34]:

$$W_{adhesion} = 2\sqrt{\gamma_b^d \gamma_{lim}^d} + \sqrt{\gamma_b^p \gamma_{lim}^p} \tag{9}$$

where $\gamma_{lim}^p$ is the polar component of the limestone cube; $\gamma_{lim}^d$ is the dispersed component of the limestone cube; $\gamma_b^p$ is the polar component of the asphalt; $\gamma_b^d$ is the dispersed component of the asphalt.

The calculated adhesion works of asphalts and limestone are shown in Figure 15. It can be seen that the adhesion work of the WWMA and limestone is higher than that of the matrix asphalt and limestone. This is because the wax modifier increases the polar component of the asphalt in Figure 14. In Equation (9), it can be seen that the adhesion work of asphalt and limestone is positively related to the polar components of asphalt. Compared with matrix asphalt, the adhesion work of WWMA-3 and limestone has increased by 10.3 mJ/m². WWMA-3 has a higher adhesion work with limestone than other asphalts. It indicates that WWMA-3 has the best adhesion strength with limestone. The results of the contact angle experiment also verify the accuracy of the MD simulation calculation results.

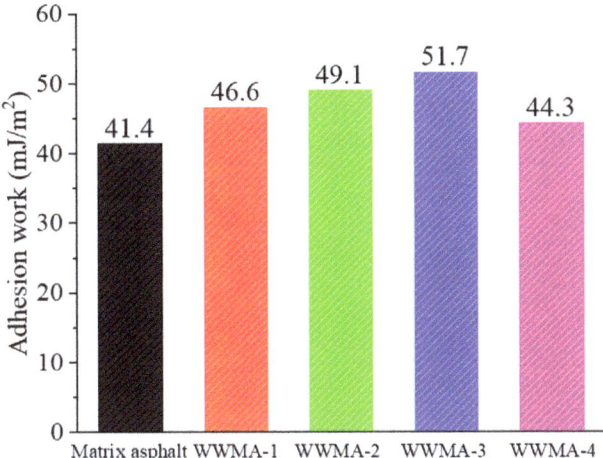

**Figure 15.** Calculated adhesion work of each asphalt and limestone by the contact angle test.

### 3.6. FTIR

The FTIR spectra of matrix asphalt, WWMA-3, and WWMA-4, are shown in Figure 16. The relative transmittance of the carbonyl group was calculated according to the reference method [40]. The relative transmittance of the carbonyl group can be calculated by Equation (10):

$$R_{C=O} = \frac{Tr_{C=O}}{Tr_{-CH2-}} \tag{10}$$

where $Tr_{C=O}$ is the transmittance of the carbonyl group; $Tr_{-CH2-}$ is the transmittance of the methylene group.

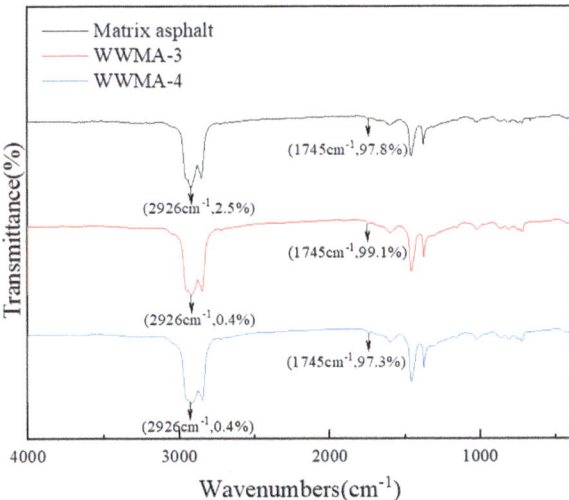

**Figure 16.** The FTIR spectra of matrix asphalt, WWMA-3 and WWMA-4.

The peak at 1745 cm$^{-1}$ and 2926 cm$^{-1}$ is ascribed to the carbonyl group and methylene [41,42]. The relative transmittances of the carbonyl group of matrix asphalt, WWMA-3, and WWMA-4, are 39.12, 247.75, and 243.25, respectively. It indicates the relative carbonyl content of WWMA-3 and WWMA-4 are both higher than that of matrix asphalt. The electronegativity of oxygen in the carbonyl group is higher than that of carbon, so the carbonyl group is negatively charged. In $CaCO_3$, $Ca^{2+}$ is positively charged with +2 valence. This may lead to the mutual attraction of carbonyl and $Ca^{2+}$ to improve the adhesion of asphalt and aggregate. In addition, it can be seen that the relative carbonyl content in WWMA-3 is higher than those in matrix asphalt and WWMA-4. This reveals the inherent mechanism that the adhesion strength of WWMA-3 and $CaCO_3$ is higher than that of matrix asphalt and WWMA-4. Based on the above calculated adhesion work, it can be inferred that the adhesion work between asphalt and aggregate is positively correlated with the relative carbonyl content.

## 4. Conclusions

This paper uses MD simulation to study the adhesion of WWMA and aggregate. Two typical aggregate oxides ($SiO_2$ and $CaCO_3$) were selected to construct the asphalt-aggregate interface model. The molecular interaction between WWMA-aggregate was quantitatively studied with MD simulation. Pull-off and contact angle experiments verify the simulation result. The conclusions are as follows:

(1) In MD simulation, the adhesion work of the WWMA and the two aggregate oxides is larger than that of matrix asphalt and the two aggregate oxides. The adhesion works of the WWMA-3 on $SiO_2$ and $CaCO_3$ increase to the peak values. They increase by 31.2% and 14.0% compared to the matrix asphalt on $SiO_2$ and $CaCO_3$. The pull-off experiments and the contact angle experiments are also in good agreement with that of the MD simulation.

(2) The pull-off and contact angle experiments prove that WWMA can increase the tensile strength and adhesion work between the asphalt and the limestone cube. The tensile strength and the adhesion work between the aggregate and the asphalt containing 3 wt% wax modifier reach the peak values. These values are 140.7% and 124.9% compared with those between the aggregate and the matrix asphalt.

(3) FTIR reveals that the carbonyl content in WWMA is much higher than that in matrix asphalt. It explains well that carbonyl can enhance the adhesion strength of asphalt and aggregate.

In this paper, the performances of WWMA are only studied indoors, and field experiments will be a new direction to further prove the accuracy of MD simulation results.

**Author Contributions:** Data curation, H.Y.; formal analysis, H.M.; investigation, F.X. and Z.L.; methodology, A.D. and J.D.; project administration, C.P.; resources, L.L., Y.H. and Y.L.; supervision, C.P.; writing—original draft, C.P. and H.Y.; writing—review and editing, Z.Y. and L.Y. All authors have read and agreed to the published version of the manuscript.

**Funding:** This research was funded by the National Natural Science Foundation of China (52108425), China University of Geosciences (Wuhan) (CUGL150412, G1323531606 and G1323519261), National College Student Innovation and Entrepreneurship Training Program (S202010491065), Anhui Road and Bridge Engineering Group Co., LTD (2021056235) and Shandong Highway and Bridge Maintenance Co., Ltd. (2021056502).

**Institutional Review Board Statement:** Not applicable.

**Informed Consent Statement:** Not applicable.

**Data Availability Statement:** The data presented in this study are available on request from the corresponding author.

**Acknowledgments:** This research was completed at the Faculty of Engineering in China University of Geosciences, Wuhan, China.

**Conflicts of Interest:** The authors declare no conflict of interest.

## References

1. Abed, A.; Thom, N.; Grenfell, J. A novel approach for rational determination of warm mix asphalt production temperatures. *Constr. Build. Mater.* **2019**, *200*, 80–93.
2. Abd, D.M.; Al-Khalid, H.; Akhtar, R. An investigation into the impact of warm mix asphalt additives on asphalt mixture phases through a nano-mechanical approach. *Constr. Build. Mater.* **2018**, *198*, 296–306.
3. Zhang, Y.; Leng, Z.; Zou, F.; Wang, L.; Chen, S.S.; Tsang, D.C.W. Synthesis of zeolite A using sewage sludge ash for application in warm mix asphalt. *J. Clean. Prod.* **2018**, *172*, 687–695.
4. Feitosa, J.P.M.; Alencar, A.E.V.; Filho, N.W.; Souza, J.R.R. Evaluation of sun-oxidized carnauba wax as warm mix asphalt additive. *Constr. Build. Mater.* **2016**, *115*, 294–298.
5. Woszuk, A.; Zofka, A.; Bandura, L.; Franus, W. Effect of zeolite properties on asphalt foaming. *Constr. Build. Mater.* **2017**, *139*, 247–255.
6. Han, X.; Cao, Z.; Wang, R.; He, P.; Zhang, Y.; Yu, J.; Ge, Y. Effect of silane coupling agent modified zeolite warm mix additives on properties of asphalt. *Constr. Build. Mater.* **2020**, *259*, 119713.
7. Wan, Z.; Zheng, B.; Xie, X.; Yang, J.; Zhou, H. Preparation method and performance test of Evotherm pre-Wet treatment aluminum hydroxide type warm-Mixed flame-Retardant asphalt. *Constr. Build. Mater.* **2020**, *262*, 120618.
8. Wu, S.; Zhang, W.; Shen, S.; Li, X.; Muhunthan, B.; Mohammad, L.N. Field-Aged asphalt binder performance evaluation for Evotherm warm mix asphalt: Comparisons with hot mix asphalt. *Constr. Build. Mater.* **2017**, *156*, 574–583.
9. Yue, M.; Yue, J.; Wang, R.; Xiong, Y. Evaluating the fatigue characteristics and healing potential of asphalt binder modified with Sasobit and polymers using linear amplitude sweep test. *Constr. Build. Mater.* **2021**, *289*, 123054.
10. Selvadurai, S.K.; Hasan, M.R.M.; Sani, A.; Hiromitsu, N.; Poovaneshvaran, S. Improvements of TPS-Porous Asphalt Using Wax-Based Additives for the Application on Malaysian Expressway. *J. Kejuruter.* **2021**, *33*, 205–215.
11. Vaitkus, A.; Čygas, D.; Laurinavičius, A.; Vorobjovas, V.; Perveneckas, Z. Influence of warm mix asphalt technology on asphalt physical and mechanical properties. *Constr. Build. Mater.* **2016**, *112*, 800–806. [CrossRef]
12. Janshidi, A.; Golchin, B.; Hamzah, M.O.; Turner, P. Selection of type of warm mix asphalt additive based on the rheological properties of asphalt binders. *J. Clean. Prod.* **2015**, *100*, 89–106. [CrossRef]
13. Dong, Z.; Gong, X.; Zhao, L.; Zhang, L. Mesostructural damage simulation of asphalt mixture using microscopic interface contact models. *Constr. Build. Mater.* **2014**, *53*, 665–673. [CrossRef]
14. Abd, D.M.; Al-Khalid, H.; Akhtar, R. Nano-Scale properties of warm-Modified bituminous binders determined with atomic force microscopy. *Road Mater. Pavement Des.* **2017**, *18*, 189–202. [CrossRef]
15. Wen, B. Adhesion of Warm-Mixed Asphalt Aggregate Interface with Organic Wax and Evaluation of Water Stability of Mixtures. Master's Thesis, Beijing Architecture University, Beijing, China, 2016.
16. Yang, S.H.; Rachman, F.; Susanto, H.A. Effect of moisture in aggregate on adhesive properties of warm-Mix asphalt. *Constr. Build. Mater.* **2018**, *190*, 1295–1307. [CrossRef]
17. Tarefder, R.; Rafiqul, A. Molecular dynamic simulation of oxidative aging in asphaltene. *Pavements Mater.* **2010**, *1*, 16–30.
18. Qu, X.; Wang, D.; Hou, Y.; Oeser, M.; Wang, L. Influence of Paraffin on the Microproperties of Asphalt Binder Using MD Simulation. *J. Mater. Civ. Eng.* **2018**, *30*, 04018191.
19. Chu, L.; Luo, L.; Fwa, T.F. Effects of aggregate mineral surface anisotropy on asphalt-aggregate interfacial bonding using molecular dynamics (MD) simulation. *Constr. Build. Mater.* **2019**, *225*, 1–12. [CrossRef]
20. Gao, Y.; Zhang, Y.; Gu, F.; Xu, T.; Wang, H. Impact of minerals and water on bitumen-Mineral adhesion and debonding behaviours using molecular dynamics simulations. *Constr. Build. Mater.* **2018**, *171*, 214–222.
21. Xu, Z.; Wang, Y.; Cao, J.; Chai, J.; Cao, C.; Si, Z.; Li, Y. Adhesion between asphalt molecules and acid aggregates under extreme temperature: A ReaxFF reactive molecular dynamics study. *Constr. Build. Mater.* **2021**, *285*, 122882.
22. Dong, Z.; Liu, Z.; Wang, P.; Gong, X. Nanostructure characterization of asphalt-Aggregate interface through molecular dynamics simulation and atomic force microscopy. *Fuel* **2017**, *189*, 155–163. [CrossRef]
23. Sun, H. COMPASS: An ab initio force-Field optimized for condensed-Phase applications-Overview with details on alkane and benzene compounds. *J. Phys. Chem. B* **1998**, *102*, 7338–7364. [CrossRef]
24. Li, D.D.; Greenfield, M.L. Chemical compositions of improved model asphalt systems for molecular simulations. *Fuel* **2014**, *115*, 347–356. [CrossRef]
25. Jones, D.R. *SHRP Materials Reference Library, Asphalt Cements: A Concise Data Compilation*; In Strategic Highway Research Program National Research Council: Washington, DC, USA, 1993.
26. Samieadel, A.; Oldham, D.; Fini, E.H. Multi-Scale characterization of the effect of wax on intermolecular interactions in asphalt binder. *Constr. Build. Mater.* **2017**, *157*, 1163–1172. [CrossRef]
27. Xu, G.; Wang, H. Molecular dynamics study of oxidative aging effect on asphalt binder properties. *Fuel* **2017**, *188*, 1–10. [CrossRef]
28. Yalghouzaghaj, M.N.; Sarkar, A.; Hamedi, G.H.; Hayati, P. Application of the surface free energy method on the mechanism of low-Temperature cracking of asphalt mixtures. *Constr. Build. Mater.* **2021**, *268*, 121194. [CrossRef]
29. Xu, G.; Wang, H. Study of cohesion and adhesion properties of asphalt concrete with molecular dynamics simulation. *Comput. Mater. Sci.* **2016**, *112*, 161–169. [CrossRef]
30. Peng, C.; Huang, S.; You, Z.; Xu, F.; You, L. Effect of a lignin-Based polyurethane on adhesion properties of asphalt binder during UV aging process. *Constr. Build. Mater.* **2020**, *247*, 118547. [CrossRef]

31. Zhang, L.; Greenfield, M.L. Effects of Polymer Modification on Properties and Microstructure of Model Asphalt Systems. *Energy Fuels* **2008**, *22*, 3363–3375. [CrossRef]
32. Yin, Y.; Chen, H.; Kuang, D.; Song, L.; Wang, L. Effect of chemical composition of aggregate on interfacial adhesion property between aggregate and asphalt. *Constr. Build. Mater.* **2017**, *146*, 231–237. [CrossRef]
33. Guo, M. Study on Mechanism and Multiscale Evaluation Method of Interfacial Interaction between Asphalt Binder and Mineral Aggregate. Ph.D. Thesis, Harbin Institute of Technology, Harbin, China, 2015.
34. Peng, C.; Chen, P.; You, Z.; Lv, S.; Zhang, R.; Xu, F.; Zhang, H.; Chen, H. Effect of silane coupling agent on improving the adhesive properties between asphalt binder and aggregates. *Constr. Build. Mater.* **2018**, *169*, 591–600. [CrossRef]
35. Zhang, H.; Li, H.; Abdelhady, A.; Jia, M.; Xie, N. Investigation on surface free energy and moisture damage of asphalt mortar with fine solid waste. *Constr. Build. Mater.* **2020**, *231*, 117140. [CrossRef]
36. Zhang, F.; Muhammad, Y.; Liu, Y.; Han, M.; Yin, Y.; Hou, D.; Li, J. Measurement of water resistance of asphalt based on surface free energy analysis using stripping work between asphalt-Aggregate system. *Constr. Build. Mater.* **2018**, *176*, 422–431. [CrossRef]
37. Tan, Y.; Guo, M. Using surface free energy method to study the cohesion and adhesion of asphalt mastic. *Constr. Build. Mater.* **2013**, *47*, 254–260. [CrossRef]
38. Zhu, J.; Zhang, K.; Liu, K.; Shi, X. Adhesion characteristics of graphene oxide modified asphalt unveiled by surface free energy and AFM-Scanned micro-Morphology. *Constr. Build. Mater.* **2020**, *244*, 118404. [CrossRef]
39. Hu, J.; Zhang, L.; Zhang, X.; Guo, Y.; Yu, X. Comparative evaluation of moisture susceptibility of modified/foamed asphalt binders combined with different types of aggregates using surface free energy approach. *Constr. Build. Mater.* **2020**, *256*, 119429. [CrossRef]
40. Peng, C.; Chen, P.; You, Z.; Lv, S.; Xu, F.; Zhang, W.; Yu, J.; Zhang, H. The anti-Icing and mechanical properties of a superhydrophobic coating on asphalt pavement. *Constr. Build. Mater.* **2018**, *190*, 83–94. [CrossRef]
41. Yaylayan, V.A.; Ismail, A.A. Investigation of the enolization and carbonyl group migration in reducing sugars by FTIR spectroscopy. *Carbohydr. Res.* **1995**, *276*, 253–265. [CrossRef]
42. Peng, C.; Jiang, G.; Lu, C.; Xu, F.; Yu, J.; Dai, J. Effect of 4,4′-Stilbenedicarboxylic acid-Intealated layered double hydroxides on UV aging resistance of bitumen. *Rsc Adv.* **2015**, *5*, 95504–95511. [CrossRef]

Article

# Characterization of Desulfurized Crumb Rubber/Styrene–Butadiene–Styrene Composite Modified Asphalt Based on Rheological Properties

Jingyao Yang, Gang Xu, Peipei Kong and Xianhua Chen *

School of Transportation, Southeast University, Nanjing 211189, China; yangjingyao19@foxmail.com (J.Y.); xugang619@hotmail.com (G.X.); kppwwy@163.com (P.K.)
* Correspondence: chenxh@seu.edu.cn

**Abstract:** With the growing interest in bituminous construction materials, desulfurized crumb rubber (CR)/styrene–butadiene–styrene (SBS) modified asphalts have been investigated by many researchers as low-cost environmental-friendly road construction materials. This study aimed to investigate the rheological properties of desulfurized CR/SBS composite modified asphalt within various temperature ranges. Bending beam rheometer (BBR), linear amplitude sweep (LAS), and multiple stress creep recovery (MSCR) tests were performed on conventional CR/SBS composite modified asphalt and five types of desulfurized CR/SBS modified asphalts. Meanwhile, Burgers' model and the Kelvin–Voigt model were used to derive nonlinear viscoelastic parameters and analyze the viscoelastic mechanical behavior of the asphalts. The experimental results indicate that both the desulfurized CR/SBS composite modifier and force chemical reactor technique can enhance the crosslinking of CR and SBS copolymer, resulting in an improved high-, intermediate-, and low-temperature performance of desulfurized CR/SBS composite modified asphalt. Burgers' model was found to be aposite in simulating the creep stages obtained from MSCR tests for CR/SBS composite modified asphalts. The superior high-temperature performance of desulfurized CR/SBS modified asphalt prepared with 4% SBS, 20% desulfurized rubber, and a force chemical reactor time of 45 min contributes to the good high-temperature elastic properties of the asphalt. Therefore, this combination is recommended as an optimal preparation process. In summary, the desulfurization of crumb rubber and using the force chemical reactor technique are beneficial to composite asphalt performance and can provide a new way of utilizing waste tire rubber.

**Keywords:** CR/SBS modified asphalt; desulfurized rubber; rheology; Burgers' model; multiple stress creep recovery

## 1. Introduction

With the rapid and large-scale construction and maintenance of highways, the demand for high-performance asphalt binder has increased sharply over the past several years. Fatigue, rutting, and other diseases of asphalt pavement are closely related to the elastic and viscous deformation behaviors of asphalt under various working conditions [1,2]. At low temperatures, elasticity dominates, and asphalt behaves as an elastomer. As the ambient temperature drops, the asphalt pavement shrinks, which can generate cracking due to the increased stiffness of asphalt binder limiting this shrinkage [3,4]. Whereas during construction and at high temperatures, viscous properties dominate, and it behaves as a fluid. When the traffic load is applied to the asphalt pavement, the asphalt flows and resulting in unstable rutting.

Traditional asphalt does not have sufficient rheological properties to alleviate the issues mentioned above. To meet the performance demands of bituminous materials caused by the rapid increase in traffic volumes, a series of studies by various researchers have been conducted in search of innovative asphalt preparation methods. One of the effective methods is to use various modifiers to decrease the stiffness modulus of asphalt at low temperatures and increase the complex modulus at high temperatures. Therefore, modified asphalt is being vigorously developed as an alternative road construction material [5]. Styrene–butadiene–styrene (SBS) copolymers are the most common modifiers used to enhance the performance and extend the service life of asphalt pavements [6,7]. Compared with traditional asphalt, SBS-modified asphalts exhibit better high- and low-temperature performance owing to three-dimensional crosslinking networks formed by SBS copolymer. However, the high cost of SBS can significantly increase construction costs when large quantities are required. In recent years, researchers have gradually paid more attention to waste rubber, which is responsible for serious environmental pollution. Many researchers have investigated the potential use of crumb rubber (CR) from waste tires in modified asphalt. Compared with SBS-modified asphalts, large yields of crumb rubber (CR) modified asphalts can be produced at relatively low cost. Moreover, dispersed CR can enhance the mechanical and rheological properties of asphalt, specifically by reducing fatigue and low-temperature cracking. Nevertheless, the high viscosity and limited high-temperature performance of CR modified asphalts has restricted their widespread use.

To take full advantage of SBS and CR modified asphalt and to compensate for their respective weaknesses, SBS copolymer has been incorporated in CR modified asphalt [8,9]. Zhang et al. [10] showed that SBS copolymer is an effective physical modifier for enhancing the high-temperature stability of conventional CR modified asphalt, mainly due to the interlaced copolymer network structure formed between SBS and CR. Liang et al. [11] explored the effect of SBS content on the rheological properties of CR/SBS modified asphalt and found that as a composite modifier, CR and SBS copolymer can improve the deformation resistance of asphalt under various ambient temperatures. However, CR/SBS modified asphalt has some drawbacks; in particular, its low storage stability and high viscosity. The desulfurization of CR was shown to solve these problems [12]. Wang et al. [13] proposed the desulfurization of CR to break the stable crosslinking network structure, resulting in decreased viscosity and increased elasticity. The chain scission in rubber is responsible for improving the storage stability and processing ability of modified asphalt [14].

However, conventional prepared methods of desulfurized CR/SBS modified asphalt are generally the same as that of CR/SBS modified asphalt—that is, mixing SBS and CR with asphalt successively and shearing at high temperature and high speed. Therefore, few chemical reactions occurred in the modification process, which would mitigate the modifications effect. Researchers [15,16] found that external energy can overcome the nonpolar characteristic of CR that limits its dispersibility in asphalt, thereby improving the incompatibility problem of asphalt.

Based on the considerations above, this study used desulfurized CR and SBS polymer to modify rubber asphalt. A type of desulfurized CR/SBS modifier with a force chemical preparation method was proposed. Conventional tests, bending beam rheometer (BBR) tests, and DSR tests were performed to acquire the rheological properties of desulfurized CR/SBS composite modified asphalt at different frequencies and temperatures. Furthermore, two existing rheological models were conducted to derive the nonlinear viscoelastic parameters and high-temperature performance of the asphalts.

## 2. Materials and Methods

### 2.1. Materials

Pen-70 asphalt was used as the base asphalt for modification. The CR (#40 mesh), desulfurized CR, and SBS copolymer (LG 501, which is linear and produced in Korea) were provided by Jiangsu Zhonghong Environment Technology Co, Ltd. (Nanjing, China). Desulfurized CR/SBS composite modifiers with desulfurized CR (20% and 25% by weight of base asphalt binder) and 4% SBS copolymer were pre-prepared in a force chemical reactor with mixing times of 15 min, 30 min, and 45 min.

## 2.2. Preparation of Asphalt Samples

Desulfurized CR/SBS composite modified asphalt was prepared in a laboratory at 180 °C. First, the desulfurized CR/SBS composite modifier was added to the base asphalt and the blend was sheared at a speed of 6000 r/min for 1.5 h. Then, the speed was reduced to 1500 r/min for 1 h until fully swelled. The preparation process of the desulfurized CR/SBS composite modified asphalt is illustrated in Figure 1. Conventional CR/SBS composite modified asphalt with 20% CR and 4% SBS was prepared using the same method. Six types of asphalt were used in this study, as listed in Table 1.

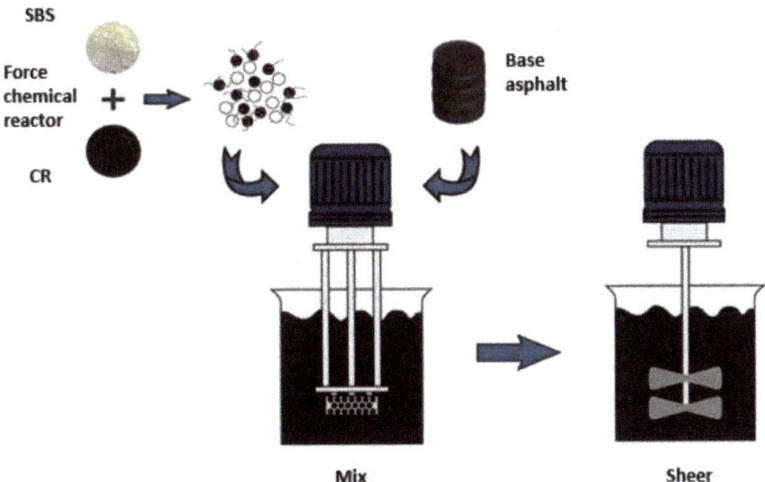

**Figure 1.** Flowchart of material preparation.

**Table 1.** Types of asphalt samples.

| Asphalt Type | Proportion of Modifier | Force Chemical Reactor Time |
| --- | --- | --- |
| 70# | - | - |
| J-20 | 4% SBS + 20% rubber | - |
| M-20 | 4% SBS + 20% desulfurized rubber | 30 min |
| M-25 | 4% SBS + 25% desulfurized rubber | 30 min |
| M-15MINS | 4% SBS + 20% desulfurized rubber | 15 min |
| M-45MINS | 4% SBS + 20% desulfurized rubber | 45 min |

## 2.3. Laboratory Tests

Prior to testing, all asphalt binders underwent short-term aging using the rolling thin film oven test (RTFOT) and were further aged in a pressure aging vessel (PAV) for 20 h to simulate aging. Rheological tests were repeated three times in each temperature range to obtain reliable results.

### 2.3.1. Bending Beam Rheometer Test

The BBR test was performed on five PAV-aged asphalt samples at three different temperatures (−18 °C, −24 °C, and −30 °C), according to the AASHTO M320 standard. The stiffness modulus (S) and creep rate (m) were obtained and used to evaluate the low-temperature rheological characteristics of the asphalts. Here, the stiffness modulus characterizes the ability to resist deformation under constant load, and the m value describes the stress relaxation ability and sensitivity of stiffness to time. A larger S value and smaller m indicate poorer anti-cracking properties of asphalt.

2.3.2. Linear Amplitude Sweep Test

The linear amplitude sweep test was performed on PAV-aged asphalt samples at 25 °C using an 8 mm dynamic shear rheometer (DSR, Anton Paar, Germany), according to the AASHTO TP101-12 [17] standard. First, a frequency sweep was performed at 0.1% strain in a frequency range of 0.2 to 30 Hz. The data were used to derive the slope of the relationship between frequency and shear modulus. Next, a linear amplitude sweep was carried out at a frequency of 10 Hz, with a linearly increasing strain amplitude from 0.05% to 30%. Then, the viscoelastic continuum damage (VECD) model was applied to obtain the damage characteristic and fatigue life curves at each strain amplitude. The fatigue damage can be characterized by damage intensity (D) and damage ratio C. When D is given, a smaller C indicates more serious damage. According to AASHTO TP 101-12, D = 0.35 was selected as the fatigue criterion.

2.3.3. Multiple Stress Creep Recovery Test

The MSCR test was performed on RTFOT-aged asphalt samples using a 25 mm DSR geometry according to AASHTO T350-14 [18]. The specimen was loaded at two stress levels of 0.1 kPa and 3.2 kPa for twenty cycles and ten cycles, respectively. Each cycle consisted of 1 s of creep stage, followed by 9 s of recovery stage for a total of 10 s. The temperature range was 70–82 °C. The creep recovery rate (R) and non-recoverable creep compliance ($J_{nr}$) were obtained and used to evaluate anti-rutting performance.

2.4. Rheological Models for Characterizing Asphalt Binders

2.4.1. Burgers' Model

One of the limitations of desulfurized CR/SBS composite modified asphalt is its poor high-temperature performance [13]. To further characterize the asphalt binder in the high-temperature range, this study investigated changes in nonlinear viscoelastic parameters. Taking the conventional CR/SBS composite modified asphalt as a control, creep curves from the MSCR test were fitted to Burgers' model to obtain the viscoelastic parameters [19].

$$\varepsilon_t = \varepsilon_e + \varepsilon_v + \varepsilon_d \\ = \sigma\left[\frac{1}{E} + \frac{1}{\eta_1}t + \frac{1}{E_2}(1 - e^{-\frac{E_2}{\eta_2}t})\right] \tag{1}$$

$$\varepsilon(t) = \sigma\left[\frac{1}{E_1} + \frac{1}{\eta_1}t + \frac{1}{E_2}(1 - e^{-\frac{E_2}{\eta_2}t})\right] \ (t < t_1) \tag{2}$$

$$\varepsilon(t) = \sigma\left[\frac{t_1}{\eta_2} + \frac{1}{E_2}(1 - e^{-\frac{E_2}{\eta_2}t_1})e^{-\frac{E_2}{\eta_2}(t-t_1)}\right] \ (t \geq t_1) \tag{3}$$

$$AEE = \frac{1}{N}\sum_{1}^{N}\frac{\varepsilon_{predicted}(t_i) - \varepsilon_{measured}(t_i)}{\varepsilon_{measured}(t_i)} \tag{4}$$

where $\varepsilon_e$ is the instantaneous elastic strain, $\varepsilon_v$ is the viscous flow, and $\varepsilon_d$ is the viscoelastic strain when the creep stress ($\sigma$) is applied; $E_1$, $E_2$, $\eta_1$, and $\eta_2$ are the four model parameters represented, respectively.

Liu [20] proposed a method for determining the four Burgers' model parameters based on the constitutive equation of Burgers' model, as presented in Equation (1). First, the time at the end of creep load (t = $t_1$) is taken as the boundary, and the strain response in the creep section is expressed using Equation (2). The strain response in the recovery section is expressed by Equation (3). The initial viscous strain is taken as the measured strain at the end of a recovery cycle. The model parameters are obtained by fitting the recovery data and initial viscous strain to Equations (2) and (3). Finally, average absolute errors are calculated to evaluate the fitted results.

### 2.4.2. Kelvin–Voigt Model

However, the burgers parameters were obtained mainly based on the creep stage of the MSCR test within a certain period. A recent study by Prashant et al. [21] proposed that three components (linear viscoelastic, nonlinear viscoelastic, and permanent strain) from the entire creep and recovery curve would help to better understand the behavior of modified binders. To verify the accuracy of Burgers' model fitting results and improve the rigor of this research, the nonlinear viscoelastic characterization of the MSCR curve was performed.

The specified procedure is briefly described as follows:

Step 1: Hypothesis: The strain data of 10 cycles at 0.1 kPa were the same, ignoring the elastic component A.

Step 2: Determine the linear viscoelastic parameters.

Fit the recovery strain data of cycle number one at 0.1 kPa using Equation (5) to obtain the calculated linear parameters B, C, D, and E.

$$\varepsilon(t) = 0.1\left[A + B\left(1 - e^{\frac{-t}{C}}\right) + D\left(1 - e^{\frac{-t}{E}}\right)\right] - 0.1\left[A + B\left(1 - e^{\frac{-(t-1)}{C}}\right) + D\left(1 - e^{\frac{-(t-1)}{E}}\right)\right] \quad (5)$$

Step 3: Determine the nonlinear viscoelastic parameters.

To determine the nonlinear viscoelastic parameter $G_1$, $G_2$, and $G_3$, 1 to 10 s strain data points of the recovery stage at 3.2 kPa was analyzed. The parameter $G_1$ is derived from fitting the first 10 data points (1 to 2 s) using Equation (6):

$$\varepsilon(t)_{NLVE-G_1} = G_1\left[B\left(1 - e^{\frac{-t}{C}}\right) + D\left(1 - e^{\frac{-t}{E}}\right)\right] - \left[B\left(1 - e^{\frac{-(t-1)}{C}}\right) + D\left(1 - e^{\frac{-(t-1)}{E}}\right)\right] \quad (6)$$

Parameters $G_2$ and $G_3$ are derived from fitting the remaining 80 data points (2 to 10 s) with Equation (7):

$$\varepsilon(t)_{NLVE-G_2-G_3} = G_2\left[3.2\left(J_{LVE(t=2)} - J_{LVE(t=1)}\right)\right] + G_3\left\{3.2\left[B\left(1 - e^{\frac{-t}{C}}\right) + D\left(1 - e^{\frac{-t}{E}}\right)\right] - \left[B\left(1 - e^{\frac{-(t-1)}{C}}\right) + D\left(1 - e^{\frac{-(t-1)}{E}}\right)\right]\right\} \quad (7)$$

where $J_{LVE(t=i)} = B\left(1 - e^{\frac{-i}{C}}\right) + D\left(1 - e^{\frac{-i}{E}}\right)$, i from 1 to 2.

Step 4: Calculate permanent strain.

$$\text{Permanent Strain, \%} = 100\{(Measured\ Strain) - (Calculated\ Strain)\}_{at\ 10} \quad (8)$$

## 3. Results and Discussion

### 3.1. Traditional Performance Test

Different modifiers and production technologies lead to complex physicochemical processes involving volatilization, oxidation, and condensation. Accordingly, the physical properties of asphalt including the softening point, penetration, and ductility will change. Three tests were carried out on each sample according to the relevant specifications. The results are presented in Table 2.

Compared with 70# base asphalt, the ductility and softening point of the composite modified asphalts were greatly improved, as expected for this type of material. Rubber chains released from the destroyed CR are inserted into the polymer SBS crosslinking network, which improves the crosslinking effect. At the same time, the low flow capacity of rubber restricts the asphalt flow. Both effects resulted in a high softening point and higher ductility. However, the penetration degree decreased. Previously, both CR and SBS copolymer were shown to selectively absorb the light fraction of the base asphalt during the swelling process, resulting in an increase in asphaltene and resin content [22,23]. While asphaltene and resin have a hardening effect on asphalt, the hardening effect was also enhanced with increasing CR/SBS composite modifier dosage and swelling degree. In general, the M series modified asphalts had higher ductility and softening points than J series modified asphalts, indicating that both pre-preparation in a force chemical reactor and desulfurized CR can improve the high and low temperature performance. However, increasing the dosage of desulfurized CR to 25% reduced its penetration and ductility at 5 °C.

Table 2. Technical indexes of test samples.

| Test Index | 70# | J-20 | M-20 | M-25 | M-15MINS | M-45MINS |
|---|---|---|---|---|---|---|
| 25 °C Penetration/mm | 61 | 57.5 | 56 | 55.43 | 55.67 | 58.7 |
| 5 °C Ductility/cm | 25 | 38.2 | 59.6 | 56.3 | 57.8 | 62 |
| Softening point/°C | 54.8 | 65.1 | 80.75 | 86.5 | 75.5 | 87.1 |

### 3.2. Rheological Properties

Traditional performance tests can characterize asphalt performance from a macro perspective at a given temperature but fail to reflect the microstructural changes. In contrast, characterizing desulfurized CR/SBS composite modified asphalt based on rheological properties can effectively bridge the gap between the microstructure and macroscopic motion. Three rheological tests were conducted to evaluate the effect of desulfurized CR and pre-preparation time of desulfurized CR/SBS composite modifiers in a force chemical reactor on the rheological properties of modified asphalt.

#### 3.2.1. Bending Beam Rheometer Test Results

Figure 2 presents the evolution of stiffness modulus (S) and creep rate (m) with loading time for TFOT-PAV aged CR/SBS modified asphalts at −18, −24, and −30 °C. As the temperature decreased, S values of the five modified asphalts increased; however, the creep rate (M) exhibited a decreasing trend. When the temperature dropped to −30 °C, all samples reached the failure point. This could be because as the temperature drops, the elastic component in the asphalt increases gradually, and the asphalt becomes harder and more brittle, thereby reducing its low-temperature relaxation ability [22]. At −30 °C, the asphalt has already exceeded its low-temperature limit. It is worth noting that the S and m values of the M-25 and M-45MINS modified asphalts are significantly better than those of the other modified asphalts. This may be because the high resilience and viscoelasticity of rubber can facilitate low-temperature deformation of the rubber asphalt. The increase in CR content resulted in greater elastic deformation. However, the test data also revealed an oddity. The S value at a certain temperature is not higher than 300 MPa, while the m value is lower than 0.3, suggesting that the conclusions made from S and m values as indexes are not sufficiently convincing.

As in previous studies [24,25], the S and m values were fitted in the semi-logarithmic and arithmetic coordinate systems ($T_{L,S}$, $T_{L,m}$) to obtain the low-temperature grade temperature ($T_{LG}$) of the modified asphalts, as shown in Table 3. The $T_{LG}$ value indicates the low-temperature crack resistance, with a lower $T_{LG}$ indicating superior crack resistance. The $T_{LG}$ value is higher than −30 °C, which is consistent with the failure points of all the asphalt samples of above −30 °C within the tested temperature range. Therefore, the $T_{LG}$ of the BBR test can be used as an index for evaluating the low-temperature crack resistance of desulfurized CR/SBS composite modified asphalts. The $T_{LG}$ values of the five asphalts can be ordered as follows: M-45MINS<M-25<M-20<J-20<M-15MINS. The results show that the desulfurization of CR, increase in CR content, and application of the force chemical reactor technique can improve the low-temperature crack resistance of modified asphalts.

Table 3. Low-temperature grade temperature ($T_{LG}$) of asphalt samples.

| Asphalt Type | J-20 | M-20 | M-25 | M-15MINS | M-45MINS |
|---|---|---|---|---|---|
| $T_{L,s}$ | −25.58 | −24.73 | −24.32 | −24.77 | −25.43 |
| $T_{L,m}$ | −21.44 | −21.49 | −22.84 | −21.23 | −23.97 |
| $T_{LG}$ | −21.44 | −21.49 | −22.84 | −21.23 | −23.97 |

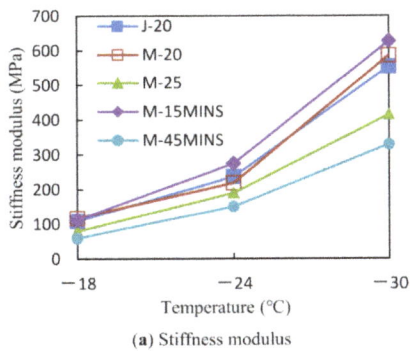
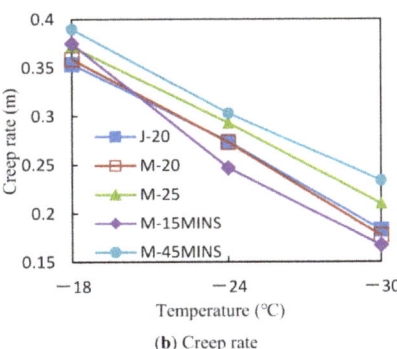

(a) Stiffness modulus　　　　　　　　(b) Creep rate

**Figure 2.** Bending beam rheometer (BBR) test results.

### 3.2.2. Linear Amplitude Sweep Test Results

Damage characteristic curves of the asphalt samples are shown in Figure 3. The damage degrees of the M-30MINS and M-45MINS modified asphalts were similar when C = 0.35, whereas M-30MINS appears to have better integrity than M-45MINS when C > 0.35. In addition, the damage rate of the M-20 and M-25 modified asphalts are similar, indicating that CR content has little effect on the fatigue performance of desulfurized CR/SBS composite modified asphalt.

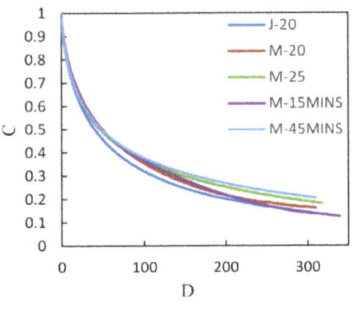
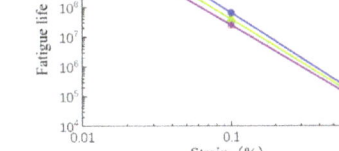

(a) Damage characteristic curve　　　　(b) Strain–fatigue life relationship curve

**Figure 3.** Linear amplitude sweep (LAS) test results.

The LAS test data were further processed using the viscoelastic continuum damage (VECD) model to obtain the relationship between strain and fatigue life. As shown in Figure 2b, the fatigue life of asphalt decreases with increasing strain and reached $10^{11}$ ESALs under the smallest strain. The effects on fatigue life may be due to the addition of carbon black to rubber, which can improve the aging resistance of asphalt. The fatigue resistance of M-45MINS modified asphalt was consistently superior to those of other modified asphalts, and an entire order of magnitude higher than the J-20 modified asphalt at low strain levels. The change in microstructure of CR as a result of the desulfurization process, which breaks the internal sulfur bonds resulting in an increase in rubber transferase activity, makes it easier to absorb the light components of the asphalt [13,14]. At the same time, increasing time in the force chemical reactor may cause devulcanized CR to release more light rubber molecules, which re-crosslink with SBS in the asphalt, thereby improving the fatigue resistance of the asphalt.

### 3.2.3. Multiple Stress Creep Recovery Test Results

Figures 4 and 5 show the MSCR test results for five composite modified asphalts at different temperatures. A lower $J_{nr}$ value indicates greater resistance to permanent deformation as well as better performance against rutting. Figure 4 shows that the $J_{nr}$ value of the composite modified asphalt gradually increased with increasing temperature at both stress levels. However, at 3.2 kPa, the $J_{nr}$ of J-20 modified asphalt was significantly higher than that of the M series modified asphalts. Under high stress, the creep recovery ability of the desulfurized CR/SBS composite modified asphalts was better than that of the conventional CR composite modified asphalt, which was likely due to the desulfurization of CR. The desulfurization process can strengthen the molecular network structure of CR, and consequently improve the high-temperature stability of asphalt to a certain extent [26]. The $J_{nr}$ value did not noticeably change among the same M series modified asphalts at 0.1 kPa. However, the $J_{nr}$ values of the M-25 and M-45MINS modified asphalts significantly deviated at 3.2 kPa stress and 82 °C. In summary, modified asphalt is not sensitive to temperature and CR content at low stress levels; however, an appropriate increase in CR content and force chemical reactor time can significantly improve the deformation resistance under high temperature and high stress. Increasing the reactor time promotes chain fracture in rubber and decreases the degree of crosslinking, leading to the swelling of lighter components into CR.

Furthermore, higher R values of binders indicate better strain recovering ability. From Figure 5, the curve of J-20 modified asphalt is significantly different from that of the M series modified asphalts, and its R value is lower. This also indicates that the crosslinking network structures formed by CR and between CR and SBS copolymer can complement each other to comprehensively improve the asphalt performance. At low stress, changes in temperature and CR content have little effect on the R value of M series modified asphalts, which also indicates that desulfurized CR/SBS composite-modified asphalt is not sensitive to temperature and CR content at low stress levels, and its high-temperature stability is excellent. Under high stress, the increase in temperature significantly reduces the R value, whereas M-20 modified asphalt showed better creep recovery. This may be due to the increase in viscosity and weakened resistance to deformation caused by the high stress and high temperature. Since the modulus of rubber is large and the asphalt's resistance to deformation is mainly provided by the CR, as the CR content increases, it can more effectively preserve the elastic properties.

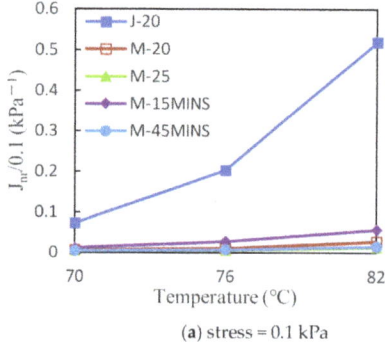
(a) stress = 0.1 kPa

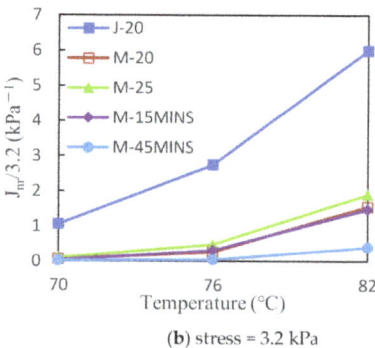
(b) stress = 3.2 kPa

**Figure 4.** Influence of temperature on unrecovered strain ($J_{nr}$) of asphalt binder at different stress levels.

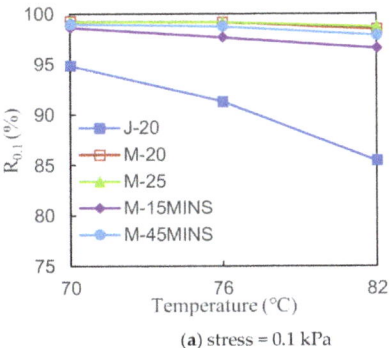 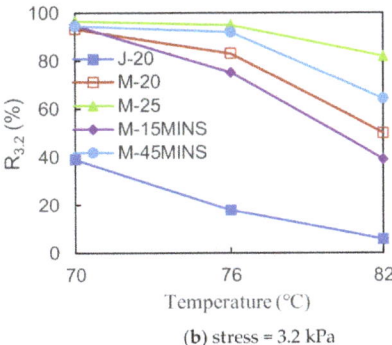

(a) stress = 0.1 kPa  (b) stress = 3.2 kPa

**Figure 5.** Influence of temperature on recovered strain (R) of asphalt binder at different stress levels.

### 3.3. Viscoelasticity Analysis Based on Multiple Stress Creep Recovery Test
#### 3.3.1. Burgers' Model Fitting Results

According to the MSCR test results, the high-temperature stability of M-45MINS modified asphalt is better than the other modified asphalts. For brevity, only the creep recovery data of the 10th cycle of the M-45MINS modified asphalt were analyzed, and J-20 modified asphalt was used as a control, as shown in Figure 6.

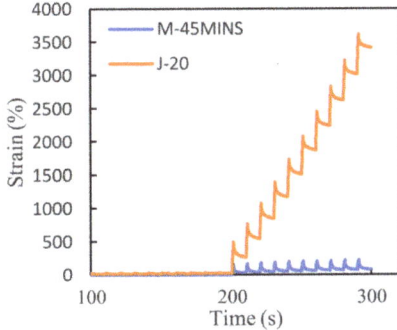

**Figure 6.** Typical creep and recovery cycle during multiple stress creep recovery (MSCR) test.

Table 4 shows the average absolute error (AAE) values of the MSCR test data at two loading levels fitted to Burgers' model. Burgers' model fits better with the creep stage of conventional CR and desulfurized CR/SBS composite modified asphalts at high temperature, with an AAE below 10%. However, the J-20 modified asphalt fits Burgers' model significantly better than M-45MINS modified asphalt [27]. Even at high stress and high temperature, the AAE value of the J-20 modified asphalt was approximately zero. This may be explained by the technical limitations of the method and the intrinsic limitations of the model, such as the number of spring and dashpot elements and the presence of only relaxation time in the equation [28].

The Burgers' model parameters in Table 5 suggest that an increase in temperature and stress level reduces the four model parameter values, which is manifested as a decrease in the high-temperature stability of the asphalt. Under the same stress level and temperature, the instantaneous elastic modulus ($E_1$) of desulfurized CR/SBS composite modified asphalt is significantly higher than that of conventional CR/SBS composite modified asphalt. Thus, the desulfurized CR can restore some elasticity and improve the deformation resistance of modified asphalt under instantaneous loading. Out of the four Burgers' model parameters, those characterizing the non-recoverable permanent deformation ($\eta_1$) of asphalt are most important, because viscous flow has a large impact on asphalt binder performance. A higher $\eta_1$ value indicates better anti-rutting properties. In general, $\eta_1$ is found to decrease with increasing temperature [29]. The $\eta_1$ value of M-45MINS is higher than that of the J-20 modified asphalt, indicating that the high-temperature performance of M-45MINS modified asphalt is better. At low stress levels, the $\eta_1$ values of the two asphalts change very little, but at high stress levels, they decrease significantly with increasing temperature. This finding is consistent with the variation of non-recoverable creep compliance observed in Figure 4.

**Table 4.** Average absolute error value for fitted multiple stress creep recovery (MSCR) test data.

| Items | Stress | Temperature | | |
|---|---|---|---|---|
| | | 70 °C | 76 °C | 82 °C |
| M-45MINS | 0.1 kPa | 8.49% | 7.33% | 6.26% |
| | 3.2 kPa | 8.12% | 6.36% | 0.68% |
| J-20 | 0.1 kPa | 2.16% | 0.14% | 0.02% |
| | 3.2 kPa | 0.01% | 0 | 0 |

**Table 5.** Burgers' model parameters.

| Stress | Temperature | M-45MINS | | | | J-20 | | | |
|---|---|---|---|---|---|---|---|---|---|
| | | $E_1$ | $E_2$ | $\eta_1$ | $\eta_2$ | $E_1$ | $E_2$ | $\eta_1$ | $\eta_2$ |
| 0.1 kPa | 70 °C | 36.06 | 16.22 | 30.37 | 71.33 | 4.41 | 3.16 | 7.17 | 28.67 |
| | 76 °C | 32.01 | 16.40 | 2.18 | 60.55 | 2.68 | 1.99 | 4.5 | 22.02 |
| | 82 °C | 13.58 | 7.94 | 15.85 | 59.45 | 1.37 | 1.37 | 3.09 | 20.77 |
| 3.2 kPa | 70 °C | 24.69 | 13.06 | 24.90 | 51.23 | 1.02 | 7.83 | 16.78 | 8.33 |
| | 76 °C | 14.26 | 8.61 | 17.70 | 33.24 | 0.40 | 10.93 | 21.84 | 3.28 |
| | 82 °C | 2.60 | 8.55 | 17.06 | 19.67 | 0.18 | 16.28 | 30.50 | 1.53 |

### 3.3.2. Kelvin–Voigt Model Fitting Results

Table 6 shows the linear viscoelastic parameters (B, C, D, E), nonlinear viscoelastic parameters ($G_1$, $G_2$, $G_3$), and PS for two types of asphalt.

Nonlinear viscoelastic parameters can be used to characterize the recovery behavior for asphalt [20]. Parameter $G_1$ represents the vertical shift of the recovery curve from 1 to 2 s. The $G_1$ value of M-45MINS modified asphalt is larger than that of J-20 modified asphalt, indicating that M-45MINS modified asphalt has better deformation recovery under instantaneous loading. It is consistent with the change of $E_1$ value of Burgers' model, since desulfurized CR restores elasticity. Parameters $G_2$ and $G_3$ represent the vertical shift and slope changes of the recovery curve from 2 to 10 s. The derived $G_3$ value of two asphalts are negative and increase closely to zero. The fitting variation of $G_3$ accurately reflects the change of recovery curve, indicating that the Kelvin–Voigt model can be used to fit the viscoelastic mechanism of CR modified asphalt. As the recovery time increases, the delayed elastic deformation of asphalt gradually recovered completely. With the increase of temperature, the PS value of M-45MINS asphalts increase sharply, indicating that high temperature impairs the asphalt's ability to resist deformation. Meanwhile, when the PS value of J-20 asphalt under 82 °C decreases, it may due to the fitting errors.

Table 6. Viscoelastic parameters and PS.

| Parameters | M-45MINS | | | J-20 | | |
|---|---|---|---|---|---|---|
| | 70 °C | 76 °C | 82 °C | 70 °C | 76 °C | 82 °C |
| B | 0.04188 | 0.06221 | 0.08895 | 0.36835 | 0.60998 | 0.84986 |
| C | 0.58801 | 0.60414 | 0.62757 | 1.01302 | 1.02523 | 0.98769 |
| D | 2.51056 | 3.8495 | 9.94901 | 59.20465 | 187.36224 | 514.839 |
| E | 18.41396 | 18.29115 | 24.90884 | 32.00404 | 46.03671 | 65.2637 |
| $G_1$ | 1.03109 | 0.74513 | 0.75977 | 0.52863 | 0.50518 | 0.57031 |
| $G_2$ | 0.67731 | 0.44075 | 0.4745 | 0.33085 | 0.31722 | 0.36266 |
| $G_3$ | −0.11469 | −0.11445 | −0.08411 | −0.02759 | −0.01106 | −0.0026 |
| PS | 3.20959 | 4.5752 | 6.36363 | 8.36741 | 8.83956 | 4.97893 |

## 4. Conclusions

A type of desulfurized CR/SBS modifier with a force chemical preparation method was proposed. The rheological properties of desulfurized CR/SBS composite modified asphalt were investigated through rheological tests and rheological models. The conclusions can be summarized as follows:

1. As a result of the CR chain inserted into the SBS crosslinking network and the low flow capacity of rubber, all of the modified asphalts were found to have higher ductility and softening points than the base asphalt. The force chemical reactor may strengthen this crosslinking effect. However, both CR and SBS copolymer were shown to selectively absorb the light fraction of the base asphalt, resulting in the decrease of penetration degree.
2. The BBR test showed that the high resilience and viscoelasticity of rubber facilitates low-temperature deformation of CR/SBS composite modified asphalt. Desulfurized CR/SBS modifier and its force chemical preparation method were found to be more effective than conventional CR/SBS modifier in improving the asphalt behavior at low temperatures. $T_{LG}$ can be used as an index for evaluating the low-temperature crack resistance of desulfurized CR/SBS composite modified asphalts.
3. The results of LAS tests indicated that both the desulfurization of CR and force chemical reactor time successfully enhanced the fatigue resistance of asphalt, because the change in the microstructure of CR makes it absorb more light components, which is beneficial for resisting fatigue disease. The fatigue resistance of M-45MINS modified asphalt was superior to those of other modified asphalts.
4. Comparing the MSCR tests in different temperatures and at different addition levels of CR, the desulfurization of CR and force chemical reactor time mitigate the $J_{nr}$ and R values. However, modified asphalt is not sensitive to temperature and CR content at low stress levels.
5. Rheological modeling of the creep-recovery behavior of desulfurized CR/SBS composite modified asphalt suggests that an increase in both temperature and stress level significantly affect the four Burgers' model parameters. Consequently, the high-temperature stability of the asphalt decreases. In addition, the nonlinear viscoelastic characterization of the MSCR curve based on the Kelvin–Voigt model verify the Burgers' model fitting results' accuracy.

**Author Contributions:** Conceptualization, J.Y. and X.C.; methodology, J.Y.; formal analysis, P.K.; investigation, J.Y.; resources, G.X.; writing—original draft preparation, J.Y.; writing—review and editing, G.X.; supervision, X.C. All authors have read and agreed to the published version of the manuscript.

**Funding:** This research was supported by the Fundamental Research Funds for the Central Universities under Grant 3221002122D, in part by the Postgraduate Research & Practice Innovation Program of Jiangsu Province, grant number SJCX20_0047 and the National Natural Science Foundation of China (No. 51778136).

**Institutional Review Board Statement:** Not applicable.

**Informed Consent Statement:** Not applicable.

**Data Availability Statement:** Data is contained within the article.

**Conflicts of Interest:** The authors declare no conflict of interest.

## References

1. Wang, L.; Chang, C. Rheological Evaluation of Polymer Modified Asphalt Binders. *J. Wuhan Univ. Technol. Mater. Sci. Ed.* **2015**, *30*, 695–702. [CrossRef]
2. Fernandes, M.R.; Forte, M.M.; Leite, L.F. Rheological Evaluation of Polymer-Modified Asphalt Binders. *Mater. Res. Ibero Am. J. Mater.* **2008**, *11*, 381–438. [CrossRef]
3. Hassanpour-Kasanagh, S.; Ahmedzade, P.; Fainleib, A.M.; Behnood, A. Rheological properties of asphalt binders modified with recycled materials: A comparison with Styrene-Butadiene-Styrene (SBS). *Constr. Build. Mater.* **2020**, *230*, 117047. [CrossRef]
4. Behnood, A.; Olek, J. Rheological properties of asphalt binders modified with styrene-butadiene-styrene (SBS), ground tire rubber (GTR), or polyphosphoric acid (PPA). *Constr. Build. Mater.* **2017**, *151*, 464–478. [CrossRef]
5. Wang, L.; Razaqpur, G.; Xing, Y.; Chen, G. Microstructure and rheological properties of aged and unaged polymer-modified asphalt binders. *Road Mater. Pavement Des.* **2015**, *16*, 592–607. [CrossRef]
6. Lo Presti, D. Recycled Tyre Rubber Modified Bitumens for road asphalt mixtures: A literature review. *Constr. Build. Mater.* **2013**, *49*, 863–881. [CrossRef]
7. Bamigboye, G.O.; Bassey, D.E.; Olukanni, D.O.; Ngene, B.U.; Adegoke, D.; Odetoyan, A.O.; Kareem, M.A.; Enabulele, D.O.; Nworgu, A.T. Waste materials in highway applications: An overview on generation and utilization implications on sustainability. *J. Clean. Prod.* **2021**, *283*, 124581. [CrossRef]
8. Kok, B.V.; Colak, H. Laboratory comparison of the crumb-rubber and SBS modified bitumen and hot mix asphalt. *Constr. Build. Mater.* **2011**, *25*, 3204–3212. [CrossRef]
9. Dong, F.; Yu, X.; Liu, S.; Wei, J. Rheological behaviors and microstructure of SBS/CR composite modified hard asphalt. *Constr. Build. Mater.* **2016**, *115*, 285–293. [CrossRef]
10. Zhang, F.; Hu, C. The research for structural characteristics and modification mechanism of crumb rubber compound modified asphalts. *Constr. Build. Mater.* **2015**, *76*, 330–342. [CrossRef]
11. Liang, M.; Xin, X.; Fan, W.; Luo, H.; Wang, X.; Xing, B. Investigation of the rheological properties and storage stability of CR/SBS modified asphalt. *Constr. Build. Mater.* **2015**, *74*, 235–240. [CrossRef]
12. Sienkiewicz, M.; Borzędowska-Labuda, K.; Wojtkiewicz, A.; Janik, H. Development of methods improving storage stability of bitumen modified with ground tire rubber: A review. *Fuel Process. Technol.* **2017**, *159*, 272–279. [CrossRef]
13. Rasool, R.T.; Song, P.; Wang, S. Thermal analysis on the interactions among asphalt modified with SBS and different degraded tire rubber. *Constr. Build. Mater.* **2018**, *182*, 134–143. [CrossRef]
14. Song, P.; Zhao, X.; Cheng, X.; Li, S.; Wang, S. Recycling the nanostructured carbon from waste tires. *Compos. Commun.* **2018**, *7*, 12–15. [CrossRef]
15. Fu, Q.; Xu, G.; Chen, X.; Zhou, J.; Sun, F. Rheological properties of SBS/CR-C composite modified asphalt binders in different aging conditions. *Constr. Build. Mater.* **2019**, *215*, 1–8. [CrossRef]
16. Attia, M.; Abdelrahman, M. Enhancing the performance of crumb rubber-modified binders through varying the interaction conditions. *Int. J. Pavement Eng.* **2009**, *10*, 423–434. [CrossRef]
17. American Association of State and Highway Transportation Officials. *AASHTO TP 101-12 (2018) Standard Method of Test for Estimating Fatigue Resistance of Asphalt Binders Using the Linear Amplitude Sweep*; American Association of State and Highway Transportation Officials: Washington, DC, USA, 2018.
18. American Association of State and Highway Transportation Officials. *AASHTO T 350-14 (2018) Standard Method of Test for Multiple Stress Creep Recovery (MSCR) Test of Asphalt Binder Using a Dynamic Shear Rheometer (DSR)*; American Association of State and Highway Transportation Officials: Washington, DC, USA, 2018.
19. Liu, H.; Zeiada, W.; Al-Khateeb, G.G.; Shanableh, A.; Samarai, M. Use of the multiple stress creep recovery (MSCR) test to characterize the rutting potential of asphalt binders: A literature review. *Constr. Build. Mater.* **2021**, *269*, 121320. [CrossRef]
20. Liu, Y.; You, Z. Determining Burger's Model Parameters of Asphalt Materials Using Creep-Recovery Testing Data. In *Proceedings of the Symposium on Pavement Mechanics and Materials at the Inaugural International Conference of the Engineering Mechanics Institute-Pavements and Materials 2008: Modeling Testing, and Performance, Minneapolis, MN, USA, 18–21 May 2008*; American Society of Civil Engineers (ASCE): Minneapolis, MN, USA, 2008; pp. 26–36.
21. Shirodkar, P.; Mehta, Y.; Nolan, A.; Dahm, K.; Dusseau, R.; McCarthy, L. Characterization of creep and recovery curve of polymer modified binder. *Constr. Build. Mater.* **2012**, *34*, 504–511. [CrossRef]

22. Dong, D.; Huang, X.; Li, X.; Zhang, L. Swelling process of rubber in asphalt and its effect on the structure and properties of rubber and asphalt. *Constr. Build. Mater.* **2012**, *29*, 316–322. [CrossRef]
23. Xiang, L.; Cheng, J.; Que, G. Microstructure and performance of crumb rubber modified asphalt. *Constr. Build. Mater.* **2009**, *23*, 3586–3590. [CrossRef]
24. Shen, J.; Arnirkhanian, S.; Tang, B. Effects of rejuvenator on performance-based properties of rejuvenated asphalt binder and mixtures. *Constr. Build. Mater.* **2007**, *21*, 958–964. [CrossRef]
25. You, Z.; Mills-Beale, J.; Fini, E.; Goh, S.W.; Colbert, B. Evaluation of Low-Temperature Binder Properties of Warm-Mix Asphalt, Extracted and Recovered RAP and RAS, and Bioasphalt. *J. Mater. Civ. Eng.* **2011**, *23*, 1569–1574. [CrossRef]
26. Yu, H.; Leng, Z.; Gao, Z. Thermal analysis on the component interaction of asphalt binders modified with crumb rubber and warm mix additives. *Constr. Build. Mater.* **2016**, *125*, 168–174. [CrossRef]
27. Lagos-Varas, M.; Movilla-Quesada, D.; Arenas, J.P.; Raposeiras, A.C.; Castro-Fresno, D.; Calzada-Perez, M.A.; Vega-Zamanillo, A.; Maturana, J. Study of the mechanical behavior of asphalt mixtures using fractional rheology to model their viscoelasticity. *Constr. Build. Mater.* **2019**, *200*, 124–134. [CrossRef]
28. Domingos, M.D.; Faxina, A.L. Rheological behaviour of bitumens modified with PE and PPA at different MSCR creep-recovery times. *Int. J. Pavement Eng.* **2015**, *16*, 771–783. [CrossRef]
29. Kumar, R.; Saboo, N.; Kumar, P.; Chandra, S. Effect of warm mix additives on creep and recovery response of conventional and polymer modified asphalt binders. *Constr. Build. Mater.* **2017**, *138*, 352–362. [CrossRef]

Article

# Investigation into Rheological Behavior of Warm-Mix Recycled Asphalt Binders with High Percentages of RAP Binder

Hui Xu [1], Yiren Sun [1,*], Jingyun Chen [1,*], Jiyang Li [1], Bowen Yu [1], Guoqing Qiu [1], Yan Zhang [2] and Bin Xu [3]

1 School of Transportation and Logistics, Dalian University of Technology, Dalian 116024, China
2 City Institute, Dalian University of Technology, Dalian 116600, China
3 Research and Development Center of Transport Industry of New Materials, Technologies Application for Highway Construction and Maintenance (Zhong Lu Gao Ke (Beijing) Road Technology Co., Ltd.), Ministry of Transport, Beijing 100088, China
* Correspondence: sunyiren@dlut.edu.cn (Y.S.); chenjy@dlut.edu.cn (J.C.)

Citation: Xu, H.; Sun, Y.; Chen, J.; Li, J.; Yu, B.; Qiu, G.; Zhang, Y.; Xu, B. Investigation into Rheological Behavior of Warm-Mix Recycled Asphalt Binders with High Percentages of RAP Binder. *Materials* 2023, *16*, 1599. https://doi.org/10.3390/ma16041599

Academic Editor: Giovanni Polacco

Received: 10 January 2023
Revised: 9 February 2023
Accepted: 13 February 2023
Published: 14 February 2023

Copyright: © 2023 by the authors. Licensee MDPI, Basel, Switzerland. This article is an open access article distributed under the terms and conditions of the Creative Commons Attribution (CC BY) license (https://creativecommons.org/licenses/by/4.0/).

**Abstract:** The rheological properties of warm-mix recycled asphalt binders are critical to enhancing design quality and interpreting the performance mechanisms of the corresponding mixtures. This study investigated the rheological behavior of warm-mix recycled asphalt binders with high percentages of RAP binder. The effects of two warm-mix additives [wax-based Sasobit (S) and surfactant-based Evotherm-M1 (E)], a rejuvenating aging [ZGSB (Z)], four RAP binder contents (0%, 30%, 50% and 70%), and three aging states (unaged, short-term aged and long-term aged) were evaluated in detail using the dynamic shear rheometer (DSR), bending beam rheometer (BBR) and Brookfield rotational viscometer tests as well as conventional performance tests over the whole range of temperatures. The results showed that the rejuvenating agent Z effectively alleviated the aging effect of the RAP binder; however, it could hardly eliminate entirely this negative impact, especially at higher RAP binder contents. The addition of S remarkably lowered the apparent viscosity of the warm-mix recycled binders by up to 35.0%, whereas E had little influence on the binder viscosity due to its surfactant nature. Besides, S performed much better in improving rutting resistance (with the increase of up to 411.3% in $|G^*|/\sin\delta$) than E, while E exhibited superior fatigue performance (with the reduction of up to 42.3% in $|G^*|\cdot\sin\delta$) to that of S. In terms of the thermal cracking resistance, E had very slight influence and S even yielded an adverse impact (with the increase of up to 70.2% in $S_a$ and the decrease of up to 34.1% in m-value). Further, S broadened the ranges of pavement service temperatures by about 12 °C, whereas E almost did not change the PG grades of the binders. Finally, regarding the characteristics of viscoelastic master curves, S considerably improved the dynamic modulus and lowered the phase angle of the binders over a wide range of frequencies and temperatures but led to the failure of the time-temperature superposition principle due to its thermorheologically complex nature. Nevertheless, in this regard, the effect of E was found very mild.

**Keywords:** asphalt binder; warm-mix asphalt (WMA); reclaimed asphalt pavement (RAP); rheological behavior; performance

## 1. Introduction

The reuse of reclaimed asphalt pavement (RAP) in new asphalt mixtures is becoming increasingly prevalent in the asphalt paving industry [1–3]. RAP milled up or ripped off from old worn-out pavements contains valuable crushed aggregates and aged/stiff binder, and its recycling can conserve natural resources and save money, thus contributing to significant environmental and economic benefits. The aging of asphalt binder occurs owing to the loss of volatiles and oxidation during the whole process of production, construction, and service of asphalt mixture. The aged binder in RAP becomes significantly stiffer and, therefore, can furnish more desirable rutting resistance for the recycled asphalt mixture. Nevertheless, high RAP contents may cause a few typical problems, e.g., inferior fatigue cracking resistance, workability, and blending efficiency [4–6]. To eliminate these potential deficiencies, the percentages of RAP applied to the pavement surface layers are commonly limited below 25% by many highway agencies.

Over the past few years, applying warm-mix asphalt (WMA) technologies to RAP mixtures has attracted numerous interests from the asphalt pavement community [7–12]. Currently, three principal types of WMA technologies are involved, i.e., the chemical additives, organic additives, and foaming processes [13,14]. As low-carbon sustainable products/processes, WMA technologies enable asphalt mixtures to possess superior workability by reducing binder viscosity or raising lubrication. The mixing and compaction temperatures of WMA can thus be lowered by about 20~40 °C compared with those of the traditional hot-mix asphalt (HMA) [14]. This advantage not only helps decrease fuel consumption but also curbs the emissions of asphalt fumes and greenhouse gas in construction. Due to the lower degree of asphalt aging caused by reduced temperatures in production and compaction, WMA technologies tend to provide better fatigue performance but poorer rutting resistance for asphalt mixtures [14]. Taking into account these factors, the combined utilization of the WMA and RAP can rationally avoid the drawbacks of both technologies and effectively improve the percentages of RAP in the recycled mixtures.

Some research attempts have been made to investigate the performance of warm-mix recycled asphalt mixtures with high RAP contents. Zhao et al. [7] comparatively assessed the joint influence of two WMA technologies (foaming and Evotherm) and high RAP contents (up to 40%) on the rutting, moisture, and fatigue susceptibility of 15 mixes and found that rutting might be a potential issue for WMA-high RAP mixtures. Mogawer et al. [15] evaluated the performance of asphalt rubber surface mixtures with RAP percentages up to 40% and a wax-based WMA additive. They found that high-content RAP had an adverse effect on the fatigue and reflective cracking resistance of the mixtures; however, the WMA technology could mitigate this impact. Xiao et al. [16] analyzed Superpave mix design characteristics of various high percentages (up to 50%) of RAP in terms of two WMA technologies (Evotherm additive and foaming technology). Song et al. [17] compared the influence of a rejuvenating agent and a foaming technology commonly used in the United States on the performance of asphalt mixtures containing 50% RAP and observed that both technologies improved the moisture susceptibility, but they exhibited opposite effects on rutting performance. Wang et al. [18] assessed multiple performances, including moisture susceptibility, low-temperature, high-temperature, and fatigue, of WMA recycled mixtures with two RAP contents (50% and 70%), two WMA additives (wax R and surfactant M) and two gradations (AC-16 and AC-13).

Another important issue lies in evaluating the rheological properties of warm-mix recycled asphalt binders, which can facilitate enhancing the design quality and interpreting the performance mechanisms of the corresponding mixtures. Xiao et al. [19] analyzed the rutting and fatigue performance of asphalt binders with high contents (up to 50%) of RAP binders and a warm-mix additive. Yu et al. [20] evaluated the rheological properties of foamed warm-mix recycled asphalts with different RAP binder contents (up to 80%) using the Brookfield rotational viscometer and dynamic shear rheometer (DSR) at both high- and intermediate-temperatures. Sun et al. [21] assessed the impacts of two WMA additives on the rutting and fatigue performance of binders with high contents (30~70%) of RAP binder by the multiple stress creep and recovery (MSCR) and linear amplitude sweep (LAS) tests.

Despite these studies involving this topic, they were mostly concentrated on the rheological properties over only part of the service temperature range (e.g., high- or intermediate-temperature). Besides, very limited research efforts have been devoted to exploring the combined influence of the rejuvenating agent, WAM additives/processes, and high-content RAP binder on the rheological properties of warm-mix recycled asphalt binders. To this end, this study investigated the rheological behavior of warm-mix recycled asphalt binders with high percentages of RAP binder over the entire service temperature range (low-, intermediate-, and high-temperature). The influence of a rejuvenating agent, two warm-mix additives, four RAP binder contents (up to 70%), and three aging states (unaged, short-term aged, and long-term aged) were evaluated.

## 2. Materials

A #90 asphalt (penetration grade 80/100) was selected as the virgin binder. An actual RAP binder was considered in this study, which was extracted and recovered from a RAP collected from a highway in Dalian, China. To have a grasp of the two binders, their basic properties were measured, as displayed in Table 1. It can be seen that the actual RAP binder exhibited inferior low- and intermediate-temperature performance but superior high-temperature performance to that of the virgin binder (#90).

**Table 1.** Properties of the #90 and actual RAP binders.

| Binder | Original | | | | RTFO | | PAV | | | |
|---|---|---|---|---|---|---|---|---|---|---|
| | Penetration at 25 °C (0.1 mm) | Softening Point (°C) | Ductility at 15 °C (mm) | Apparent Viscosity at 135 °C (Pa·s) | $\|G^*\|/\sin\delta$ (58 °C) (kPa) | $\|G^*\|/\sin\delta$ at 58 °C (kPa) | $\|G^*\|\cdot\sin\delta$ at 25 °C (kPa) | $S_a$ at −18 °C (MPa) | m-Value at −18 °C |
| #90 | 85.7 | 46.7 | >1000 | 0.329 | 2.207 | 3.766 | 2278 | 260.5 | 0.324 |
| Actual RAP | 42.1 | 55.1 | 147.8 | 0.605 | 8.13 | — | 1456 | 288.5 | 0.325 |

Given that laboratory-aged asphalts always display relatively more reproducible and stable rheological behaviors, an artificial RAP binder was produced as an alternative to the actual RAP binder for the purpose of this study, which was fabricated by subjecting the #90 asphalt to successive rolling thin-film oven (RTFO) aging [22] and pressurized aging vessel (PAV) aging [23]. The short-term RTFO aging was performed at 163 °C for 85 min, and the long-term PAV aging was performed at 100 °C. To ensure similar performance and aging degree of the artificial RAP binder to the actual one, three different exposure durations in the PAV, 10, 15, and 20 h, were applied, and accordingly, the artificial RAP binders were denoted as PAV-10, PAV-15, and PAV-20, respectively. In this study, four RAP binder percentages, 0%, 30%, 50%, and 70%, were adopted to analyze the effect of high RAP binder contents.

To investigate the influence of WMA additives, two different WMA additives, Sasobit and Evotherm-M1, were employed. Sasobit is a white organic synthetic hard wax produced from coal gasification through the Fischer-Tropsch (FT) method, which can reduce the binder viscosity and thus decrease the production and placement temperatures. Sasobit is described as an "asphalt flow improver", and its use enables production temperatures to be reduced by 20~30 °C [24]. Evotherm-M1 is an amber chemical liquid surfactant that can enhance the coating of asphalt binder to aggregates and thus improve the workability of asphalt concrete. The Evotherm-M1 belongs to the Evotherm 3G product, which is a water-free warm-mix technology [14]. The dosages of 3% and 0.5% by weight of the total binder were used for Sasobit and Evotherm-M1, respectively, according to the suppliers' recommendations. Figure 1 shows the images of the two WMA additives.

Besides, a rejuvenating agent ZGSB (Figure 1), was used to restore the performance of the RAP binder. Table 2 presents the basic properties of ZGSB. To determine the optimal content of ZGSB, four dosages, 2%, 4%, 6%, and 8% by weight of the RAP binder, were taken into account.

**Table 2.** Basic properties of the rejuvenating agent ZGSB.

| Saturates Fraction (%) | Aromatics Fraction (%) | Viscosity by Vacuum Capillary Viscometer at 60 °C (mm²/s) | Density at 15 °C (g/cm³) |
|---|---|---|---|
| 23.9 | 59.9 | 153.5 | 0.987 |

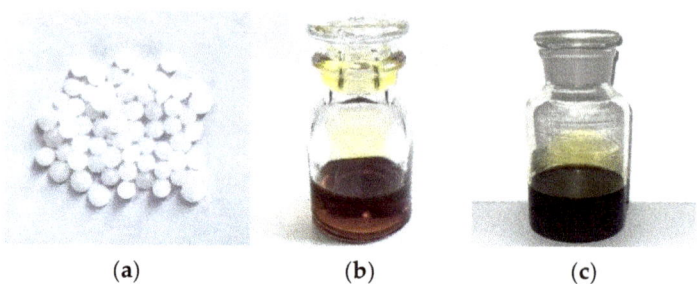

**Figure 1.** Images of the WMA additives and rejuvenating agent: (**a**) Sasobit; (**b**) Evotherm-M1; (**c**) ZGSB.

For brevity, the two WMA additives, Sasobit and Evotherm-M1, and the rejuvenating agent, ZGSB, are respectively designated as S, E, and Z in the following analysis. Thus, three categories of recycled binders are involved, in which the Z binders represent those containing only the rejuvenator ZGSB, the S + Z binders represent those containing both Sasobit and ZGSB and the E + Z binders represent those containing both Evotherm-M1 and ZGSB.

## 3. Methodology

### 3.1. Conventional Performance Tests

According to the standard test methods of China JTG E20-2011 [25], three conventional performance tests, i.e., the penetration, softening point, and ductility tests, were conducted to evaluate the basic performance of the binders.

### 3.2. Apparent Viscosity Test by Brookfield Rotational Viscometer

The apparent viscosity was measured at four temperatures, 115, 135, 155, and 175 °C, in accordance with ASTM D4402 [26] using a Brookfield rotational viscometer. The apparent viscosity can well reflect the workability of the binders during production and construction.

### 3.3. High- and Intermediate-Temperature Performance Test

The high- and intermediate-temperature performance of asphalt binders can be characterized using a DSR (Lab⁺, Malvern) according to ASTM D7175 [27]. In this method, the dynamic shear modulus $|G^*|$ and phase angle $\delta$ at the loading frequency of 10 rad/s are obtained.

Further, the high-temperature rutting performance can be assessed using the rutting parameter $|G^*|/\sin\delta$, which is derived from the dissipated energy density per oscillatory loading cycle $W_d$ at a constant stress amplitude $\tau_a$:

$$W_d = \pi(\tau_a)^2/(|G^*|/\sin\delta) \tag{1}$$

In the rutting performance testing, the binders under both the unaged original state and the short-term RTFO aged state are required.

Similarly, the intermediate-temperature fatigue performance can be evaluated using the fatigue parameter $|G^*|\cdot\sin\delta$, which is calculated from the dissipated energy density per oscillatory loading cycle $W_d$ at a constant strain amplitude $\gamma_a$:

$$W_d = \pi(\gamma_a)^2(|G^*|\cdot\sin\delta) \tag{2}$$

In the fatigue performance testing, the binder subjected to successive short-term RTFO aging and long-term PAV aging is required.

Two types of plates were adopted in testing. The 8-mm plates with a gap of 2 mm were used at intermediate temperatures for evaluating the fatigue performance, while the

25-mm plates with a gap of 1 mm were employed at high temperatures for evaluating the rutting performance.

### 3.4. Low-Temperature Performance Test

The thermal cracking resistance of asphalt binders at low temperatures can be assessed using a BBR (Cannon Instrument) in accordance with ASTM D6648 [28]. In this method, the flexural-creep stiffness $S_a(t)$, creep compliance $D_a(t) = 1/S_a(t)$ and m-value of asphalt binder are determined through a creep process. The fixed contact load $P$ around 980 mN is applied to the mid-point of the prismatic test specimen for 240 s, and the mid-span deflection $\xi(t)$ was monitored at an interval of 0.5 s. The flexural-creep stiffness $S_a(t)$ and compliance $D_a(t)$ can be calculated by the following:

$$S_a(t) = \frac{1}{D_a(t)} = \frac{Pl^3}{4bh^3\xi(t)} \tag{3}$$

where $l$ is the span length between the two supports; $h$ is the thickness of the prismatic specimen; $b$ is the width of the prismatic specimen; $t$ is the loading time. Moreover, the m-value can be determined using the slope of the stiffness versus the time on the log-log scale. In the low-temperature performance testing, the binder undergoing both RTFO and PAV aging is required. In this study, the temperatures of $-12$, $-18$, and $-24$ °C were used for testing.

### 3.5. Frequency Sweep Test

To further evaluate the linear viscoelastic behavior over a wide range of frequencies and temperatures, the frequency sweep test was performed on the binders utilizing the DSR. The tests were carried out in a strain-controlled mode. At intermediate temperatures, the 8-mm plates with a gap of 2 mm were applied, and the angular strain amplitudes were set to 0.5%, while at high temperatures, the 25-mm plates with a gap of 1 mm were applied, and the angular strain amplitudes were set to 1%. The testing was conducted at $-2$, 4, 16, 28, 40, 52, and 64 °C, and at each temperature, the complex modulus $G^*$ was measured at the frequencies of 0.1~100 rad/s as follows [29]:

$$G^* = G' + i \cdot G'', \; G' = |G^*|\cos\delta, \; G'' = |G^*|\sin\delta \tag{4}$$

where $i = \sqrt{-1}$ is the imaginary unit; $G'$ is the storage modulus; $G''$ is the loss modulus.

With the measurements of $G^*$ obtained at different temperatures and frequencies, the corresponding master curves can be constructed following the time-temperature superposition principle. The time-temperature shift factor ($\alpha_T$) is represented on the frequency domain by the following [30]:

$$\alpha_T = \frac{\omega_r}{\omega} \tag{5}$$

where $\omega$ is the angular frequency; $\omega_r$ is the reduced angular frequency.

The master curves of $G^*$ for asphalt binders can be modeled by the 2S2P1D model [31]. The 2S2P1D model is a complex-valued model consisting of two spring elements, two parabolic elements, and a dashpot element; thus, it can analytically represent the dynamic modulus and phase angle components as well as the storage modulus and loss modulus components. Besides, its continuous relaxation and retardation spectra can also be analytically derived [32]. The 2S2P1D model has the following form:

$$G^*(\omega_r) = G_e + \frac{G_g - G_e}{1 + \alpha(i\omega_r\tau_0)^{-k} + (i\omega_r\tau_0)^{-h} + (i\omega_r\beta\tau_0)^{-1}} \tag{6}$$

where $G_e$ is the equilibrium modulus; $G_g$ is the instantaneous modulus; $\alpha$, $k$, $h$, $\beta$, and $\tau_0$ are the model parameters. Following a common practice, $G_e$ and $G_g$ were set to 0 Pa and $10^9$ Pa in this study.

In the development of the master curves, the parameters of the 2S2P1D model and the time-temperature shift factors can be calculated simultaneously. The target error function to minimize, $f$, is given by the following:

$$f = \frac{1}{N}\sqrt{\sum_{i=1}^{N}\left(\frac{|G^*|_{\text{mea},i} - |G^*|_{\text{cal},i}}{|G^*|_{\text{mea},i}}\right)^2} + \frac{1}{N}\sqrt{\sum_{i=1}^{N}\left(\frac{\delta_{\text{mea},i} - \delta_{\text{cal},i}}{\delta_{\text{mea},i}}\right)^2} \qquad (7)$$

where $N$ is the number of the data points; the subscripts mea and cal refer to measured and calculated values, respectively.

### 3.6. Experimental Procedure

Figure 2 shows the experimental procedure of this study. First, an artificial RAP binder was fabricated as an alternative to the actual RAP binder. Then, the rejuvenating agent content was determined to generate the recycled binders with different RAP binder contents. Further, the two WMA additives were added to produce the warm-mix recycled binders. Finally, a variety of rheological tests were carried out to analyze the rheological behavior and determine the continuous grading temperatures and PG grades of the binders.

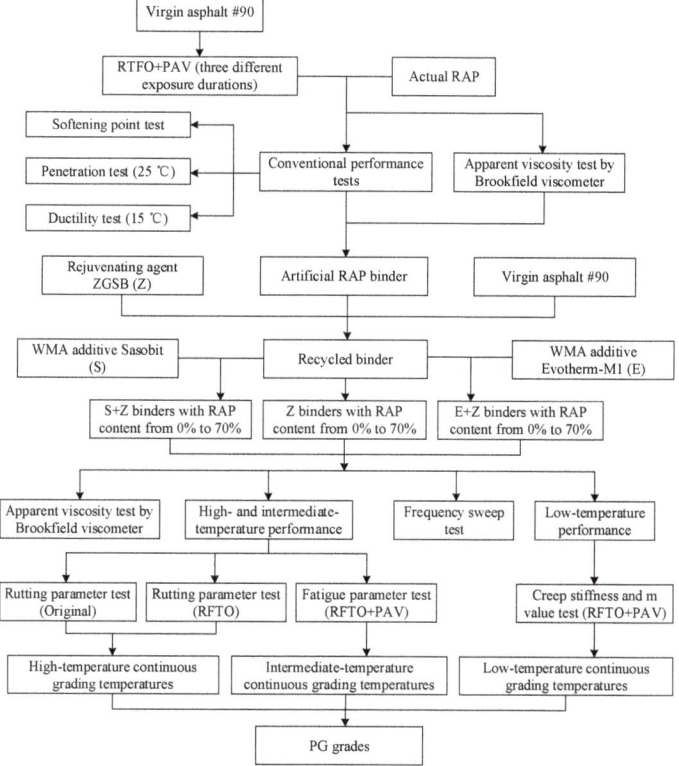

**Figure 2.** Flowchart of the experimental procedure.

## 4. Results and Discussion

### 4.1. Determination of PAV Aging Time for Artificial RAP Binder

As mentioned above, three PAV exposure times, 10, 15, and 20 h, were applied to manufacturing the artificial RAP binder to achieve a similar performance and aging degree to those of the actual one. Figure 3 presents the basic properties of the artificial RAP binders aged at different PAV exposure times. As can be seen, the penetration and ductility (15 °C) reduced, and the softening point and apparent viscosity rose with the increasing exposure time, indicating that the artificial RAP binder became increasingly stiffer. The artificial RAP binder aged for 10 h, PAV-10, exhibited very similar basic properties to those of the actual RAP binder; thus, the RAP-10 binder was used as an alternative to the actual RAP binder in this study.

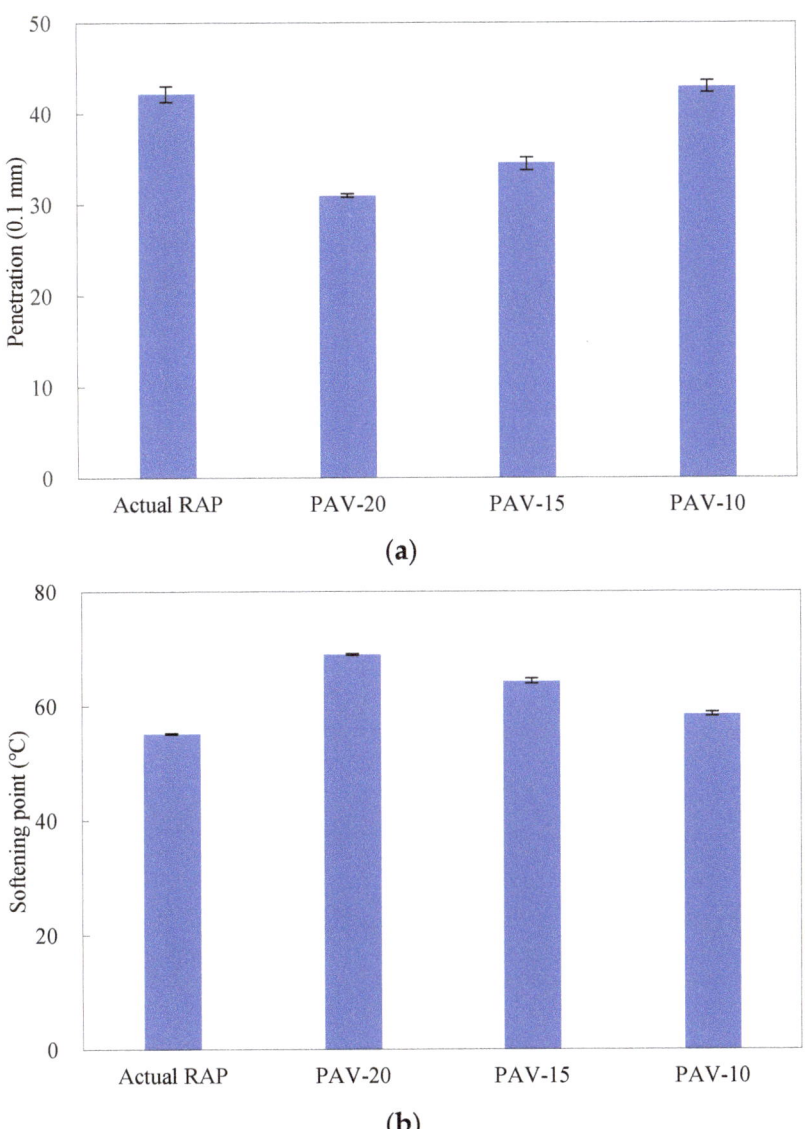

Figure 3. *Cont.*

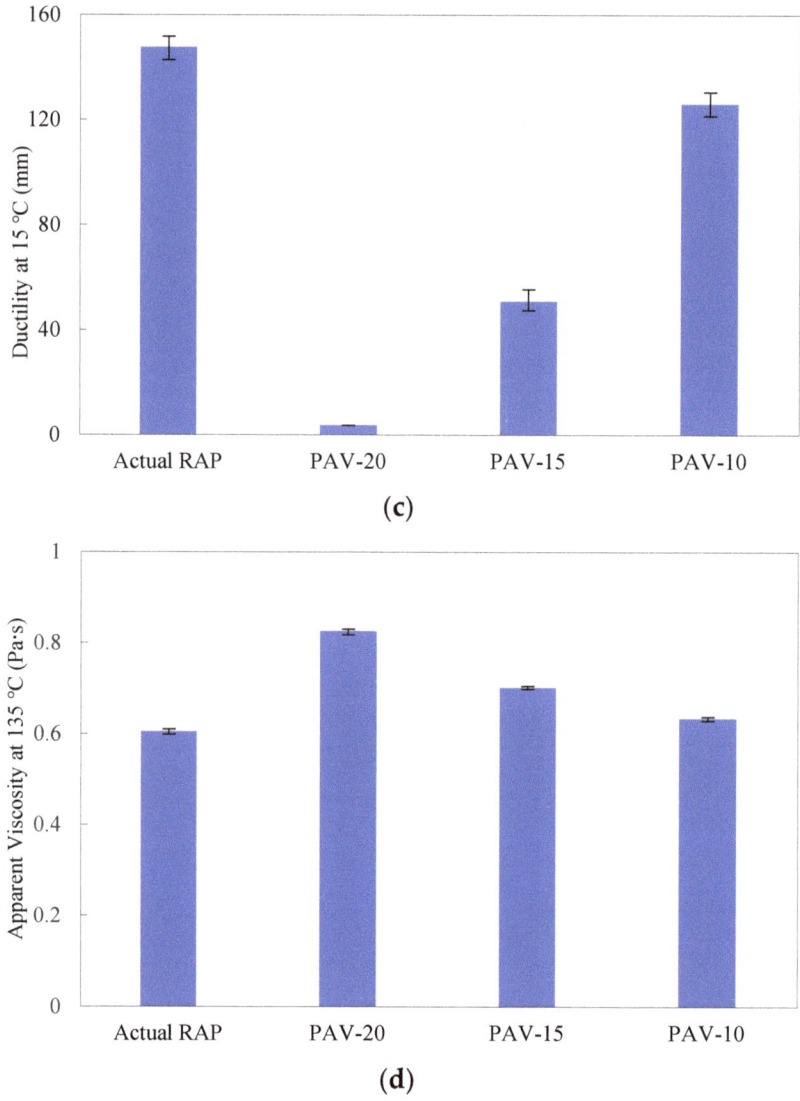

**Figure 3.** Basic properties of artificial RAP binders aged at different PAV exposure times: (**a**) penetration; (**b**) softening point; (**c**) ductility; (**d**) apparent viscosity.

### 4.2. Determination of Rejuvenating Agent Content

To recycle the RAP binder, the rejuvenating agent content needs to be appropriately determined. In this study, the basic properties of the binders, including the penetration, softening point, ductility, and apparent viscosity, were used to this end, as shown in Figure 4. It should be mentioned that since the ductility values of all the binders were larger than 100 cm at 15 °C, the ductility tests were conducted at 5 °C to more clearly reveal the effect of the rejuvenating agent content herein. It can be observed that as the rejuvenating agent content increased, the recycled binder became softer significantly. At the content of 6%, the properties of the recycled binder were found most close to those of the original #90 binder; therefore, the rejuvenating agent dosage of 6% by weight of the RAP binder was employed in the present study.

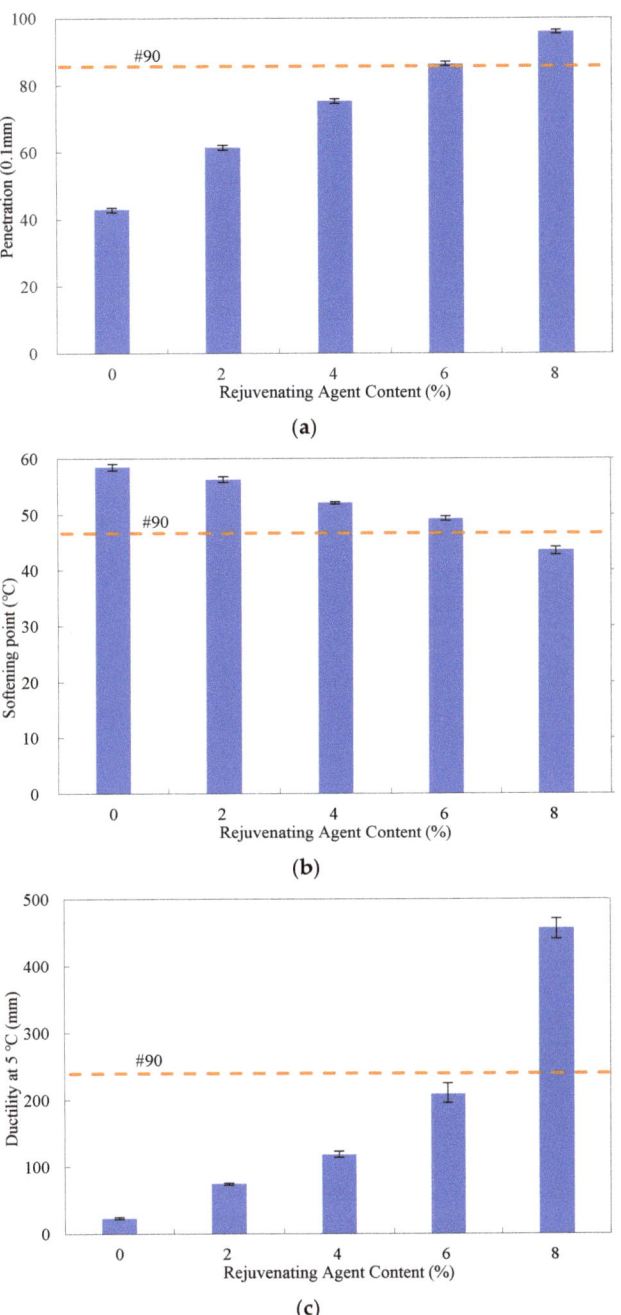

**Figure 4.** *Cont.*

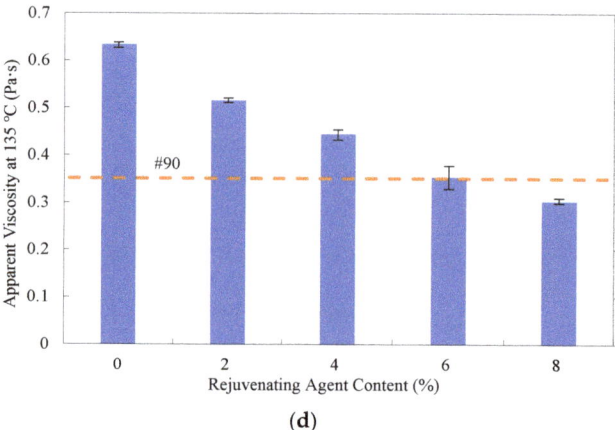

(**d**)

**Figure 4.** Basic properties of rejuvenated asphalt binders at different rejuvenating agent contents: (**a**) penetration; (**b**) softening point; (**c**) ductility; (**d**) apparent viscosity.

*4.3. Analysis of Apparent Viscosity Test Results*

Figure 5 shows the apparent viscosity test results of the warm-mix recycled binders at different RAP binder contents and testing temperatures. As observed, the apparent viscosity values of all three types of recycled binders (i.e., Z, S + Z, and E + Z) increased linearly with the RAP binder content. This indicates that although a relatively accurate rejuvenating agent dosage by weight of the RAP binder (6%) was used, it was still difficult to completely offset the aging effect, especially for higher RAP binder contents.

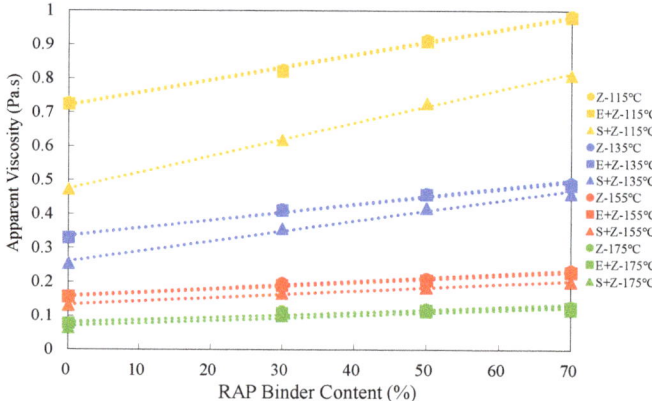

**Figure 5.** Apparent viscosity test results of the warm-mix recycled binders at different RAP binder contents and testing temperatures.

It can be seen from Table 3 that the S + Z binders exhibited the highest percentage increase of apparent viscosity, e.g., 71.2% and 81.5%, respectively, at 115 °C and 175 °C for the RAP binder content of 70%, whereas the E + Z binders presented the lowest percentage increase of apparent viscosity, e.g., only 35.3% and 60.3%, respectively, at 115 °C and 175 °C for the RAP binder content of 70%. This indicates that the WMA additive S can raise the RAP binder content susceptibility of apparent viscosity while E has the opposite effect.

Table 3. Percentage increase in apparent viscosity with the RAP binder content.

| RAP Binder Content (%) | Percentage Increase (%) | | | | | |
|---|---|---|---|---|---|---|
| | 115 °C | | | 175 °C | | |
| | Z | E + Z | S + Z | Z | E + Z | S + Z |
| 0 | 0 | 0 | 0 | 0 | 0 | 0 |
| 30 | 13.5 | 13.2 | 31.1 | 59.7 | 39.7 | 55.4 |
| 50 | 25.7 | 25.4 | 53.9 | 68.1 | 51.3 | 73.8 |
| 70 | 35.4 | 35.3 | 71.2 | 79.2 | 60.3 | 81.5 |

Moreover, Figure 5 shows that at the four different temperatures, the viscosity values of the S + Z binders were lower than those of the E + Z and Z binders, and as the temperature decreased, the viscosity differences between the S + Z binders and the other two types of binders became more remarkable. This indicates that the WMA additive S plays a critical role in reducing the binder viscosity. Thus, it allows a higher RAP binder content. On the other hand, the viscosity values of the E + Z and Z binders were very close to each other, and this indicates that the WMA additive E has little influence on the binder viscosity. Clearly, the WMA additive E operates by a mechanism different from that of S. Actually, as a surfactant, E affects the chemical bonding between binder and aggregate rather than the binder viscosity.

Table 4 shows that the addition of S could decrease the apparent viscosity by up to 35.0% and 12.1%, whereas the use of E could only decrease the apparent viscosity by up to 0.7% and 5.2%, respectively, at the temperatures of 115 °C and 175 °C.

Table 4. Percentage increase in apparent viscosity caused by the WMA additive.

| Binder Type | Percentage Increase (%) | | | | | | | |
|---|---|---|---|---|---|---|---|---|
| | 115 °C | | | | 175 °C | | | |
| | RAP-0% | RAP-30% | RAP-50% | RAP-70% | RAP-0% | RAP-30% | RAP-50% | RAP-70% |
| Z | 0 | 0 | 0 | 0 | 0 | 0 | 0 | 0 |
| E + Z | −0.4 | −0.6 | −0.7 | −0.5 | 8.3 | −5.2 | −2.5 | −3.1 |
| S + Z | −35.0 | −24.9 | −20.4 | −17.9 | −9.7 | −12.1 | −6.6 | −8.5 |

*4.4. Analysis of High-Temperature Performance Test Results*

Typically, a higher $|G^*|/\sin\delta$ value represents a superior rutting resistance. Figure 6 presents the rutting parameter test results of the warm-mix recycled binders at different RAP binder contents under the unaged condition. As can be seen, the rutting parameter values exhibited linear increasing trends with the RAP binder content on the logarithmic scale regardless of WMA additives used, indicating an improvement in rutting performance. It is seen that the difference of $|G^*|/\sin\delta$ between the binders with and without the RAP binder increased considerably with the rising RAP binder content. Thus, accurately controlling the rejuvenating agent content is critical to guaranteeing the performance of recycled asphalt mixtures with high RAP contents.

Table 5 shows that S could better alleviate the increase in $|G^*|/\sin\delta$ with the RAP binder content than E under the unaged condition. For instance, at the RAP binder content of 70% and 70 °C, the percentage increase of $|G^*|/\sin\delta$ for the S + Z binder was 56.2% while that for the E + Z binder was 67.5%, both of which were less than 72.3% for the Z binder.

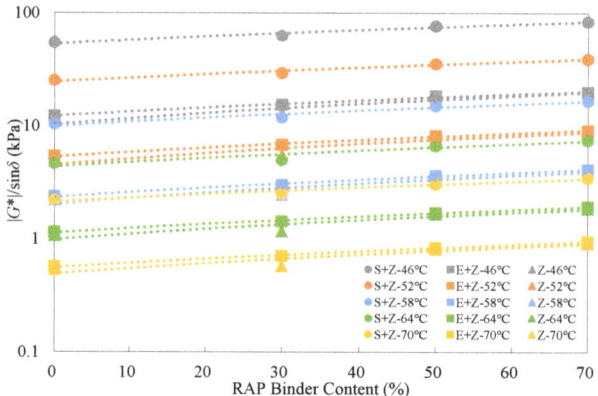

**Figure 6.** Rutting parameter test results of the warm-mix recycled binders at different RAP binder contents under the unaged condition.

**Table 5.** Percentage increase in $|G^*|/\sin\delta$ with the RAP binder content under the unaged condition.

| RAP Binder Content (%) | Percentage Increase (%) | | | | | |
|---|---|---|---|---|---|---|
| | 46 °C | | | 70 °C | | |
| | Z | E + Z | S + Z | Z | E + Z | S + Z |
| 0 | 0 | 0 | 0 | 0 | 0 | 0 |
| 30 | 16.8 | 27.2 | 15.5 | 8.1 | 25.2 | 14.3 |
| 50 | 59.4 | 46.1 | 39.8 | 53.4 | 37.7 | 37.9 |
| 70 | 82.1 | 63.7 | 53.0 | 72.3 | 67.5 | 56.2 |

Besides, Figure 6 presents that at any given RAP content and temperature, the S + Z binders displayed the best performance, followed by the E + Z binders, and the Z binders ranked the lowest. Although both WMA additives contributed to the increasing $|G^*|/\sin\delta$, S performed significantly better than E. This may be because wax crystallization in the additive S reinforces the permanent deformation resistance of asphalt binders at elevated temperatures that are lower than the melting point of S (about 100 °C).

Table 6 shows that the inclusion of S could increase $|G^*|/\sin\delta$ by up to 399.5% and 340.8%, but the use of E could only increase $|G^*|/\sin\delta$ by up to 20.6% and 23.0%, respectively, at the temperatures of 46 °C and 70 °C under the unaged condition.

**Table 6.** Percentage increase in $|G^*|/\sin\delta$ caused by the WMA additive under the unaged condition.

| Binder Type | Percentage Increase (%) | | | | | | | |
|---|---|---|---|---|---|---|---|---|
| | 46 °C | | | | 70 °C | | | |
| | RAP-0% | RAP-30% | RAP-50% | RAP-70% | RAP-0% | RAP-30% | RAP-50% | RAP-70% |
| Z | 0 | 0 | 0 | 0 | 0 | 0 | 0 | 0 |
| E + Z | 10.8 | 20.6 | 1.5 | −0.4 | 6.2 | 23.0 | −4.7 | 3.3 |
| S + Z | 399.5 | 393.9 | 338.3 | 319.7 | 317.0 | 340.8 | 274.8 | 278.0 |

According to the Superpave PG specification ASTM D6373 [33], unaged binders should satisfy the requirement that $|G^*|/\sin\delta$ values are larger than 1 kPa. It can be observed from Figure 7 that the rutting parameter test results of the warm-mix recycled binders under the unaged condition were all greater than 1 kPa at testing temperatures below 64 °C; however, when the temperature rose to 70 °C, only the S + Z binders met the requirement, implying a more desirable rutting resistance for the S + Z binders than the E + Z and Z ones.

The role of E in enhancing rutting performance was found to be quite slight at different temperatures, and this can be ascribed to its nature as a chemical surfactant.

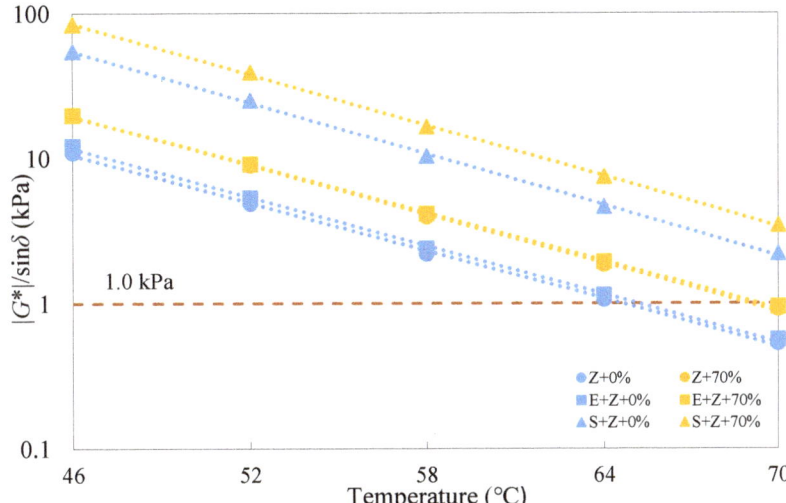

**Figure 7.** Rutting parameter test results of the warm-mix recycled binders at different testing temperatures under the unaged condition.

Figures 8 and 9 give the rutting parameter test results of the warm-mix recycled binders at different RAP binder contents and testing temperatures under the short-term RTFO aged condition. In this case, the $|G^*|/\sin\delta$ values should be larger than 2.2 kPa. It is evident that similar observations to those under the unaged condition can be made for all the recycled binders under the short-term aged condition.

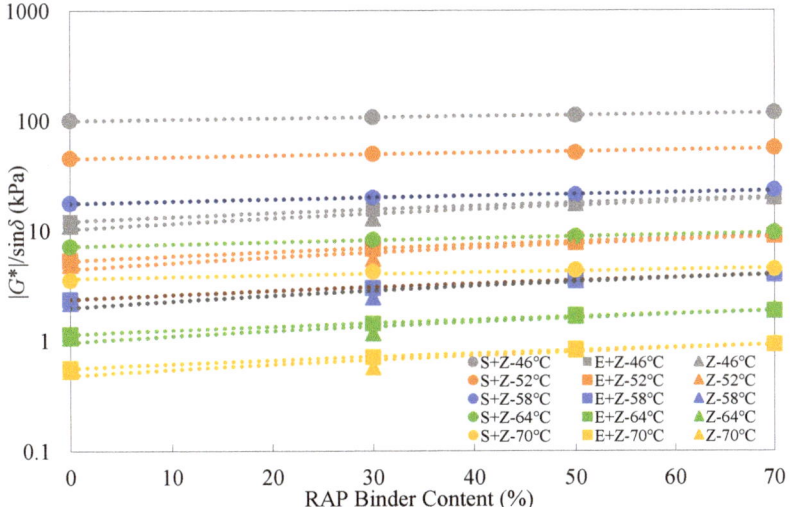

**Figure 8.** Rutting parameter test results of the warm-mix recycled binders at different RAP binder contents under the short-term aged condition.

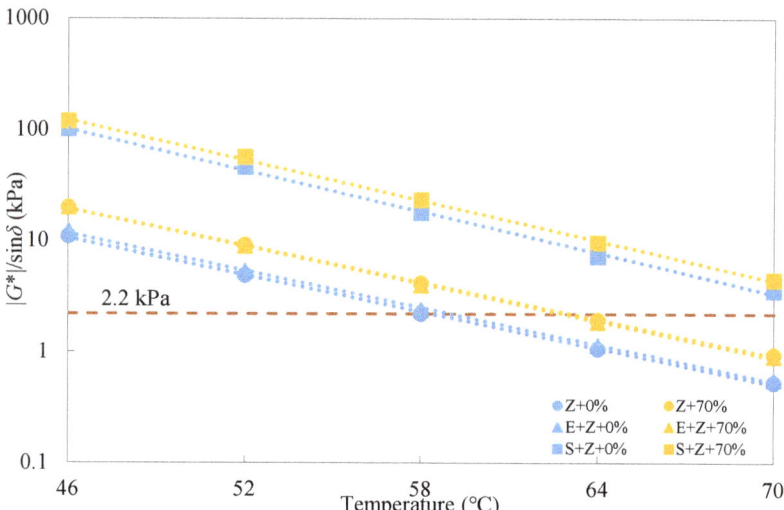

**Figure 9.** Rutting parameter test results of the warm-mix recycled binders at different testing temperatures under the short-term aged condition.

Table 7 shows that S could alleviate the increase in $|G^*|/\sin\delta$ with the RAP binder content, but E intensified this percentage increase under the short-term aged condition. This observation was different from that obtained under the unaged condition.

**Table 7.** Percentage increase in $|G^*|/\sin\delta$ with the RAP binder content under the short-term aged condition.

| RAP Binder Content (%) | Percentage Increase (%) | | | | | |
|---|---|---|---|---|---|---|
| | 46 °C | | | 70 °C | | |
| | Z | E + Z | S + Z | Z | E + Z | S + Z |
| 0 | 0 | 0 | 0 | 0 | 0 | 0 |
| 30 | 8.1 | 16.4 | 6.5 | 20.8 | 24.9 | 18.5 |
| 50 | 25.9 | 32.5 | 10.6 | 38.9 | 42.6 | 22.9 |
| 70 | 46.6 | 52.5 | 17.2 | 68.8 | 71.6 | 24.5 |

Table 8 indicates that the inclusion of S could increase $|G^*|/\sin\delta$ by up to 411.3% and 335.3%, but the use of E could only increase $|G^*|/\sin\delta$ by up to 29.9% and 26.0%, respectively, at the temperatures of 46 °C and 70 °C under the short-term aged condition.

**Table 8.** Percentage increase in $|G^*|/\sin\delta$ caused by the WMA additive under the short-term aged condition.

| Binder Type | Percentage Increase (%) | | | | | | | |
|---|---|---|---|---|---|---|---|---|
| | 46 °C | | | | 70 °C | | | |
| | RAP-0% | RAP-30% | RAP-50% | RAP-70% | RAP-0% | RAP-30% | RAP-50% | RAP-70% |
| Z | 0 | 0 | 0 | 0 | 0 | 0 | 0 | 0 |
| E + Z | 20.6 | **29.9** | 26.8 | 25.4 | 21.9 | **26.0** | 25.2 | 23.9 |
| S + Z | **411.3** | 403.8 | 349.1 | 308.9 | **335.3** | 327.1 | 285.2 | 221.0 |

## 4.5. Analysis of Intermediate-Temperature Performance Test Results

Generally, a higher fatigue parameter $|G^*|\cdot\sin\delta$ represents an inferior fatigue performance. Figure 10 displays the fatigue parameter test results of the warm-mix recycled binders at different RAP binder contents under the long-term aged condition. As the RAP binder content increased, the $|G^*|\cdot\sin\delta$ values for all the binders increased linearly, suggesting that the RAP binder has an adverse impact on the fatigue resistance. Moreover, it is noticed that the rejuvenating agent was unable to entirely compensate for the negative effect of the aged binder despite a relatively accurate determination of the rejuvenating agent dosage. In addition, both the WMA additives S and E lowered the $|G^*|\cdot\sin\delta$ values, which indicates that both S and E are capable of improving the fatigue performance of the recycled binders. Specifically, E performed better than S in this aspect.

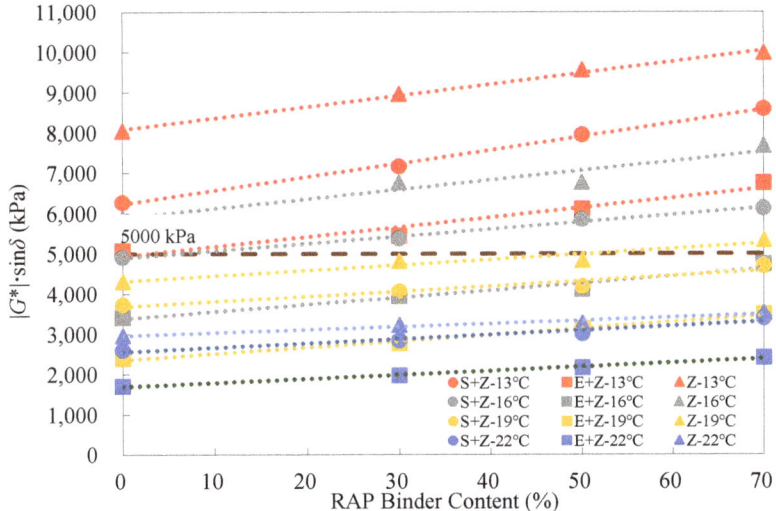

**Figure 10.** Fatigue parameter test results of the warm-mix recycled binders at different RAP binder contents and testing temperatures under the long-term aged condition.

Table 9 shows that both S and E could promote the increase in $|G^*|\cdot\sin\delta$ with the RAP binder content, and the two WMA additives seemed to exhibit a comparable effect. Table 10 indicates that the addition of S could reduce $|G^*|\cdot\sin\delta$ by up to 22.1%, and 12.1% and the use of E could even reduce $|G^*|\cdot\sin\delta$ by up to 39.3% and 42.3%, respectively, at the temperatures of 13 °C and 22 °C under the long-term aged condition.

**Table 9.** Percentage increase in $|G^*|\cdot\sin\delta$ with the RAP binder content under the long-term aged condition.

| RAP Binder Content (%) | Percentage Increase (%) | | | | | |
|---|---|---|---|---|---|---|
| | 13 °C | | | 22 °C | | |
| | Z | E + Z | S + Z | Z | E + Z | S + Z |
| 0 | 0 | 0 | 0 | 0 | 0 | 0 |
| 30 | 8.2 | 11.7 | 14.2 | 8.8 | 15.5 | 9.1 |
| 50 | 13.5 | 24.9 | 26.7 | 10.4 | 26.8 | 15.5 |
| 70 | 23.4 | 33.8 | 37.0 | 19.0 | 41.1 | 30.1 |

**Table 10.** Percentage increase in $|G^*|\cdot\sin\delta$ caused by the WMA additive under the long-term aged condition.

| Binder Type | Percentage Increase (%) | | | | | | | |
|---|---|---|---|---|---|---|---|---|
| | 13 °C | | | | 22 °C | | | |
| | RAP-0% | RAP-30% | RAP-50% | RAP-70% | RAP-0% | RAP-30% | RAP-50% | RAP-70% |
| Z | 0 | 0 | 0 | 0 | 0 | 0 | 0 | 0 |
| E + Z | −36.9 | **−39.3** | −36.2 | −32.4 | **−42.3** | −38.8 | −33.8 | −31.7 |
| S + Z | **−22.1** | −20.0 | −16.9 | −13.9 | **−12.1** | −11.9 | −8.0 | −3.9 |

In accordance with the Superpave PG specification ASTM D6373 [33], long-term aged binders should meet the requirement that the $|G^*|\cdot\sin\delta$ values must be less than 5000 kPa. Figure 11 gives the $|G^*|\cdot\sin\delta$ test results of the binders at different testing temperatures. At temperatures above 22 °C, the $|G^*|\cdot\sin\delta$ values of all the binders could satisfy the criterion. At 19 °C, only the Z binders with the RAP content of 70% (Z + 70%) did not meet the requirement, whereas, at 16 °C, only the E + Z binders were desirable. When the temperature dropped below 13 °C, all the test results exceeded the limiting value (5000 kPa). It can be observed from the slopes of the fitted lines that in terms of fatigue resistance, the addition of E slightly elevated the temperature susceptibility of the recycled binders, but the inclusion of S had a very limited effect on the temperature susceptibility.

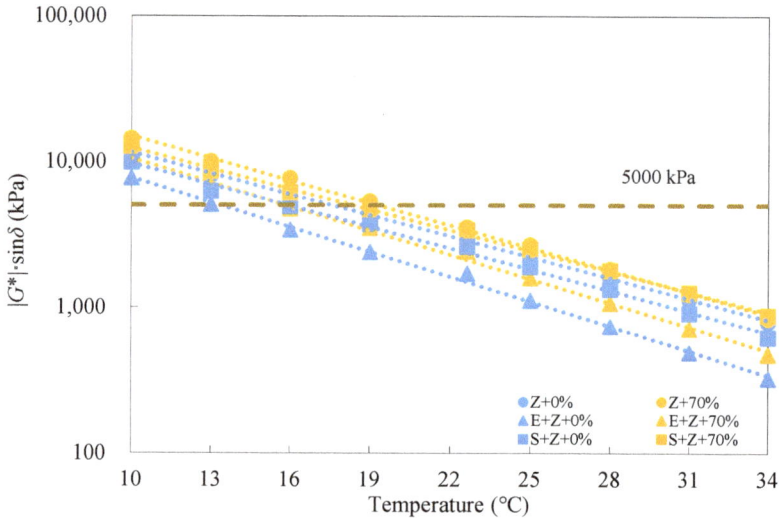

**Figure 11.** Fatigue parameter test results of the warm-mix recycled binders at different testing temperatures under the long-term aged condition.

*4.6. Analysis of Low-Temperature Performance Test Results*

In general, the binder with a lower stiffness modulus $S_a$ and a higher m-value has a more satisfactory low-temperature thermal cracking resistance. Figure 12 presents the creep stiffness and m-value test results of the warm-mix recycled binders at different RAP binder contents and testing temperatures. As can be seen, compared to the stiffness moduli at the RAP binder content of 0%, those at 30% to 70% were even lower at the three temperatures. This abnormal observation can actually be attributed to the addition of the rejuvenating agent Z, which has a softening effect on the aged binder. Additionally, with the growth of the RAP binder content, the m-values of the Z binders did not exhibit noticeable change, and this also demonstrates the effect of the rejuvenating agent. Table 11 shows no remarkable trends for the percentage increases of $S_a$ and m-value with the RAP binder content.

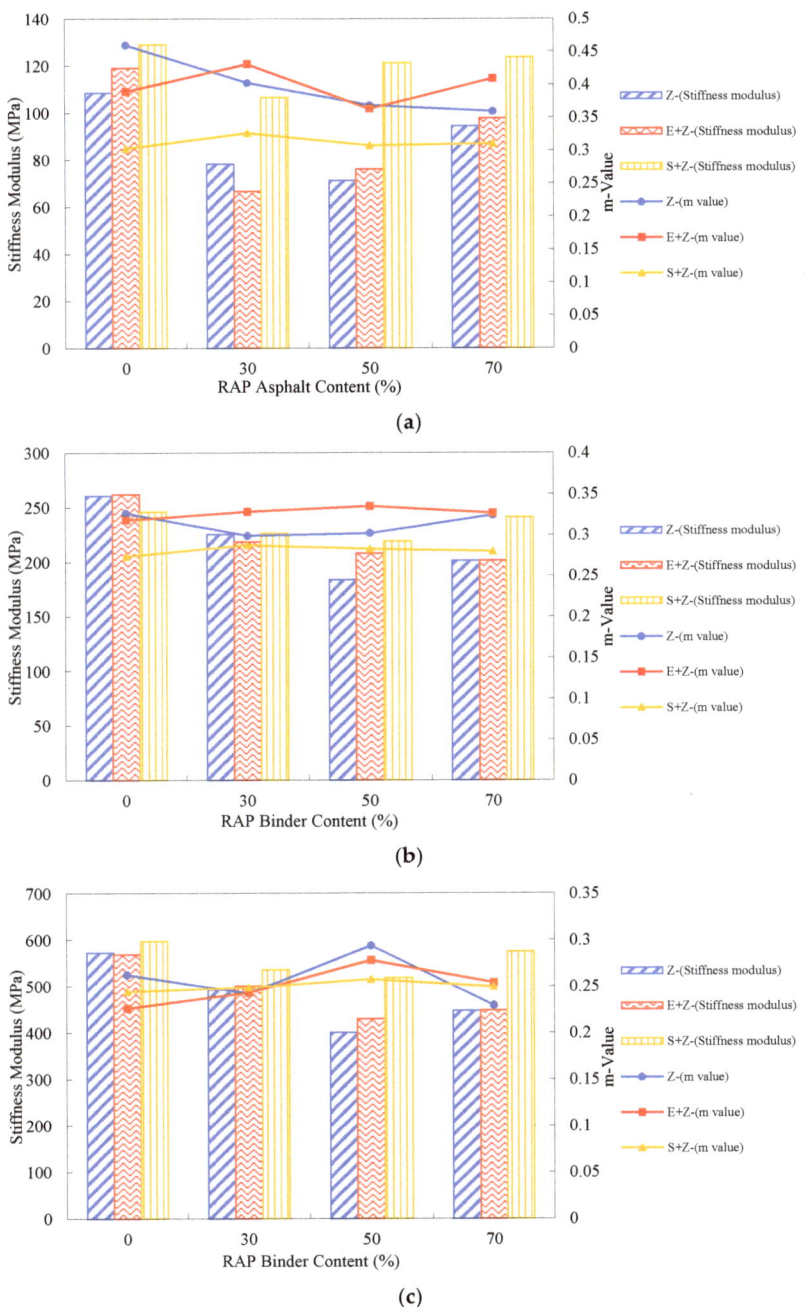

**Figure 12.** Creep stiffness and m value test results of the warm-mix recycled binders at different RAP binder contents and testing temperatures under the long-term aged condition: (**a**) −12 °C; (**b**) −18 °C; (**c**) −24 °C.

Table 11. Percentage increases of $S_a$ and m-value with the RAP binder content under the long-term aged condition.

| Quantity | RAP Binder Content (%) | Percentage Increase (%) | | | | | | | | |
|---|---|---|---|---|---|---|---|---|---|---|
| | | −24 °C | | | −18 °C | | | −12 °C | | |
| | | Z | E + Z | S + Z | Z | E + Z | S + Z | Z | E + Z | S + Z |
| $S_a$ | 0 | 0 | 0 | 0 | 0 | 0 | 0 | 0 | 0 | 0 |
| | 30 | −14.1 | −12 | −10.2 | −13.4 | −16.6 | −7.9 | −27.9 | −44.2 | −17.4 |
| | 50 | −30.1 | −24.5 | −13.2 | −29.6 | −20.6 | −11 | −34.5 | −36.3 | −6.2 |
| | 70 | −21.9 | −21.2 | −3.7 | −22.6 | −23.1 | −2 | −13.1 | −17.9 | −4.3 |
| m-value | 0 | 0 | 0 | 0 | 0 | 0 | 0 | 0 | 0 | 0 |
| | 30 | −7.6 | 7.5 | 1.8 | −8.4 | 3.1 | 5.1 | −12.5 | 10.7 | 7.8 |
| | 50 | 12 | 23 | 5.5 | −7.5 | 5.2 | 3.3 | −19.9 | −6.8 | 1.5 |
| | 70 | −12.6 | 12.4 | 2.3 | −0.8 | 2.7 | 2.4 | −21.8 | 5 | 2.5 |

Further, Figure 12 shows that regardless of the RAP percentage, the stiffness modulus values of the S + Z binders were all higher than those of the Z and E + Z binders, and the m-values of the S + Z binders were mostly lower than those of the other two types of binders. This implies that the use of S has a negative impact on the thermal cracking resistance. For the E + Z and Z binders, both stiffness moduli and m-values were very close to each other; thus, E has a very mild influence on the low-temperature rheological properties of the recycled binders. Table 12 presents that the addition of S could increase the $S_a$ by up to 29.3%, 19.6%, and 70.2% and reduce the m-values by up to 12.4%, 16.2%, and 34.1%, respectively, at the temperatures of −24 °C, −18 °C, and −12 °C under the long-term aged condition.

Table 12. Percentage increases of $S_a$ and m-value caused by the WMA additive under the long-term aged condition.

| Quantity | Binder Type | Percentage Increase (%) | | | | | | | | | | | |
|---|---|---|---|---|---|---|---|---|---|---|---|---|---|
| | | −24 °C | | | | −18 °C | | | | −12 °C | | | |
| | | RAP-0% | RAP-30% | RAP-50% | RAP-70% | RAP-0% | RAP-30% | RAP-50% | RAP-70% | RAP-0% | RAP-30% | RAP-50% | RAP-70% |
| $S_a$ | Z | 0 | 0 | 0 | 0 | 0 | 0 | 0 | 0 | 0 | 0 | 0 | 0 |
| | E + Z | −0.8 | 1.7 | 7.1 | 0.1 | 0.6 | −3.1 | 13.4 | 0 | 9.7 | −15.1 | 6.7 | 3.6 |
| | S + Z | 4.2 | 8.9 | 29.3 | 28.5 | −5.6 | 0.4 | 19.3 | 19.6 | 18.9 | 36.2 | 70.2 | 31 |
| m-value | Z | 0 | 0 | 0 | 0 | 0 | 0 | 0 | 0 | 0 | 0 | 0 | 0 |
| | E + Z | −13.9 | 0.2 | −5.4 | 10.7 | −2.6 | 9.7 | 10.8 | 0.8 | −15.3 | 7.1 | −1.5 | 13.8 |
| | S + Z | −7 | 2.5 | −12.4 | 8.7 | −16.2 | −3.8 | −6.5 | −13.6 | −34.1 | −18.9 | −16.6 | −13.6 |

In terms of the Superpave PG specification ASTM D6373 [33], the stiffness moduli of the long-term aged binders should be less than 300 MPa, but the m-values should be greater than 0.3. Obviously, at −12 °C, the stiffness moduli and m-values of all the binders were acceptable, but at −24 °C, were ineligible. At −18 °C, due to the limitation of the m-value, the Z + S binders at all the RAP binder contents and the Z binders at the RAP binder content of 30% were undesirable for thermal cracking resistance.

### 4.7. Analysis of Continuous Grading Temperatures and PG Grades

In terms of the specification ASTM D7643 [34], the continuous grading temperatures of asphalt binder can be calculated by interpolating between test data measured at two adjacent specification temperatures so that high-, intermediate- and low-temperature grades can be more accurately evaluated. Figure 13 displays the resulting high-temperature continuous grading temperatures of the binders at different RAP binder contents. It is seen that as the RAP binder percentage rose, the continuous grading temperatures merely increased moderately (by about 4 °C from 0% to 70%) for both the original and RTFO-aged binders. This is because the addition of the rejuvenating agent (6% by weight of the RAP binder) alleviated the hardening effect of the aged binder to a certain extent. Compared to the continuous grading temperatures of the Z binders, the results for the S + Z binders increased by more than two high-temperature grades (12 °C), while the results for the E + Z binders solely increased slightly. This demonstrates the advantage of S over E in enhancing the rutting resistance.

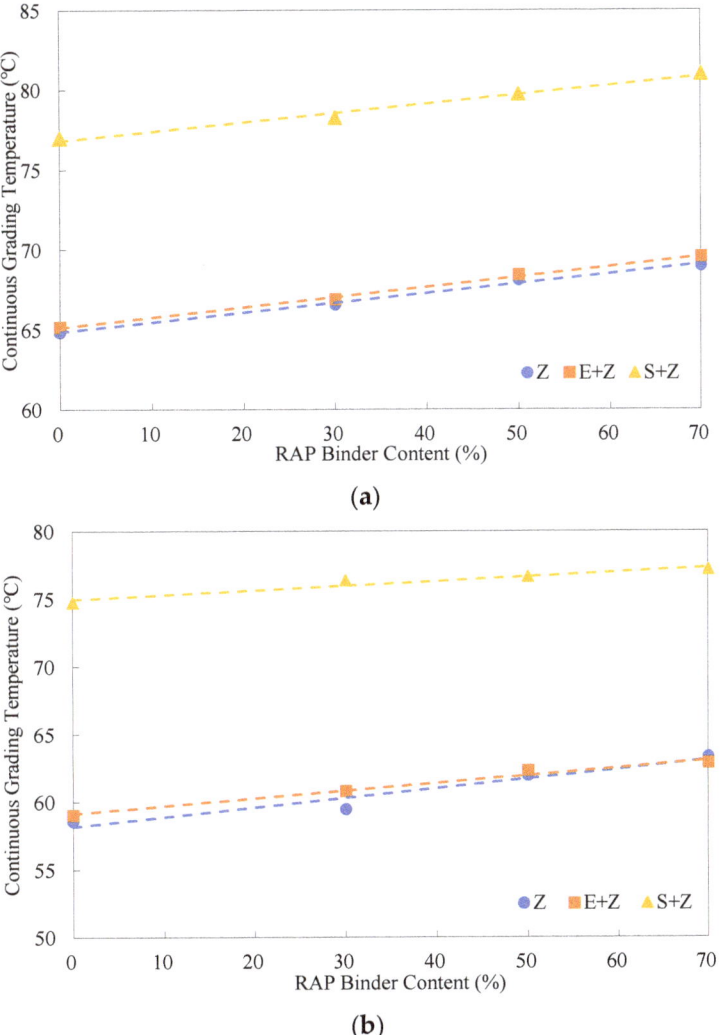

**Figure 13.** High-temperature continuous grading temperatures of the warm-mix recycled binders at different RAP binder contents: (**a**) original; (**b**) RTFO-aged.

Figure 14 gives the intermediate-temperature continuous grading temperatures of the warm-mix recycled binders at different RAP binder contents. The continuous grading temperatures of the three types of binders presented linear trends with the RAP binder content. The addition of E led to the greatest reduction of the continuous grading temperatures (more than a grade for fatigue performance, 3 °C), followed by the use of S.

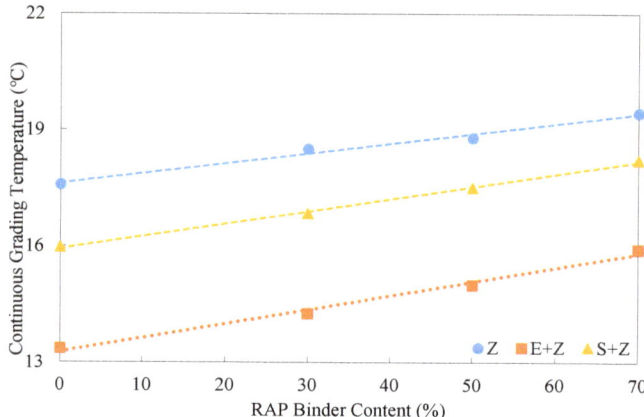

**Figure 14.** Intermediate-temperature continuous grading temperatures of the warm-mix recycled binders at different RAP binder contents.

Figure 15 shows the low-temperature continuous grading temperatures determined using the stiffness modulus $S_a$ and m-value, respectively. The final low-temperature grade is dependent on the upper of the two continuous grading temperatures from $S_a$ and the m-value. Evidently, the inclusion of S considerably elevated the continuous grades, thus having a negative impact on the thermal cracking resistance. Unlike the WMA additive S, E almost did not change the low-temperature continuous grading temperatures of the Z binders.

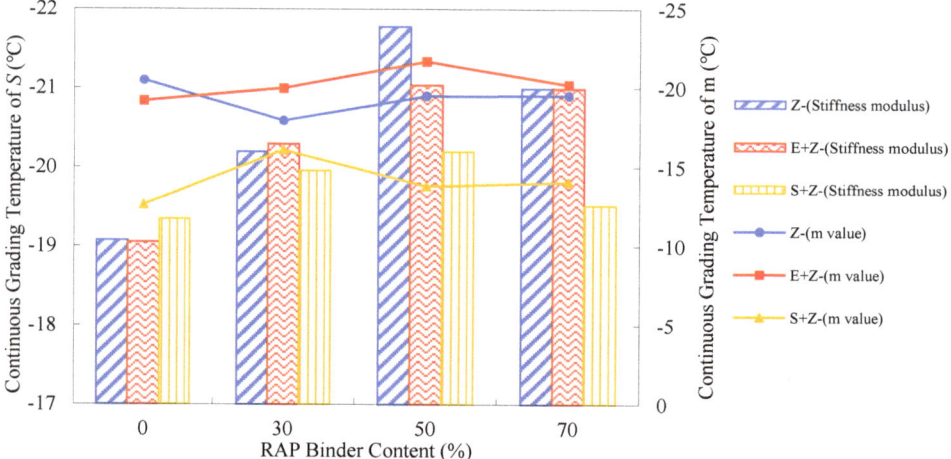

**Figure 15.** Low-temperature continuous grading temperatures of the warm-mix recycled binders at different RAP binder contents.

Table 13 shows the resulting PG grades of all the warm-mix recycled binders according to the specification ASTM D6373 [33]. As can be seen, when subjected to short-term aging, the high-temperature grades of both the E + Z and Z binders declined by one grade (6 °C), whereas the S + Z binders mostly maintained the original grades, which implies the role of the WMA additive S in aging resistance. Additionally, the S + Z binders possessed wider ranges of pavement service temperature (wider by about two PG grades, 12 °C) than those of the other two types of binders though S lowered the low-temperature performance of the binders. In this regard, the use of E almost did not change the PG grades of the binders.

Table 13. Performance grading (PG) grades of the warm-mix recycled binders.

| Binder | PG High (Original) | PG High (RFTO) | PG Intermediate (RTFO + PAV) | PG Low (RTFO + PAV) | PG Grade |
|---|---|---|---|---|---|
| S + Z + 0% | 76 | 70 | 16 | −12 | 70–22 |
| S + Z + 30% | 76 | 76 | 19 | −12 | 76–22 |
| S + Z + 50% | 76 | 76 | 19 | −12 | 76–22 |
| S + Z + 70% | 76 | 76 | 19 | −12 | 76–22 |
| E + Z + 0% | 64 | 58 | 16 | −18 | 58–28 |
| E + Z + 30% | 64 | 58 | 16 | −18 | 58–28 |
| E + Z + 50% | 64 | 58 | 16 | −18 | 58–28 |
| E + Z + 70% | 64 | 58 | 19 | −18 | 58–28 |
| Z + 0% | 64 | 58 | 19 | −18 | 58–28 |
| Z + 30% | 64 | 58 | 19 | −12 | 58–22 |
| Z + 50% | 64 | 58 | 19 | −18 | 58–28 |
| Z + 70% | 64 | 58 | 22 | −18 | 58–28 |

*4.8. Analysis of Linear Viscoelastic Master Curves*

In this study, the reference temperature $T_r$ was selected as 28 °C for the construction of the master curves for the binders. Figure 16 presents the calculated master curves of dynamic modulus and phase angle for the three types of binders. As observed, the 2S2P1D model well simulated the dynamic modulus master curves for all the binders. For the Z and E + Z binders, the phase angle master curves of the 2S2P1D model were also in good agreement with the test data. However, due to the addition of S, the phase angle test data of the S + Z binders obtained at different test temperatures were incapable of forming single curves by horizontal shift, which led to the failure of the time-temperature superposition principle and unsatisfactory fitting of the 2S2P1D model. This may be because the WMA additive S is a thermorheologically complex material in essence. Table 14 summarizes the resulting parameters of the 2S2P1D model and fitting errors.

Table 14. Resulting parameters of the 2S2P1D model and fitting errors for the binders.

| Sample | $\alpha$ | $k$ | $h$ | $\beta$ | $\tau_0$ (s) | $f$ (%) |
|---|---|---|---|---|---|---|
| Actual RAP | 8.22 | 0.33 | 0.68 | $9.20 \times 10^1$ | $4.18 \times 10^{-6}$ | 1.92 |
| Pav-10 | 4.46 | 0.21 | 0.56 | $2.12 \times 10^3$ | $2.28 \times 10^{-7}$ | 3.68 |
| Z + 0% | 7.50 | 0.36 | 0.69 | $8.37 \times 10^1$ | $8.98 \times 10^{-7}$ | 1.37 |
| Z + 30% | 6.78 | 0.35 | 0.70 | $4.87 \times 10^1$ | $1.75 \times 10^{-6}$ | 1.77 |
| Z + 50% | 8.87 | 0.37 | 0.71 | $6.58 \times 10^1$ | $2.63 \times 10^{-6}$ | 1.63 |
| Z + 70% | 9.21 | 0.36 | 0.70 | $1.01 \times 10^2$ | $1.75 \times 10^{-6}$ | 1.01 |
| E + Z + 0% | 8.69 | 0.32 | 0.69 | $6.97 \times 10^1$ | $9.76 \times 10^{-7}$ | 1.31 |
| E + Z + 30% | 9.47 | 0.33 | 0.69 | $9.36 \times 10^1$ | $9.31 \times 10^{-7}$ | 1.46 |
| E + Z + 50% | 9.36 | 0.35 | 0.70 | $8.70 \times 10^1$ | $1.75 \times 10^{-6}$ | 1.39 |
| E + Z + 70% | 8.93 | 0.34 | 0.69 | $1.08 \times 10^2$ | $1.75 \times 10^{-6}$ | 1.70 |
| S + Z + 0% | 14.19 | 0.34 | 0.69 | $1.84 \times 10^3$ | $2.68 \times 10^{-5}$ | 4.00 |
| S + Z + 30% | 14.16 | 0.31 | 0.67 | $3.22 \times 10^3$ | $2.22 \times 10^{-5}$ | 3.78 |
| S + Z + 50% | 23.46 | 0.31 | 0.70 | $8.36 \times 10^2$ | $3.86 \times 10^{-5}$ | 3.21 |
| S + Z + 70% | 16.19 | 0.28 | 0.65 | $4.39 \times 10^3$ | $1.18 \times 10^{-5}$ | 3.24 |

Figure 17 gives the master curves of dynamic modulus and phase angle for the actual RAP binder and the PAV-10 binder. As can be seen, the master curves of the two aged binders were very close to each other. This indicates that their linear viscoelastic properties were very similar, and the use of the PAV-10 binder as an alternative to the actual RAP binder was rational and effective.

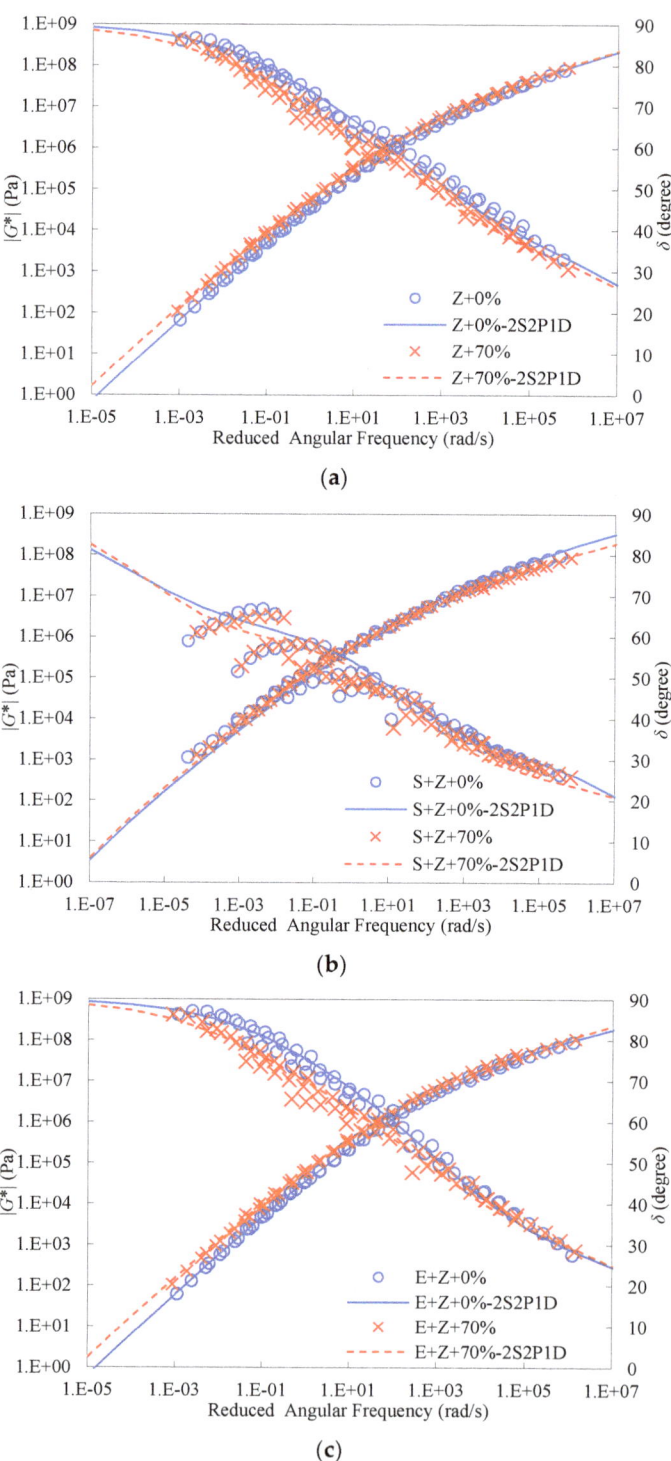

**Figure 16.** Calculated master curves of dynamic modulus and phase angle for: (**a**) Z binders; (**b**) S + Z binders; (**c**) E + Z binders.

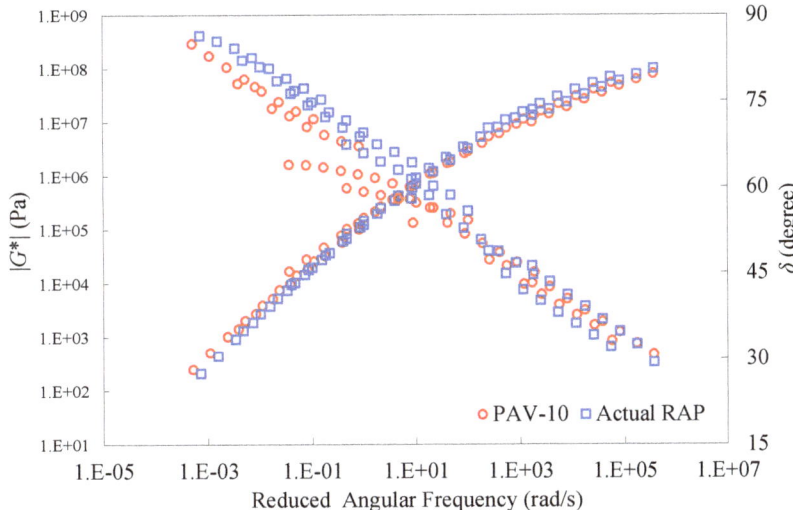

**Figure 17.** Master curves of dynamic modulus and phase angle for the actual RAP binder and the PAV-10 binder.

It can be seen from Figure 18 that both the master curves of dynamic modulus and phase angle at different RAP binder contents were very close for the Z binders. This further verifies the recycling effect of the rejuvenating agent. Even so, there still existed differences between the master curves, in particular when the percentage of the RAP binder became high. This also indicates that it is difficult for the rejuvenating agent to completely balance the negative impact of the aged binder.

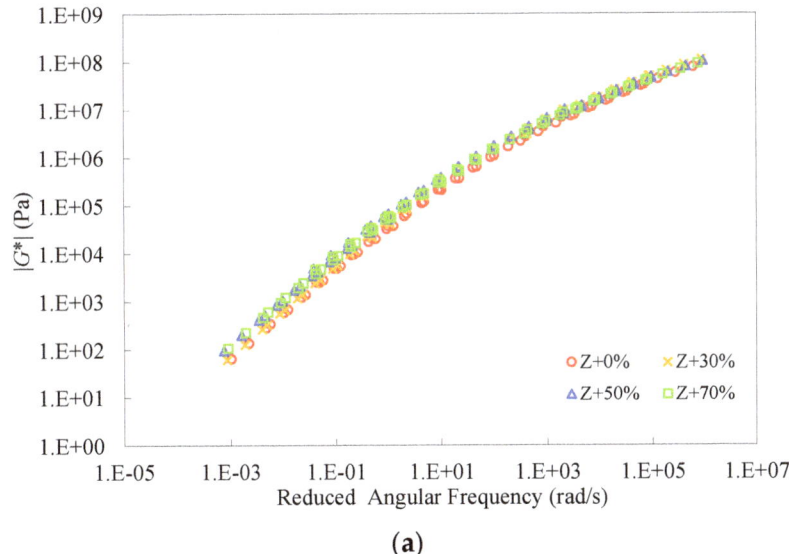

(**a**)

**Figure 18.** *Cont.*

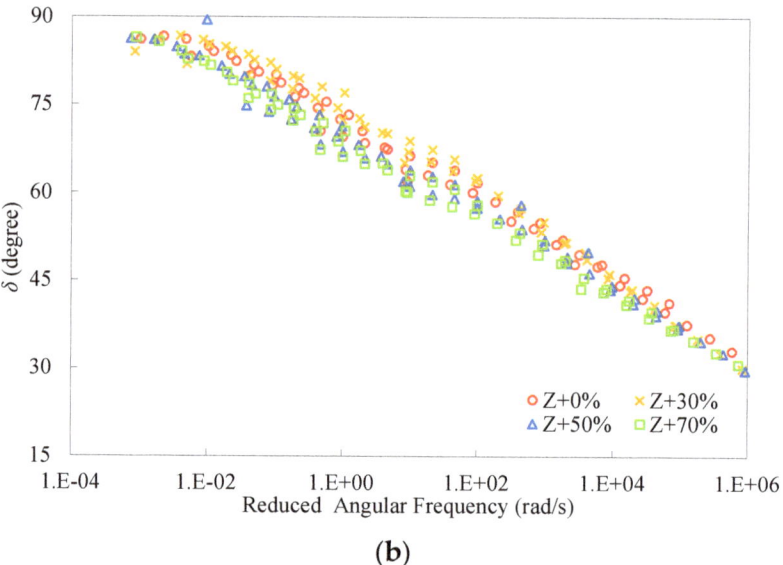

(**b**)

**Figure 18.** Master curves of dynamic modulus and phase angle of the Z binders at different RAP binder contents.

Figure 19 shows the master curves of dynamic modulus and phase angle of the warm-mix recycled binders at the RAP binder content of 70%. It can be seen that the inclusion of S considerably improved the dynamic modulus and lowered the phase angle of the binders over a wide range of frequencies and temperatures, indicating the reinforcement effect on the stiffness and the enhancement effect on the elasticity. Moreover, the difference between the master curves of the S + Z binder and the other two binders became more pronounced at lower frequencies. Unlike the additive S, E had a very slight influence on the master curves of dynamic modulus and phase angle of the recycled binder at the RAP binder content of 70%. Similar observations can be attained for the binders at the other RAP binder contents, 0%, 30%, and 50%.

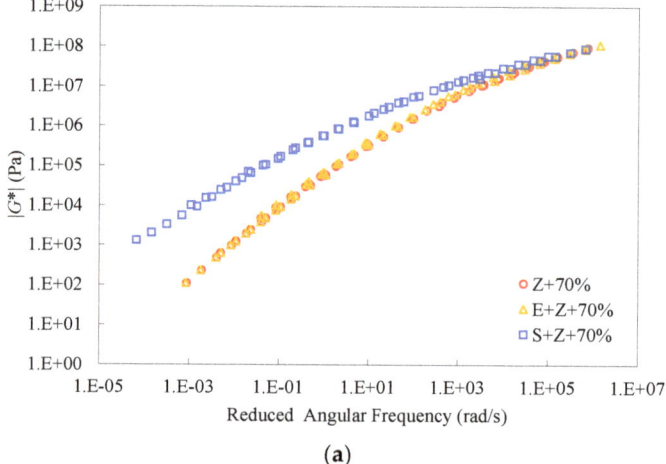

(**a**)

**Figure 19.** *Cont.*

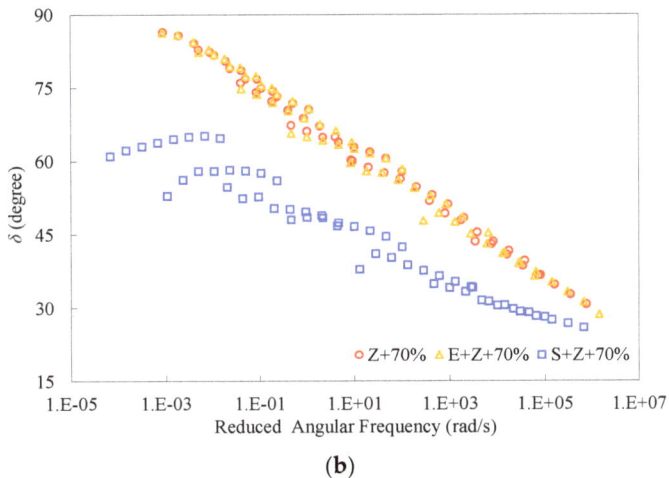

(**b**)

**Figure 19.** Master curves of dynamic modulus and phase angle of the warm-mix recycled binders at the RAP binder content of 70%.

## 5. Conclusions

The present study investigated the rheological behavior of warm-mix recycled asphalt binders with high percentages of RAP binder. The combined impacts of two WMA additives (wax-based S and surfactant-based E), a rejuvenating aging (Z), four RAP binder contents (0~70%), and three aging states (unaged, short-term aged, and long-term aged) were assessed in detail using various rheological and performance tests over the entire range of service temperatures. On the basis of the results and analyses, the following conclusions can be drawn:

1. In terms of the conventional performance tests and the apparent viscosity test, the artificial RAP binder that was obtained by subjecting the virgin binder to successive RTFO aging and PAV aging with a specific exposure duration could achieve similar performance and aging degree to those of the actual one;
2. The rejuvenating agent Z could effectively alleviate the aging effect of the RAP binder; however, it was still difficult to completely eliminate this negative impact, in particular at higher RAP binder contents, despite a relatively accurate rejuvenating agent dosage by weight of the RAP binder (6%) used;
3. The addition of S remarkably lowered the apparent viscosity of the warm-mix recycled binders (by up to 35.0%), and with the decreasing temperature, this effect of viscosity reduction became more significant (within the range of 115 °C to 175 °C); however, the WMA additive E had little influence (with the reduction of only up to 5.2% in apparent viscosity) on the binder viscosity due to its surfactant nature;
4. Under both unaged and RTFO short-term aged conditions, the WMA additive S could considerably enhance the rutting performance (respectively with the increases of up to 399.5% and 411.3% in $|G^*|/\sin\delta$) due to the wax crystallization in S, whereas E only enhanced the rutting resistance very slightly (respectively with the increases of up to 23.0% and 29.9% in $|G^*|/\sin\delta$);
5. Both S and E were capable of improving the fatigue performance of the recycled binders, but E performed better than S in this aspect (respectively with the reductions of up to 42.3% and 22.1% in $|G^*|\cdot\sin\delta$);
6. The use of S had an adverse impact on the thermal cracking resistance (with an increase of up to 70.2% in $S_a$ and a decrease of up to 34.1% in m-value between −24 °C and −12 °C), but the inclusion of E had a very mild influence on the low-temperature performance of the recycled binders;

7. The addition of S broadened the ranges of pavement service temperature by about two PG grades, 12 °C though S lowered the low-temperature performance of the binders (about one PG grade, 6 °C), whereas the use of E almost did not change the PG grades of the binders. Moreover, S could enhance the short-term aging resistance of the binders;
8. The inclusion of S considerably improved the dynamic modulus and lowered the phase angle of the binders over a wide range of frequencies and temperatures, indicating the reinforcement effect on the stiffness and the enhancement effect on the elasticity, but E had a very slight influence on the two master curves of the warm-mix recycled binders;
9. Due to the thermorheologically complex nature of S, the phase angle test data of the S + Z binders measured at different temperatures were incapable of forming single curves by horizontal shift, which led to the failure of the time-temperature superposition principle and unsatisfactory fitting of the 2S2P1D model.

This study served as a preliminary investigation of the rheological behavior of warm-mix recycled asphalt binders with high percentages of RAP binder. Additional research efforts are needed to evaluate the effects of other WMA techniques, like the foaming-based technology and other RAP sources. Besides, the relationships between the performance of the warm-mix recycled asphalt binders with high contents of RAP binder and the corresponding WMA-high RAP asphalt mixtures are recommended to develop. More advanced rheological tests, like the MSCR and LAS tests, are needed for the rheological evaluation of the materials in future studies.

**Author Contributions:** Methodology, H.X., Y.S. and J.C.; Investigation, H.X., J.L., B.Y., G.Q., Y.Z. and B.X.; Writing—original draft, H.X.; Writing—review & editing, Y.S.; Supervision, J.C. All authors have read and agreed to the published version of the manuscript.

**Funding:** This study was funded by the National Natural Science Foundation of China (51808098 and 51878122), the Natural Science Foundation of Liaoning Province (2022-MS-140), and Fundamental Research Funds for the Central Universities (DUT22JC22).

**Institutional Review Board Statement:** Not applicable.

**Informed Consent Statement:** Not applicable.

**Data Availability Statement:** Data sharing is not applicable to this article.

**Acknowledgments:** This study was sponsored by the National Natural Science Foundation of China (51808098 and 51878122), the Natural Science Foundation of Liaoning Province (2022-MS-140), and Fundamental Research Funds for the Central Universities (DUT22JC22). The supports are gratefully acknowledged.

**Conflicts of Interest:** The authors declare no conflict of interest.

## References

1. Gong, H.; Huang, B.; Shu, X. Field performance evaluation of asphalt mixtures containing high percentage of RAP using LTPP data. *Constr. Build. Mater.* **2018**, *176*, 118–128. [CrossRef]
2. Zhu, J.; Ma, T.; Fan, J.; Fang, Z.; Chen, T.; Zhou, Y. Experimental study of high modulus asphalt mixture containing reclaimed asphalt pavement. *J. Clean. Prod.* **2020**, *263*, 121447. [CrossRef]
3. Behnood, A. Application of rejuvenators to improve the rheological and mechanical properties of asphalt binders and mixtures: A review. *J. Clean. Prod.* **2019**, *231*, 171–182. [CrossRef]
4. Song, W.; Xu, Z.; Xu, F.; Wu, H.; Yin, J. Fracture investigation of asphalt mixtures containing reclaimed asphalt pavement using an equivalent energy approach. *Eng. Fract. Mech.* **2021**, *253*, 107892. [CrossRef]
5. Hettiarachchi, C.; Hou, X.; Xiang, Q.; Yong, D.; Xiao, F. A blending efficiency model for virgin and aged binders in recycled as-phalt mixtures based on blending temperature and duration. *Resour. Conserv. Recycl.* **2020**, *161*, 104957. [CrossRef]
6. McDaniel, R.; Anderson, R. *Recommended Use of Reclaimed Asphalt Pavement in the Superpave Mix Design Method: Technician's Manual*; Transportation Research Broad-National Research Council: Washington, DC, USA, 2001.
7. Zhao, S.; Huang, B.; Shu, X.; Woods, M. Comparative evaluation of warm mix asphalt containing high percentages of reclaimed asphalt pavement. *Constr. Build. Mater.* **2013**, *44*, 92–100. [CrossRef]

8. Alsalihi, M.; Faheem, A. Compaction and Performance of Warm-Mix Asphalt with High-Reclaimed Asphalt Pavement. *J. Mater. Civ. Eng.* **2020**, *32*, 04020055. [CrossRef]
9. Goli, H.; Latifi, M.; Sadeghian, M. Moisture characteristics of warm mix asphalt containing reclaimed asphalt pavement (RAP) or steel slag. *Mater. Struct.* **2022**, *55*, 53. [CrossRef]
10. Xu, H.; Chen, J.; Sun, Y.; Zhu, X.; Wang, W.; Liu, J. Rheological and physico-chemical properties of warm-mix recycled asphalt mastic containing high percentage of RAP binder. *J. Clean. Prod.* **2020**, *289*, 125134. [CrossRef]
11. Zhu, X.; Sun, Y.; Du, C.; Wang, W.; Liu, J.; Chen, J. Rutting and fatigue performance evaluation of warm mix asphalt mastic containing high percentage of artificial RAP binder. *Constr. Build. Mater.* **2020**, *240*, 117860. [CrossRef]
12. Guo, N.; You, Z.; Tan, Y.; Zhao, Y. Performance evaluation of warm mix asphalt containing reclaimed asphalt mixtures. *Int. J. Pavement Eng.* **2016**, *18*, 981–989. [CrossRef]
13. Behnood, A. A review of the warm mix asphalt (WMA) technologies: Effects on thermo-mechanical and rheological properties. *J. Clean. Prod.* **2020**, *259*, 120817. [CrossRef]
14. Bonaquist, R. *Mix Design Practices for Warm Mix Asphalt*; Transportation Research Board: Washington, DC, USA, 2011.
15. Mogawer, W.; Austerman, A.; Mohammad, L.; Kutay, M.E. Evaluation of high RAP-WMA asphalt rubber mixtures. *Road Mater. Pavement Des.* **2013**, *14*, 129–147. [CrossRef]
16. Xiao, F.; Hou, X.; Amirkhanian, S.; Kim, K.W. Superpave evaluation of higher RAP contents using WMA technologies. *Constr. Build. Mater.* **2016**, *112*, 1080–1087. [CrossRef]
17. Song, W.; Huang, B.; Shu, X. Influence of warm-mix asphalt technology and rejuvenator on performance of asphalt mixtures containing 50% reclaimed asphalt pavement. *J. Clean. Prod.* **2018**, *192*, 191–198. [CrossRef]
18. Wang, W.; Cheng, H.; Sun, L.; Sun, Y.; Liu, N. Multi-performance evaluation of recycled warm-mix asphalt mixtures with high reclaimed asphalt pavement contents. *J. Clean. Prod.* **2022**, *377*, 134209. [CrossRef]
19. Xiao, F.; Putman, B.; Amirkhanian, S. Rheological characteristics investigation of high percentage RAP binders with WMA technology at various aging states. *Constr. Build. Mater.* **2015**, *98*, 315–324. [CrossRef]
20. Yu, X.; Dong, F.; Xu, B.; Ding, G.; Ding, P. RAP Binder Influences on the Rheological Characteristics of Foamed Warm-Mix Recy-cled Asphalt. *J. Mater. Civ. Eng.* **2017**, *29*, 04017145. [CrossRef]
21. Sun, Y.; Wang, W.; Chen, J. Investigating impacts of warm-mix asphalt technologies and high reclaimed asphalt pavement binder content on rutting and fatigue performance of asphalt binder through MSCR and LAS tests. *J. Clean. Prod.* **2019**, *219*, 879–893. [CrossRef]
22. *ASTM D2872*; Standard Test Method for Effect of Heat and Air on a Moving Film of Asphalt (Rolling Thin-Film Oven Test). ASTM International: West Conshohocken, PA, USA, 2022.
23. *ASTM D6521*; Standard Practice for Accelerated Aging of Asphalt Binder Using a Pressurized Aging Vessel (PAV). ASTM International: West Conshohocken, PA, USA, 2022.
24. Jamshidi, A.; Hamzah, M.O.; You, Z. Performance of Warm Mix Asphalt containing Sasobit®: State-of-the-art. *Constr. Build. Mater.* **2013**, *38*, 530–553. [CrossRef]
25. Ministry of Transport of the People's Republic of China. *JTG E20-2011*; Standard Test Methods of Bitumen and Bituminous Mixtures for Highway Engineering. China Communications Press: Beijing, China, 2011.
26. *ASTM D4402*; Standard Test Method for Viscosity Determination of Asphalt at Elevated Temperatures Using a Rotational Viscometer. ASTM International: West Conshohocken, PA, USA, 2022.
27. *ASTM D7175*; Standard Test Method for Determining the Rheological Properties of Asphalt Binder Using a Dynamic Shear Rheometer. ASTM International: West Conshohocken, PA, USA, 2015.
28. *ASTM D6648*; Standard Test Method for Determining the Flexural Creep Stiffness of Asphalt Binder Using the Bending Beam Rheometer (BBR). ASTM International: West Conshohocken, PA, USA, 2016.
29. Tschoegl, N. *The phenomenological Theory of Linear Viscoelastic Behavior: An Introduction*; Springer: New York, NY, USA, 1989.
30. Ferry, J. *Viscoelastic Properties of Polymers*; Wiley: New York, NY, USA, 1980.
31. Olard, F.; Di Benedetto, H. General 2S2P1D Model and Relation Between the Linear Viscoelastic Behaviours of Bituminous Binders and Mixes. *Road Mater. Pavement Des.* **2003**, *4*, 185–224. [CrossRef]
32. Zhang, Y.; Sun, Y. Fast-Acquiring High-Quality Prony Series Parameters of Asphalt Concrete through Viscoelastic Continuous Spectral Models. *Materials* **2022**, *15*, 716. [CrossRef] [PubMed]
33. *ASTM D6373*; Standard Specification for Performance Graded Asphalt Binder. ASTM International: West Conshohocken, PA, USA, 2016.
34. *ASTM D7643*; Standard Practice for Determining the Continuous Grading Temperatures and Continuous Grades for PG Graded Asphalt Binders. ASTM International: West Conshohocken, PA, USA, 2022.

**Disclaimer/Publisher's Note:** The statements, opinions and data contained in all publications are solely those of the individual author(s) and contributor(s) and not of MDPI and/or the editor(s). MDPI and/or the editor(s) disclaim responsibility for any injury to people or property resulting from any ideas, methods, instructions or products referred to in the content.

Article

# Reinforcement Effect of Different Fibers on Asphalt Mastic

Tingting Xie [1], Kang Zhao [1] and Linbing Wang [1,2,*]

[1] National Center for Materials Service Safety, University of Science and Technology Beijing, Beijing 100083, China
[2] School of Environmental, Civil, Mechanical and Agricultural Engineering, University of Georgia, Athens, GA 30602, USA
* Correspondence: linbing.wang@uga.edu; Tel.: +86-178-0105-2381

**Abstract:** Fiber materials as an asphalt mixture additive and stabilizer can effectively improve the performance index of asphalt pavement. In this study, lignin and carbon fiber were used as modifiers to study their effects on the road performance of asphalt mastic. Based on the frequency sweep, linear amplitude sweep (LAS) and multi-stress creep recovery (MSCR) experiments were conducted to test the high-temperature rutting and medium-temperature fatigue resistance of asphalt mastic with different fiber incorporation and low-temperature performance tests based on bending beam rheometer (BBR). The results indicate that adding fibers increased the stiffness of the asphalt mastic, and the modification effect of lignin fibers was better than that of carbon fibers. Meanwhile, the characteristic flow index of the asphalt mastic gradually increased with the increase in temperature, indicating that it gradually became a near-Newtonian fluid at higher temperatures. The addition of fibers also improved the high temperature rutting resistance of the asphalt mastic but did not have an advantageous effect on fatigue and low temperature cracking resistance. Additionally, the fitting results of the four-parameter Burgers model show that the use of fiber modification decreases the proportion of elasticity and viscous creep compliance but increases the delayed elasticity part.

**Keywords:** asphalt mastic; lignin fiber; carbon fiber; rheology; burgers model

## 1. Introduction

Different fibers have been used in asphalt binders and mix to address mechanical performance issues. Fibers provide this material with higher modulus, resistance, durability, deformability, and excellent ductility [1]. In addition, fibers have been used as reinforcement for polymer matrices in other industries [2,3]. These fibers give stiffness and strength to the composite, thus allowing the matrix to better transfer loads between fibers. Similarly, adding fibers to asphalt can improve the performance of the mix and contribute to sustainability by extending its service life and reducing road maintenance. Many studies have been conducted on fiber-reinforcement asphalt mixtures or mastics [4–7]. There are two main categories of fibers: synthetic fibers and natural fibers [8]. Synthetic fibers include polyester, carbon, and glass fibers. It has been demonstrated that applying synthetic fibers to asphalt mixtures can improve their fatigue life [9–11]. Wu et al. studied the voids and low-temperature performance of asphalt mixes with different polyester fiber contents. It was discovered that with the increase of polyester fiber content, both voids and connected voids decreased first and then increased, with the best performance in low-temperature crack resistance produced given a 0.4% polyester fiber content [12]. Qin et al. characterized the performance of different fibers (including basalt fibers, polyester fibers, and lignin fibers) on asphalt biners, and the results showed that asphalt binders reinforced with basalt fibers showed the best overall performance [13]. In light of moisture damage and low temperatures, Khater et al. evaluated the efficiency of lignin fiber and glass fiber-modified asphalt mixes. Fibers significantly improved the water resistance, low-temperature stability, and quality of asphalt mixtures [14]. Fu et al. evaluated the fracture resistance of different fiber-modified asphalt mixes based on acoustic emission techniques and found that the incorporation of fibers greatly improved the strength of the control asphalt mixture and that steel fiber-reinforced asphalt mixes exhibited more favorable fracture resistance than glass fiber-reinforced asphalt mixes and basalt fiber-reinforced asphalt mixtures [15]. Abtahi et al. investigated the strengthening effect of asphalt mixtures incorporated with both glass and polypropylene fibers in the range of 0.05% to 0.2% by weight of aggregate for glass fibers and 2% to 6% by weight of asphalt for polypropylene fibers, and proposed an optimum amount of 0.1% glass fibers in combination with 6% polypropylene fibers [16].

Since natural fibers usually exhibit considerable mechanical properties and are more environmentally friendly and cheaper than synthetic fibers, they are often used to reinforce asphalt mixtures such as cotton, hemp, wool, and silk fibers. Chen et al. conducted a systematic study and analysis of several types of fiber-modified asphalt mixes through penetration tests and dynamic shear tests and pointed out that the fiber surface area should be taken into account when increasing asphalt amount and when mixing asphalt mixes in order to meet the needs of asphalt-coated fibers [17]. Miao et al. investigated the effect of interfacial properties on the performance of fiber-reinforced asphalt. They found that fiber reinforcement was significantly more effective when fiber surface energy was high, and the fiber surface energy correlated positively with the fiber reinforcement effects [18]. Stone Matrix Asphalt (SMA) was investigated by Zhang et al. to determine its rheological behavior and strengthening mechanism. They found that the fibers in the asphalt mastic were well strengthened and that the creep recovery rate of asphalt mastic increased significantly, while the residual creep value decreased under high stress [19]. Research from Noorvand et al. found that fiber-reinforced asphalt mixes with higher micro fibrillation had a better dispersion uniformity and were more resistant to high-temperature rutting [20]. Tanzadeh et al. modified asphalt binders with 4.5% styrene-butadiene-styrene (SBS) and 2% and 4% nano-silica to improve the performance properties of the modified porous asphalt mixes by adding 0.5% and 1% lime powder and blended synthetic fibers to 0.4% and 0.5% of the asphalt mix as filler types, resulting in weight loss, while improving tensile strength and rutting resistance [21]. Liu and Xia et al. found that bamboo fibers outperformed lignin fibers regarding high-temperature stability, low-temperature crack resistance, and moisture stability of asphalt mixtures with good road properties [22,23]. Sheng et al. reported that adding bamboo fibers to asphalt mixes improved their water damage resistance, rutting resistance, and cracking resistance at low temperatures. It was determined from Marshall mix design calculations that bamboo fiber content should be 0.2–0.3% for dense grade asphalt (DG) and 0.4% for stone matrix asphalt (SMA) [24]. Yu et al. found that bamboo fiber incorporation improved the stability and tensile strength of the asphalt mixture, and SEM images revealed a strong bond between the fiber modified asphalt mixture and the asphalt binder [25].

As fiber reinforcement technology is currently used mainly in asphalt mixtures, and asphalt mastic is an essential binder, the interaction between it and fibers has been less studied. Most studies were conducted on a single fiber type, and not enough research has been conducted on the reinforcing effect of different fiber types. Therefore, two different fibers, carbon fiber and lignin fiber, were selected for this study to select a better-performing fiber and promote its application in asphalt pavements. Frequency scan (FS) tests were used to evaluate the linear viscoelastic rheological properties; fatigue properties were evaluated using linear amplitude scan (LAS) tests coupled with viscoelastic continuum damage (VECD) theory; rutting resistance was evaluated using multiple stress creep recovery (MSCR) tests; bending beam rheometer (BBR) tests were used to evaluate the crack resistance of asphalt mastics; and the Burgers model was used to analyze the viscoelastic composition of the fiber-modified asphalt mastic.

## 2. Materials and Methods

### 2.1. Materials

#### 2.1.1. Asphalt Binder

The asphalt binder (penetration grade of pen-70) provided by Beijing Changping Aphalt Plant was selected as the base asphalt for this study. The main physical characteristics of the base asphalt are shown in Table 1.

**Table 1.** Physical characteristics of base asphalt.

| Properties | Testing Standards | Results |
| --- | --- | --- |
| Penetration (0.1 mm) at 25 °C | JTG E20-2011/T0604 | 75.2 |
| Softening point (°C) | JTG E20-2011/T0606 | 49.2 |
| Ductility (cm) at 5 °C | JTG E20-2011/T0605 | 35.5 |
| Viscosity (Pa·s) at 135 °C | JTG E20-2011/T0625 | 0.35 |

#### 2.1.2. Filler

The plain asphalt mastics used in this study were prepared in a 1:1 ratio of base asphalt binder and filler. The physical properties of filler are shown in Table 2.

**Table 2.** Physical properties of filler.

| Apparent Density g/cm$^3$ | Water Content (%) | Hydrophilic Coefficient | Appearance | Screening Test (%) | | |
| --- | --- | --- | --- | --- | --- | --- |
| | | | | 100 | 90–100 | 75–100 |
| 2.792 | 0.33 | 1.0 | No agglomeration | 100 | 95.8 | 83.1 |

#### 2.1.3. Fibers

Considering that lignin fibers tend to agglomerate in asphalt mastics with high content, the rheological performance of lignin-modified asphalt mastics at lower content was investigated. In this study, 0.5 mm carbon fiber and lignin fiber were selected as modifiers to prepare modified asphalt mastic, and the carbon fiber content was selected as 3%, 6%, and 9%, and the lignin fiber was selected as 1%, 3%, and 6% in order to adsorb enough asphalt and so that the fiber would not agglomerate. The performance of fibrous asphalt mastics at 3% and 6% content can also be compared with that of carbon fiber modified asphalt mastics and the results are equally valid. The macroscopic forms of carbon and lignin fibers are shown in Figure 1.

**Figure 1.** Fiber morphology.

## 2.2. Methods

Since there is a lack of methodological specifications for assessing the performance of asphalt mastic materials, this study refers to test specifications for asphalt for testing and analysis.

### 2.2.1. Preparation of Fiber-Asphalt Mastic

The preparation of fiber-modified asphalt binder is divided into the following four steps: firstly, fibers must be heated separately for 24 h at 105 °C to remove water from their surfaces; secondly, the stored solid asphalt (600 g) is preheated at 160 °C for 2 h to make it liquid for mixing with the filler and fiber; then, the fiber and filler are weighed in proportion and slowly added to the asphalt at 2000 r/min stirring in order to prevent the fibers from clumping; and finally, the asphalt binder is continuously stirred with the heated filler and fibers at 160 °C for about 30 min to produce a homogeneous fiber-modified asphalt mastic. There are 7 groups of research materials in this study, as shown in Table 3.

**Table 3.** Overview of all tested asphalt mastics.

| NO. | Percent Weight of Lignin Fiber (%) | Percent Weight of Carbon Fiber (%) |
|---|---|---|
| 1 | 0 | 0 |
| 2 | 1 | 0 |
| 3 | 3 | 0 |
| 4 | 6 | 0 |
| 5 | 0 | 3 |
| 6 | 0 | 6 |
| 7 | 0 | 9 |

### 2.2.2. Frequency Sweep Test

The rheological testing in this study was done based on an Anton Par MCR 102 dynamic shear rheometer (DSR) from Graz, Austria. A 25 mm parallel plate geometry with a 1 mm gap setup was used when the test temperature was above 40 °C, and a 2 mm gap geometry with an 8 mm parallel plate was used when the temperature was below 40 °C. In the present study, frequency sweep tests from 0.1 rad/s to 100 rad/s at seven temperatures of 10 °C, 20 °C, 30 °C, 40 °C, 50 °C, 60 °C, and 70 °C were selected, and the experimental results were analyzed by fitting the master curve according to the Christenson–Anderson–Marasteanu (CAM) model [26,27].

Based on the time–temperature superposition principle, the dynamic shear modulus at multiple temperatures is shifted to construct the modulus master curve, which is then matched to the Christensen–Andersen–Marastanou (CAM) model; Equation (1) illustrates the process of fitting dynamic modulus master curves:

$$|G^*| = \frac{\left|G_g^*\right|}{\left[1 + (f_c/f')^k\right]^{m/k}} \quad (1)$$

where $|G^*|$ is the dynamic shear modulus; $G_g^*$ is the glassy modulus taken as 1 GPa [28]; $f_c$ is the crossover angular frequency; $k$ and $m$ are dimensionless shape parameters referred to as the rheological index; $f'$ is reduced angular frequency, $f' = \varphi_T \times f$, where $f$ is physical angular frequency; and $\varphi_T$ is time–temperature shift factor fitted via a polynomial function, as shown in Equation (2):

$$Log \varphi_T = -\frac{D_1 \cdot (T - T_0)}{D_2 + (T - T_0)} \quad (2)$$

where $T_0$ id reference temperature, and $D_1$ and $D_2$ are regression coefficients.

2.2.3. Linear Amplitude Sweep Test

Testing was conducted using the AASHTO TP 101 linear amplitude sweep (LAS) procedure to determine the asphalt binder's performance [29]. In this study, the theory is also applied to the fatigue performance analysis of asphalt mastic. The linear strain sweeps with amplitudes ranging from 0.1% to 30% over 5 min (hereafter referred to as LAS-5) was used in the LAS test at the desired intermediate temperature. LAS test results are shown in Figure 2. LAS test data is interpreted using the simplified-viscoelastic continuum damage (S-VECD) model developed for asphalt concrete fatigue modeling [30,31]. As the Beijing region typically experiences intermediate temperatures of around 20 °C, this study used 20 °C as the temperature for the LAS test. Taking the artificial failure definition of 35% reduction in $|G^*|\cdot\sin\delta$ estimates the binder fatigue life [32].

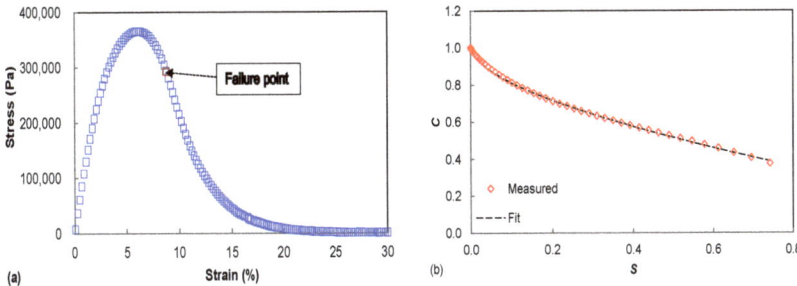

**Figure 2.** Typical fatigue model of base asphalt binder based on LAS at 20 °C (**a**) Stress–strain curve (**b**) Damage characteristic curve.

2.2.4. Multiple Sweep Recovery Test

The MSCR test based on AASHTO TO 70 [33] protocol was used to determine the asphalt binder's capacity for permanent deformation under high-temperature circumstances. During creep-recovery loading mode, the DSR applies a lower stress level of 0.1 kPa after 90 s of recovery for ten cycles. Afterward, the stress is raised to 3.2 kPa, which is repeated continuously for another 10 cycles. Performance parameters for MSCR tests include recovery rate ($R$) and non-recoverable compliance ($J_{nr}$). According to the Equations (3) and (4), the $R$ and $J_{nr}$ for any particular creep-recovery cycle are calculated, in which $\gamma_0$ indicates the shear strain at the initiation of this cycle, $\gamma_p$ represents Peak strain after creep duration of 1 s, and $\gamma_n$ represents Unrecoverable strain after 9 s recovery. $\tau$ indicates the creep stress level in each cycle. $R$ and $J_{nr}$ are the average values of 10 creep-recovery modes at each strain level, respectively, thus giving four parameters $R_{0.1}$, $J_{nr0.1}$, $R_{3.2}$ and $J_{nr3.2}$. MSCR tests for all asphalt mastics were performed at 60 °C instead of the corresponding PG temperature, in order to compare the rutting resistance equally.

$$R = (\gamma_p - \gamma_{nr})/(\gamma_p - \gamma_0) \tag{3}$$

$$J_{nr} = (\gamma_{nr} - \gamma_0)/\tau \tag{4}$$

2.2.5. Bending Beam Rheometer (BBR) Test

The asphalt mastic was tested for its creep resistance at low temperatures using a bending beam rheometer. The BBR samples (125 mm × 12.5 mm × 6.25 mm) were cooled in an ethanol bath for 60 min at constant temperatures of −6 °C, −12 °C, and −18 °C, respectively. Two stainless steel supports were then used to support the beam and 100 g were loaded onto them. Stiffness was calculated as a function of time by monitoring deflection over time. An investigation was carried out at a loading time of 60 s to determine the creep stiffness (S) and creep rate (m) of the mastics. The creep stiffness (S) and creep rate (m) were used to assess the asphalt mastics' low-temperature performance.

## 3. Results and Discussion

### 3.1. Effect of Frequencies on Modulus and Phase Angle

From Figure 3, it is clear that the dynamic shear modulus of the asphalt mastic increases with increasing loading frequency, and the logarithmic value of the modulus has a good linear relationship with the logarithmic value of the loading frequency. The reason for this relationship between modulus and loading frequency is that the greater the loading frequency, the shorter the contact time between the asphalt material and the applied loading at a single load. Furthermore, asphalt materials are viscoelastic materials whose deformation generally includes elastic deformation, recoverable viscoelastic deformation, and irrecoverable viscous deformation. The modulus of asphalt material increases with the increase of loading frequency because the higher the loading frequency, the shorter the loading action time at each cycle, and the smaller the deformation produced by asphalt material will be, resulting in the increase of the modulus. At low frequencies, the loading time becomes longer, and the deformation of the asphalt material increases, resulting in a decrease in modulus. The effect of loading frequency on the phase angle of the asphalt material has an opposite trend compared to the dynamic shear modulus, as shown in Figure 4. When the fiber content is less than 3%, the phase angle of asphalt mastic tends to decrease with increasing loading frequency, because the greater the loading frequency, the more there is an elastic deformation component in the deformation produced by each loading cycle, and thus the phase angle of asphalt material is reduced.

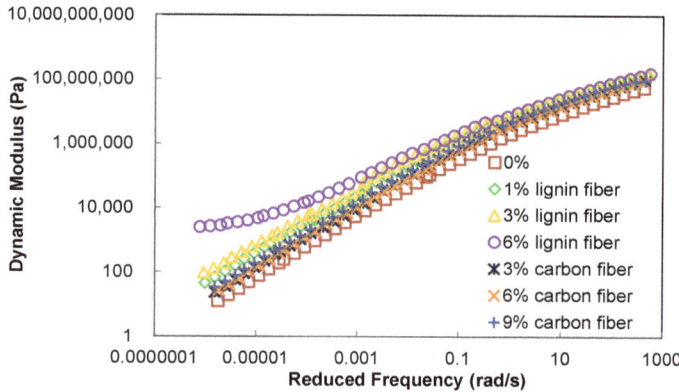

**Figure 3.** Dynamic modulus master curves.

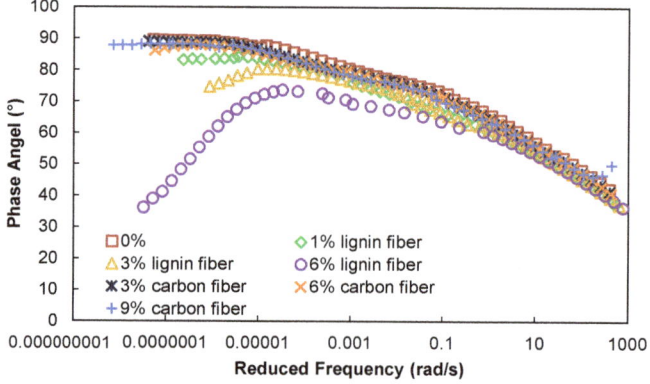

**Figure 4.** Phase angle master curves.

It can also be seen from Figure 3 that the addition of lignin fibers increases the modulus value of asphalt mastic, while the carbon fiber enhancement effect is not obvious, indicating that lignin fibers are more effective than carbon fibers in improving the stiffness and modulus of asphalt mastic. Especially at high temperatures, the stiffening effect of lignin fiber is more obvious. Compared with carbon fiber, the effect of lignin fiber on the phase angle of asphalt mastic is more apparent; whenever the fiber dose increases by 1%, it can reduce the phase angle of fiber asphalt mastic by about 5°, while when the fiber dose is higher than 3% in the high-temperature range, the phase angle of asphalt mastic shows a trend of first rising and then falling, indicating that its elastic properties first decline and then enhance. An analysis of the addition of fibers in ordinary asphalt mastic, fibers which can play the role of "macromolecular soft chain," is shown by the schematic diagram in Figure 5. The disorderly distribution of fibers can absorb most of the concentrated stress so that the asphalt material, to withstand the load capacity while making its elastic properties, increases.

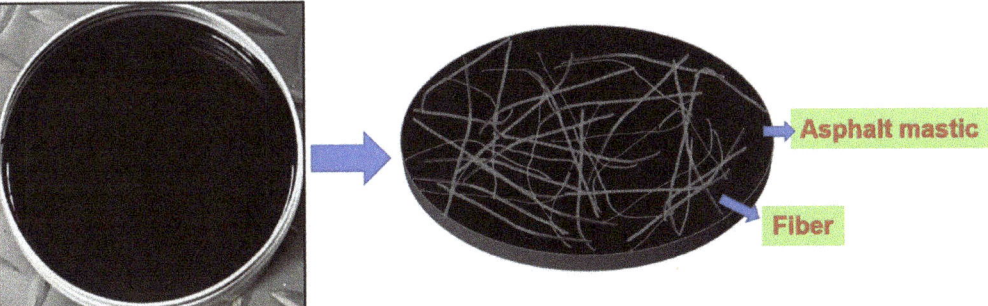

**Figure 5.** Schematic diagram of asphalt mastic sample and fiber distribution.

*3.2. Effect of Temperatures on Modulus and Phase Angle*

The dynamic shear modulus and phase angle $\delta$ at each temperature at a frequency of 10 rad/s were selected for a comparative analysis in this study. Figure 6 illustrates the effects of experimental temperature on the complex shear modulus and phase angle $\delta$ of asphalt mastic. According to Figure 6, asphalt mastic's dynamic shear modulus decreases exponentially as the experimental temperature increases until it finally converges when experiment temperature increases, indicating that both fiber asphalt mastic and plain asphalt mastic are sensitive to temperature; furthermore, the incorporation of fiber can substantially increase the dynamic shear modulus of asphalt mastic and make the material more resistant to stress, i.e., produce an enhancement effect. The phase angle $\delta$ of different types of asphalt mastic increases gradually with the increase of experimental temperature, and the phase angle of plain asphalt mastic is larger than that of fiber asphalt mastic. The larger the value of $\delta$, indicating that the larger the viscous component of asphalt mastic, the more likely it is to produce high-temperature permanent deformation. Therefore, different types of asphalt mastic gradually transformed from elastic to viscous states as the experimental temperature increased, and the viscous state of plain asphalt mastic was the most obvious and most likely to produce high-temperature permanent deformation. It can be seen that the incorporation of fibers can substantially improve the high-temperature deformation resistance of asphalt mastic.

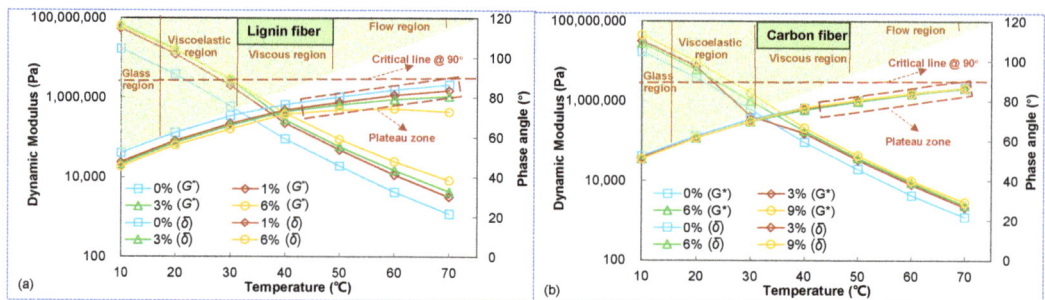

**Figure 6.** Effect of temperature on dynamic modulus and phase angle (**a**) dynamic modulus (**b**) phase angle.

### 3.3. Initial Self-Healing Temperature

The asphalt mastics with different fiber contents were subjected to frequency sweep experiments at temperatures of 10–70 °C and frequencies in the range of 0.1–10 rad/s. The flowability of fiber asphalt mastics can be inferred from the composite viscosity and frequency, and the values of the obtained composite viscosity and frequency were fitted to a power function according to the Equation (5) to obtain the flow characteristic index $n$.

$$\eta* = m|w|^{n-1} \tag{5}$$

where: $w$ indicates the frequency; $\eta^*$ indicates the composite viscosity; $m$ and $n$ indicate the fitting parameters. Where the fitted parameter $n$ also becomes the flow characteristic index, according to which the initial self-healing temperature of the fiber asphalt mastic is analyzed and thus its self-healing ability is studied.

The test results of the composite viscosity of asphalt mastics with different fiber contents obtained from the frequency sweep test are shown in Figure 7. The composite viscosity of fiber-modified asphalt mastics will be increased more significantly with the increase of fiber contents, which is due to the formation of a space-randomly distributed network structure of fibers in the mastic impeding the flow of asphalt, which is expressed in the macroscopic increase of asphalt mastic viscosity.

Figure 7 also shows that with the increase of frequency, the composite viscosity of the fiber asphalt mastics has an obvious decreasing trend. When the temperature rises to 50 °C, the correlation between the composite viscosity and frequency becomes worse, and the viscosity curve almost becomes a horizontal straight line, indicating that at this time, the asphalt mastics compound the characteristics of the Newtonian fluid, the viscosity value is constant. After the lignin content reaches 6%, the composite viscosity at 50 °C still shows a decreasing trend with the increase of frequency, which indicates that the flow of fiber asphalt mastic is seriously hindered at higher fiber content, and it is difficult to reach the state of Newtonian fluid ultimately.

According to the results of Figure 8, the flow characteristics index of the fiber asphalt mastic was obtained from Table 4. The fitting results showed that the flow characteristics index of plain asphalt mastic and fiber-modified asphalt mastic tended to increase gradually with the increase in temperature, which indicated that the asphalt mastic gradually became a near-Newtonian fluid at higher temperatures.

The flow characteristic index of plain asphalt mastic increased from 0.571 to 0.98 in the range of 10–70 °C. The flow characteristic index of fiber asphalt mastic with lignin fibers decreased slightly, and the flow characteristic index increased from 0.519 to 0.924 and 0.5 to 0.856 and 0.485 to 0.639 in the range of 10–70 °C for 1%, 3%, and 6% of lignin fiber asphalt mastic, respectively. The carbon fiber modified asphalt mastic with 3%, 6%, and 9% of carbon fiber modified asphalt mastic increased from 0.547 to 0.97, 0.557 to 0.952, and 0.522

to 0.961 in the range of 10–70 °C, respectively, which shows that the carbon fiber modified asphalt mastic has better flow properties than the lignin modified asphalt mastic.

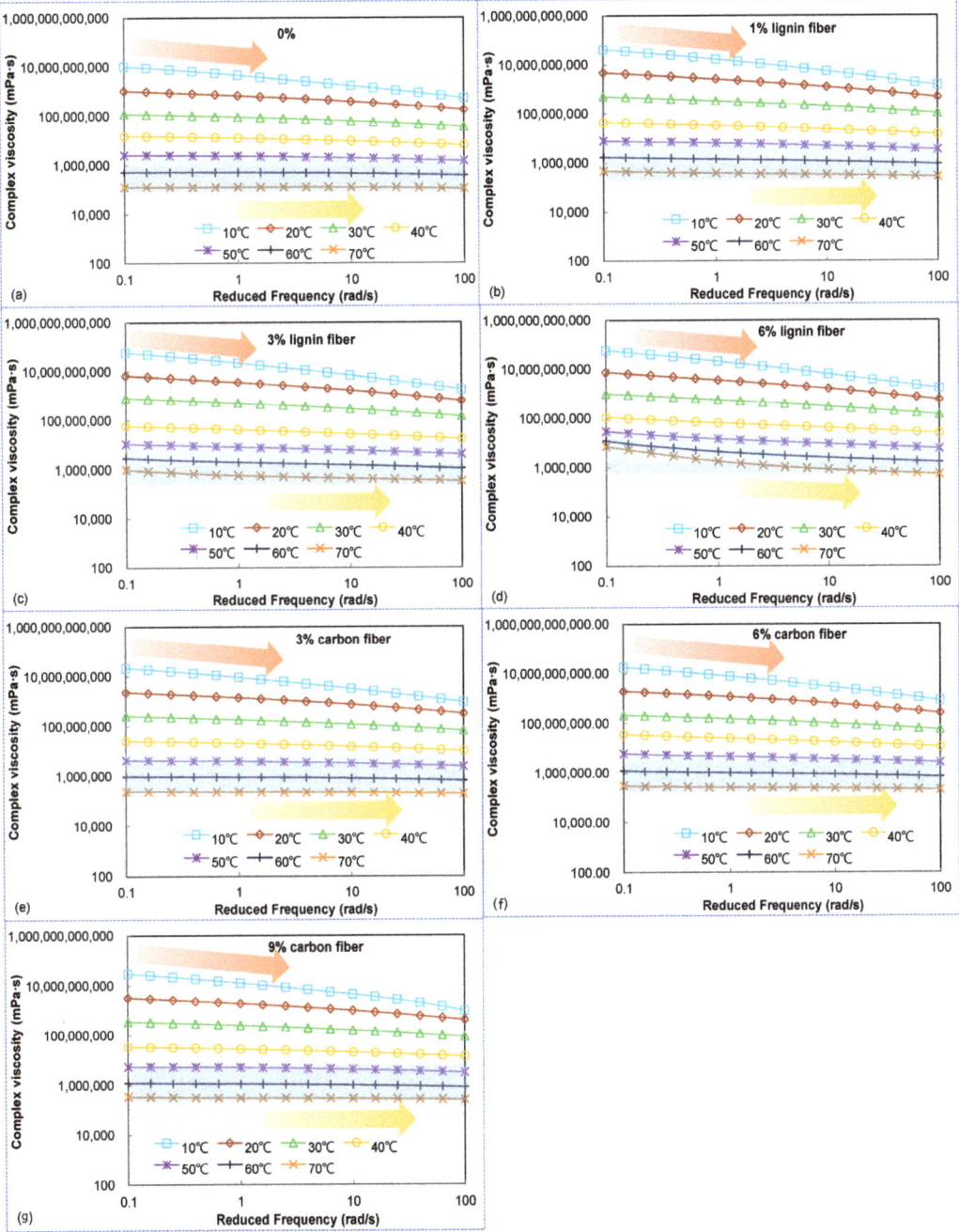

**Figure 7.** Complex viscosity of fiber modified asphalt mastics (**a**) 0% (**b**) 1% lignin fiber (**c**) 3% lignin fiber (**d**) 6% lignin fiber (**e**) 3% carbon fiber (**f**) 6% carbon fiber (**g**) 9% carbon fiber.

**Table 4.** Flow characteristic index of tested asphalt mastics.

| T (°C) | 0% | Lignin Fiber Content (%) | | | Carbon Fiber Content (%) | | |
|---|---|---|---|---|---|---|---|
| | | 1% | 3% | 6% | 3% | 6% | 9% |
| 10 | 0.571 | 0.519 | 0.5 | 0.485 | 0.547 | 0.557 | 0.522 |
| 20 | 0.74 | 0.672 | 0.656 | 0.629 | 0.716 | 0.712 | 0.703 |
| 30 | 0.828 | 0.769 | 0.756 | 0.719 | 0.808 | 0.806 | 0.802 |
| 40 | 0.874 | 0.836 | 0.825 | 0.785 | 0.866 | 0.836 | 0.869 |
| 50 | 0.92 | 0.882 | 0.86 | 0.78 | 0.916 | 0.885 | 0.92 |
| 60 | 0.956 | 0.909 | 0.868 | 0.729 | 0.947 | 0.927 | 0.945 |
| 70 | 0.924 | 0.924 | 0.856 | 0.639 | 0.97 | 0.952 | 0.961 |

Figure 8 shows the tendency of the flow characteristics index with temperature for different fiber-modified asphalt mastics. As can be seen from Figure 8, fiber-modified asphalt mastic has a higher initial self-healing temperature when fibers are added, which indicates that the fiber incorporation reduces the flowability of the asphalt mastic and needs to be increased to a higher temperature to achieve the same flow state. The initial self-healing temperature of plain asphalt mastic is 46 °C at 0.9, and 57 °C at 1% lignin content. The initial self-healing temperatures for carbon fiber blending of 3%, 6%, and 9% are 47 °C, 54 °C, and 46 °C, respectively, indicating that there is a peak in the content of uniformly distributed fibers in asphalt when lower than this value. The fiber incorporation reduces the asphalt mobility and requires an increase in temperature to achieve the same flow state, and when the content is exceeded, the fibers tend to coalesce into clumps, the self-healing asphalt content increases, the asphalt mobility increases, and the temperature needed to achieve the same flow state is reduced.

As mentioned above, the initial self-healing temperature of asphalt mastic increased significantly with the increase of fiber content, indicating that the interaction between fibers affects the main factor of asphalt mastic fluidity performance when the fiber content is higher. Therefore, in the actual construction, the appropriate temperature should be set according to the different fiber types and amounts for crack self-healing, which can achieve the healing effect and reduce consumption at the same time.

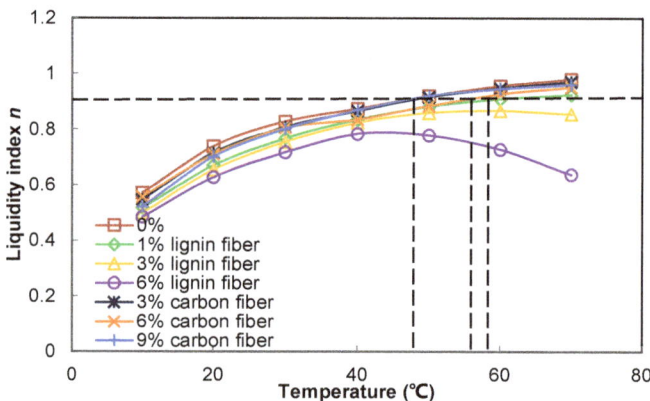

**Figure 8.** The flow characteristics index of different fiber modified asphalt mastics.

*3.4. Fatigue Performance*

This paper assumes that asphalt mastic is a homogeneous material, so the S-VECD model for asphalt is also applicable to asphalt mastic [34–36]. The stress–strain curves obtained based on the linear amplitude sweep test are shown in Figure 9, where the viscous damage points are identified, and the data points after the damage have been removed. The stress–strain curve of the plain asphalt mastic was lower than that of the fiber-modified asphalt mastic. The fibers increase the breaking stress while decreasing the strain, indicating that the hardener mastic consistently exhibits greater strength but has a smaller deformation limit. The highest yield strain of the asphalt material corresponds to its better elastic properties, so the increased fiber admixture adversely affects the elastic properties of the asphalt mastic. Nevertheless, the failure strain indicator only depicts the capability of the asphalt mastic under repeated loading, and it may indicate fatigue resistance in some cases. In order to obtain a more accurate fatigue evaluation, more fatigue damage and failure characteristics must be evaluated.

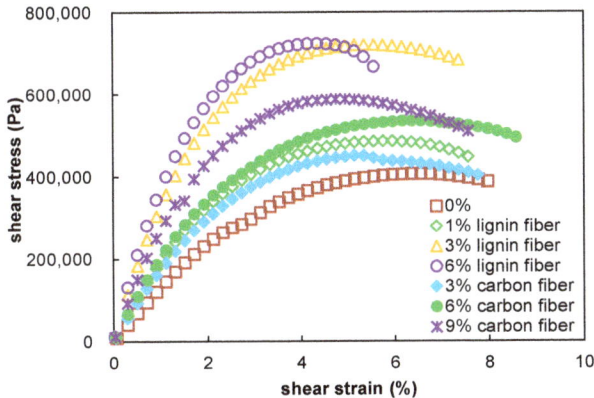

**Figure 9.** Strain–stress curves of fiber modified asphalt mastics.

The damage characteristic curves (DCC) of plain asphalt mastic and lignin fiber-modified asphalt mastic presented in Figure 10 were calculated using the S-VECD model. The observed distinct DCCs suggest that the lignin fibers significantly affected the fatigue damage evolution of the base asphalt mastic. Using the C vs. S relationship as an input to fatigue performance prediction, the fitting results of each DCC represent the unique damage property of each asphalt mastic. The C(S) curve then gradually increases with increasing lignin content. The relevant position of the C(S) curve is mainly determined by the stiffness of the material, and considering the role of $|G^*|$ in the calculation of damage S, lower stiffness usually yields a lower curve.

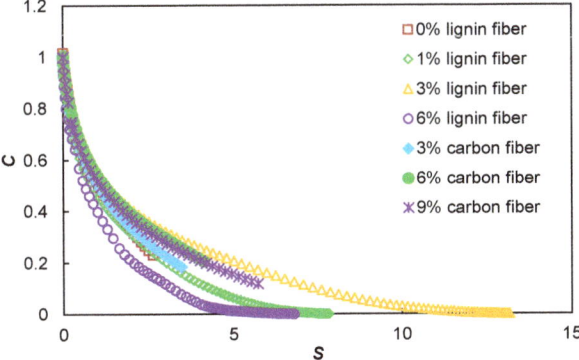

**Figure 10.** The damage characteristic curves of fiber modified asphalt mastics.

In this study, an artificial definition of failure, i.e., a 35% reduction in $|G^*|\text{-sin }\delta$, is adopted to estimate the fatigue life of asphalt mastic [32]. Using the calculated material properties, the fatigue life of the fiber-modified asphalt mastic was simulated and predicted under cyclic strain-controlled fatigue loading, as shown in Figure 11. Although the lignin fiber modified asphalt mastics show smaller fatigue life than plain asphalt mastic, the fiber content increase improves the fatigue lives of fiber-modified asphalt. Therefore, higher fiber content may exhibit greater fatigue life than normal asphalt mastic and is worthy of further investigation. However, carbon fiber incorporation significantly reduced the fatigue life, indicating that carbon fiber has an adverse effect on the fatigue resistance of asphalt mastic.

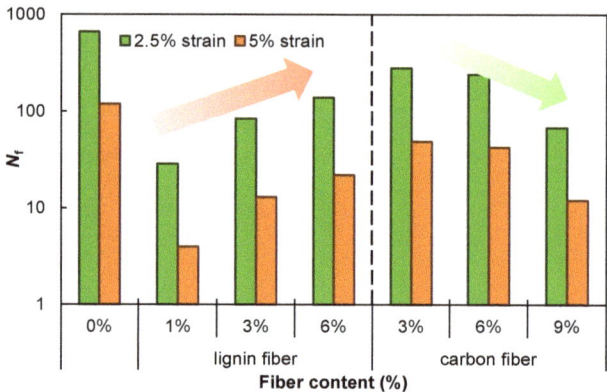

**Figure 11.** Fatigue lives of fiber modified asphalt mastics.

### 3.5. Rutting Resistance

There is a lack of tests and indicators to evaluate the high-temperature performance of asphalt mastic materials and there are limitations of using rutting factors to evaluate the high-temperature performance of asphalt mastic. Therefore, this study used the creep test to evaluate the high-temperature performance of asphalt mastic materials. The non-recoverable creep flexibility $J_{nr}$ and the non-recoverable creep flexibility difference $J_{nr\text{-diff}}$ were used as the evaluation indexes for the high-temperature performance of asphalt mastic.

The time–strain curves measured based on the MSCR test are summarized and compared in Figure 12, which evidences that fiber-modified asphalt mastics have higher high-temperature stability than plain asphalt mastics. Increased fiber content gradually enhances fiber-modified asphalt mastic's high-temperature stability. $J_{nr}$ is used as a standard index to evaluate the irrecoverable creep flexibility of asphalt materials. Its value can more accurately reflect the high-temperature rutting resistance of asphalt materials. A lower value indicates a better high-temperature performance. As shown in Figure 13, the change pattern of $J_{nr}$ is consistent with the change law of the time–strain curve, i.e., adding fiber improves its high-temperature performance. In addition, the lignin fibers have a stiffening and viscosity-enhancing effect, while the carbon fiber-modified asphalt mastic is brittle, prone to fracture, and has poor temperature resistance. Therefore, lignin fibers are chosen for the modification of asphalt mastic, which will result in a better modification effect and reduce the cost of the experiment.

The results of the high-temperature creep stress sensitivity analysis of the fiber-modified asphalt mastic based on the $J_{nr\text{-diff}}$ index are presented in Figure 14. As the lignin content of the modified asphalt mastic rises, the stress sensitivity of the modified asphalt mastic appears to rise sharply, especially when the lignin content reaches 6%. In asphalt mastic modified with carbon fiber, the stress sensitivity increases with fiber content and then decreases. AASHTO MP 19 [37] specifies that the limit for $J_{nr\text{-diff}}$ index is 75%, which means that asphalt mastic mixed with 6% fiber has reached creep damage stage and is incompatible with the technical standards, which means that modified asphalt mastic is subject to the stress sensitivity requirements.

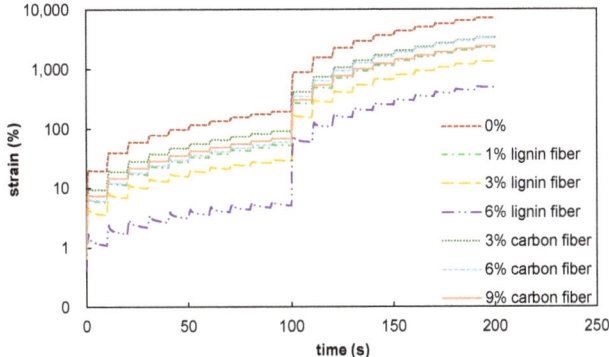

**Figure 12.** Time–strain curves of fiber-modified asphalt mastics.

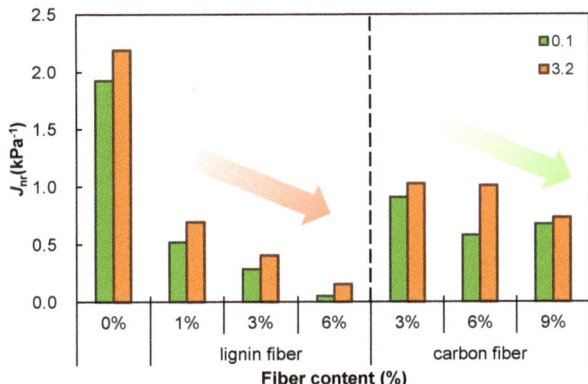

**Figure 13.** $J_{nr}$ values of fiber modified asphalt mastics.

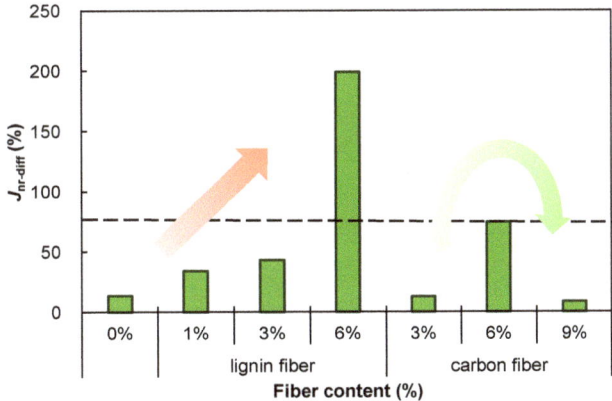

**Figure 14.** $J_{nr\text{-diff}}$ of fiber modified asphalt mastics.

### 3.6. Low Temperature Cracking Resistance

Two indicators of the BBR test: bending creep modulus of strength $S$ and creep curve slope $m$ (slope of the curve of strength modulus to load time) are used to evaluate the low-temperature properties of asphalt mastic. Asphalt mastic with a low $S$ value is more flexible, has a greater deformation tolerance, and is more resistant to low-temperature cracking [38,39]. It can be seen from Figure 15 that the creep modulus $S$ of asphalt mastic increases with the increase of fiber content. It shows that the low-temperature cracking resistance of asphalt mastic becomes worse, so the increase of fiber content is not conducive to improving the low-temperature characteristics of asphalt mastic. Meanwhile, the experimental results also reveal that the $S$ value decreases rapidly with the increase in temperature, so increasing the temperature is beneficial to improving the low-temperature performance of asphalt mastic. The slope of creep curve $m$ characterizes the relaxation performance of asphalt mastic. In general, increasing the value of $m$ will result in faster stress release, more substantial relaxation capacity, and better crack resistance at low temperatures. Figure 16 shows the effect of fiber content on the slope of creep curve $m$ of asphalt mastic. The value of the slope of creep curve $m$ of asphalt mastic decreases slightly with the increase of asphalt content, but the changing trend is not apparent, which indicates that the fiber content has less effect on the stress accumulation ability of asphalt mastic, so the increase of fiber content has a negative effect on the low-temperature performance of asphalt mastic, but the effect is weak. Furthermore, asphalt mastic's $m$ value increases quickly with increasing temperature, which has a beneficial effect on crack resistance at low temperatures.

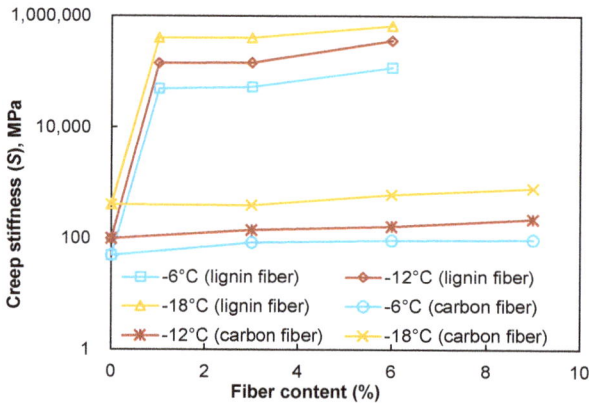

**Figure 15.** Creep stiffness of fiber modified asphalt mastics.

### 3.7. Four-Parameter Burgers Model Fitting Analysis

#### 3.7.1. Four-Parameter Burgers Model

The four-parameter Burgers model is a widely used viscoelastic mechanics model, and it can better reflect the viscoelastic properties of asphalt-based materials. The viscoelastic model consists of a set of Maxwell models in series with a set of Kelvin models, which can respond to the instantaneous elastic strain, viscoelastic strain, and viscous strain of viscoelastic materials [40–42]. The Burgers model and its creep curve are shown in Figure 17. The model generally contains two equations, one is the creep loading equation with constant stress input, and the other is the stress relaxation mode equation with constant strain input, and the two equations can be obtained by Laplace transformation and inversion. The mode of creep loading with constant stress is used in this study, and its instanton equation is shown in Equation (6):

$$\frac{\varepsilon(t)}{\sigma_0} = \frac{1}{E_m} + \frac{t}{\eta_m} + \frac{1}{E_k}(1 - e^{-tE_k/\eta_k}) \tag{6}$$

The creep flexibility of viscoelastic asphalt materials under creep loading $J$ is generally divided into three components, as shown in Equation (7):

$$J(t) = J_e + J_{ev}(1 - e^{-t/J_{ev}}) + J_v \qquad (7)$$

where $J_e = 1/E_m$ is the elastic flexibility; $J_{ev} = \eta_k/E_k$ is the delayed elastic flexibility or elastic flexibility; $J_v = t/\eta_m$ is the viscous flexibility.

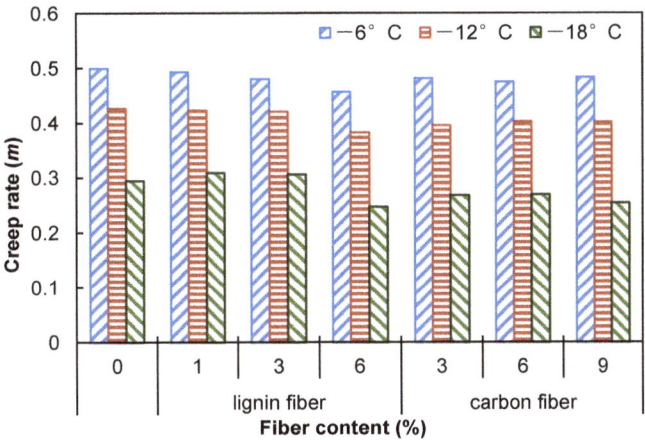

Figure 16. Creep rate of fiber modified asphalt mastics.

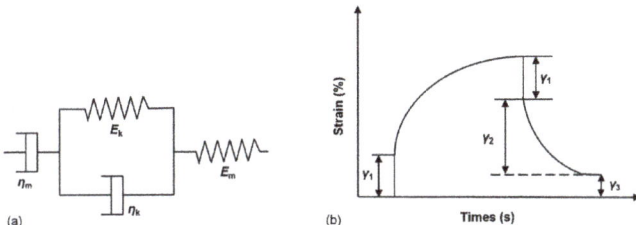

Figure 17. Burgers model graphs and creep curves (a) Burgers model (b) creep curves.

3.7.2. Viscous Part of Creep Stiffness $G_v$

The NCHRP 9–10 group has proposed to evaluate the high-temperature rutting resistance of asphalt binders based on the "viscous component of creep stiffness ($G_v$, and $G_v$, =1/$J_v$)", and $G_v$ is a parameter indicating the resistance of the asphalt binder to deformation at high temperatures [43]. The creep recovery behavior of the tested asphalt mastic samples was analyzed in this paper based on the four-parameter Burgers model, and the 10th creep-recovery cycle at two creep stress levels were selected for simulation fitting, and the $G_v$ index of the viscous component of the creep stiffness was calculated and summarized in Figure 18. In the presence of different types of fibers, asphalt mastic's $G_v$ value is enhanced to some extent, giving fiber-modified asphalt mastic a significantly better rutting resistance compared to plain asphalt mastic. This is generally consistent with the $J_{nr}$ index test results in Section 3.5 based on MSCR testing; therefore, the high temperature rutting resistance of the asphalt mastic evaluated by the $J_{nr}$ index is consistent with the fitted analysis based on the four-parameter Burgers model.

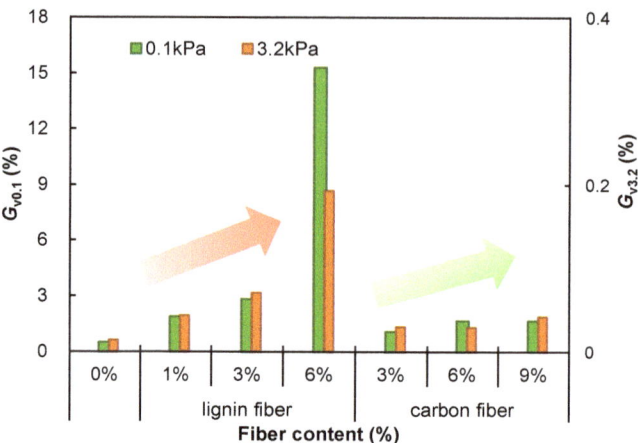

**Figure 18.** Viscous part of creep stiffness $G_v$ of fiber modified asphalt mastics.

3.7.3. Model Reliability Verification

This study was conducted by using Origin's own formula editor to custom edit the required functions and then to fit the data, and finally the fitted parameters of the Burgers model were obtained based on the convergence of the data. Figure 19 shows the results of back-calculating the data to verify the fitness of the fitted parameters. A good correlation exists between measured and predicted values, and the correlation parameter is close to one. Therefore, it can be demonstrated that the parameters obtained from the model fitting can be used for the subsequent study of viscoelastic component analysis.

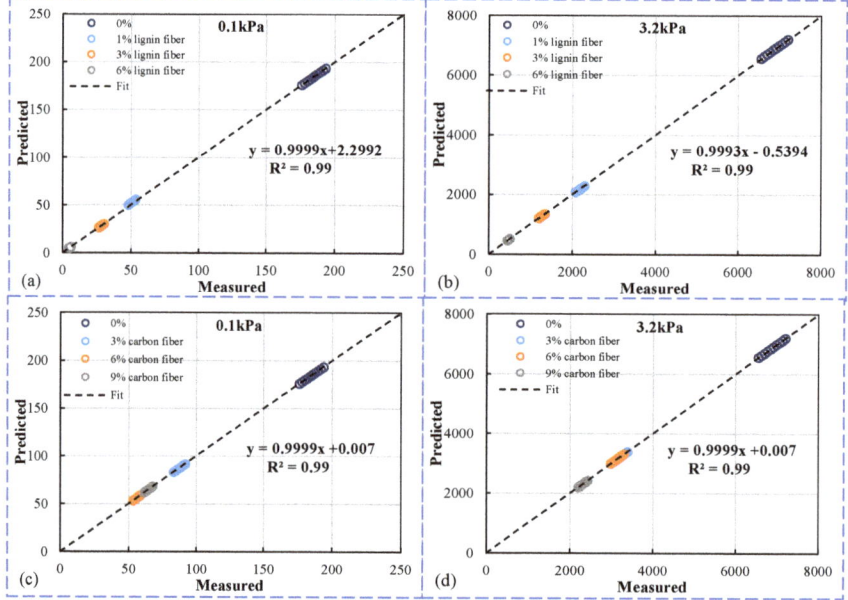

**Figure 19.** Measured values vs. Predicted values of $J_{nr}$ under two stresses. (**a**) lignin fiber modified under 0.1 kPa (**b**) lignin fiber modified under 3.2 kPa (**c**) carbon fiber modified under 0.1 kPa (**d**) carbon fiber modified under 3.2 kPa.

### 3.7.4. Viscoelastic Component Comparison

Based on the results of the four-parameter Burgers model fitting, a comparative analysis of the instantaneous elastic compliance, delayed elastic compliance, and viscous compliance of the asphalt during creep recovery in the MSCR test can be carried out, as shown in Figure 20. It can be seen that under the lower creep stress of 0.1 kPa, the percentage of elastic compliance and viscous compliance gradually decreases and the percentage of delayed elastic compliance gradually increases with the addition of fibers; however, under the higher creep stress of 3.2 kPa, the asphalt mastic samples basically only reflect the elastic compliance and viscous compliance, and the delayed elastic compliance generally accounts for a relatively low percentage, indicating that the creep stress level is crucial for the analysis of the viscoelastic component of the creep recovery process of fiber-modified asphalt mastic.

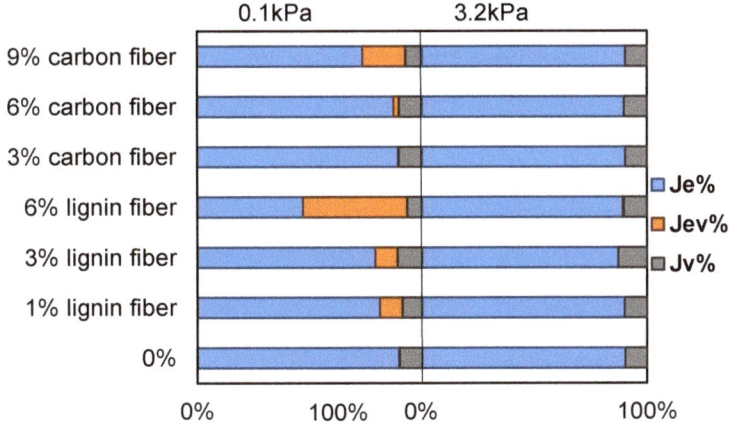

**Figure 20.** Viscoelastic component comparison of fiber modified asphalt mastics.

## 4. Conclusions

A series of studies on lignin fiber and carbon fiber-modified asphalt mastic based on DSR rheometer and BBR rheometer were conducted in this study and the following findings were obtained:

1. The addition of fibers enhances the stiffness of the asphalt, strengthens its resistance to deformation, and raises its self-healing temperature.
2. Fibers positively enhance the rutting resistance of asphalt mastics but negatively affect their fatigue resistance.
3. The low-temperature crack resistance of asphalt mastics is adversely affected by adding fibers, while the low-temperature crack resistance is positively affected by the increase in temperature.
4. Analysis of the viscoelastic component by the Burger model revealed that adding fibers increased the percentage of delayed elasticity and decreased the percentage of elastic compliance and viscous compliance of the asphalt mastic.
5. Lignin fibers are chosen for the modification of asphalt mastic, which will result in a better modification effect and reduce the cost of the experiment.
6. Natural plant fibers have a large surface area, a rough surface compared to other fibers, are highly oleophobic, and are mainly used as oil-holding, stabilizing anti-leakage, and reinforcement for asphalt mixtures. However, natural plant fibers have defects such as poor compatibility, hydrophilicity, and poor thermal stability, and the surface modification of plant fibers can be further investigated to solve these problems. Although the performance of carbon fiber is excellent, the price is too high to replace the existing engineering materials on a large scale, so how to obtain

relatively low production costs has become the direction of the future development of carbon fibers. In addition, the conclusion that fiber incorporation adversely affects the low-temperature performance of asphalt binders is contrary to the fiber cracking mechanism, and perhaps fiber-aggregate interactions may change this. Therefore, it is worthwhile to follow up with more in-depth research and analyses of fiber asphalt mixtures.

**Author Contributions:** Author Contributions: data curation, T.X. and K.Z.; formal analysis, T.X.; funding acquisition, L.W.; methodology, T.X., K.Z. and L.W.; project administration, L.W.; resources, L.W.; software, T.X. and K.Z.; supervision, L.W.; validation, T.X.; writing—original draft, T.X., K.Z. and L.W.; writing—review and editing, T.X. and L.W. All authors have read and agreed to the published version of the manuscript.

**Funding:** This study was funded by The National Key Research and Development Program of China (No. 2019YFE0117600).

**Institutional Review Board Statement:** The study did not require ethical approval.

**Informed Consent Statement:** The study did not involve humans.

**Data Availability Statement:** The study did not report any data.

**Conflicts of Interest:** The authors declare no conflict of interest.

## References

1. Carlos, J.S.A.; Pedro, L.G.; Pablo, P.M.; Daniel, C.F. Mechanical performance of fibers in hot mix asphalt: A review. *Constr. Build. Mater.* **2019**, *200*, 756–769.
2. Elanchezhiana, J.C.; Ramnathb, B.V.; Ramakrishnanc, G.; Rajendrakumard, M.; Naveenkumare, V.; Saravanakumarf, M.K. Review on mechanical properties of natural fiber composites. *Materials* **2018**, *5*, 785–1790. [CrossRef]
3. Sood, M.; Dwivedi, G. Effect of fiber treatment on flexural properties of natural fiber reinforced composites: A review. *Egypt J. Pet.* **2018**, *27*, 775–783. [CrossRef]
4. Luo, D.; Khater, A.; Yue, Y.; Abdelsalam, M.; Zhang, A.; Li, Y.; Li, J.; Iseley, D.T. The performance of asphalt mixtures modified with lignin fiber and glass fiber: A review. *Constr. Build. Mater.* **2019**, *209*, 377–387. [CrossRef]
5. Mansourian, A.; Razmi, A.; Razavi, M. Evaluation of fracture resistance of warm mix asphalt containing jute fibers. *Constr. Build. Mater.* **2016**, *117*, 37–46. [CrossRef]
6. Oda, S.; Leomar Fernandes, J.; Ildefonso, J.S. Analysis of use of natural fibers and asphalt rubber binder in discontinuous asphalt mixtures. *Construct. Build. Mater.* **2011**, *26*, 13–20. [CrossRef]
7. Vale, A.C.; Casagrande, M.D.T.; Soares, J.B. Behavior of natural fiber in Stone Matrix Asphalt mixtures using two design methods. *J. Mater. Civil. Eng.* **2013**, *26*, 457–465. [CrossRef]
8. Abiola, O.S.; Kupolati, W.K.; Sadiku, E.R.; Ndambuki, J.M. Utilization of natural fiber as modifier in bituminous mixes: A review. *Constr. Build. Mater.* **2014**, *54*, 305–312. [CrossRef]
9. Carlos, J.; Slebi-Acevedo, P.L.-G.; Irune, I.-V.; Daniel, C.-F. Laboratory assessment of porous asphalt mixtures reinforced with synthetic fibers. *Constr. Build. Mater.* **2020**, *234*, 117224.
10. Enieb, M.; Diab, A.; Yang, X. Short- and long-term properties of glass fiber reinforced asphalt mixtures. *Int. J. Pavement Eng.* **2019**, *22*, 64–76. [CrossRef]
11. Kim, M.-J.; Kim, S.; Yoo, D.-Y.; Shin, H.-O. Enhancing mechanical properties of asphalt concrete using synthetic fibers. *Constr. Build. Mater.* **2018**, *178*, 233–243. [CrossRef]
12. Wu, J.; Li, F.; Ma, Q. Effect of Polyester Fiber on Air Voids and Low-Temperature Crack Resistance of Permeable Asphalt Mixture. *Adv. Civ. Eng.* **2020**, *2020*, 2381504. [CrossRef]
13. Qin, X.; Shen, A.; Guo, Y.; Li, Z.; Lv, Z. Characterization of asphalt mastics reinforced with basalt fibers. *Constr. Build. Mater.* **2018**, *159*, 508–516. [CrossRef]
14. Khater, A.; Luo, D.; Abdelsalam, M.; Yue, Y.; Hou, Y.; Ghazy, M. Laboratory evaluation of asphalt mixture performance using composite admixtures of lignin and glass fibers. *Appl. Sci.* **2021**, *11*, 364. [CrossRef]
15. Fu, L.; Jiao, Y.; Chen, X. Reinforcement evaluation of different fibers on fracture resistance of asphalt mixture based on acoustic emission technique. *Constr. Build. Mater.* **2022**, *314*, 125606. [CrossRef]
16. Abtahi, S.M.; Esfandiarpour, S.; Kunt, M.; Hejazi, S.M.; Ebrahimi, M.G. Hybrid reinforcement of asphalt-concrete mixtures using glass and polypropylene fibers. *J. Eng. Fibers Fabr.* **2013**, *8*, 2. [CrossRef]
17. Chen, Z.; Yi, J.; Chen, Z.; Feng, D. Properties of asphalt binder modified by corn stalk fiber. *Constr. Build. Mater.* **2019**, *212*, 225–235. [CrossRef]

18. Miao, Y.; Wang, T.; Wang, L. Influences of Interface Properties on the Performance of Fiber-Reinforced Asphalt Binder. *Polymers* **2019**, *11*, 542. [CrossRef]
19. Zhang, X.; Gu, X.; Lv, J.; Zhu, Z.; Ni, F. Mechanism and behavior of fiber-reinforced asphalt mastic at high temperature. *Int. J. Pavement Eng.* **2018**, *19*, 407–415. [CrossRef]
20. Noorvand, H.; Salim, R.; Medina, J.; Stempihar, J.; Underwood, B.S. Effect of synthetic fiber state on mechanical performance of fiber reinforced asphalt concrete. *Trans. Res. Rec.* **2018**, *2672*, 42–51. [CrossRef]
21. Tanzadeh, J.; Shahrezagamasaei, R. Laboratory Assessment of Hybrid Fiber and Nano-silica on Reinforced Porous Asphalt Mixtures. *Constr. Build. Mater.* **2017**, *44*, 260–270. [CrossRef]
22. Liu, K.; Li, T.; Wu, C.; Jiang, K.; Shi, X. Bamboo fiber has engineering properties and performance suitable as reinforcement for asphalt mixture. *Constr. Build. Mater.* **2021**, *290*, 123240. [CrossRef]
23. Xia, C.; Wu, C.; Liu, K.; Jiang, K. Study on the durability of bamboo fiber asphalt mixture. *Materials* **2021**, *14*, 1667. [CrossRef]
24. Sheng, Y.; Zhang, B.; Yang, Y.; Li, H.; Chen, Z.; Chen, H. Laboratory investigation on the use of bamboo fiber in asphalt mixtures for enhanced performance. *Arabian J. Sci. Eng.* **2019**, *44*, 4629–4638. [CrossRef]
25. Yu, D.; Jia, A.; Feng, C.; Liu, W.; Fu, T.; Qiu, R. Preparation and mechanical properties of asphalt mixtures reinforced by modified bamboo fibers. *Constr. Build. Mater.* **2021**, *286*, 122984. [CrossRef]
26. Marateanu, M.; Anderson, D. Time-temperature dependency of asphalt binders—An improved model. *J. Assoc. Asphalt Paving Technol.* **1996**, *65*, 408–448.
27. Williams, M.L.; Landel, R.F.; Ferry, J.D. The temperature dependence of relaxation mechanisms in amorphous polymers and other glass-forming liquids. *J. Am. Chem. Soc.* **1955**, *77*, 3701–3707. [CrossRef]
28. Underwood, B.; Kim, Y.R.; Guddati, M.N. Improved calculation method of damage parameter in viscoelastic continuum damage model. *Int. J. Pavement Eng.* **2009**, *11*, 459–476. [CrossRef]
29. *AASHTP TP 101*; Standard Method of Test for Estimating Damage Tolerance of Asphalt Binders Using the Linear Amplitude Sweep. AASHTO: Washington, DC, USA, 2014.
30. Johnson, C.M. Estimating Asphalt Binder Fatigue Resistance Using an Accelerated Test Method. Ph.D. Thesis, University of Wisconsin-Madison, Madison, WI, USA, 2010.
31. Hintz, C.; Bahia, H. Simplification of linear amplitude sweep test and specification parameter. *Transp. Res. Record.* **2013**, *2370*, 10–16. [CrossRef]
32. Wang, C.; Chen, Y.; Xie, W. A comparative study for fatigue characterization of asphalt binder using the linear amplitude sweep test. *Mater. Struct.* **2020**, *53*, 1–12. [CrossRef]
33. *AASHTO TP 70*; Standard Method of Test for Multiple Stress Creep Recovery (MSCR) Test of Asphalt Binder Using a Dynamic Shear Rheometer (DSR). AASHTO: Washington, DC, USA, 2010.
34. Kutay, M.E.; Lanotte, M. Viscoelastic continuum damage (VECD) models for cracking problems in asphalt mixtures. *Int. J. Pavement Eng.* **2018**, *19*, 231–242. [CrossRef]
35. Daniel, J.S.; Kim, Y.R. Development of a simplified fatigue test and analysis procedure using a viscoelastic, continuum damage model. *J. Assoc. Asphalt Paving Technol.* **2002**, *71*, 619–650.
36. Underwood, B.S. Multiscale Constitutive Modeling of Asphalt Concrete. Master's Thesis, North Carolina State University, Raleigh, NC, USA, 2011.
37. *AASHTO MP10*; Standard Specification for Performance Graded Asphalt Binder Using Multiple Stress Creep Recovery (MSCR) Test. AASHTO: Washington, DC, USA, 2010.
38. Wu, M.; Li, R.; Zhang, Y.; Wei, J.; Lv, Y.; Ding, X. Reinforcement effect of fiber and deoiled asphalt on high viscosity rubber/SBS modified asphalt mortar. *Pet. Sci.* **2014**, *11*, 454–459. [CrossRef]
39. Wu, M.M.; Li, R.; Zhang, Y.; Fan, L.; Lv, Y.; Wei, J.M. Stabilizing and reinforcing effects of different fibers on asphalt mortar performance. *Pet. Sci.* **2015**, *12*, 189–196. [CrossRef]
40. Wang, C.; Ji, X.; Xie, T. Evaluation rutting resistance and performance of compound modified bio-asphalt based on viscoelastic analysis. *J. Beijing Univ Technol.* **2022**, *48*, 6. (In Chinese)
41. Rodrigo, D.; Hussain, U.; Bahia, R.L. A nonlinear constitutive relationship for asphalt binders. *Mater. Struct.* **2012**, *45*, 457–473.
42. Kumar, R.; Saboo, N.; Kumar, P. Effect of warm mix additive on creep and recovery response of conventional and polymer modified asphalt binders. *Constr. Build. Mater.* **2017**, *138*, 352–362. [CrossRef]
43. Bahia, U.H.; Hanson, I.D.; Zeng, M. *Characterization of Modified Asphalt Binders in Super Pave Mix Design*; Transportation Research Board: Washington, DC, USA, 2001.

Article

# The Influence of Aeolian Sand on the Anti-Skid Characteristics of Asphalt Pavement

Jingsheng Pan [1,†], Hua Zhao [1,†], Yong Wang [1,*] and Gang Liu [2,*]

1. College of Water and Architectural Engineering, Shihezi University, Shihezi 832003, China; pjs@stu.shzu.edu.cn (J.P.); jackerry_zhao@163.com (H.Z.)
2. State Key Laboratory of Silicate Materials for Architectures, Wuhan University of Technology, Wuhan 430070, China
* Correspondence: wyong@shzu.edu.cn (Y.W.); liug@whut.edu.cn (G.L.)
† These authors contributed the paper equally.

**Abstract:** The influence of sand accumulation on the skid resistance of asphalt pavement was studied. Many scholars have researched the anti-skid performance of conventional asphalt pavements. However, there is a lack of research on the anti-skid performance of desert roads under the condition of sand accumulation. In this study, AC-13 and AC-16 asphalt mixtures were used. The British Pendulum Number (BPN) under different sand accumulations was measured with a pendulum friction coefficient meter, and the Ames engineering texture scanner was used to obtain different sand accumulations. The texture index of asphalt mixture was used to study the macro and micro texture of asphalt pavement under different amounts of sand accumulation, and the degree of influence of different particle sizes on BPN was obtained through gray correlation analysis. The test results show that the presence of aeolian sand has a significant impact on the macro and micro texture of the asphalt pavement and will cause the anti-skid performance to decrease. Moreover, there is an apparent positive linear correlation between the road surface texture index and BPN. The research results may provide reference and reference for the design and maintenance of desert highways.

**Keywords:** asphalt pavement; sand accumulation conditions; skid resistance; British Pendulum Number (BPN); texture index

## 1. Introduction

The anti-skidding of the pavement is of great significance to traffic safety [1]. The anti-skid performance of the road surface depends on the macro-structure and micro-texture of the road surface, which provides sufficient adhesion for vehicle driving [2,3]. The lithology and angular characteristics of aggregate minerals affect the texture of the road surface [4–6]. Specifically, the lithology of the aggregate determines the resistance to abrasion of the road surface under traffic loads, and the angularity affects the micro/macro texture of the road surface. Different rocks have different resistance to abrasion.

The skid resistance of pavement surface is mainly controlled by microstructure and macrostructure, in which microstructure mainly affects the friction at low speed, while macro structure mainly affects the friction at high speed. Many scholars have done relevant research and discussion on the specific quantitative relationship between these two structures and road surface friction coefficient. Friel et al. [7] obtained the surface texture image of coarse aggregate through the scanning electron microscope, analyzed its digital image and fractal dimension by computer, and obtained the influence of micro-texture on pavement skid resistance. Serigost [8] collected many texture data and friction coefficient of Texas highway surface, compared and regressed them, obtained the relationship between pavement friction coefficient and microtexture, and realized the purpose of predicting BPN and friction coefficient with pavement microtexture. Based on the comprehensive analysis of surface characteristics such as friction, texture, and anti-sliding performance

of asphalt mixture in laboratory and field, Rezaei [9] proposed a model to characterize anti-sliding performance by mixture gradation, aggregate texture, and traffic volume level. Li [10] studied the changes of aggregate area, volume, and texture parameters of different aggregates before and after microscale wear and polishing with the help of a portable three-dimensional scanner. Ling [11] obtained the pavement surface texture elevation by using the image processing method and analyzed the relationship between different wavelength textures and anti-sliding performance.

In reality, the road surface texture is inevitably affected by external environmental factors, such as rainfall, snowfall, ice, and other pollutants, which affect the friction between the tire and the road [12,13]. In addition, changes in climate and traffic volume significantly impact the texture of the road surface [14,15]. Rain has seriously affected the anti-skid performance of the road surface [16–19]. The presence of rainwater becomes a lubricant between the tires and the road surface. The membrane generates hydrodynamic pressure on the wheels, which reduces the adhesion of the tire to the road surface. As the thickness of the water film increases, the anti-skid performance decreases more [20]. Under freezing conditions, ice and snow fill the gap on the road surface, reduce the adequate depth of the macrotexture, hinder the direct contact between the tire and the road. The anti-sliding performance is seriously reduced [21–24], resulting in the loss of the original stability of vehicle braking and the increase of braking distance, which brings hidden dangers for driving safety. Other pollutants such as sand, salt, and oil leakage will also affect the anti-skid performance of the road [25,26]. It blocks the gaps of the asphalt pavement, reduces the texture of the road surface, and makes the pavement lose its original anti-skid ability. In addition, the change of temperature will also affect the anti-sliding performance of pavement [27,28]. In the case of clean road surface, the anti-skid performance in summer is generally lower than in winter. The above are the problems faced by the anti-skid performance of asphalt pavement in conventional areas. The anti-skid performance of the pavement is easily disturbed by the external environment.

Xinjiang is a typical arid and semi-arid area in China, with severe desertification. In the desert area, the harm of wind sand to roads is particularly prominent [29]. Among them, aeolian sand covering the road surface is one of the typical hazards. The research shows that the contact area will affect the anti-sliding performance in a particular proportion [30]. Aeolian sand reduces the effective contact area between the tire and the road surface, in which the larger particles cover the road surface and hinder the direct contact between the tire and the road surface, and the smaller particles fill the macrostructure of the road surface and reduce the friction coefficient. As a result, the skid resistance of asphalt pavement is seriously reduced, and sand accumulation has become an important reason for weakening the function of the road surface.

Desert roads are a unique form of the road in extreme climate environments, which accounts for a relatively small number globally, but the harm of wind sand to the highway can not be ignored. Compared with the research on the anti-sliding performance of conventional asphalt pavement, the literature on the anti-sliding performance of desert highways under sand accumulation is more minor. This study took the sand accumulation into account on desert highways to analyze the morphological characteristics of aeolian sand. The texture change trend of asphalt pavements and the evolution of anti-skid performance were studied under different amounts of sand accumulation. Meanwhile, the impact of aeolian sand granularity was considered. The research results are expected to provide a reference for the prevention and control of desert highway diseases, especially the anti-skid performance.

## 2. Materials and Methods

*2.1. Materials*

Based on the Test Methods of Aggregate for Highway Engineering [31], the properties of aggregate were tested, and their results are given in Table 1. The aggregate used in this study is a conglomerate from Xinjiang of China.

Table 1. Aggregate properties.

| Test Items | Results | Technical Requirements |
|---|---|---|
| Crushed value (%) | 12.8 | ≤26 |
| Weared value (%) | 10.5 | ≤28 |
| Polished value (PSV) | 38 | ≥36 |
| Angularity (s) | 31.8 | ≥30 |

In this paper, two types of asphalt mixtures were prepared, AC-13 and AC-16. The aggregates are both conglomerates and have the same specific gravity. The passing rate of each sieve is shown in Figure 1. The optimal binder content of AC-13 is 4.7%, the air voids content 4.3%, the optimal binder content of AC-16 4.2%, and the air voids content is 4.3%.

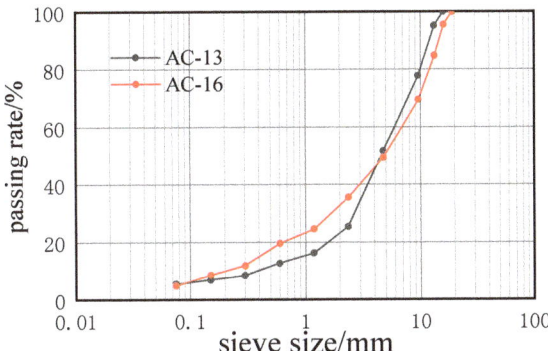

Figure 1. Aggregate gradation.

The sand sample comes from aeolian sand-covered on the road surface in the desert area of Xinjiang. In order to explore the different particle size content and microscopic morphology of aeolian sand, the particle size analysis of aeolian sand was carried out, and the particle size distribution of the sample was tested with a laser particle size analyzer (instrument model: Master-sizer 2000) (Malvern Instruments Ltd., Worcestershire, UK). The results are shown in Figure 2 by using the field emission scanning electron microscope (Quanta FEG 250) (FEI Co., Hillsboro, OR, USA). The morphology of aeolian sand under different magnifications is shown in Figure 3.

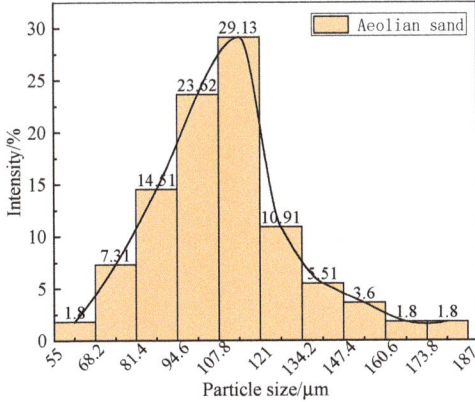

Figure 2. Particle size distribution of aeolian sand.

**Figure 3.** SEM images of aeolian sand. (**a**) SEM image of 100 times; (**b**) SEM image of 500 times.

According to the results obtained by the laser particle size analyzer, the maximum particle size of aeolian sand is 187 μm, and the minimum is 55 μm. Among them, the particle size distribution of aeolian sand between 94.6–121 μm accounts for the most significant proportion, which is 52.8%. From Figure 3, it can be observed that the aeolian sand particles are relatively granular, which can provide a rolling medium for the rolling friction between the wheel and the road surface and reduce the anti-skid performance of the road surface when it exists on the road surface.

2.2. Test Method

The pendulum friction coefficient instrument was used to measure the British Pendulum Number (BPN) of the surface of the laboratory compacted slab (300 mm × 300 mm × 50 mm) intended for wheel track rutting test with different sand accumulation [32].

In order to study the impact of sand accumulation on the anti-skid performance of the pavement, a rut plate specimen was made indoors, and the sand on the road area was simulated through different amounts of sand, and the BPN value under different conditions was tested, as shown in Figure 4. Moreover, the different amounts of sand accumulation were 0 g, 10 g, 20 g, 30 g, 40 g, 50 g, 60 g, 70 g, and 80 g, respectively. Before the test, clean the surface of the test piece with a brush and then spread the sand weighed in advance evenly over the whole test piece. After the test, clean the sand on the surface of the test piece with a brush and repeat this method many times until the test was completed.

**Figure 4.** The accumulation patterns of different aeolian sands are 30 g (**a**) and 70 g (**b**), respectively.

As shown in Figures 5 and 6, we prepared aeolian sand samples with moisture contents of 5% and 10%, respectively, and used the pendulum friction coefficient meter to determine the BPN value under different sand accumulation in order to study the influence of the moisture content of aeolian sand on the anti-sliding performance.

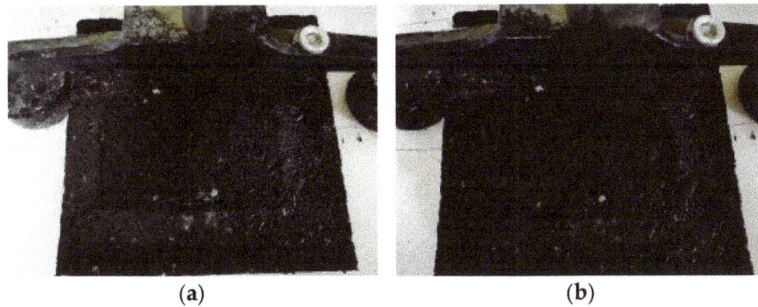

**Figure 5.** Specimens with water content of 5% (**a**) and 10% (**b**) respectively before adding aeolian sand.

**Figure 6.** Specimens with a water content of 5% (**a**) and 10% (**b**) respectively after adding aeolian sand.

In order to study the influence of different particle diameters of aeolian sand on the anti-skid performance of two types of asphalt mixtures, the aeolian sand samples were first screened. The particle size distribution of aeolian sand was mainly divided into three ranges, namely less than 0.075 mm, 0.075–0.15 mm, and greater than 0.15 mm, as shown in Figure 7. The mass percentages are 12.5%, 83.6%, and 3.9%, respectively. Then, the BPN values of the three particle sizes of aeolian sand were tested separately on the road surface.

**Figure 7.** Aeolian sand with different particle size ranges less than 0.075 mm (**a**), 0.075–0.15 mm (**b**), and greater than 0.15 mm (**c**), respectively.

The Ames engineering texture scanner (LTS 9400) (Ames Engineering Co., Ames, IA, USA) was used to detect the texture indicators of the specimens under different amounts of sand. As shown in Figure 8, select 30 scan lines among them, and select the three texture indicators of MPD, Ra, and Rq. Among them, MPD is the mean profile depth of the specimen surface, Ra is the average deviation of the contour arithmetic, and Rq is the root mean square deviation of the profile.

**Figure 8.** Ames texture scanner.

### 3. Results and Discussion

*3.1. BPN Test*

As shown in Figure 9, the BPN values of the two types of specimens change under different amounts of sand. The amount of sand is inversely proportional to the BPN value on the surface of the specimen. The BPN value decreases with the increase in the amount of sand. The BPN values of the two asphalt mixture types show the same change law, which can be divided into three stages. In increasing the initial sand volume from 0 to 10 g, the BPN value decreased slowly. The BPN value of AC-13 decreased by 4.1% and 3.8%, and the BPN value of AC-16 decreased by 2.3% and 1.7%, respectively. Then, when the sediment volume increased from 10 g to 50 g, the decreasing range of BPN value of the two mixtures increased obviously, in which the BPN value of AC-13 decreased by 27.5% and 27.6%, and the BPN value of AC-16 decreased by 31% and 30.2%. In the later stage, when the amount of sedimentation reached 60 g, the BPN value decreased slowly, and the final change curve gradually became flat. The BPN value of the two mixtures showed a slight rebound, of which AC-13 rebounded by 0.6% and 0.7%, and AC-16 rebounded 0.9% and 1.1%.

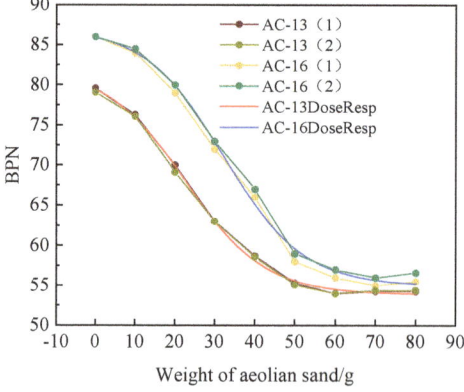

**Figure 9.** BPN value under different aeolian sand weight.

The DoseResp model is used to fit the BPN value change curves of the two mixtures. The model equations and various parameters are shown in Table 2. The goodness of fit of the two mixtures is above 0.99, indicating that the changes in the BPN value of the two mixtures under different amounts of sedimentation conform to the model. The model can predict the trend of the BPN value of AC-13 and AC-16 well under the conditions of sedimentation.

**Table 2.** Fitting curve of BPN value change under different aeolian sand weight.

| Mixture Type | Fitting Equation | Fitting Precision |
| --- | --- | --- |
| AC-13 | $y = 53.911 + \dfrac{28.182}{1 + 10^{0.044x - 0.996}}$ | 0.998 |
| AC-16 | $y = 54.994 + \dfrac{32.141}{1 + 10^{0.044x - 1.420}}$ | 0.995 |

The analysis believes that due to the generally small particle size of aeolian sand, it enters the gaps of the asphalt mixture at the initial stage and blocks the gaps, which reduces the depth of the structure and decreases the anti-skid performance. Then, as the quality of the aeolian sand increases, the gap is gradually filled, and part of the aeolian sand covers the road surface. Due to its round appearance, the tire and the road surface form a micro-bearing system, which changes sliding friction into rolling friction, and the friction force drops sharply. The main factor causing the decrease of friction is that the aeolian sand is not adhered to the adhesive asphalt, resulting in rolling friction under the impact force of the pendulum. Finally, when the amount of accumulated sand increases to a certain level (60 g), the friction is controlled by the accumulated sand. The accumulated sand produces resistance to the rubber slider of the pendulum friction coefficient meter, causing its BPN value to rise. The reason for this phenomenon may be the limitation of the test equipment, which does not conform to the actual situation.

As shown in Figure 10a, when the amount of aeolian sand with the water content of 5% increases to 10 g, the BPN values of the two mixtures decrease by 4.9% and 4.5%, respectively. At this time, the AC-13 decreases. The amplitude is more significant than AC-16. The analysis believes that the air voids content and texture depth of AC-13 is less than AC-16. Under the interference of water and sand, the decrease of BPN value is slightly more significant than AC-16. With the increase of sand accumulation, the anti-sliding performance of the two mixtures decreases and finally tends to be stable. During the whole process, the BPN values of the two mixtures decrease by 28.5% and 35.7%, respectively. When the water content of aeolian sand is 10%, as shown in Figure 10b, the initial BPN value of the two mixtures decreases overall. The initial value of AC-13 is significantly lower than AC-16. When the amount of sand increased to 30 g, the BPN value droped to the lowest point. The BPN value of AC-16 reached the lowest point when the amount of sand increased to 50 g. There was a slight rebound. Among them, the BPN value of AC-13 rebounded by 3.1% and 4.4%, and the BPN value of AC-16 rebounded. The margins were 1.9% and 2.4%. The analysis shows that water becomes the lubricant between the sand, which makes the micro-bearing system formed by the wheel sand road more significant. The wet sand exists between the wheel and the road like a ball, and the anti-sliding ability drops sharply at this time.

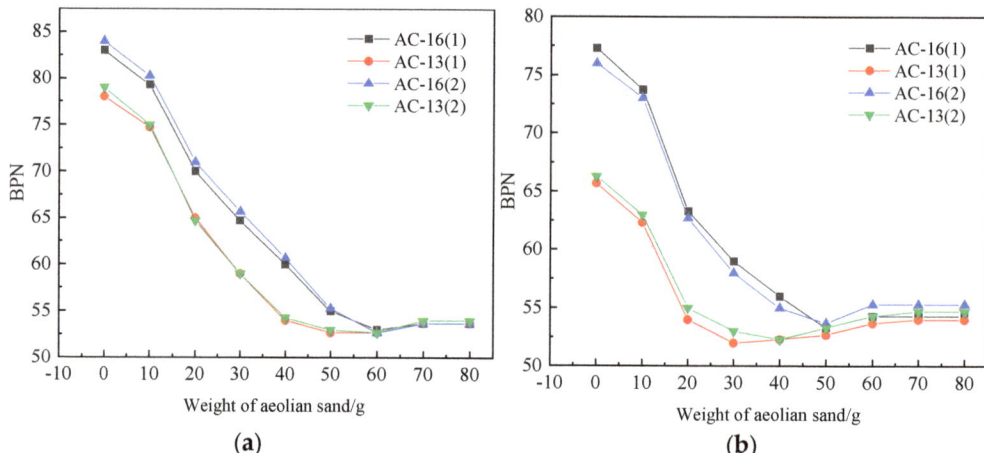

**Figure 10.** BPN under different sediment accumulation conditions in the wet state. (**a**) The moisture content of accumulated sand is 5%; (**b**) The moisture content of accumulated sand is 10%.

Compared with the dry aeolian sand acting on the road surface, the aeolian sand in the wet state makes the BPN value of the two mixtures generally lower. At this time, the pavement voids and macrostructures are not only filled with sand but filled with sand and water. This process accelerates the decrease in the depth of the pavement structure, allowing more sand and water to act on the road surface and hinder the tires and the pavement. The direct contact between the sand and the moisture exists in the sand surface and the gaps, which provides sufficient lubrication between the sand and between the sand and the wheel and the road surface, so the BPN value can quickly drop to the lowest point.

Figure 11a shows the variation law of BPN values of three particle sizes of AC-16 under different sediment volumes. On the whole, the variation trend of three particle sizes is the same as that of conventional aeolian sand samples. However, the slope of the BPN value curve for aeolian sand less than 0.075 mm in the initial stage is the smallest, while the curve greater than 0.15 mm has the most significant variation range, and the curve decreases rapidly in the whole process. In the end, there was a significant recovery. In increasing the amount of sand from 0 to 10 g, the BPN corresponding to the three particle sizes decreased by 4.1%, 5.0%, and 6.5%, respectively. When the amount of sand increased from 10 g to 50 g, the three particle sizes, the corresponding BPN dropped by 29.6%, 31.6%, and 30.2%, respectively. The overall decline in Figure 11b is smaller than that in Figure 11a, indicating that AC-16 is more disturbed by wind and sand. However, the slope of the BPN curve at the initial stage was less than that of AC-13, and three particle sizes decreased by 1.2%, 2.3%, and 3.4%, respectively. The analysis shows that the macro structure of AC-16 is larger than AC-13, and it can contain aeolian sand particles. Relatively strong, the decrease in BPN value is slight in the initial stage. Then, when the macro structure was filled with aeolian sand particles, the BPN value dropped sharply, and the three particle sizes dropped by 34.7%, 34.2%, and 35.7%, respectively. When the BPN value of the two mixtures fell to the lowest point in the later period, because the surface texture of AC-16 was slightly larger than AC-13, the BPN value of AC-16 rose slightly.

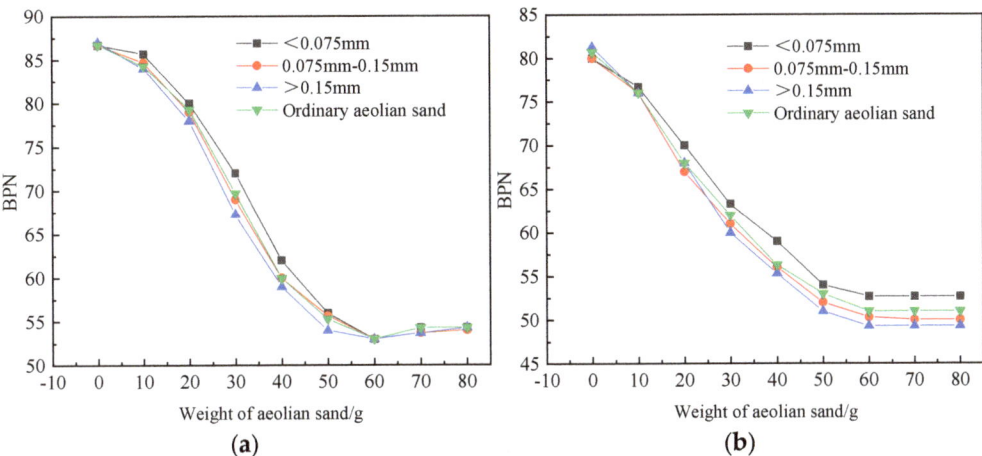

**Figure 11.** BPN value of aeolian sand with three particle sizes existing alone on the road surface. (**a**) BPN value of aeolian sand with three particle sizes existing alone on the AC-16 pavement surface; (**b**) BPN value of aeolian sand with three particle sizes existing alone on the AC-13 pavement surface.

The decrease of the BPN value on the surface of the test piece is related to the particle size. The larger the particle size, the more significant the decrease of the BPN value on the surface of the test piece. The results indicated that the larger the particle size of the aeolian sand, the greater the ability to fill the gap and the surface structure of the specimen. It reduced the macrostructure of the specimen surface and its anti-sliding ability. The BPN value decreases more rapidly. Therefore, under the same filling rate, the larger the particle size of the aeolian sand, the faster the anti-skid performance of the pavement decreases.

In order to determine the degree of influence of the three particle sizes on the BPN value, the test results were analyzed with a gray correlation. The BPN value under the three particle sizes was used as the comparison sequence. The BPN value under the conventional aeolian sand was used as the reference sequence for comparative analysis. The results are shown in Figure 12.

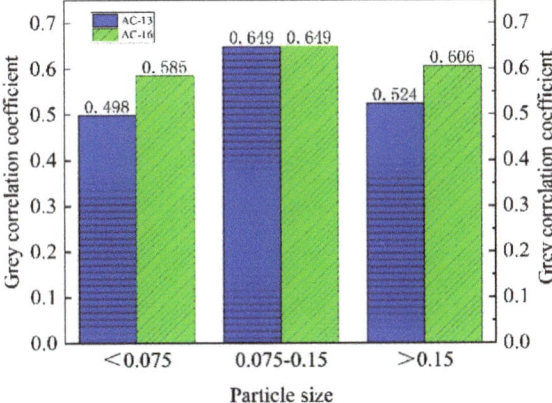

**Figure 12.** Grey correlation analysis results.

It can be seen from Figure 12 that the gray correlation coefficient of particles less than 0.075 mm in the two kinds of asphalt mixture is the smallest, which means that the particle size has the least influence on the BPN value. In the conventional aeolian sand, the particle size less than 0.075 mm accounts for a relatively small proportion. In addition, its particle size is small, which mainly fills the gaps in the road surface and contributes little to the rolling medium between the tire and the road surface. The 0.075–0.15 mm particles have an essential impact on the BPN value. They had a large mass proportion, and acted as a rolling medium for the rolling friction between the tire and the road surface. Although the particles larger than 0.15 mm had the smallest proportion, they played a significant supporting role between the wheel and the road surface, disturbing the BPN value. The gray correlation coefficient corresponding to the particle larger than 0.15 mm was in the middle level.

### 3.2. Texture Index

The most intuitive impact of aeolian sand on the pavement is to change its original macroscopic structure. Meanwhile, smaller particles enter the pavement voids to fill the macroscopic structure and reduce the structure depth, thereby reducing the friction coefficient. On the other hand, larger particles with diameters covering the road surface hinder the direct contact between the tire and the road surface and reduce the effective contact area between the tire and the road surface, thereby reducing the anti-skid performance.

In order to further understand the road texture under different sand accumulation, the Ames engineering texture scanner (LTS 9400) was used to scan the surface of specimens under different sand accumulation. Thirty scanning lines were selected for each specimen to extract three texture indexes, namely MPD, Ra, and Rq. Figure 13a,b are the change trends of texture indicators of AC-13 and AC-16 mixtures under different sand accumulations. The sizes of the three indicators are inversely proportional to the sand accumulation, and Ra, Rq. The similarity of the change curves of the two indicators is relatively high, indicating that the two are affected by the same degree of sand accumulation.

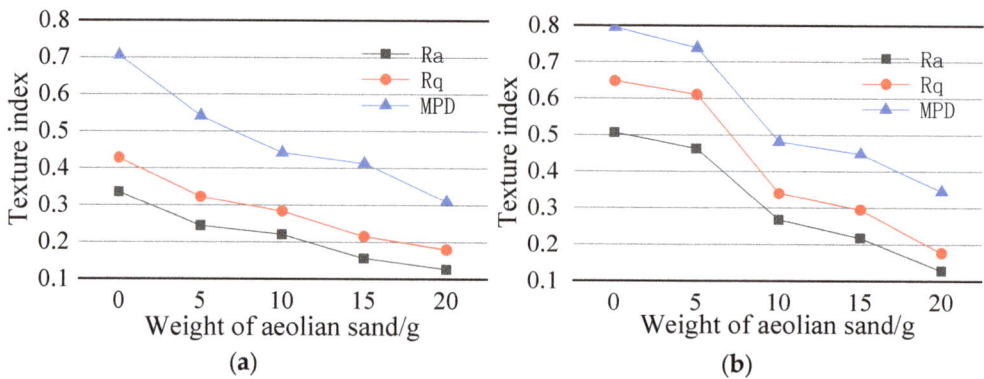

**Figure 13.** Texture index under different sand accumulation. (**a**) Texture index under different sand accumulation on the AC-13 pavement surface; (**b**) Texture index under different sand accumulation on the AC-16 pavement surface.

MPD reflects the average section depth of the specimen surface. With the increasing amount of aeolian sand, the texture depth of the specimen surface is gradually reduced, and the surface texture changes passively. At this time, MPD decreases with the trend, and the BPN value greatly correlates with the surface texture of the specimen, which also decreases. The MPD of AC-13 in Figure 13a and AC-16 in Figure 13b decreased by 23% and 7.2%, respectively, when the initial sand accumulation increased from 0 g to 5 g. Then, with the gradual increase of the amount of aeolian sand on the specimen surface, the MPD decreased. During the whole process, the MPD of the two specimens decreased by 52% and 56.4%, respectively.

Ra is the average deviation of the contour arithmetic and is used to characterize the texture of the road surface in the scanning range. It reflects the dispersion degree of the change of the pavement structure profile concerning the reference line. The increase of aeolian sand makes the surface part of the specimen to be filled. As the depth decreases, the texture decreases. When aeolian sand increased from 0 g to 5 g, the Ra of the two mixtures decreased by 27.4% and 8.6%, respectively. During the whole process, the Ra of the two mixtures decreased by 62.2% and 74.6%, respectively. As shown in Figure 13, the Ra value of AC-16 is more significant than that of AC-13, and it sharply drops with the increase of sediment accumulation.

Rq is the root mean square deviation of the profile and is used to indicate the surface texture of the specimen. As the amount of aeolian sand gradually increases, the internal voids of the specimen are occupied by aeolian sand particles, the macrostructure is reduced, and the surface is rough. When the temperature decreases, the Rq index decreases, and then the BPN value of the specimen surface decreases. In the initial stage, Rq of AC-16 decreased slowly, compared with AC-13. The decrease of Rq index was 24.5% and 5.7%, respectively, and the decline of AC-16 in the late stage was significantly increased, and the overall decline was more than AC-13. During the whole process, the Rq of AC-13 and AC-16 decreased by 57.9% and 72.5%, respectively.

On the whole, when the number of sand increases from 0 g to 5 g, the three texture indicators of MPD, Ra, and Rq decrease by AC-13 larger than AC-16. Meanwhile, the voids and macro-structure of AC-13 are smaller than AC-16. In the presence of a small amount of aeolian sand, the texture index of the AC-13 surface is more sensitive. AC-16 can contain a small amount of sand and has little effect on its texture in the initial stage.

Aeolian sand covers the road surface, which directly affects the macroscopic structure of the road surface, which in turn makes the road surface texture change significantly. The result of the change in the road sign texture is the corresponding decrease in the BPN value. Therefore, it is believed that the relationship between the road surface texture index and anti-skid performance certainly exist.

The correlation analysis between the BPN value and three texture indicators of MPD, Ra, and Rq was performed. As shown in Figure 14, BPN and Ra values have an excellent linear relationship, indicating that the increase of aeolian sand make the contour of the pavement structure. The slope of the trend line for AC-16 is more significant than that for AC-13.

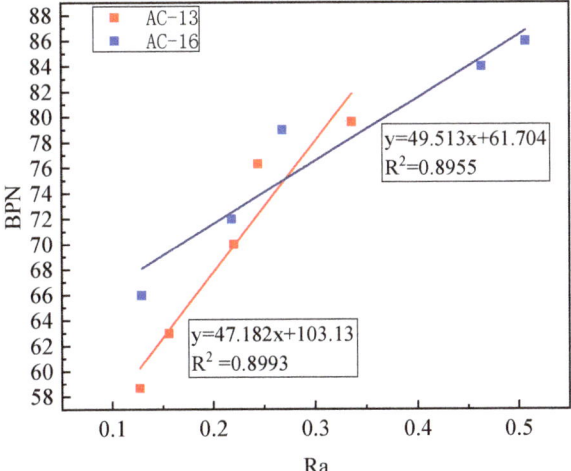

**Figure 14.** Relationship between BPN and Ra.

The BPN value and Rq of the two mixtures were fitted, and the results are shown in Figure 15. There was also an excellent linear relationship between the BPN value and Rq. The goodness of fit for AC-13 is 0.9088, and for AC-16, 0.8727. The increase of the contoured root means square deviation Rq will cause the increase of the BPN value. The two are in direct proportion, indicating that the more the amount of aeolian sand on the road surface, the stronger the filling capacity of the road surface. At this time, the road surface texture decreases, and the BPN value also increases. It decreases accordingly.

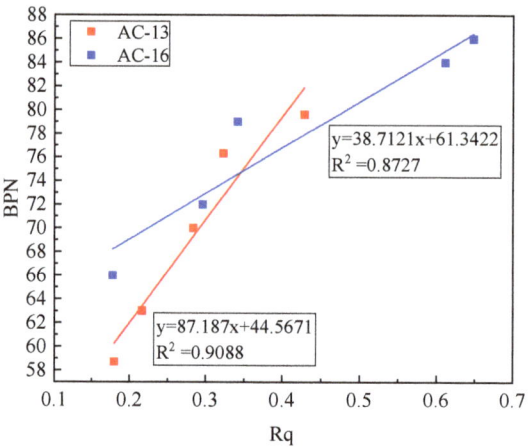

**Figure 15.** Relationship between BPN and Rq.

Figure 16 shows the results of fitting the BPN value and MPD of the two mixtures. The results show that the BPN value and MPD also have a good linear relationship. The goodness of fit of the two mixtures is above 0.84. The average section depth of MPD decreases with the decrease of the BPN value. The analysis shows that as the number of aeolian sand increases, the texture depth of the specimen surface gradually decreases, and the surface texture changes passively. At this time, the MPD decreases accordingly.

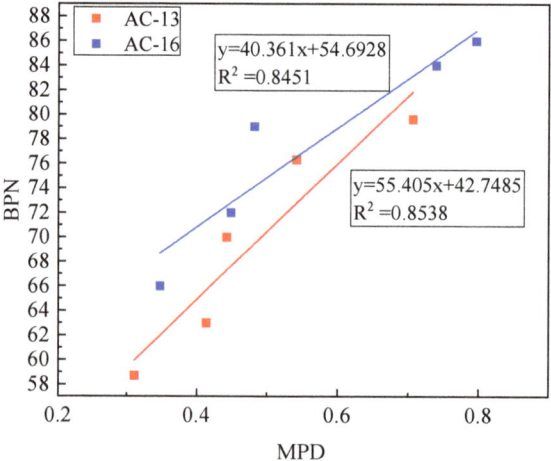

**Figure 16.** Relationship between BPN and MPD.

## 4. Conclusions

This paper studies the influence of aeolian sand on the anti-sliding performance and texture index of asphalt mixture, which is a study based on the macro perspective of the anti-sliding performance changes caused by different amounts of sand and asphalt mixture types.

- As the number of sand accumulation increases, the BPN values of AC-13 and AC-16 mixtures show a downward trend. By fitting the DoseResp model, the later changes can be predicted based on the changes in the initial BPN value.
- By studying the anti-sliding performance of pavement surface whether aeolian sand contains water, we find that the sand with water is generally lower than that of dry sand, and the skid resistance decreases with the increase of water volume.
- Through the grey correlation analysis, it is found that the aeolian sand particles between 0.075–0.15 mm have the most significant influence on the BPN value of the two mixtures because its mass percentage is the largest.
- Ames engineering texture scanner was used to obtain texture indicators under different amounts of sand. It is found that all three texture indicators are correlated with the BPN value. With the increase of the sand deposition within 20 g, the effective structural depth of pavement macrotexture decreases, resulting in the decrease of BPN value of road surface, which means that a certain amount of aeolian sand has a significant impact on the anti-skid performance of the mixture at the macro level.

**Author Contributions:** Conceptualization, J.P. and H.Z.; methodology, J.P.; literature survey, J.P. and H.Z.; validation, J.P., H.Z. and G.L.; formal analysis, J.P. and H.Z.; investigation, J.P., H.Z. and Y.W.; resources, Y.W. and G.L.; data curation, J.P. and H.Z.; writing—original draft preparation, J.P.; writing—review and editing, G.L. and H.Z.; supervision, Y.W. and G.L.; project administration, Y.W. and G.L.; funding acquisition, Y.W. and G.L. All authors have read and agreed to the published version of the manuscript.

**Funding:** This research was funded by National Natural Science Fund of China, grant numbers 51768062 and 52068061.

**Institutional Review Board Statement:** Not applicable.

**Informed Consent Statement:** Not applicable.

**Data Availability Statement:** No new data were created or analyzed in this study, data sharing is not applicable to this article.

**Conflicts of Interest:** The authors declare no conflict of interest.

## References

1. Anupam, K.; Tang, T.; Kasbergen, C.; Scarpas, A.; Erkens, S. 3-D Thermomechanical Tire–Pavement Interaction Model for Evaluation of Pavement Skid Resistance. *Transp. Res. Rec.* **2020**, *2675*, 65–80. [CrossRef]
2. Chu, L.J.; Fwa, T.F. Pavement skid resistance consideration in rain-related wet-weather speed limits determination. *Road Mater. Pavement Des.* **2016**, *19*, 334–352. [CrossRef]
3. Dan, H.-C.; He, L.-H.; Xu, B. Experimental investigation on skid resistance of asphalt pavement under various slippery conditions. *Int. J. Pavement Eng.* **2015**, *18*, 485–499. [CrossRef]
4. Do, M.; Cerezo, V.; Zahouani, H. Laboratory test to evaluate the effect of contaminants on road skid resistance. *Proc. Inst. Mech. Eng. Part J* **2014**, *228*, 1276–1284. [CrossRef]
5. Eisenberg, D. The mixed effects of precipitation on traffic crashes. *Accid. Anal. Prev.* **2004**, *36*, 637–647. [CrossRef]
6. Geedipally, S.R.; Das, S.; Pratt, M.P.; Lord, D. Determining Skid Resistance Needs on Horizontal Curves for Different Levels of Precipitation. *Transp. Res. Rec.* **2020**, *2674*, 358–370. [CrossRef]
7. Friel, J.J.; Pande, C.S. A direct determination of fractal dimension of fracture surfaces using scanning electron microscopy and stereoscopy. *J. Mater. Res.* **1993**, *8*, 100–104. [CrossRef]
8. Serigos, P.A.; de Fortier Smit, A.; Prozzi, J.A. Incorporating surface microtexture in the prediction of skid resistance of flexible pavements. *Transp. Res. Rec.* **2014**, *2457*, 105–113. [CrossRef]
9. Rezaei, A.; Masad, E. Experimental-based model for predicting the skid resistance of asphalt pavements. *Int. J. Pavement Eng.* **2013**, *14*, 24–35. [CrossRef]

10. Li, Q.J.; Zhan, Y.; Yang, G.; Pittenger, D.M.; Wang, K.C.P. 3D Characterization of Aggregates for Pavement Skid Resistance. *J. Transp. Eng. Part B Pavements* **2019**, *145*, 04019002. [CrossRef]
11. Chen, D.; Ling, C.; Wang, T.; Su, Q.; Ye, A. Prediction of tire-pavement noise of porous asphalt mixture based on mixture surface texture level and distributions. *Constr. Build. Mater.* **2018**, *173*, 801–810. [CrossRef]
12. Gierasimiuk, P.; Wasilewska, M.; Gardziejczyk, W. A Comparative Study on Skid Resistance of Concrete Pavements Differing in Texturing Technique. *Materials* **2021**, *14*, 178. [CrossRef] [PubMed]
13. Mataei, B.; Zakeri, H.; Zahedi, M.; Nejad, F.M. Pavement Friction and Skid Resistance Measurement Methods: A Literature Review. *Open J. Civ. Eng.* **2016**, *6*, 537–565. [CrossRef]
14. Kienle, R.; Ressel, W.; Götz, T.; Weise, M. The influence of road surface texture on the skid resistance under wet conditions. *Proc. Inst. Mech. Eng. Part J.* **2018**, *234*, 313–319. [CrossRef]
15. Kogbara, R.B.; Masad, E.A.; Kassem, E.; Scarpas, A. Skid Resistance Characteristics of Asphalt Pavements in Hot Climates. *J. Transp. Eng. Part B Pavements* **2018**, *144*, 04018015. [CrossRef]
16. Li, C.; Wang, Y.; Lei, J.; Xu, X.; Wang, S.; Fan, J.; Fan, S. Damage by wind-blown sand and its control measures along the Taklimakan Desert Highway in China. *J. Arid. Land* **2020**, *13*, 98–106. [CrossRef]
17. Li, P.; Yi, K.; Yu, H.; Xiong, J.; Xu, R. Effect of Aggregate Properties on Long-Term Skid Resistance of Asphalt Mixture. *J. Mater. Civ. Eng.* **2021**, *33*, 04020413. [CrossRef]
18. Lubis, A.S.; A Muis, Z.; Gultom, E.M. The effect of contaminant on skid resistance of pavement surface. *IOP Conf. Ser. Earth Environ. Sci.* **2018**, *126*, 012040. [CrossRef]
19. Nicolosi, V.; D'Apuzzo, M.; Evangelisti, A. Cumulated frictional dissipated energy and pavement skid deterioration: Evaluation and correlation. *Constr. Build. Mater.* **2020**, *263*, 120020. [CrossRef]
20. Pomoni, M.; Plati, C.; Loizos, A.; Yannis, G. Investigation of pavement skid resistance and macrotexture on a long-term basis. *Int. J. Pavement Eng.* **2020**, *1*, 1–10. [CrossRef]
21. Iii, H.R.; Nassiri, S.; AlShareedah, O.; Yekkalar, M.; Haselbach, L. Evaluation of skid resistance of pervious concrete slabs under various winter conditions for driver and pedestrian users. *Road Mater. Pavement Des.* **2019**, *22*, 1350–1368. [CrossRef]
22. Tang, T.; Anupam, K.; Kasbergen, C.; Scarpas, A.; Erkens, S. A finite element study of rain intensity on skid resistance for permeable asphalt concrete mixes. *Constr. Build. Mater.* **2019**, *220*, 464–475. [CrossRef]
23. Wang, D.; Zhang, Z.; Kollmann, J.; Oeser, M. Development of aggregate micro-texture during polishing and correlation with skid resistance. *Int. J. Pavement Eng.* **2018**, *21*, 629–641. [CrossRef]
24. Wang, H.; Wang, C.; Bu, Y.; You, Z.; Yang, X.; Oeser, M. Correlate aggregate angularity characteristics to the skid resistance of asphalt pavement based on image analysis technology. *Constr. Build. Mater.* **2020**, *242*, 118150. [CrossRef]
25. Wang, Y.; Liu, Y.; Cheng, Y.; Xue, J. Evaluation of the Decay Characteristics of Pavement Skid Resistance Using Three-Dimensional Texture from Accelerated Abrasion Test. *J. Transp. Eng. Part B Pavements* **2020**, *146*, 04020073. [CrossRef]
26. Wu, J.; Wang, X.; Wang, L.; Zhang, L.; Xiao, Q.; Yang, H. Temperature Correction and Analysis of Pavement Skid Resistance Performance Based on RIOHTrack Full-Scale Track. *Coatings* **2020**, *10*, 832. [CrossRef]
27. Yan, B.; Mao, H.; Zhong, S.; Zhang, P.; Zhang, X. Experimental Study on Wet Skid Resistance of Asphalt Pavements in Icy Conditions. *Materials* **2019**, *12*, 1201. [CrossRef] [PubMed]
28. Yu, M.; You, Z.; Wu, G.; Kong, L.; Liu, C.; Gao, J. Measurement and modeling of skid resistance of asphalt pavement: A review. *Constr. Build. Mater.* **2020**, *260*, 119878. [CrossRef]
29. Yun, D.; Hu, L.; Tang, C. Tire-Road Contact Area on Asphalt Concrete Pavement and Its Relationship with the Skid Resistance. *Materials* **2020**, *13*, 615. [CrossRef]
30. Zhu, X.; Yang, Y.; Zhao, H.; Jelagin, D.; Chen, F.; Gilabert, F.A.; Guarin, A. Effects of surface texture deterioration and wet surface conditions on asphalt runway skid resistance. *Tribol. Int.* **2020**, *153*, 106589. [CrossRef]
31. Ministry of Transport of the People's Republic of China. *Test. Methods of Aggregate for Highway Engineering (JTG E42-2005)*; Ministry of Transport of the People's Republic of China: Beijing, China, 2005.
32. Ministry of Transport of the People's Republic of China. *Standard Test Methods of Bitumen and Bituminous Mixtures for Highway Engineering (JTG E20-2011)*; Ministry of Transport of the People's Republic of China: Beijing, China, 2011.

Article

# Anti-Skid Characteristics of Asphalt Pavement Based on Partial Tire Aquaplane Conditions

Miao Yu [1,2,*], Yao Kong [3], Zhanping You [4,*], Jue Li [2], Liming Yang [5] and Lingyun Kong [2]

1. Key Laboratory of Road Structure and Material of Transport Ministry, Chang'an University, Xi'an 710064, China
2. School of Civil Engineering, Chongqing Jiaotong University, 66 Xuefu Blvd, Chongqing 400074, China; lijue1207@cqjtu.edu.cn (J.L.); konglingyun@cqjtu.edu.cn (L.K.)
3. Geotechnical Engineering Institute, CCTEG Chongqing Engineering Co., Ltd., Chongqing 400042, China; kongyao@cqmsy.com
4. Department of Civil and Environmental Engineering, Michigan Technological University, 1400 Townsend Drive, Houghton, MI 49931, USA
5. Guangxi Communications Design Group Co., Ltd., Nanning 530028, China; luemingyang@163.com
* Correspondence: yumiaoym@126.com (M.Y.); zyou@mtu.edu (Z.Y.)

**Abstract:** This study presented a finite element model of radial tire–asphalt pavement interaction using ABAQUS 6.14 software to investigate the skid resistance properties of asphalt pavement under partial tire aquaplane conditions. Firstly, the pavement profile datum acquired by laser scanning were imported to Finite Element Analysis (FEA) software to conduct the pavement modeling. Secondly, a steady state rolling analysis of a tire on three types of asphalt pavements under drying conditions was carried out. Variation laws of the friction coefficient of the radial tire on different pavements with different pavement textures, tire pressures, and loads on the tire were examined. Subsequently, calculation results of the steady state rolling analysis were transmitted to dynamic explicit analysis, and an aquaplane model of a radial tire on asphalt pavements was built by inputting the flow Euler grids. The tire–pavement adhesive characteristics under partial aquaplane conditions are discussed regarding the aquaplane model. Influences of the thickness of water film, the texture of asphalt pavement, and the rolling speed of the tire on the vertical pavement-tire contact force are analyzed. It is found that the vertical contact force between open graded friction course (OGFC) pavement and tire is the highest, followed by stone mastic asphalt (SMA) pavement and dense graded asphalt concrete (AC) pavement surface. The vertical contact force between tire and pavement will be greatly reduced, even with increasing speed or water film thickness. As tire speed increases from 70 km/h to 130 km/h, the tire–pavement contact force is reduced by about 25%. Moreover, when the thickness of water film increases from 0 (dry condition) to 4 mm and then to 12 mm, the vertical contact force reduced 50% and 15%, respectively, compared with under the dry contact condition. This study provided a key theoretical reference for safe driving on wet pavements.

**Keywords:** skid resistance; tire-pavement friction; pavement textures; partial tire aquaplane conditions; water film; vertical contact force

**Citation:** Yu, M.; Kong, Y.; You, Z.; Li, J.; Yang, L.; Kong, L. Anti-Skid Characteristics of Asphalt Pavement Based on Partial Tire Aquaplane Conditions. *Materials* **2022**, *15*, 4976. https://doi.org/10.3390/ma15144976

Academic Editor: Francesco Canestrari

Received: 19 June 2022
Accepted: 15 July 2022
Published: 17 July 2022

**Publisher's Note:** MDPI stays neutral with regard to jurisdictional claims in published maps and institutional affiliations.

**Copyright:** © 2022 by the authors. Licensee MDPI, Basel, Switzerland. This article is an open access article distributed under the terms and conditions of the Creative Commons Attribution (CC BY) license (https:// creativecommons.org/licenses/by/ 4.0/).

## 1. Introduction

The anti-skid performance of pavement is one of the major influencing factors of traffic accidents. Inadequate skid-resistance may decrease the braking efficiency of vehicles, contributing to accidents, including collisions [1,2]. Whether exhibiting the pure skidding behavior of lock slip or the usual simultaneous rolling-skidding state of vehicles, the tire–pavement friction behavior is the control factor of the braking effect since the braking force of the tire on the pavement is significantly higher than the rolling resistance in the braking process [3,4]. Under severe weather conditions like heavy rain, the pavement surface is covered by a layer of water film, as the rainwater cannot be discharged quickly enough. When vehicles drive on the pavement covered with water film, the tires are lifted by water flow and cannot contact the pavement sufficiently. The adhesive characteristics and braking characteristics of vehicles decrease significantly compared with those on dry pavement, thus greatly increasing the risk of traffic accidents [5]. In addition, traffic accidents induced by weather conditions not only have a high frequency of occurrence, but also are accompanied by secondary accidents [6,7]. Weather-related traffic accidents have attracted much attention from the traffic safety industry [8,9]. Once the tire aquaplane phenomenon occurs, adhesive characteristics of tires and the anti-skid characteristics of pavement under wet conditions is crucial to ensuring driving safety [10,11]. However, anti-skid performance of asphalt pavement under tire aquaplane conditions is mainly discussed through simulation analysis due to the great differences between test simulation and practical vehicle-pavement coupling friction conditions.

Ong et al. [12] built a finite element model to evaluate anti-skid performance of trucks on wet pavement. The model covers basic structural mechanics and fluid mechanic theory, as well as tire contact and tire–fluid interaction. The model is used to analyze the effects of speed, tire load, road thickness, and water film thickness on the skid resistance of trucks under zero load and full load. With consideration to the influences of the geometric design of pavement, tire surface design, and various working conditions of tires, a finite element model of tire–water–pavement interaction was established by Tang Tianci et al. [13] to evaluate the slipping performance of vehicles under different rainfall intensities. The real pavement surface texture and the internal microstructure of the asphalt mixture were integrated into the pavement grid by CT scanning. The established model could not only simulate water flow on a pavement surface, but also capture the process of water penetration into the porous structure of pavement, thus enabling the simulation of the tire–water–pavement interaction effectively. Furthermore, Anupam [14] also conducted an investigation on the tire–pavement heat transfer by modeling the coupling contact between tire and asphalt pavement. The textured pavement surface submodel were modeled by the combination of CT scanning and the commercial software Simpleware, which was incorporated for the post-process of surface images. Feng et al. [15] constructed finite element models of tires with smooth, longitudinal grooves and tires with complete texture in ABAQUS by using the three-dimensional solid modeling software CATIA. These models were used for water-skidding simulations of tires under static inflation loading, drive braking, free rolling, and transient water-skidding states, respectively, which further verified the validity of the CEL method in studying water-skidding behaviors of tires. Zhu et al. [16] built an aquaplane model of rib tires on asphalt pavement by using ABAQUS. The solid model of asphalt pavement was established by CT scanning, accompanied by the coupled tire model and fluid model. Furthermore, the effect of tire pressure, thickness of water film, and macrotexture of asphalt pavement on a tire aquaplane were analyzed. Zhu et al. [17] built an aquaplane model of rib tires on runways in airports by using the CEL algorithm with consideration to the macrotexture of asphalt pavement, and they used the model to discuss influences of slip ratio, flow distribution, and tire pressure on a tire aquaplane. Numerical simulation results were verified by calculating the braking distance and trailer test.

To sum up, there are various influencing factors of tire–water–pavement coupling friction. For simplification of calculation in road engineering, pavement is usually simplified as a rigid plane [18]. That is, the investigation of wet friction performance of pavement is insufficient in comparison. With further deepening studies, some scholars have attempted to study the tire–pavement coupling friction behaviors in humid environments through CT scanning and have achieved good evaluation effects [19,20]. However, these studies often have high test costs and require much time for calculation, which brings some objective limitations against further investigation.

Based on the above brief review of the research status of tire–pavement contact behaviors on tire aquaplane conditions, it can be inferred that the research on pavement skid resistance, especially when tires rolling on partial aquaplane conditions is insufficient. In addition, the texture characteristics are acquired mainly by CT scanning, which also incurs high cost. Therefore, the inadequate investigation of tire rolling on wet pavement surfaces with different thicknesses of water film has prevented further study on pavement frictional mechanism under the wet condition of pavement design, which also plays a key role in traffic safety.

The objective of this study is to investigate the frictional properties of asphalt pavement on partial tire aquaplane conditions at a relatively low cost. The pavement texture characteristics can be acquired by laser scanning, and then the textured pavement can also be modeled in ABAQUS by data processing. Variation laws of the friction coefficient of the radial tire on various pavements with the associated pavement textures, tire pressures, and loads on the tire are further investigated. Moreover, the tire–pavement contact characteristics under partial aquaplane conditions are discussed in explicit analysis mode. Influences of the thickness of water film, the texture of asphalt pavement, and the rolling speed of the tire on the vertical pavement-tire contact force are analyzed. It seeks a relatively good computer analysis target at a low computational cost and provides application references as well as theoretical references to the anti-skid performance of the wet pavement.

## 2. Methodology

Based on the finite element model of radial tire–asphalt pavement interaction, which was built with ABAQUS, a steady state rolling analysis of a tire on three drying asphalt pavements was carried out first to explore variation laws of the friction coefficient of radial tires with pavement texture, tire pressure, and loads on tire. Second, calculation results of the steady state rolling analysis were transmitted to a dynamic explicit analysis, and an aquaplane model of a radial tire on asphalt pavements was built by inputting the flow Euler grids. The model is used to analyze the adhesion characteristics of tire and pavement under partial aquaplane conditions and explore the effects of water film thickness, asphalt pavement texture, and tire movement speed on the vertical contact force between pavement and tire, so as to provide some theoretical guidance for the safety of vehicle driving on rainy days.

### 2.1. Construction of the Tire–Pavement Contact Model
#### 2.1.1. Tire Modeling and Assembly with Pavement

In this study, the tire–pavement contact simulation analysis was carried out using ABAQUS finite element simulation software. The tire model used the 175SR14 radial tire. Initially, the axisymmetric model is constructed according to the geometry and material properties of a radial tire. Subsequently, the partial 3D model and an integrated three-dimensional model of the tire can be generated in turn in ABAQUS, as shown in Figure 1.

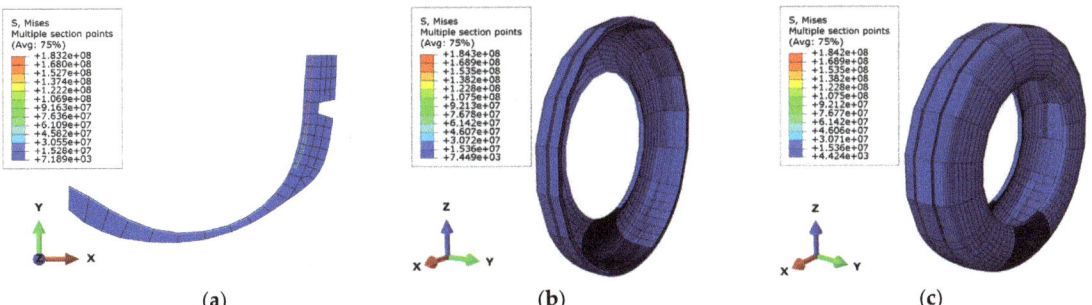

**Figure 1.** Tire modeling. (**a**) Axisymmetric model; (**b**) partial 3D model; (**c**) full 3D model.

Based on the Input file of the completed 3D tire model, the keyword *INCLUDE, INPUT was adopted for importing the Input file of the asphalt pavement model. Moreover, the tire–pavement contact behaviors were defined, accompanied with the set of loading and boundary conditions, to realize the tire–pavement assembly, as illustrated in Figure 2. More details on modeling and assembly can be found in Reference [21]. In addition, the construction steps of the pavement texture model are illustrated as follows.

**Figure 2.** Tire–pavement assembly.

2.1.2. Acquisition of Pavement Texture Properties

The laser profile scanner was assembled to acquire relative altitude information on surfaces of asphalt concrete specimens. This laser scanner was assembled with the cross-skidding table and double-shaft connected control cabinet, as shown in Figure 3a. To acquire sufficient texture properties of asphalt pavement, a 100 mm × 100 mm rectangular zone of each specimen was scanned, with a 1 mm points-acquisition interval in the scanning area, as shown in Figure 3b,c. That is, 10,000 effective altitude data were collected in each scanning region.

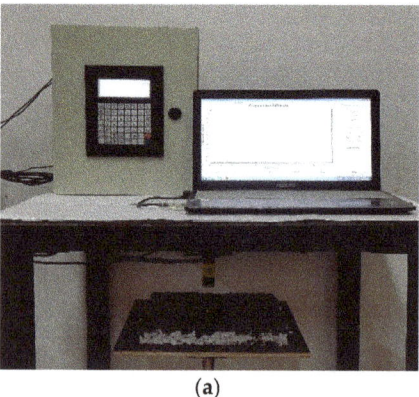

(**a**)

**Figure 3.** *Cont.*

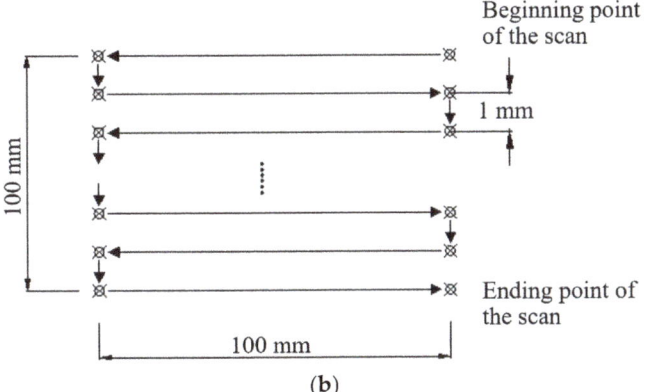

**Figure 3.** Laser scanning of asphalt pavement profile. (**a**) Basic configuration of laser scanner; (**b**) scanning path; (**c**) scanning region of texture.

The laser displacement sensor moves continuously along the path in Figure 3b, according to the programming of the control cabinet. The output data of the supporting software are continuous two-dimensional altitude data, and the continuous altitude data were, thereafter, divided into different contour lines. Since the standard scanning range of the laser scanning head is ±10 mm, the data were uniformly displayed as 12.5 mm. Therefore, the singular points should be deleted and interpolated prior to data processing. Following that, in order to differentiate the contour lines, the transverse length of the scanning path was extended to get the significant height difference of the altitude, and the boundary of the contour line could be easily verified according to the data mutation. Moreover, altitude data of the even-order contour lines were reverted around the central point and then arranged according to the relative positional relations of the contour lines. Thus, 100 contour altitude lines with an interval of 1 mm along the same direction were obtained.

To eliminate the uneven phenomena of the asphalt pavement surface originating from specimen compaction, linear fitting was conducted to each contour line. Moreover, 100 altitude data of each contour line were arranged in order from small to high, and the absolute value of the median was set as the elevation zero point. Subsequently, the whole contour line was moved vertically to eliminate influences of average altitude differences among specimens. Following that, the processed texture data were input into the MATLAB software to reconstruct three-dimensional profile maps of the three types of asphalt pavement, namely, Asphalt Concrete (AC), Stone Matrix Asphalt (SMA), and Open Graded Friction Course (OGFC), as shown in Figure 4.

2.1.3. Modeling of Asphalt Pavement with Profile Textures

The collected three-dimensional coordinates of asphalt pavement were transformed into a three-dimensional grid model through Python programming to generate the INP file of pavement for further simulation analysis in ABAQUS.

In general, the reconstructed three-dimensional geometric models of asphalt pavement can be categorized as wireframe models, curved shell models, and solid models. Only the undulating state of pavement surface is needed in the modeling of curved shell; that is, the whole rigid surface can be constituted by a series of triangular or quadrilateral rigid slices so as to realize the accurate characterization of road surface texture. Hence, this type of shell model is adopted in this research to build the 3D model of asphalt pavement with the following steps.

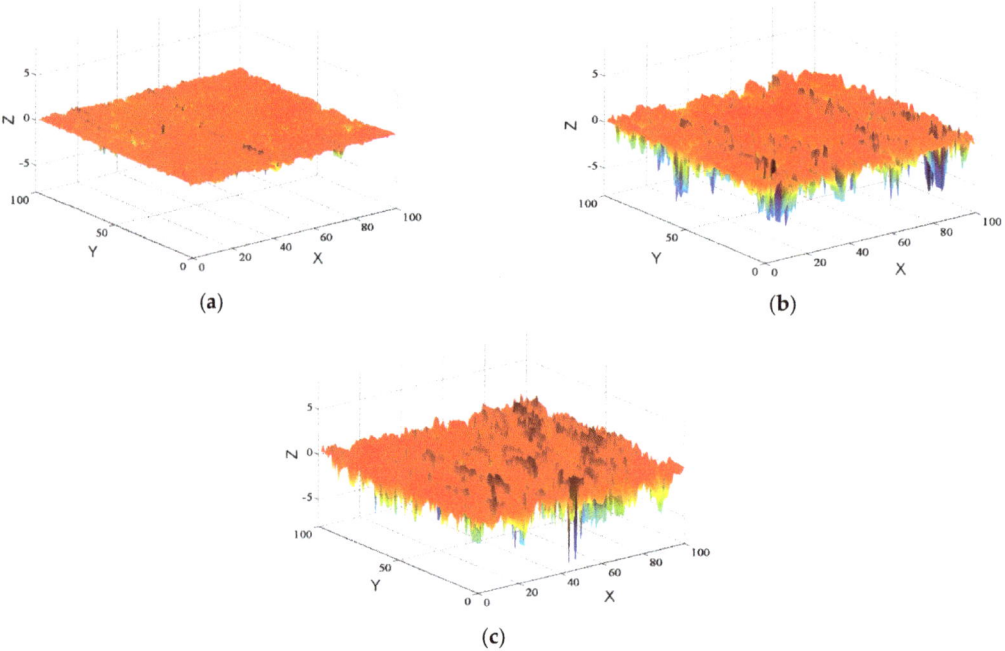

**Figure 4.** Three-dimensional profile maps of different types of asphalt pavement (Unit: mm). (**a**) AC pavement; (**b**) SMA pavement; (**c**) OGFC pavement.

(1) Pavement filtering and smoothing

Considering that a large number of sharp angles can be generated from the curved shell modeling based on the data of pavement texture, significant stress concentration often occurs in the areas that contact those sharp peaks of the pavement surface. This phenomenon always leads to the penetration of the road model to the tire model exceeding the tolerance range, thus triggering the termination of model calculation. In addition, the tire–pavement contact still cannot converge if the meshing of the pavement model is too fine. As a result, it was necessary to implement pavement filtering and smoothing processing first to eliminate sharp points on the pavement surface. Next, the pavement was amplified to some extent along the plane *xoy* to make the grid size of the pavement larger than that of the tire. Furthermore, mean filtering and Grid data function were applied in combination to conduct the smoothing process to the texture data, as illustrated in Figure 5.

(2) Extension and amplification of pavement

Compared with the dimensions of the tire model, the scanning region of the asphalt pavement was too small to develop normal contact with the tire. Due to the self-similarity and self-affinity of asphalt pavement, the pavement model could be expanded both in longitudinal and lateral directions. To ensure a smooth transition of altitude between the original pavement and the expanded pavement, the altitude data of the original pavement were expanded asymmetrically along the pavement boundary. Subsequently, the original pavement was expanded by three times along the longitudinal and transverse directions, respectively. Meanwhile, local coordinate systems were defined through the keyword of *SYSTEM, and the center of the expanded pavement was moved to the place below the tire.

**Figure 5.** Grid data diagram of pavement profile in 150 × 150 area.

The dimensions of the pavement grid model after extension were 300 mm × 300 mm, which covered 150 × 150 altitude data points, with 2 mm precision. In contrast, the size of the texture unit of the tire was about 2.5 mm. Since the grid size of the pavement was smaller than the grid size of the tire, the tire–pavement contact analysis could not reach convergence. Hence, it was necessary to expand the pavement coordinates by 1.5 times along $x$, $y$, and $z$ directions under the premise of an unchanged number of altitude data points. Therefore, the size of the pavement grid was set to 3 mm, and the whole size of the pavement model became 450 mm × 450 mm, as shown in Figure 6.

(3) Generation of the grid model

Altitude data in Microsoft EXCEL were extracted by Python language programming and then converted into 3D coordinates. Four adjacent nodes formed a quadrilateral element, and corresponding relations between node number and element number were established, as presented in Figure 7. All nodes and elements were written into the keyword *NODE of nodes in turn, and the keyword *ELEMENT for element was defined. The element type was set as R3D4. Finally, the INP format file of the finite element model of pavement was generated, which contained 22,500 nodes and 22,200 elements. The finite element grid models of a 100 mm × 100 mm pavement after mean filtering and smoothing before extension and amplification are shown in Figure 8.

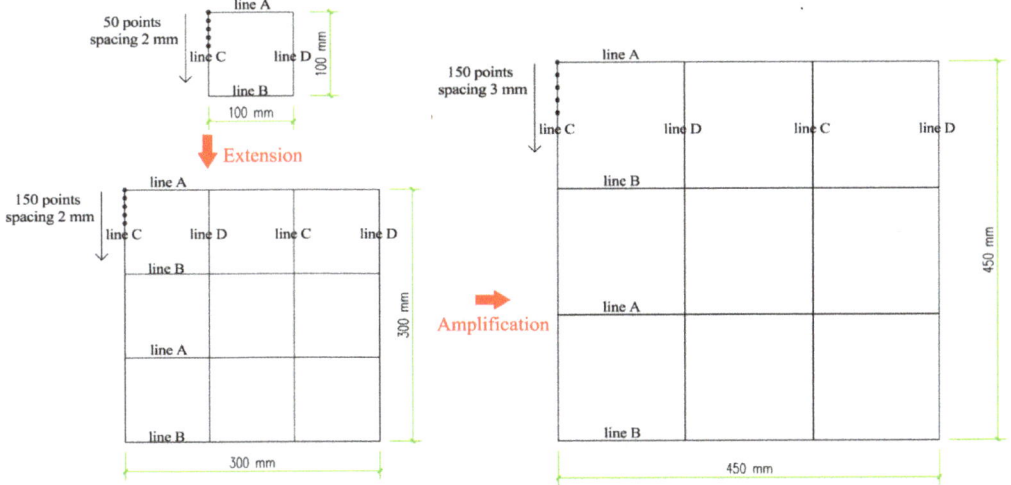

**Figure 6.** Pavement grid setting after the extension and the amplification processing.

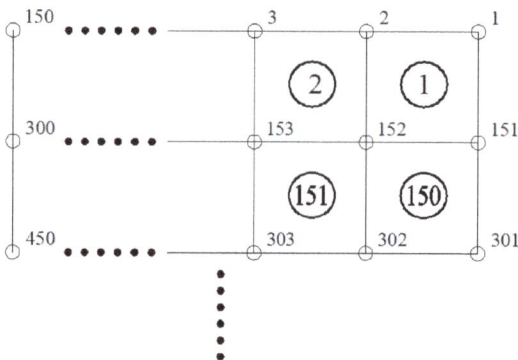

**Figure 7.** Schematic diagram of node number and element number.

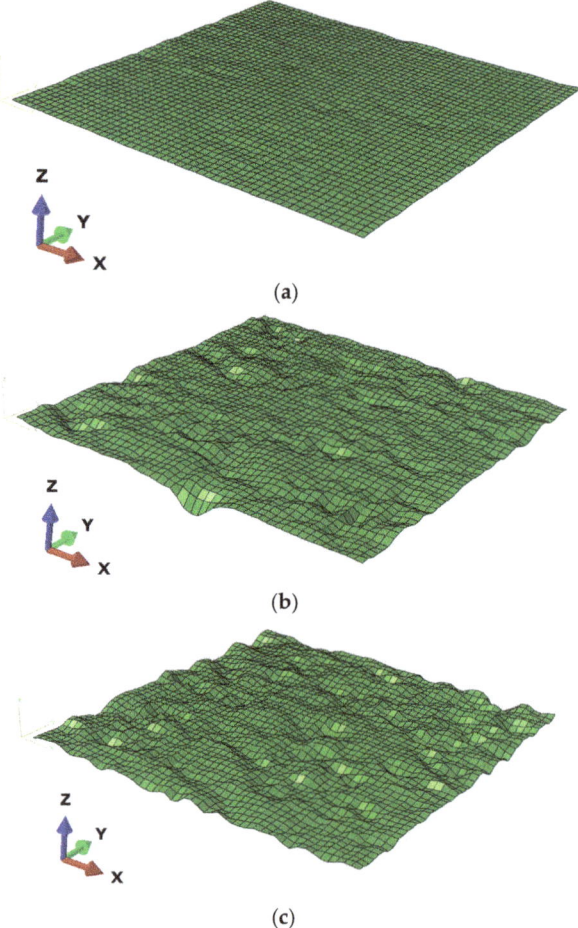

**Figure 8.** Finite element grid models of asphalt pavements. (**a**) AC pavement; (**b**) SMA pavement; (**c**) OGFC pavement.

## 2.2. Steady Rolling Analysis of the Radial Tire

### 2.2.1. Modeling of Tire–Pavement Frictional Contact

(1) Definition of rolling speed of the tire

Since the rolling speed of the tire cannot be defined directly in steady-state rolling analysis, it is necessary to determine the rolling angular velocity and linear velocity of the tire through the keywords *TRANSPORT VELOCITY and *MOTION. Specifically, *TRANSPORT VELOCITY is used to define the angular velocity of tire material running through the model grid. *MOTION is used to define the kinematic velocity of the moving reference coordinates, which is the translational velocity of the tire. As the tire rolls freely, the linear velocity of the tire (V) is just equal to the product of the rolling angular velocity ($\omega_0$) and rolling radius (r). In addition, when the rolling angular velocity of the tire ($\omega$) is smaller than $\omega_0$, the tire is in the braking state. By contrast, while $\omega > \omega_0$, the tire is in the driving state. Hence, the rolling contact analyses of the tire under braking, free rolling, and driving states can be realized by defining $\omega$ and V. It should be noted that the rolling radius of the tire is not the radius of the tire after inflation. In other words, static trace analysis is needed before the rolling analysis. Following that, the rolling radius of the tire can be calculated according to the inflating radius of the tire and the radial deformation under the corresponding load. More details about the establishment of the tire–pavement frictional contact simulation model, such as the tire modeling and the definition of tire–pavement contact, can be found at the reference [19].

### 2.2.2. Influencing Factors of Anti-Skid Performance of Asphalt Pavements

(1) Effects of pavement texture on the frictional force of pavement

The tire pressure was set to 2.5 bar, and a load of 3900 N was applied onto the tire. On this basis, the rolling radii of the tire on AC, SMA, and OGFC were calculated to be 297.91 mm, 297.77 mm, and 297.21 mm, respectively. The steady state rolling analysis when the velocity and slip rate of the tire was 30 km/h and 20% was carried out by defining the rotating angular velocity of the tire and the pavement-to-tire average velocity. The keywords *TRANSPORT VELOCITY valued 22.4 rad/s and *MOTION,TYPE = VELOCITY valued 8.3334 m/s.

Figure 9 shows the distribution of the shear force on the rolling direction of the tire on different types of pavements, namely, AC, SMA, and OGFC. The shear force on the tire–pavement interface shows significant differences among the different types of pavement. Specifically, the maximum frictional force that the OGFC provides to the tire element is higher than those of SMA and AC. The frictional shear force of AC pavement is the smallest, at nearly 10 times lower than that of OGFC. Additionally, regions where the tire generates frictional force are closely related to texture features of asphalt pavements. When the pavement texture is relatively flat, frictional force shows a relatively uniform distribution, and it is generated in most regions at the tire surface and tire shoulder in the contact area. Moreover, frictional force at the area with protruding pavement texture is relatively high, indicating that frictional force when the tire rolls on asphalt pavements is mainly provided by protruding textures of the asphalt surface. Furthermore, as pavement texture becomes relatively rough, frictional forces usually show the trend of the dispersing distribution, as illustrated in Figure 9b,c. Since a few coarse aggregates in OGFC are too protruding, frictional forces of the tire at these nodes are negative, thus resulting in the direction of friction force being opposite to the direction of resultant friction force in these areas.

(2) Effects of tire pressure on friction coefficient

In the process of vehicle braking, controlling the slip rate of the tire is conducive to increasing the braking efficiency. The slip rate of the vehicle tire is generally controlled in the range between 15% and 20% to avoid locking of tires during emergency braking and thereby increase braking efficiency. In this study, the variation laws of friction coefficient with tire pressure and loads under free rolling and braking conditions were analyzed by defining the rotating angular velocity of the tire and translational velocity of the pavement.

The testing conditions in this simulation were set as follows: tire pressure was set in four ranges, that is, 1.5 bar, 2.0 bar, 2.5 bar, and 3.0 bar, and the vertical loads and translational velocity of pavement were 3900 N and 30 km/h, accompanied by a constant 20% slip ratio. Furthermore, the shear force and vertical reaction force on the tire were output through the keyword *CONTACT PRINT from the DAT file, thus obtaining the friction coefficient of the tire under different motion states.

It can be seen from Figure 10 that under the free rolling state, all the friction coefficients of the tire on the three types of pavement decrease to some extent with the increase of tire pressure. Additionally, the SMA and OGFC show a significantly greater decreasing trend than the AC, which can be attributed to the rapid reduction of the actual contact area between the tire and the pavement with a rough profile. By contrast, the effects of tire pressure on the friction coefficient under the braking state at a slip rate of 20% are opposite those under the free rolling state. In other words, the friction coefficients of the tire on all three pavements increase with the increase of tire pressure. Moreover, it is found that the friction coefficients of the OGFC and SMA are significantly higher than that of AC pavement under braking conditions. This also sheds light on the fact that pavement texture influences anti-skid performance of pavement significantly.

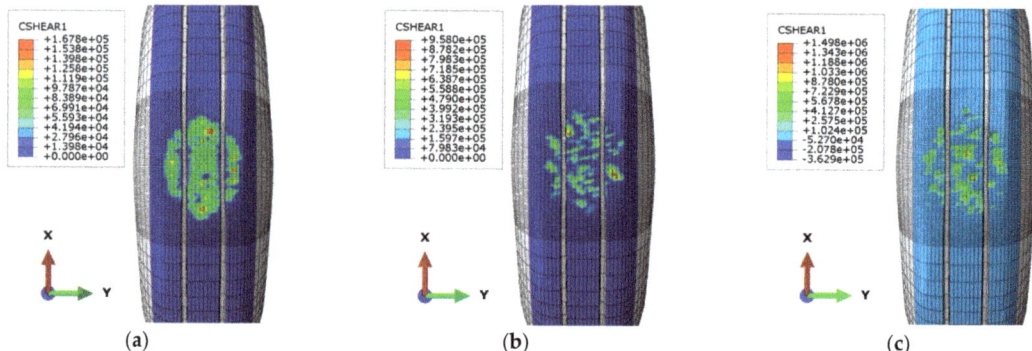

**Figure 9.** Cloud chart of shear force distribution on different pavements. (**a**) AC Pavement (**b**) SMA Pavement (**c**) OGFC Pavement.

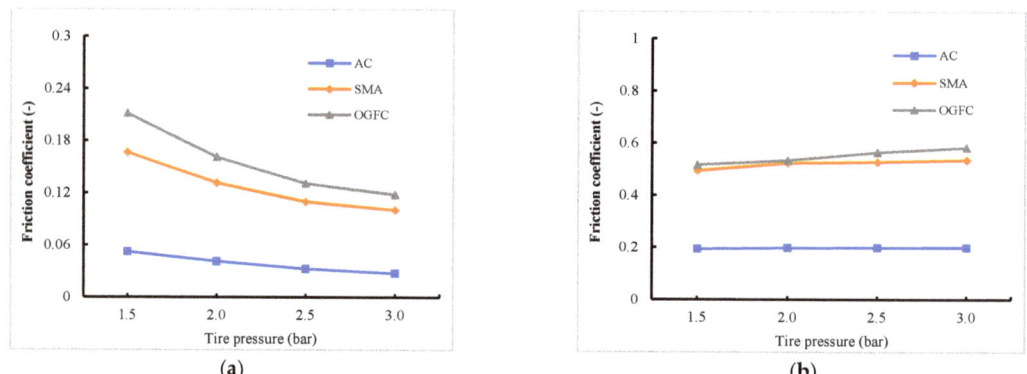

**Figure 10.** Variation laws of friction coefficient with tire pressure. (**a**) Tire free rolling state; (**b**) tire braking state.

(3) Effects of loads on the friction coefficient

In this analyzing phase, the simulation conditions were set as follows. Tire pressure and the translational speed of the road relative to the tire were set at 2.5 bar and 30 km/h. Different loads of 1700 N, 2200 N, 2700 N, and 3200 N were applied onto the tire under the free rolling state and braking state with a slip rate of 20%. The dynamic friction coefficients of the tire on different pavements were obtained.

In Figure 11, the variation trend of friction coefficients of the tire on three pavements, both under the free rolling state and under the braking state with a slip rate of 20%, shows positive trends related to the rising loads. In other words, given a certain range of loads, the tire–pavement contact is better with the increase of loads, and, furthermore, the rising trend of the frictional force of the tire provided by pavement is higher than that of the vertical reaction force. In addition, it can be seen from a comparison of the friction coefficients of different pavements that OGFC and SMA have similar friction coefficients, while AC shows a relatively lower friction coefficient, especially under the braking state. Such a difference is mainly due to the different degrees of surface roughness and the corresponding change of actual contact area.

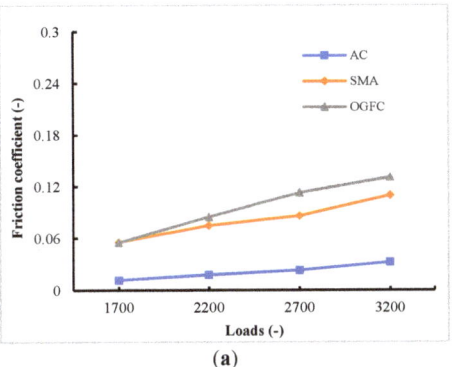
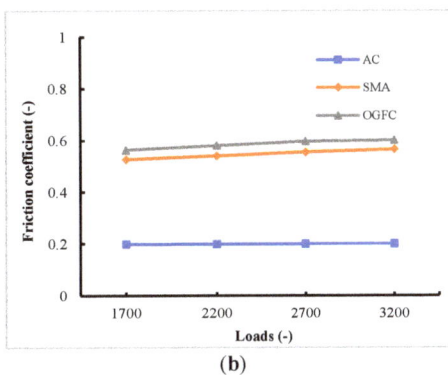

Figure 11. Variation laws of friction coefficients with loads. (a) Tire free rolling state; (b) tire braking state.

### 2.3. Finite Element Modeling of the Tire Aquaplane

#### 2.3.1. Principle of CEL Method

In ABAQUS, the fluid-solid coupling issue is mainly solved through the fluid-solid coupling algorithm (CEL), smoothed particle hydrodynamics (SPH), and computational fluid dynamics (CFD). The Euler-Language coupling (CEL) method is adopted to solve the tire aquaplane problem. In CEL, the tire and pavement are defined by Lagrange elements, while the water body is defined by Euler elements, as presented in Figure 12. CEL method combines advantages of the Lagrange method and Euler method. It not only solves the great deformation of fluid grids, but also identifies coupling boundaries effectively. During modeling, the tire grids and fluid grids overlapped to some extent. On fluid free boundaries, pressure on the fluid surface was applied onto the tire surface as intensity of pressure. Meanwhile, the volume fraction and velocity boundary of the fluid grids were updated according to the displacement and velocity of the tire on the coupling surface.

#### 2.3.2. Euler Grid Modeling Based on Flow Model

(1) Selection of tire rolling model and flow model

The tire aquaplane model can be categorized as "tire rolling model" and "flow model" according to different selections of reference coordinate systems. It can be seen from Figure 13 that the tire rolling model constrains all degrees of freedom (DOFs) of the pavement model, and it applies the rolling velocity and translational velocity onto the tire to make it roll on the water film model. The flow model constrains the DOF of the tire in the translational direction, and it applies rotating angular velocity onto the tire only. Meanwhile, translational velocities opposite the advancing direction of the tire were applied onto the pavement model and flow model simultaneously to simulate a tire driving through the water film on pavements. The "tire rolling model" and "flow model" have been compared by some scholars [16]. It is found that the tire rolling model has advantages in simulating flow traces, and it can simulate water traces completely. However, the quantity scale of grids can be decreased significantly during modeling and thereby increases the solving efficiency. Since the tire rolling model and flow model show almost equal calculation accuracies and the flow trace analysis is not the research focus of this paper, the flow model was adopted for the modeling analysis of the tire aquaplane.

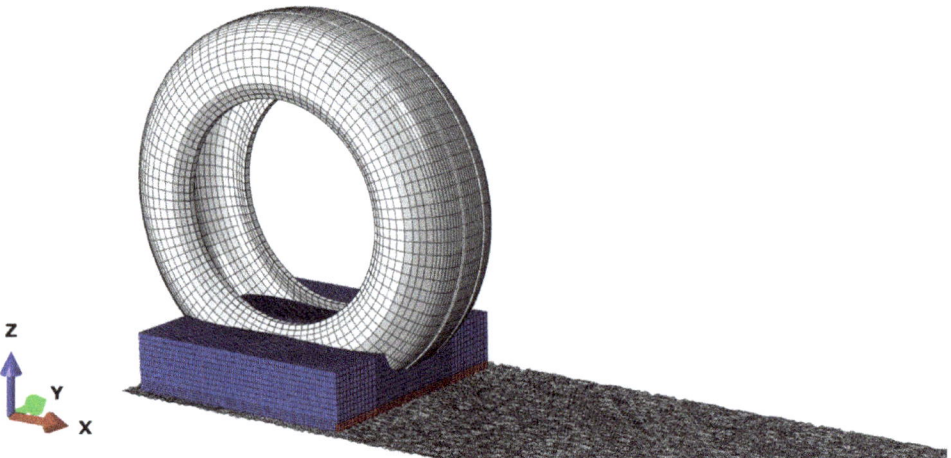

**Figure 12.** Tire aquaplane model on asphalt pavements.

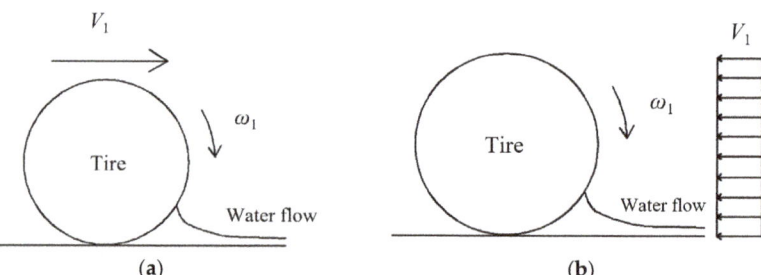

**Figure 13.** Schematic diagram of different modeling techniques. (**a**) Tire rolling model; (**b**) flow model.

(2) Flow state equation

In this study, the relations of fluid pressure with density and internal energy of fluid were described by the Mie–Gruneisen state equation. It is assumed that the fluid pressure (p) is linear with the internal energy per unit mass ($E_m$):

$$p = f(\rho) + g(\rho) E_m \quad (1)$$

where $f(\rho)$ and $g(\rho)$ are determined by the selected state equation. In the Mie–Gruneisen state equation, values of $f(\rho)$ and $g(\rho)$ are determined as follows:

$$f(\rho) = \frac{\rho_0 c_0^2 \eta}{(1-s\eta)^2}\left(1 - \frac{\Gamma_0 \eta}{2}\right) \qquad (2)$$

$$g(\rho) = \Gamma_0 \rho_0 \qquad (3)$$

where $\eta = 1 - \rho_0/\rho$, $\rho_0$ is the initial density of the water body, and $\rho$ is the density of water bodies after impacts. $c_0$ is the propagation velocity of acoustic waves in water, and s is a material constant. $c_0$ and s determine the linear relationship between the impact speed of water flow ($U_s$) and particle velocity of water flow ($U_p$): $U_s = c_0 + sU_p$. $\Gamma_0$ is the material constant. The flow parameters in this study are listed in Table 1.

**Table 1.** Parameters of flow materials.

| Initial Density (ton/mm$^3$) | Dynamic Viscosity (N·s/mm$^2$) | $c_0$ (mm/s) | s | $\Gamma_0$ |
|---|---|---|---|---|
| $1.0 \times 10^{-9}$ | $8.9 \times 10^{-10}$ | $1.5 \times 10^6$ | 0 | 0 |

(3) Construction of flow grid models

The Euler grid model established in this study was divided into air zone and flow zone. The overall dimensions were 400 mm in $x$ direction, 400 mm in $y$ direction, and 100 mm in $z$ direction, as shown in Figure 14a. To set boundary conditions of the Euler grid model, the flow zone was divided into side, bottom, and inlet of water pump, as shown in Figure 14b,c. The Inlet dimension was 4 mm in $x$ direction and 400 mm in $y$ direction. Size in the $z$ direction the dimension was determined according to the preset thickness of the water film in the model. Furthermore, regions overlapping the tire grid that might be flowed through by water were refined, and the minimum side of the minimum grid was 4 mm. The Euler elements EC3D8R were used as the flow model. The whole flow model contained 110,594 nodes and 100,799 elements.

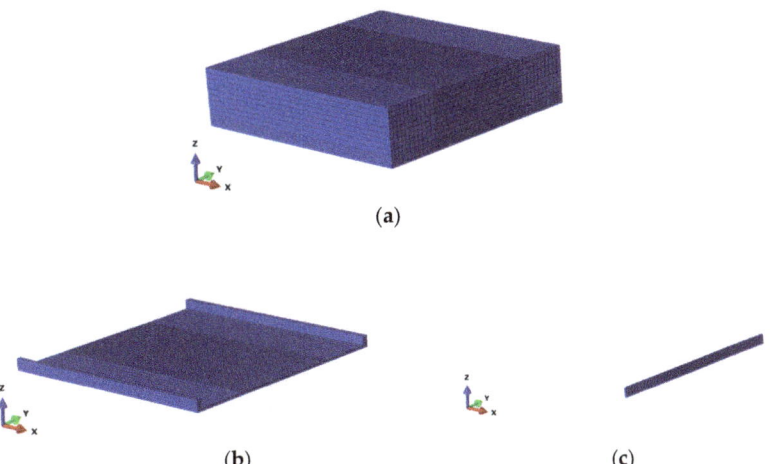

**Figure 14.** Fluid finite element model. (**a**) Euler grid model; (**b**) side and bottom of flow; (**c**) inlet of water pump (inlet).

2.3.3. Tire Aquaplane Analysis Based on ABAQUS/Standard and Explicit

To reduce the calculation expense in tire aquaplane analysis, an implicit expression was carried out first by using the Standard solver. Next, a dynamic explicit analysis was conducted. Steady state calculation results were transmitted to the explicit analysis by the keyword *IMPORT, and finally, kinetic analysis in the tire aquaplane process was performed through the Explicit solver.

(1) Implicit analysis

First, a complete 3D radial tire was built in the implicit analysis. The pavement grid files were input into the tire model through the keyword *INCLUDE, and the tire–pavement contact properties were defined. Static analysis and steady state rolling analysis of the tire on asphalt pavements were carried out. During steady state rolling analysis of the tire, only the meshing of the tire–ground contact area was refined to increase computational efficiency. However, the uniform meshing of the tire is needed in the explicit analysis. Hence, one grid was generated every 4° during construction of a complete 3D radial tire model using the keyword *SYMMETRIC MODEL GENERATION, and a total of 90 tire grids were generated. Since the tire did not achieve real rolling and only materials flow in grids during steady state rolling analysis, a 450 mm × 450 mm asphalt pavement model was constructed. However, the real relative movements were made by the tire and pavement in the explicit analysis. Considering that the asphalt pavements previously built for Standard analysis were too short in size, they had to be extended to 1800 mm × 450 mm, as shown in Figure 15.

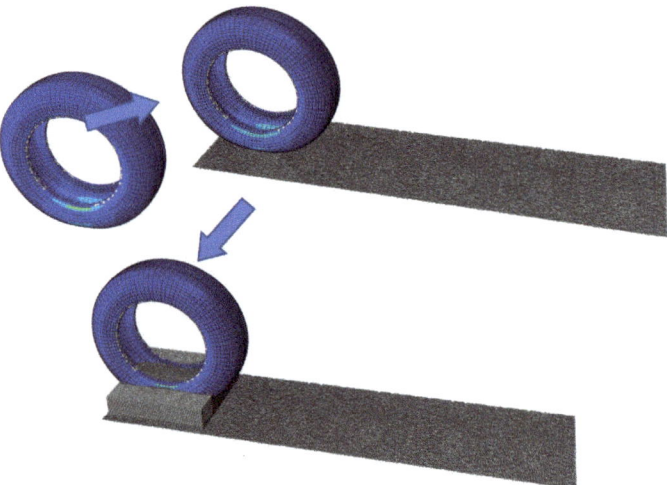

**Figure 15.** Modeling process of tire aquaplane.

(2) Explicit analysis

In the explicit analysis, calculation results and pavement grid files of steady state rolling analysis were input from Standard analysis. The concentrated mass and rotational inertia as well as initial velocity of the tire were defined. Meanwhile, the boundary conditions and amplitude of loads were defined by the keyword *AMPLITUDE. Later, Euler grid files were input through the keyword *INCLUDE, and material properties of fluid were defined. Additionally, a vertically downward gravitational field was applied to the whole Euler model. When defining the initial state of the fluid, the volume fraction of element set INLET of the water pump was set at 1, while the volume fraction of other regions was set at 0. In other words, only grids of INLET were filled with water completely, whereas other Euler grids were filled with air, as presented in Figure 16, where the red zone is fluid and the blue zone is air.

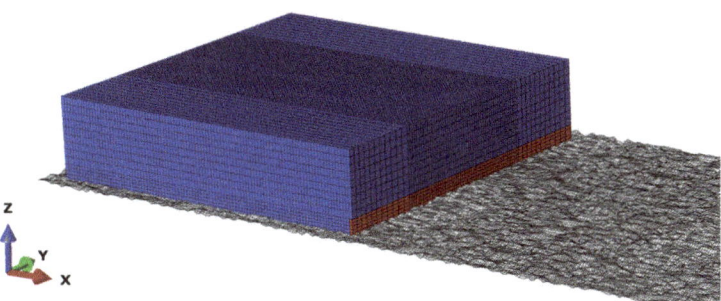

**Figure 16.** Initial state of tire aquaplane.

In the explicit analysis, the contact relation is defined by the universal contact algorithm. The tire–water contact and tire–pavement contact were built by the keyword *CONTACT INCLUSIONS. Later, tire–water contact constraint and water–pavement contact constraint were eliminated using the keyword *CONTACT EXCLUSIONS. In this way, the universal contact properties between the tire and pavement were defined. Subsequently, the contact force of the tire surface which was provided by the tire was output in the historical output through the keyword *CONTACT OUTPUT.

2.3.4. Validity Verification of Tire Aquaplane Finite Element Model

(1) Flow trace verification

In this simulation phase, first, the tire pressure and thickness of water film were set at 2.5 bar and 12 mm, and a vertical load of 3900 N was applied onto the tire. Next, the tire rolled on OGFC was defined at a constant velocity of 70 km/h. The time of analytical step was determined to be 0.03 s. As the water flow impacted the tire, the flow distribution pattern in the Euler grids was disclosed through the field output variable EVF, as presented in Figure 17.

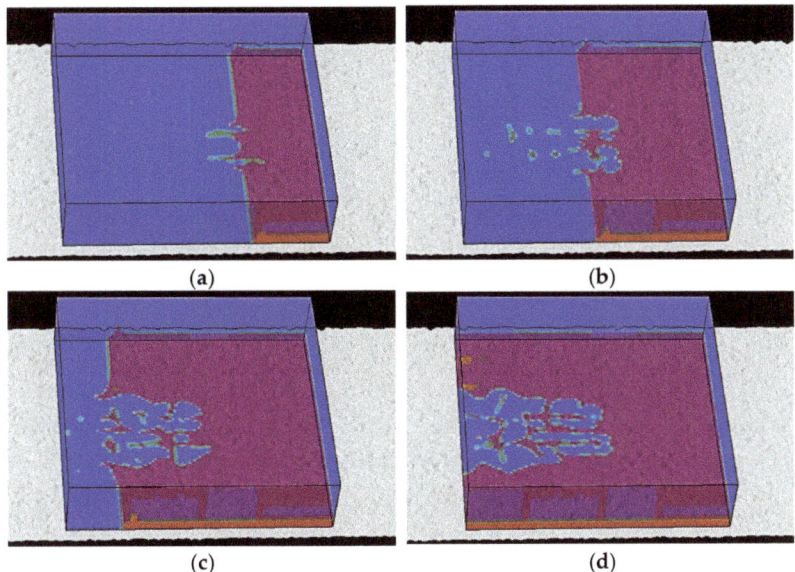

**Figure 17.** Flow trace in the tire aquaplane process. (**a**) t = 0.0060 s; (**b**) t = 0.0105 s; (**c**) t = 0.0165 s; (**d**) t = 0.0225 s.

Figure 17a presents the flow trace distribution as the time history was 0.0060 s. At this time, the tire began to contact the water flow, and the water first flew into longitudinal grooves on the tire surface. The analysis time (t) in Figure 17b was 0.0105 s, when there was no complete contact between the tire and pavement due to the support force of water flow. Specifically, the tire shoulder was in relatively full contact with the pavement, whereas there were some flows between the tire tread and pavement. In other words, loads on the tire were mainly undertaken by the tire shoulder. As the time duration reached 0.0165 s, it can be seen from Figure 17c that the tire surface began to contact the pavement gradually. The longitudinal grooves on the tire surface could not discharge the water between the tire and pavement anymore. Furthermore, the water flow in the longitudinal texture of the tire was dispersed due to the uneven pavement surface. The time duration in Figure 17d is 0.0225 s. In this phase, water flows occupied the whole Euler grid bottom and furthermore, began to splash behind the tire by inertia after tire rolling. Water flows generated the maximum support force upon the squeezing effect, which was kept unchanged. Hence, the aquaplane model in this study could display flow trace distributions in each stage of the tire aquaplane process, which was consistent with practical tire aquaplane conditions.

(2) Vertical contact force

Two tire velocities were set, while other boundary conditions were fixed. The variations of the vertical contact force between the tire and pavement under different tire speeds with time in the aquaplane process were output by defining the historical output, with the fixed boundary conditions, as shown in Figure 18. Before contact with water flows, the tire–pavement contact force fluctuated around 3900 N, which was determined by the damping structure of the tire. With the continuous contact between the tire and water, the lifting effect of the water on the tire led to the gradual decrease of the tire–road contact force. After the tire entered into the wet asphalt pavements, it reached the stress balance state. As the velocity reached 70 km/h, the tire–pavement vertical contact force reached the balance at about 0.02 s, and, finally, it was stabilized at about 1000 N. When the velocity was 150 km/h, the vertical contact force reached the balance at about 0.015 s, and, finally, it approached 0, indicating that the tire was lifted up completely by water flows. In other words, a complete aquaplane phenomenon occurred.

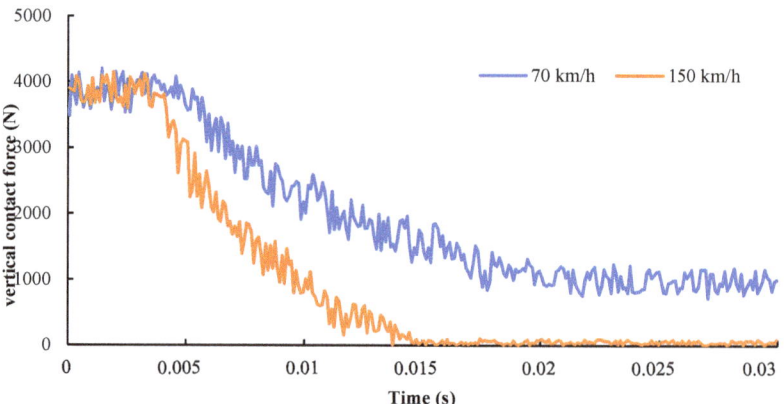

Figure 18. Vertical contact force in the tire aquaplane process.

## 3. Results and Analysis

Existing studies on tire aquaplane mainly concentrate on complete aquaplane. Experimental studies and numerical models have been conducted by scholars to solve the critical aquaplane velocity and the affecting factors under different conditions [20–22]. Nevertheless, drivers are often more cautious when driving on wet pavements. In other words, the vehicle is usually driven at a low speed, and the tires are in a partial aquaplane state [23,24], which also may often trigger low adhesive force and traffic accidents. Since the tire–pavement friction behaviors under the partial aquaplane state are closely related with vertical contact force, the effects of pavement types, tire speed, and water film thickness on the vertical contact force of the pavement are investigated as follows.

## 3.1. Effects of Asphalt Pavement Types on Partial Aquaplane Performanc

With the same testing conditions applied in the above analysis, the variation trends of vertical contact force from three types of pavement, namely, AC, SMA, and OGFC, are presented in Figure 19, based on history output. All the simulations in this phase were conducted under the same tire speed, 70 km/h.

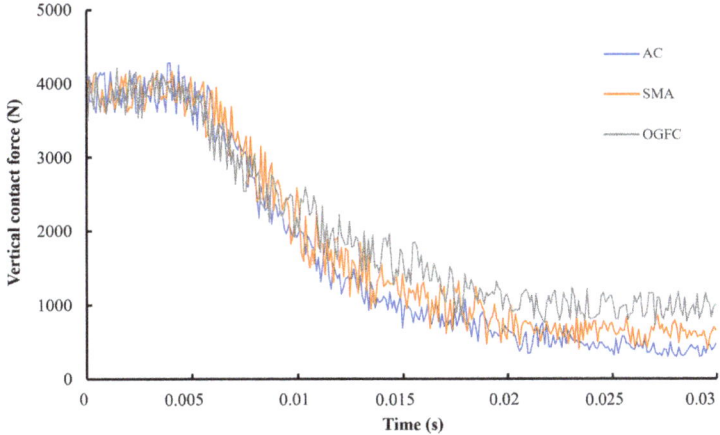

**Figure 19.** Effects of pavement type on vertical contact force.

According to the variation trend of vertical contact force, there was little difference of vertical contact force among the pavements before the tire–flow contact. However, vertical contact force decreased sharply as the tire entered the wet pavements. Specifically, vertical contact force on AC declined the most, followed by that of SMA and OGFC, which were relatively stable. As the tire reached the equilibrium state of force, the difference of vertical contact force became more significant. OGFC provides the tire with distinct vertical contact force, which is superior to SMA and AC. This difference can be attributed to water discharge capacity, which benefits from the rough surface supplied by sufficient textures, especially on OGFC and SMA pavement.

## 3.2. Effects of Tire Rolling Velocity on Partial Aquaplane Performance

The tire pressure, loads, and thickness of water film were set at 2.5 bar, 3900 N, and 12 mm, respectively. With all other parameters being constant, the tire speed was set at 70 km/h, 90 km/h, 110 km/h, and 130 km/h, for the investigation of the effect of tire rolling velocity on partial aquaplane properties. Based on historical output, the variation trends of tire–pavement vertical contact force on AC, SMA, and OGFC in the tire partial aquaplane process with velocity are presented in Figure 20.

According to variation trends of vertical contact force with velocity, the vertical contact force declined significantly after the tire made contact with water flows. When the tire speed was 70 km/h, the rolling tire had not contacted water flows yet at 0.005 s, and the vertical contact force was equivalent to the applied loads. When the velocity reached 130 km/h, the aquaplane phenomenon occurred, and the vertical contact force was about 3000 N, which was about 25% lower than that under drying conditions. Furthermore, it can be found that along with the increase of tire speed, all the pavements showed a decreasing trend. OGFC could provide the tire with greater force on the vertical contact direction, followed by SMA and AC. In other words, the water-drainage properties and pavement texture could account for the different anti-skid performance of pavement. Additionally, as the contact duration reached 0.02 s, it also can be found that the tire had basically reached a balanced state of force. When the velocity increased from 70 km/h to 130 km/h, the ultimate tire–pavement vertical contact force decreased by 40~60%.

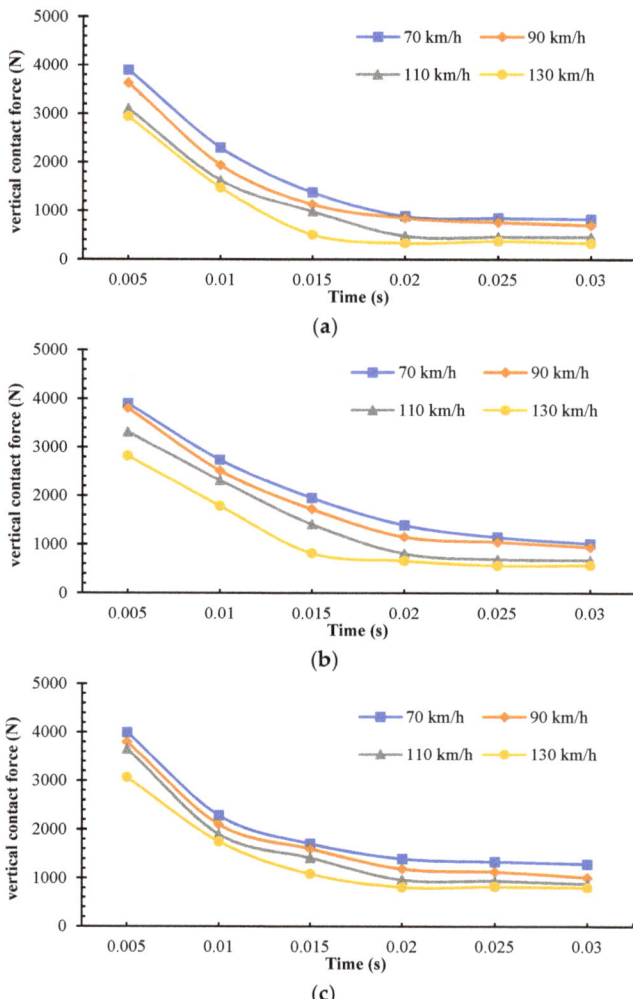

**Figure 20.** Effects of tire rolling velocity on vertical contact force. (**a**) AC pavement; (**b**) SMA pavement; (**c**) OGFC pavement.

## 3.3. Effects of the Thickness of Water Film on Partial Aquaplane Performance

With the other constant simulation parameters, the tire speed was set at 70 km/h for the investigation of the effect of water film thickness, that is, 4 mm, 8 mm, and 12 mm, on the vertical contact force under partial aquaplane circumstances. Based on historical output, the variation trends of tire–pavement vertical contact force on AC, SMA, and OGFC in partial aquaplane process with thickness of water film were obtained, as shown in Figure 21.

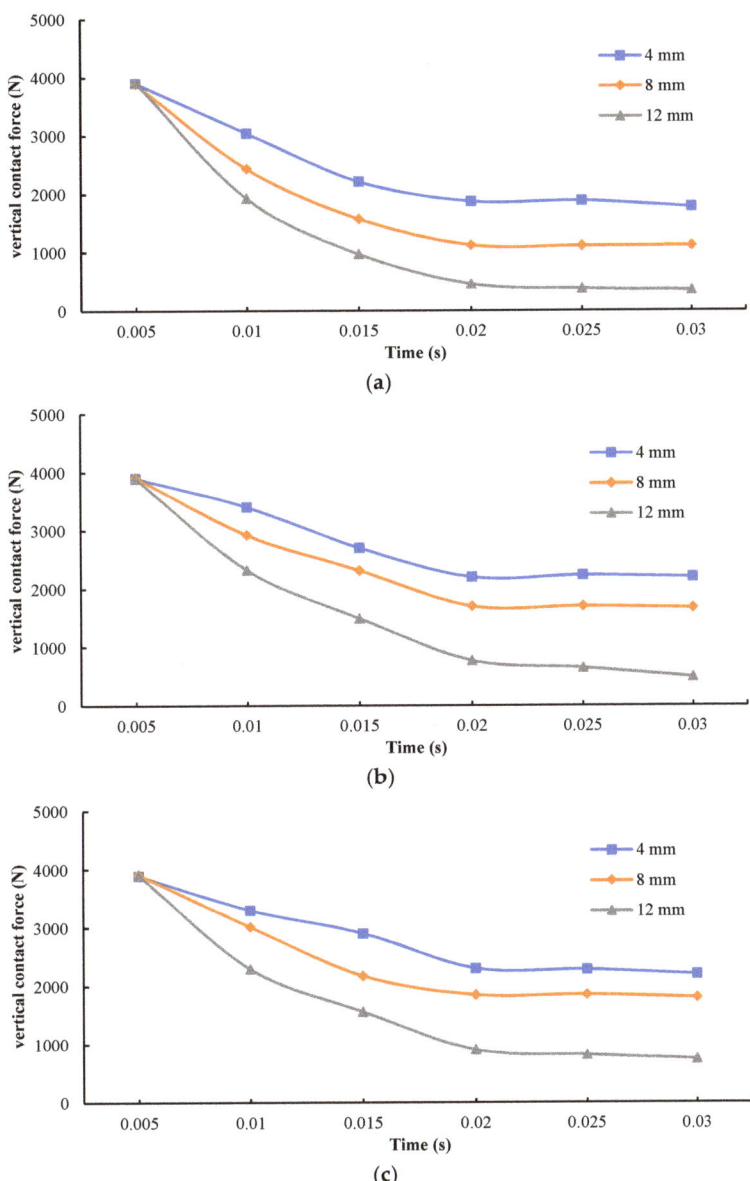

**Figure 21.** Effects of the thickness of water film on vertical contact force. (**a**) AC pavement; (**b**) SMA pavement; (**c**) OGFC pavement.

It can be seen from Figure 21 that while the thickness of the water film was 4 mm, the decreasing trends of vertical contact forces on the pavements were relatively stable. Furthermore, the ultimate vertical contact force decreased to about 50% of the applied loads onto the tire. With the increase of thickness of the water film, the decreasing trends of vertical contact force became obvious. As the thickness of water film reached 12 mm, the ultimate vertical contact force declined significantly, accounting for only about 15% of the applied vertical loads. In addition, while in the given thickness of water film, the vertical contact force of OGFC was still superior to SMA and AC, which is still due to the difference of pavement textures.

## 4. Conclusions and Future Research

In this paper, steady state rolling analyses of the tire under free rolling state and braking state with a slip rate of 20% were conducted based on the finite element analyzing method in ABAQUS. Influences of pavement texture, tire pressure, and loads on tire–pavement friction behaviors were briefly discussed under drying conditions. Next, a tire–texture pavement contact model was built. The calculation results of steady state analysis were input into the explicit analysis by compiling INP files, and then the Euler grid model was constructed. Material parameters of water were defined, and the aquaplane finite element models of a radial tire on three types of asphalt pavement were constructed using the CEL method. Finally, variation laws of the tire–pavement vertical contact force with pavement type, tire rolling velocity, and thickness of water film under the partial aquaplane condition were analyzed by the built aquaplane model. Some conclusions were drawn as follows.

(1) Tire–pavement frictional force distribution is closely related to pavement texture characteristics. The frictional force distribution between the tire and AC, which has a flat surface, is relatively uniform, but the friction forces on SMA and OGFC, with relatively rough surfaces, are mainly concentrated in the protruding aggregate particles. This conclusion also provides evidence for the perception that rough surfaces usually tend to have greater friction than smooth ones.

(2) Under the free rolling state, the tire–pavement dynamic friction coefficient decreases with the increase of tire pressure. Specifically, the tire–pavement dynamic friction coefficients on OGFC and SMA decrease more than that on AC. Under the braking state, the tire–pavement dynamic friction coefficient is positively related with a tire pressure. Moreover, whether under a free rolling state or under a braking state, the friction coefficient always increases with the increase of loads.

(3) Under the partial aquaplane state, due to the influence of the support force provided by the water flow, the vertical contact force between the tire and the pavement is significantly reduced. Finally, the tire reaches the stress balance state under the collaborative effect of the lifting force of the water flow and support force of the pavement. Under equal conditions, the vertical contact force between the OGFC and tire is the highest when the tire reaches the stress balance state, followed by that between SMA and the tire. In contrast, the vertical contact force between AC and the tire is the smallest. Moreover, the tire–pavement vertical contact force decreases more significantly under the partial aquaplane state when the velocity of tire speed or thickness of the water film increases.

Based on the conclusions drawn above, FEA method, accompanied with laser scanning for the acquirement of pavement profile, seems to be a feasible method for analyzing tire–pavement friction on partial tire aquaplane conditions. Finally, a few future research topics are recommended as follows.

(1) Tire–pavement frictional characteristics on the wet pavement should be further investigated, in combination with the coupled factor of temperature variation.

(2) Friction coefficient threshold on tire partial aquaplane conditions should be considered with the vehicle crash rates, in order to facilitate the traffic safety research. Moreover, the suggested vehicle speed should also be given on various wet conditions.

**Author Contributions:** Conceptualization, Z.Y.; Data curation, M.Y., Y.K. and L.Y.; Funding acquisition, M.Y. and L.Y.; Methodology, M.Y., J.L. and L.K.; Project administration, M.Y.; Software, Y.K. and J.L.; Writing—original draft, Y.K.; Writing—review & editing, Z.Y. and L.Y. All authors have read and agreed to the published version of the manuscript.

**Funding:** This work is financially supported by Open Fund of Key Laboratory of Road Structure and Material of Transport Ministry, Chang'an University, the Fundamental Research Funds for the Central Universities, CHD (No. 300102210506), China Postdoctoral Science Foundation (No. 2018M633444), Shaanxi Province Postdoctoral Science Foundation (No. 2018BSHEDZZ123), General Project of Chongqing Natural Science Foundation (No. cstc2021jcyj-msxmX0554), National Natural Science Foundation of China (No. 51608085), and Key R & D Program of Guangxi (No. GUI KE AB20159036).

**Institutional Review Board Statement:** Not applicable.

**Informed Consent Statement:** Not applicable.

**Data Availability Statement:** Some or all data, models, or code that support the findings of this study are available from the corresponding author upon reasonable request.

**Conflicts of Interest:** The authors declare no conflict of interest.

# References

1. Yu, M.; Xiao, B.; You, Z.; Wu, G.; Li, X.; Ding, Y. Dynamic friction coefficient between tire and compacted asphalt mixtures using tire-pavement dynamic friction analyzer. *Constr. Build. Mater.* **2020**, *258*, 119492. [CrossRef]
2. Chowdhury, A.; Kassem, E.; Aldagari, S.; Masad, E. *Validation of Asphalt Mixture Pavement Skid Prediction Model and Development of Skid Prediction Model for Surface Treatments*; Report 0-6746-01-1; Texas Department of Transportation, Research and Technology Implementation Office: Austin, TX, USA, 2017.
3. Lu, J.; Pan, B.; Liu, Q.; Sun, M.; Liu, P.; Oeser, M. A novel noncontact method for the pavement skid resistance evaluation based on surface texture. *Tribol. Int.* **2022**, *165*, 107311. [CrossRef]
4. Saghafi, M.; Abdallah, I.N.; Nazarian, S. Practical Specimen Preparation and Testing Protocol for Evaluation of Friction Performance of Asphalt Pavement Aggregates with Three-Wheel Polishing Device. *J. Mater. Civ. Eng.* **2022**, *34*, 04021397. [CrossRef]
5. Shahriar, N.; Gerardo, W. Flintsch & Alejandra MedinaLinking roadway crashes and tire–pavement friction: A case study. *Int. J. Pavement Eng.* **2017**, *18*, 119–127.
6. Hofko, B.; Kugler, H.; Chankov, G.; Spielhofer, R. Correlating Field and Lab Measurements of Skid Resistance by Skiddometer and Wehner/Schulze Device. In Proceedings of the Annual Meeting of Transportation Research Board, Washington, DC, USA, 8–12 January 2017.
7. Arce, O.D.G.; Zhang, Z. Skid resistance deterioration model at the network level using Markov chains. *Int. J. Pavement Eng.* **2019**, *22*, 118–126. [CrossRef]
8. McCarthy, R.; Flintsch, G.; de León Izeppi, E. Impact of Skid Resistance on Dry and Wet Weather Crashes. *J. Transp. Eng. Part B Pavements.* **2021**, *147*, 04021029. [CrossRef]
9. Maia, R.S.; Costa, S.L.; Cunto, F.J.C.; Branco, V.T.F.C. Relating Weather Conditions, Drivers' Behavior, and Tire-Pavement Friction to the Analysis of Microscopic Simulated Vehicular Conflicts. *J. Transp. Eng. Part B Pavements* **2021**, *147*, 04021037. [CrossRef]
10. Liu, C.; Qian, Z.; Liao, Y.; Ren, H. A Comprehensive Life-Cycle Cost Analysis Approach Developed for Steel Bridge Deck Pavement Schemes. *Coatings* **2021**, *11*, 565. [CrossRef]
11. Tang, F.; Fu, X.; Cai, M.; Lu, Y.; Zhong, S. Investigation of the Factors Influencing the Crash Frequency in Expressway Tunnels: Considering Excess Zero Observations and Unobserved Heterogeneity. *IEEE Access* **2021**, *9*, 58549–58565. [CrossRef]
12. Ong, G.P.; Fwa, T. Modeling Skid Resistance of Commercial Trucks on Highways. *J. Transp. Eng.* **2010**, *7*, 510–517. [CrossRef]
13. Tang, T.; Anupam, K.; Kasbergen, C.; Scarpas, A.; Erkens, S. A finite element study of rain intensity on skid resistance for permeable asphalt concrete mixes. *Constr. Build. Mater.* **2019**, *220*, 464–475. [CrossRef]
14. Anupam, K.; Tang, T.; Kasbergen, C.; Scarpas, A.; Erkens, S. 3-D Thermomechanical Tire–Pavement Interaction Model for Evaluation of Pavement Skid Resistance. *Transp. Res. Rec.* **2021**, *2675*, 65–80. [CrossRef]
15. Feng, X. Research on simulation technology of tire hydroplaning Performance. In Proceedings of the 19th Annual Conference of Beijing Strength Society, Beijing Mechanics Association, Beijing, China; 2013; Volume 2.
16. Zhu, S. Numerical Simulation of Tire Skid Resistance Based on Pavement Macro Texture. Ph.D. Thesis, Southeast University, Nanjing, China, 2017.
17. Zhu, X.; Pang, Y.; Yang, J.; Zhao, H. Numerical analysis of hydroplaning behavior by using a tire-water-film-runway model. *Int. J. Pavement Eng.* **2020**, *23*, 784–800. [CrossRef]
18. Yu, M.; You, Z.; Wu, G.; Kong, L.; Liu, C.; Gao, J. Measurement and modeling of skid resistance of asphalt pavement: A review. *Constr. Build. Mater.* **2020**, *260*, 119878. [CrossRef]
19. Varveri, A.; Avgerinopoulos, S.; Kasbergen, C.; Scarpas, A.; Collop, A. The Influence of Air Void Content on Moisture Damage Susceptibility of Asphalt Mixtures: A Computational Study. In Proceedings of the Annual Meeting of Transportation Research Board, Washington, DC, USA, 12–16 January 2014.
20. Tang, T.; Anupam, K.; Kasbergen, C.; Kogbara, R.; Scarpas, A.; Masad, E. Finite Element Studies of Skid Resistance under Hot Weather Condition. *Transp. Res. Rec.* **2018**, *2672*, 382–394. [CrossRef]
21. Yu, M.; Wu, G.; Kong, L.; Tang, Y. Tire-Pavement Friction Characteristics with Elastic Properties of Asphalt Pavements. *Appl. Sci.* **2017**, *7*, 1123. [CrossRef]

22. Huang, X.; Liu, X.; Cao, Q.; Yan, T.; Zhu, S.; Zhou, X. Numerical simulation of tire partial hydroplaning on flood pavement. *J. Hunan Univ. (Nat. Sci.)* **2018**, *45*, 113–121.
23. Ji, T.; Huan, X.; Liu, Q. Part hydroplaning effect on pavement friction coefficient. *J. Transp. Eng.* **2003**, *3*, 10–12.
24. Yan, Z. Simulation Study of tire Braking Performance on Wet Roads. Bachelor's Thesis, Jilin University, Jilin, China, 2017.

Article

# Characterization of Rutting Damage by the Change of Air-Void Characteristics in the Asphalt Mixture Based on Two-Dimensional Image Analysis

Kang Zhao [1], Hailu Yang [1], Wentao Wang [1] and Linbing Wang [2,*]

[1] National Centre for Materials Service Safety, University of Science and Technology Beijing, Beijing 100083, China
[2] Sensing and Perception Lab, School of Environmental, Civil, Agricultural and Mechanical Engineering, University of Georgia, Athens, GA 30602, USA
\* Correspondence: linbing.wang@uga.edu

**Abstract:** In the process of the rutting test, the air-void characteristics in asphalt mixture specimens are a dynamic change process. It is of great significance to systematically study the correlation between the change of air-void characteristics and the depth of the rutting slab and establish a relationship with damage. In this paper, the air-void information of rutting specimen sections with different loading cycles (500, 1000, 1500, 2000, 2500, and 3000 times) is obtained by two-dimensional image technology. The dynamic change process of the micro characteristics of internal air voids of two graded asphalt mixtures (AC-13 and AC-16) under cyclic wheel load is analyzed, and it is used as an index to characterize the microstructure damage of the asphalt mixture. The results show that the variation of air-void distribution, air-void shape characteristics, and air-void fractal dimension with the loading process can well characterize the permanent deformation law of the rutting slab. The fractal dimension of the air void increases with the increase in load. It is a dynamic process in which the air-void content changes with crack initiation and propagation. After rutting deformation, the total air-void area and average air-void size of the sample increase, and the total air-void number decreases. Because microcracks are formed in the specimen after rutting damage, the aspect ratio of the air void increases, and the roundness value decreases.

**Keywords:** air-void characteristics; fractal dimension; crack initiation and propagation; rutting damage

## 1. Introduction

An asphalt mixture is a composite material composed of aggregates, air voids, mastic, and other multiphase media, which is generally considered a continuous medium. However, the anisotropy of the microstructure leads to complex mechanical states and complex interactions between the components of the asphalt mixture. Among these, the air-void characteristics significantly impact the mechanical properties of the asphalt mixture. We found that two specimens with the same gradation and same air-void content may have different air-void numbers and average air-void size distributions. Specimens with the same air-void content may be composed of air voids of different sizes according to a certain combination. Therefore, two specimens with the same air-void content may also have different failures under the same load conditions. Specimens with larger air voids are more prone to relatively early failure compared with specimens with smaller air-void sizes. This is because once the specimen is subjected to load conditions, the possibility of inducing larger strains increases. In turn, it may lead to internal structural instability, and the existing air voids may continue to expand in the specimen and connect (or merge with other air voids), resulting in serious damage [1]. The increase in air void will reduce the resistance of the asphalt mixture to pavement damage. This is because the air void cannot transfer the load, and the material becomes weaker due to the decrease in the effective area (The concentrated force of each area is higher). The air-void content of the asphalt mixture shall be within a reasonable range to avoid adverse damage such as rutting [2]. Therefore, the proper compaction degree is significant for the performance of asphalt materials. The compaction index is widely used in engineering practice to ensure good service performance of asphalt mixture [3,4].

In recent years, with the progress of technology and the improvement of equipment available, the digital imaging technology of asphalt concrete is an effective tool for evaluating the internal structure. Researchers have been able to use digital image technology to monitor the internal failure process of engineering materials and connect this failure with the measured strain. Internal failure in materials can be represented in many forms, such as specific voids [5], crack surfaces [6,7], and spacing between cracks [8,9]. Air voids play an important role in characterizing the performance of asphalt mixture, and their distribution is very important for determining the overall mechanical response of asphalt mixture [10–12]. Under load, existing air voids may merge, resulting in microcracks at the interface between aggregate and mastic. Microcracks continue to expand and grow under deformation, forming macro cracks and increasing air voids [13]. Xu G et al. [14] extracted the internal void structure characteristics of asphalt mixture through an X-ray computed tomography (CT) test and three-dimensional (3D) image reconstruction technology. The change of air-void distribution before and after the freeze–thaw test is analyzed to evaluate the structural evolution of materials under freeze–thaw cycles. Yang B et al. [15] studied the correlation between the performance of porous asphalt mixture and the three pore characteristic parameters obtained by the CT. It was found that the micro pore characteristic parameters had little effect on high temperature, humidity sensitivity, and cooling performance but had a strong correlation with spalling resistance, permeability, connectivity, noise reduction, and other properties. Zhang Z et al. [16] used the connective void content of three types of asphalt mixtures was employed to characterize the damage to the corresponding asphalt mixture sample under freeze–thaw cyclic loading. The variation of connective void content revealed the nonlinear characteristics of asphalt mixture damage accumulation. Kassem Emad et al. [17] used several mechanical tests (the overlay tester, Hamburg Wheel-Tracking Test (HWTT), and a repeated tensile test) to characterize the influence of air-void distributions on mechanical properties and response of asphalt mixtures. The results show that air voids play an important role in influencing the performance of asphalt mixtures. Xu H et al. [18] used a set of image analysis programs to extract the internal structural characteristics of asphalt mixture during freeze–thaw cycles. The evolution of the internal void structure of the asphalt mixture during the freeze–thaw cycle was evaluated. Hassan et al. [19] analyzed the air void and crack characteristics caused by stress and strain by using a two-dimensional (2D) image analysis method and took them as the damage index to characterize the microstructure damage of the asphalt mixture. Therefore, it is very meaningful to analyze the damage behavior of asphalt mixture by describing the change characteristics of air voids in the deformation process of samples. However, previous researchers used image analysis to characterize the void of asphalt mixture, focusing on the change value before and after the test and less on the development of the whole process of void during the test. At the same time, the change of air-void content is more used as an evaluation indicator, and other intuitive and effective indicators are lacking. Few studies have used the rutting test to characterize the change in the internal voids of materials.

The purpose of this paper is to quantify the change of air-void characteristics as a damage characterization method by analyzing the section image information of rutting specimens during the test. This method can not only be used to quantify the changes of air-void characteristics in the process of damage development but also to determine the damage concentration area and use the determined air-void parameters to describe the severity of the damage. Although most of the current research involves 3D analysis, the internal structure of materials is obtained through various advanced scanning techniques. It is also important to understand the basic 2D measurement in damage analysis because it can provide a reference for specific problems. Using the same concept, further analysis of damage can be extended to complex problems in 3D or four-dimensional (4D) analysis [20]. At present, most of the research focuses on the change of meso-void characteristics of the whole specimen before and after wheel loading. Few studies have explained the air-void distribution of the whole wheel load area section in the whole process of loading and linked the air-void distribution with the change of air-void fractal dimension and permanent deformation of the asphalt mixture. This study is a comprehensive study on the variation law of internal air-void characteristics of two kinds of asphalt mixtures (AC-13 and AC-16) with the loading cycles of the rutting test. It is the basic work to explain the evolution of internal defects of mixtures under wheel load and evaluate the high-temperature performance of asphalt mixtures from a meso perspective.

## 2. Materials and Methods

### 2.1. Materials

In this study, the dense gradation, namely AC-13 and AC-16, with the nominal maximum aggregate size of 13 and 16 mm, was selected. Qinhuangdao 70# asphalt was selected as an asphalt binder, and the main physical properties are given in Table 1. The coarse aggregate is limestone, the fine aggregate is machine-made sand (0–5 mm), and the filler is limestone powder. Asphalt mixture samples are prepared under the optimum asphalt binder content obtained by the Marshall method [21]. The aggregate gradation, optimum asphalt binder content, and air-void content of the two asphalt mixtures are shown in Table 2. The air-void content of all types of samples is controlled within ±0.5% of the range shown in Table 2.

Table 1. The properties of asphalt binders.

| Parameter | Qinhuangdao-70# | Requirements | Test Value | Test Method |
|---|---|---|---|---|
| Penetration (25 °C, 5 s, 100 g) 0.1 mm | | 60~80 | 64 | T0604 |
| Penetration index (PI) | | −1.5~+1.0 | −0.32 | T0604 |
| Ductility (10 °C) | | ≥20 | 42 | T0605 |
| Ductility (15 °C) | | ≥100 | >100 | |
| Softening point TR&B/°C | | ≥46 | 48.0 | T0606 |
| Solubility/% | | ≥99.5 | 99.72 | T0607 |
| Flash point/°C | | ≥260 | 282 | T0611 |
| Density (15 °C) | | Measured | 1.037 | T0603 |
| Thin Film Oven Test (TFOT) | Mass loss/% | ≤±0.8 | −0.177 | T0609 |
| | Penetration ratio/% | ≥61 | 65.4 | T0604 |
| | Ductility (10 °C)/cm | ≥6 | 9.8 | T0605 |

Table 2. Aggregate gradations and mix design results of AC mixtures.

| Sieve Size (mm) | Passing Percent (%) | |
|---|---|---|
| Gradation | AC-16 | AC-13 |
| 19 | 100 | - |
| 16 | 95 | 100 |
| 13.2 | 85 | 94.8 |
| 9.5 | 71.2 | 81.5 |
| 4.75 | 35.4 | 41.2 |
| 2.36 | 23.9 | 27.6 |
| 1.18 | 20.4 | 23.3 |
| 0.6 | 16.2 | 18.2 |
| 0.3 | 13.0 | 14.2 |
| 0.15 | 11.1 | 12.0 |
| 0.075 | 7.8 | 8.0 |
| Optimum asphalt content (%) | 4.6 | 4.8 |
| Air-void content (%) | 4.0 | 4.4 |

## 2.2. Specimen Fabrication and Rutting Test

According to the Chinese code [22], the mixing temperature of asphalt mixture containing petroleum asphalt is 163 °C. The compaction temperature of the rutting specimen is 100 °C during the forming process. Therefore, the rolling equipment and test mold shall be preheated to 100 °C in advance. The roller load is set as 9 kN (line load 300 N/m). The manufacture of a rutting specimen consists of two steps: (1) rolling the rutting specimen for two cycles with a roller (moving back and forth on the sample is defined as one cycle) and (2) rolling the sample for 12 cycles after rotating it 180°. The size of the final formed rutting specimen is 300 (length) × 300 (width) × 50 mm (height). In order to accurately obtain the change of specimen section after different loading cycles, the plate is cut into three parts by an asphalt mixture cutter along the wheel loading direction before the test, 60 mm in the middle and 120 mm on both sides. Therefore, 3 specimens named 1, 2, and 3 are obtained, as shown in Figure 1.

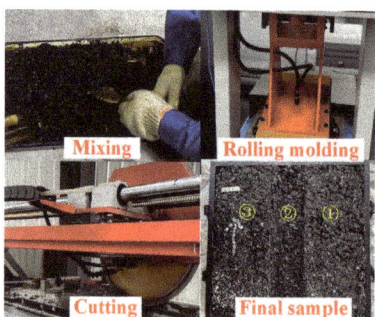

**Figure 1.** Specimen preparation and cutting.

The rutting test is carried out by using the automatic asphalt mixture rutting instrument to study the rutting development of the asphalt mixture. The device is mainly composed of rubber wheels and a fixed platform. The diameter of the rubber wheel is designed to be 200 mm, and the contact width of the wheel specimen is 50 mm. The stress applied by the rubber wheel is 0.70 ± 0.05 MPa, and it is loaded back and forth on the rutting specimen at the rolling speed of 42 r/min. The experimental temperature is set at 60 °C [22]. Due to the cutting effect, the overall size of the rutting specimen is reduced by 4 mm. In order to keep the boundary conditions unchanged, 4 mm steel sheets are inserted on one side to provide a constant lateral force. The details of the rut test are shown in Figure 2. The cumulative rutting depth changes of the specimens under different loading times (500, 1000, 1500, 2000, 2500, and 3000) are recorded by the linear variable differential transformer (LVDT) in the device. In this study, 3 rut specimens of AC-13 and AC-16 are used, respectively.

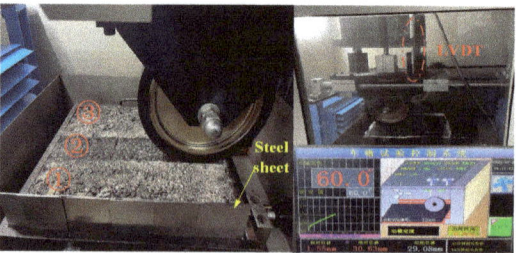

**Figure 2.** Rutting test.

*2.3. Image Acquisition and Processing*

The industrial camera is used to collect the images of four sections of each specimen after the corresponding loading times (500, 100, 1500, 2000, 2500, and 3000) to study the rutting deformation law. Because of the large difference in each specimen section, we only collected the image information of one of the three samples of each asphalt mixture. When taking images, the camera is placed on the test bench to ensure the levelness of the camera. The camera lens is located at the center of the specimen, 50 cm away from the specimen, as shown in Figure 3.

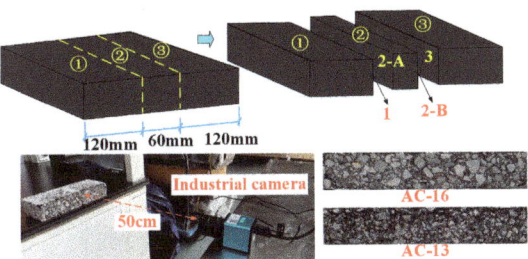

**Figure 3.** Specimen section image acquisition process.

Digital image processing uses a series of algorithms to process and analyze digital images with computers so that the images can meet the needs of human vision, other equipment, data extraction, etc. [23]. In this section, MATLAB software is used to process images, including image enhancement, image filtering, image segmentation, and feature information acquisition. (1) Image enhancement: image enhancement is mainly processed by histogram equalization, also known as gray-level equalization. The purpose is to convert the input image into an output image with the same number of pixels at each gray level through the operation of each pixel in the image. The cross-section image of the asphalt mixture specimen after gray histogram equalization is shown in Figure 4b. Histogram equalization function in MATLAB is histeq (f, 256). (2) Image filtering: In this study, median filtering method is used to filter the image, which is very effective for filtering pulse interference and image scanning noise. The cross-section of the specimens after median filtering is shown in Figure 4c. The function of the median filtering operation in MATLAB is medfilt2 (f). (3) Image segmentation: This research obtains the best threshold value through the function gray threshold value in MATLAB [24]. According to the function result, when the threshold value is set to 35, a good segmentation effect can be obtained, as shown in Figure 4d. Figure 4e shows the acquisition of air-void feature information on the segmented image using ImageJ software. The whole process is shown in Figure 4. By comparing with the air-void content of the actual mixture, we selected the air void when the intensity of the corresponding pixel is in the range of 0 to 20. The air voids are divided into small air voids (<1 mm$^2$), medium air voids (1–5 mm$^2$), and large air voids (>5 mm$^2$) according to the size.

*2.4. Air-Void Characteristics*

Some parameters, such as air-void content, air-void number, air-void shape, and average air-void size, have been successfully used to characterize the characteristics of air voids. Among them, the number and average size of air voids are very important in characterizing the characteristics of air voids. In the study of microstructure damage, the changes of these parameters (by comparing the parameters during and before deformation) provide information on the damage degree of rutting specimens with different gradations, different depths, and different loading accumulation times.

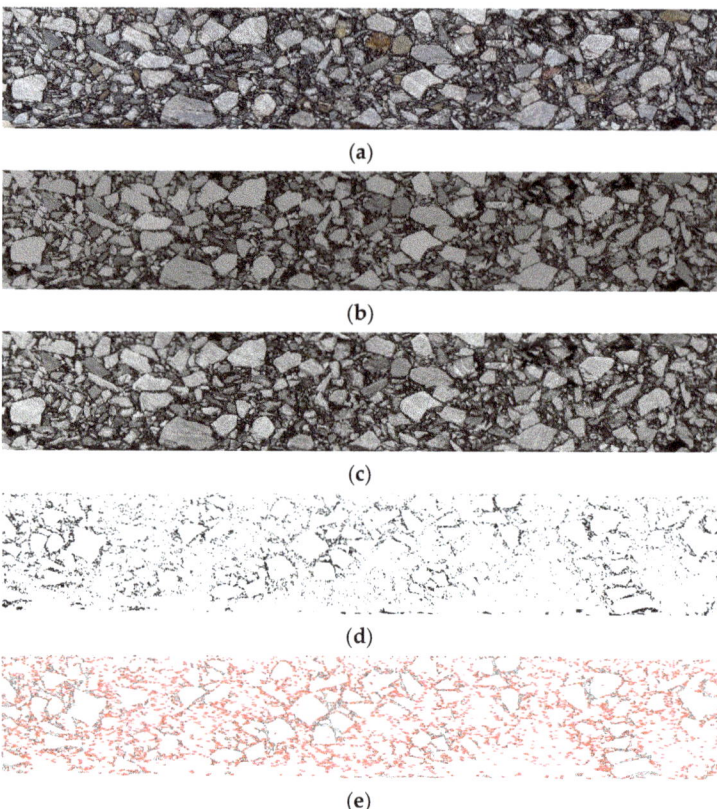

**Figure 4.** Image processing process: (**a**) original image; (**b**) section image after gray histogram equalization; (**c**) section image after median filtering; (**d**) sectional image after image segmentation; (**e**) air-void feature information.

In 2D images, the shape is usually regarded as the area surrounded by a closed contour curve. The most commonly used shape factors in image analysis are roundness, circularity, and aspect ratio. Figure 5 shows air-void shape features defined according to roundness, circularity, and aspect ratio. The shape factor is usually normalized, ranging from 0 to 1. A shape factor equal to 1 usually represents the ideal case or maximum symmetry, such as a circle and sphere [25]. Generally speaking, the crack has a high aspect ratio, and the circularity and roundness values are low (close to 0.0), showing a flat shape. In this study, the section information of the specimen before and after the rutting test is obtained through image analysis. The average air-void size, aspect ratio, circularity, and roundness of the air void in the whole process of the test are calculated by using Equations (1)–(4) to study the change of the air void.

$$\text{Average air voids size} = \frac{\text{Total air void area}}{\text{Number of air void}}, \quad (1)$$

$$\text{Aspect ratio} = \frac{\text{Length of major axis}}{\text{Length of minor axis}}, \quad (2)$$

$$\text{Circularity} = \frac{4\pi \times \text{Area}}{\text{Perimeter}^2}, \quad (3)$$

$$\text{Roundness} = \frac{4\pi \times \text{Area}}{\pi \times \text{Major axis}^2}, \tag{4}$$

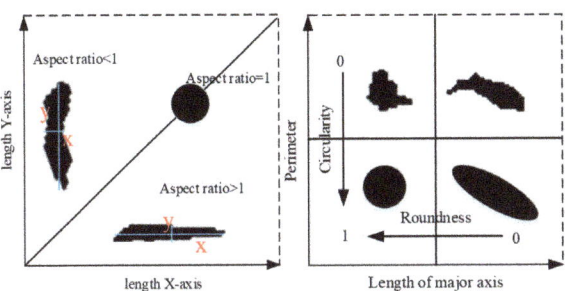

**Figure 5.** Shape features.

*2.5. Fractal Dimension Theory*

The fractal dimension reflects the validity of space occupied by complex objects. It is a measure of the complex shape and irregularity of objects, including Hausdorff dimension, box-counting method, etc. [26]. Box counting uses a set of square boxes or grids to measure the length or distance between points on the shape boundary. It can be calculated with MATLAB software [27]. Figure 6 shows an image of a box array of different sizes (r) covering the air-void image. In the curves of Log (box number, N) and Log (box size, r), the boxes containing the pixels of the air-void image are counted (N), which is used to obtain the fractal dimension, i.e., the slope of the logarithmic regression line. As shown in Figure 7, it can be seen from the fitting curve of Log N (s)–Log N that the image shows the self-similar characteristics of the fractal body at different scales. This shows that the fractal dimension based on 2D images can well reflect the changes in micro-structure. Therefore, it is feasible to quantitatively characterize the rut section by fractal dimension. We can use Equation (5) to calculate the fractal dimension. Lower fractal dimensions represent more stable air-void structures. If the measured air-void areas are the same, different air-void distributions will produce different fractal dimensions.

$$FD = -\frac{\log(N(r))}{\log(r)}, \tag{5}$$

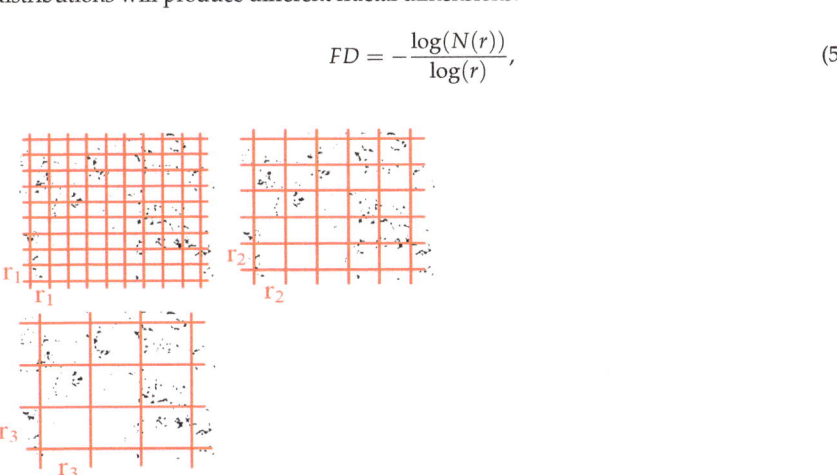

**Figure 6.** Images of box arrays of different sizes (r).

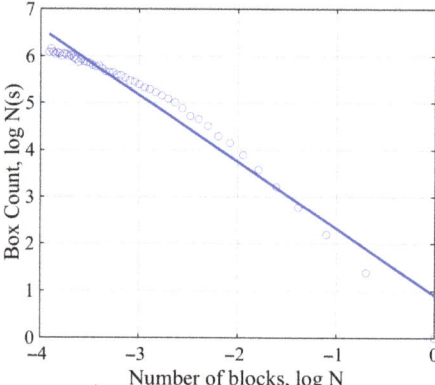

**Figure 7.** Logarithmic relationship between box count and each box size.

## 3. Results

*3.1. Correlation between Rutting Performance and Air-Void Characteristics*

3.1.1. Rutting Test Results

Rutting depth is an important index to evaluate the rutting resistance of specimens. The average value of rutting depth measured on different specimens (three samples for each grading) of the same grading is taken as the rutting test results of this study. The rutting test results are shown in Figure 8. As shown in Figure 8, the deformation of AC-16 is less than that of AC-13 during the whole flow deformation. Different asphalt mixture gradation leads to different deformation development of rutting specimens. Later, we try to explain this phenomenon from the damage analysis of the screenshot image of the rutting specimen.

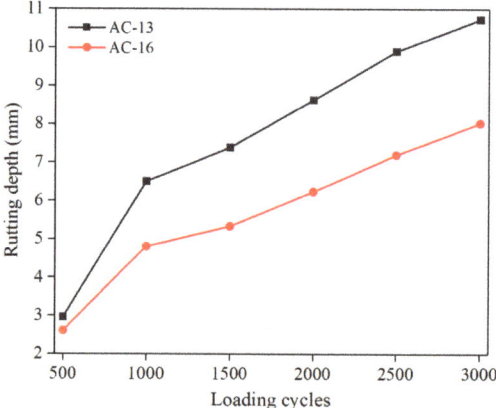

**Figure 8.** Rutting test results.

3.1.2. Correlation between Rutting Cumulative Depth and Air-Void Characteristics

Previous studies have found that the damage of the asphalt mixture under load may first occur in its internal air voids [28]. Therefore, it is of great significance to deeply understand the permanent deformation of the specimen by studying the variation characteristics of the air voids in the specimen during the rutting test. The correlation between the accumulated rut depth and the air-void characteristics (average air-void size, air-void content, circularity, and aspect ratio) is shown in Figure 9. In order not to increase the influence of other factors, only the air-void characteristics of section 2-A after different loading times are studied here.

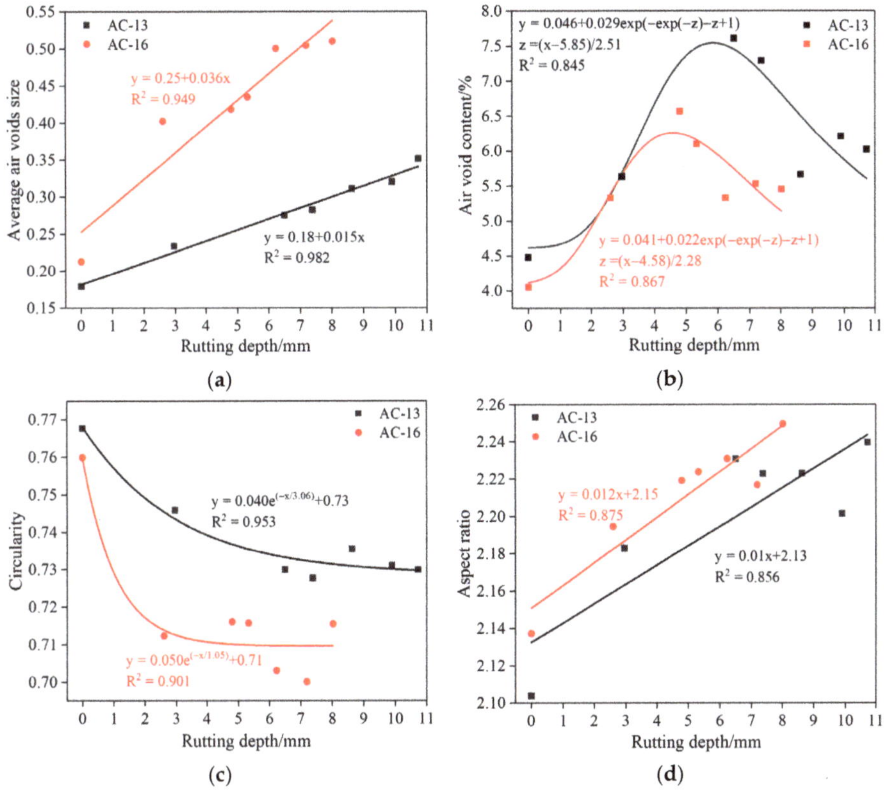

**Figure 9.** Correlation between rutting depth and air-void distribution: (**a**) average air-void size; (**b**) air-void content; (**c**) circularity; (**d**) aspect ratio.

The cumulative rut depth is linearly related to the average air-void size and aspect ratio, as shown in Figure 9a,d. With the increase in the average air-void size and aspect ratio, the cumulative rut depth also increases. At the same time, according to Figure 9c, with continuous loading, the air void inside the specimen becomes flatter, and the circularity decreases. At this time, the internal damage of the specimen increases, and it is more prone to damage. It can be seen from Figure 9b that with the increase in cumulative rut depth, the air-void content first increases, then decreases, and finally tends to a stable value. This is mainly because the number of air voids will first increase rapidly under the load. With the increase in load, the air voids in the specimen are continuously compacted and merged into larger air voids, finally forming cracks. Therefore, the air-void content decreases to a certain extent. This is consistent with Ma X et al.'s research on two possible air-void evolution laws in the mixture [29]. This indicates that under load, the change of internal air voids, along with the initiation and propagation of cracks, is a dynamic process, which is of great significance to the study of internal damage of rut specimens.

3.1.3. Correlation between Rutting Cumulative Depth and Air-Void Fractal Dimension

Based on the fractal theory, the relationship between the fractal dimension of air voids and the rutting depth of rutting specimens is studied. It solves the difficulty of measuring the deformation of each layer in the sample by a conventional test method. The measured fractal dimensions of the AC-13 and AC-16 sections vary with the loading cycles, as shown in Figure 10. For the relationship between fractal dimension and rutting depth, we only analyze the fractal results of section 2-A, and the results are shown in Figure 11. The fractal dimension is related to the complexity and disorder of surface morphology, and more complex image surfaces have higher fractal dimensions [30]. In general, the fractal dimension of air voids increases with the increase in loading times. In other words, with the increase in loading times, the air-void complexity increases. This is consistent with the formation of many irregularly shaped air voids and the increase in roughness in the specimen section during the test.

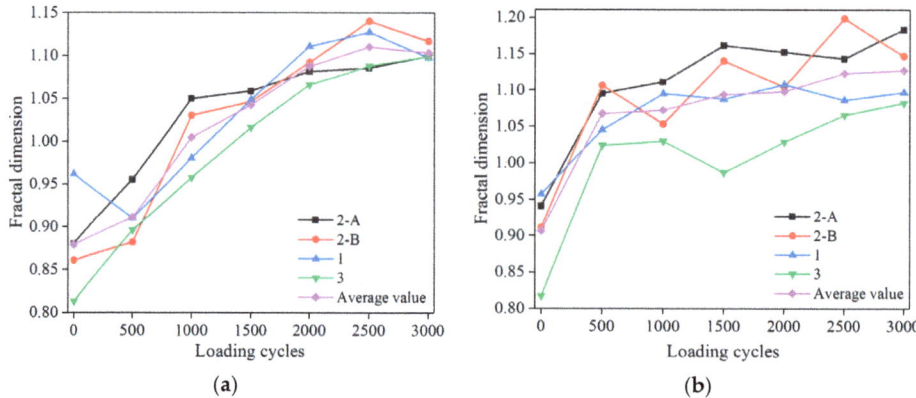

**Figure 10.** Relationship between fractal dimension and loading cycles: (**a**) AC-13; (**b**) AC-16.

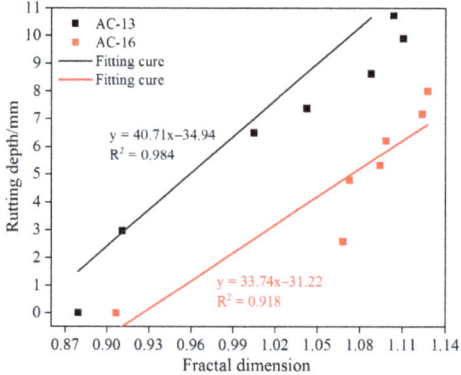

**Figure 11.** Relationship between fractal dimension and rutting depth.

By studying the correlation between rutting depth and air-void fractal dimension, it is found that the value of the fractal dimension is linearly correlated with rutting depth. Therefore, fractal theory can reliably study the deformation law of rutting samples. With the development of deformation, the fractal dimension of cross-section air voids increases gradually. During the test, air voids of all sizes are compacted under the wheel load and create new air voids, especially in the area below the load. With the development of rutting depth, new air voids appear, and the old air voids fuse and merge into micro cracks, which increases the complexity and dispersion in the section. Obviously, the fractal results of air voids accord with the physical significance of fractal theory.

## 3.2. Air-Void Structure and Rutting Specimen Damage

### 3.2.1. Change of Air-Void Content

In order to characterize the internal damage of the material by the change of the air-void content of the specimen. The changes in the air-void content of the 2-A section of the two mixtures with the loading cycles and the height along the specimen before and after loading are studied, respectively, as shown in Figure 12.

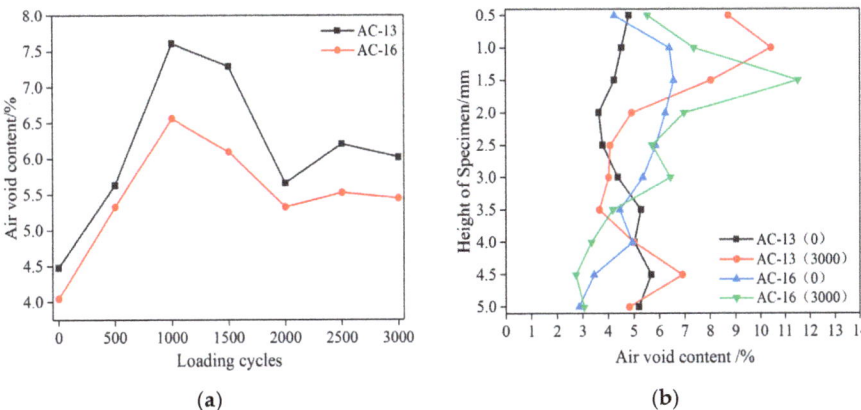

**Figure 12.** Change of air-void content: (**a**) change with loading cycles; (**b**) along the depth direction.

The air-void content of the two rutting specimens first increases, then decreases, and finally stabilizes with the increase in loading cycles, as shown in Figure 12a. This is mainly due to the formation of a large number of air voids in the loading area under compression and shear at the early stage of loading. The newly formed air voids are more than the air voids used for fusion and forming microcracks, so the air-void content will have a rising process. With the increase in loading cycles, a part of the internal air void of the rutting specimen is fused and merged to form micro cracks under load, and a part of the air void is compacted, so the internal air-void content will have a decline process. In the last stage, damage occurred inside the specimen. The change process of air voids is similar to the previous two stages, but this process is not as violent as before. Finally, the air voids will gradually tend to a stable value. Due to the lack of laboratory technology to accurately capture the air-void distribution of asphalt mixture, previous research results are based on the air-void content before and after the test [31]. It is of little significance to study the damage development process.

It can be seen from Figure 12b that before and after the test, the area with high air-void content change mainly occurs in the middle and upper part, which indicates that under the action of wheel load, this area is subject to more shear deformation, resulting in more new air voids. For AC-13, the air-void content in the middle area of the section decreases after the test. Because this area is mainly subjected to compressive stress and the air voids are continuously compacted, the air voids are reduced. The increase in bottom air-void content indicates that it is subjected to shear stress. For AC-16, the change of air-void fraction in the middle area is not obvious, and there is a certain decrease in the bottom. This is mainly due to the large particle size of aggregate in AC-16 and the good performance of the skeleton in the middle area, which can withstand large shear stress failure.

### 3.2.2. Damage Analysis of Rutting Specimen

Previous studies have shown that the load transfer capacity of asphalt mixture mainly depends on the interaction between aggregate and mortar, and macro cracks appear at the interface between aggregate and mortar [32]. Figure 13 shows the change process of air-void shape of the two mixtures under wheel load and also vividly represents the formation process of macro cracks in the mixture. As shown in Figure 13b, under the action of wheel load, the air void develops along the interface between aggregate and asphalt mortar. This process leads to the deformation, expansion, and fusion of the air void and promotes the generation of macro cracks in the mixture. Micro damage also promotes the bond failure or damage of asphalt mortar, further has a more adverse impact on the bonding performance between aggregate and asphalt mortar, and aggravates the degradation of internal structure under a wheel load. In addition, it can be seen from Figure 13a that the load also makes the aggregate with smaller particle size in the mixture crushed, which destroys the internal skeleton structure of the mixture and further accelerates the damage of the asphalt mixture. Comparing the two mixtures, it can be seen that this phenomenon is more obvious for the mixture with a smaller maximum particle size of aggregate. The damage is not only the increase in air void but also accompanied by the crushing of many aggregates and the overflow of asphalt mortar. This is different from Li P et al.'s result that only aggregate movement exists in asphalt mixture under pressure [33], which provides a new idea for the study of mixture damage.

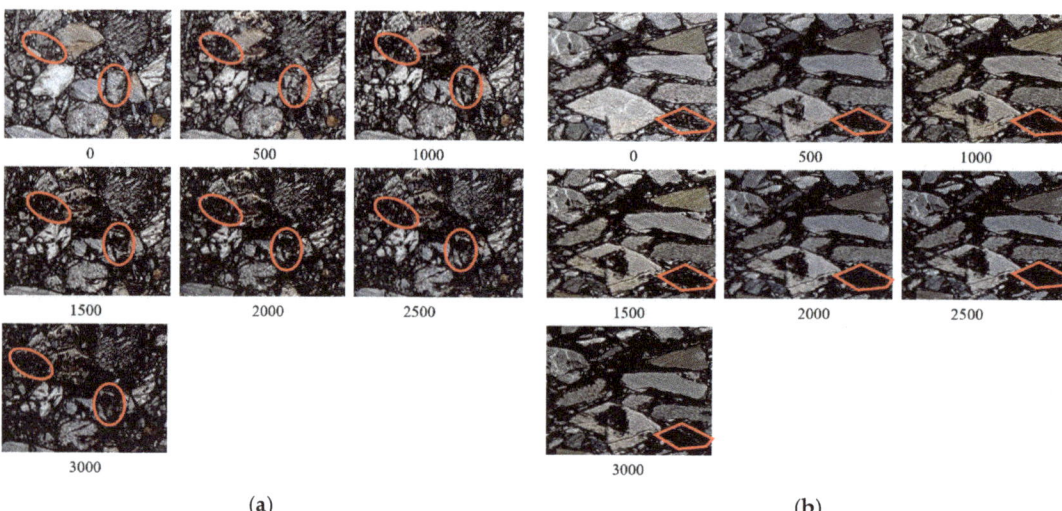

**Figure 13.** Variation of sectional air void with loading cycles: (**a**) AC-13; (**b**) AC-16.

### 3.2.3. Change of Air-Void Characteristics

Figure 14 shows the changes in total air-void number, total air-void area, and average air-void size of the two graded mixtures before and after the rutting test. These parameters have been successfully applied to characterize the characteristic changes of voids and have been well-established in previous research work [34,35]. In the study of microstructure damage, the changes of these parameters (by comparing the parameters before and after deformation) provide valuable information on the severity of damage at different depths in compacted asphalt mixture specimens. The bar graph compares changes in air-void characteristics by averaging at 5 mm intervals along the height of the specimen. Through observation, several noteworthy results were obtained along the characteristic distribution of the sample height in three main areas, namely, the bottom (0–1.5 cm), the middle (1.5–3.5 cm), and the top (3.5–5 cm).

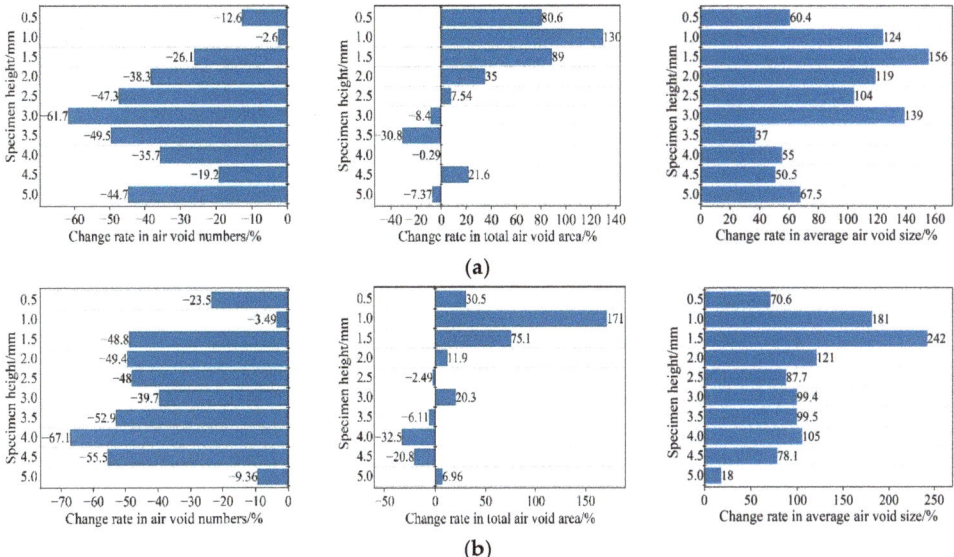

**Figure 14.** Variation of air-void characteristics in rutting test: (**a**) AC-13; (**b**) AC-16.

As shown in Figure 14, the rate of change of total air-void area and average air-void size is higher in the middle and upper region, which is consistent with the slight expansion observed in the middle and upper region of the specimen after the rutting test. Under the action of wheel load, the middle and upper mixture is seriously damaged. The total air voids of the two mixture specimens along the height direction are reduced, indicating the generation of micro cracks and macro cracks in the rutting specimen, which also promotes the expansion in the middle area of the specimen. This is mainly due to two reasons: one is that the internal air void is reduced due to the continuous compaction of the specimen under load; Second, microcracks are formed due to the fusion of small air voids.

In this study, when describing the different damage phenomena in the specimen due to deformation, the changes in total air-void number and total air-void area are determined as two key parameters. According to the previous study [36], there are two main mechanisms under permanent deformation, namely densification, and shear deformation, as shown in Figure 15. By analyzing the changes in air-void number and air-void area before and after the test, the results of this experiment are mainly related to the damage mechanism related to compression deformation. The relationship between damage mechanisms and air-void properties will be further clarified in subsequent sections, involving changes in air-void geometry.

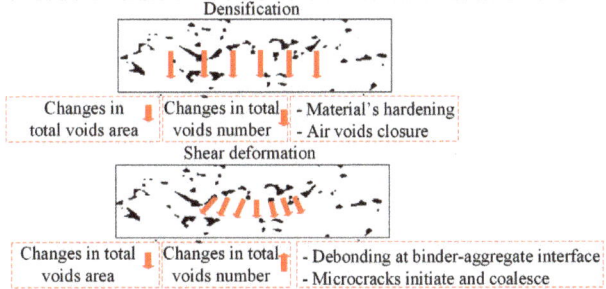

**Figure 15.** Correlation between damage mechanisms and changes in air-void content properties.

The change distribution of air-void characteristics along the height of the rutting specimen before and after the test is systematically studied in the previous part. This part will study the change of air void under load from the whole section. Firstly, the variation characteristics of air-void perimeter and area in the mixture under different loading cycles are studied. Secondly, the changes in air voids with different area sizes and their corresponding quantities before and after loading are studied. Finally, the changes in the total area and number of air voids of different types (large, medium, and small) before and after the test are studied.

Figure 16 shows the area and perimeter distribution of the air void in the rutting specimen under different loading cycles. When the polygon area is certain, the larger the ratio of major and minor axes, the longer the circumference [37]. It can be seen from Figure 16 that with the increase in loading times, the area and perimeter distribution of the internal air voids of the rutting specimen become wider, and the discreteness also increases. This shows that under the action of load, the air voids develop towards a flatter shape, preparing for further integration and macro cracks. This also confirms the rationality of the previous research on the development law of air voids before and after loading from another perspective.

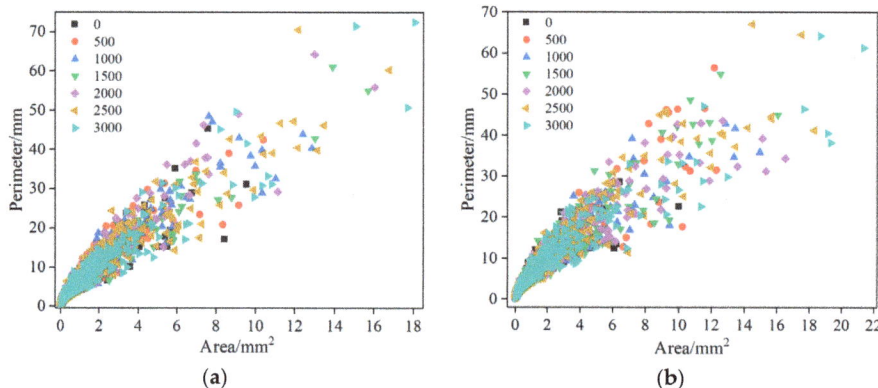

**Figure 16.** Relationship between air-void area and perimeter: (**a**) AC-13; (**b**) AC-16.

Figure 17 shows the variation law of the average area and perimeter of air voids along the height of the specimen. It can be seen from Figure 17 that before the load is applied, the average perimeter and area of the two graded mixtures change a little along the height direction, which shows that the specimen is fully compacted during the preparation process. After 3000 cycles of cyclic loading, the average area and perimeter of the specimen increase to a certain extent along the height direction, indicating that the internal air-void size of the specimen increases as a whole. For the two mixtures, the change rate of the middle and upper part of the specimen is large, which is also consistent with the obvious expansion of the middle and upper part observed after the test.

In order to further explain the change of air voids before and after the test, we calculated the area and the corresponding number of air voids in the 2-A section of the specimen before and after loading, as shown in Figure 18. Previous field tests have concluded that air voids in the pavement are compressed after permanent deformation [38]. However, it can be seen from Figure 18 that a larger area of air voids appeared in the specimen after loading. The number of air voids with smaller areas ($\leq 1$ mm$^2$) is reduced greatly, and most of them are compressed. At the same time, we also compared the change rate of the total area and quantity of different types of air voids before and after the test, as shown in Figure 19.

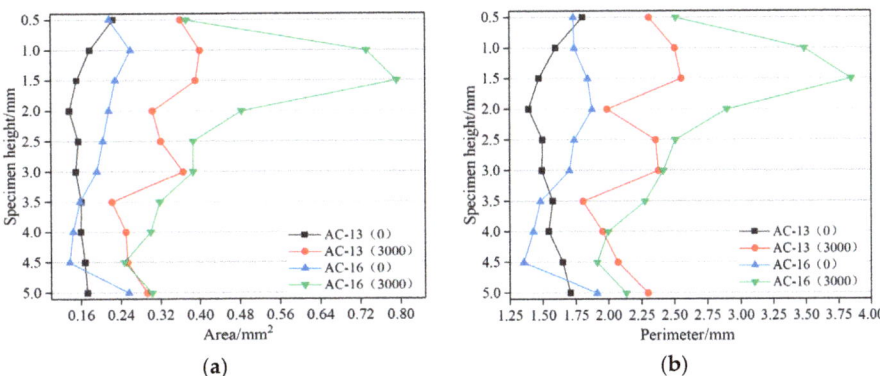

**Figure 17.** Variation of the average area and perimeter of air voids along specimen height (**a**) Average area; (**b**) Average perimeter.

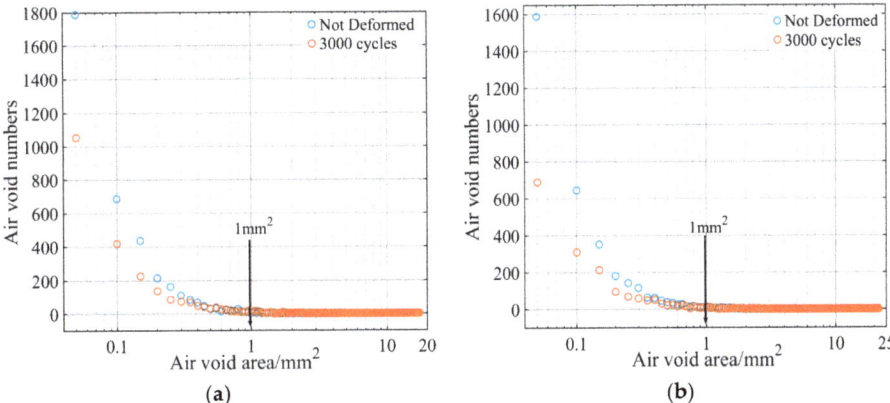

**Figure 18.** Relationship between air voids and corresponding quantities: (**a**) AC-13; (**b**) AC-16.

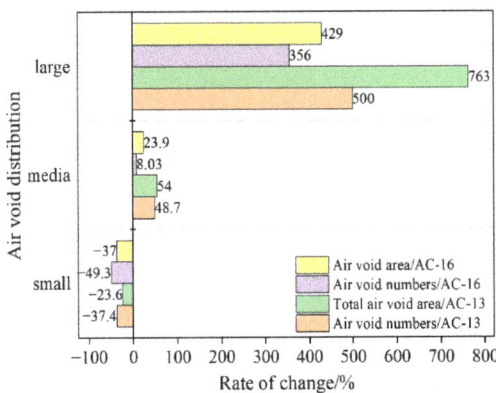

**Figure 19.** Changes in area and quantity of air voids with different sizes.

It can be seen from Figure 19 that after 3000 cycles of wheel loading, the total number and area of small air voids in the specimen decrease, while the middle air voids and large air voids increase, and the change rate of the total number and area of large air voids is the largest. This is mainly due to the small number of large air voids in the original specimen. After loading, the number of large air voids increases, making the change rate larger. The above results show that the air voids inside the specimen mainly occur through compaction and fusion of small air voids under load. The fusion between small air voids expands into micro cracks, which in turn induces damage to the asphalt mixture. At the same time, the change rate of the total number and area of the large and medium air voids in AC-13 specimens is higher than that of AC-16, indicating that the damage degree of AC-13 specimens is higher than that of AC-16. This result can be confirmed by rutting results.

### 3.2.4. Change of Air-Void Shape Characteristics

After the rutting test, not only the total number of air voids, total air-void area, and average air-void size but also the shape characteristics of air voids will change greatly. Figure 20 shows the change rate of void shape characteristics of the specimen before and after the test, i.e., aspect ratio, circularity, and roundness. The bar graph compares the changes in air-void shape characteristics by averaging at 5 mm intervals along with the height of the specimen.

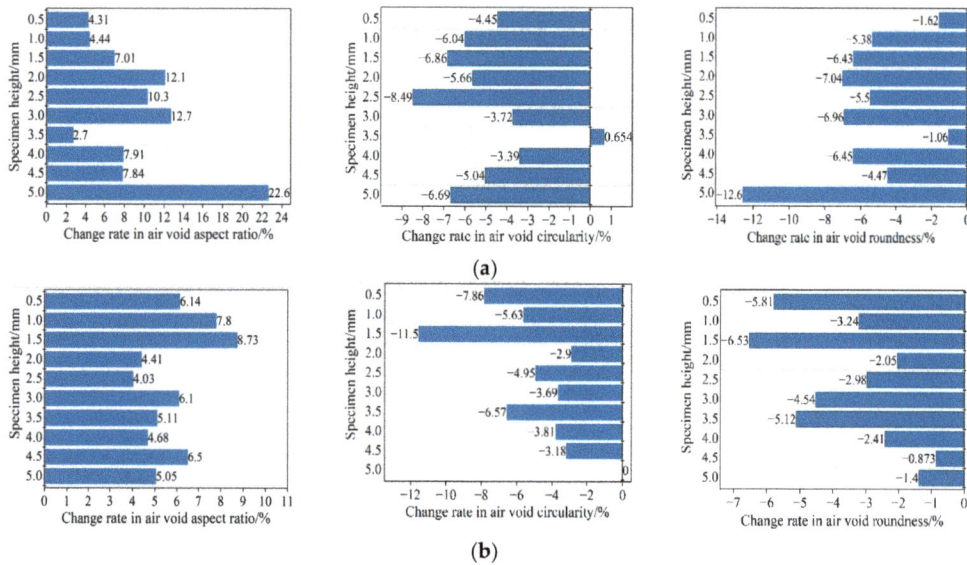

**Figure 20.** Variation of air-void shape characteristics in rutting test: (**a**) AC-13; (**b**) AC-16.

It can be seen from Figure 20 that after the test, the overall roundness of the air voids in the specimen decreased, and the shape of the air voids developed into a flat shape. At the same time, with the increase in air-void size, two adjacent air voids merge to form an air void, and its aspect ratio is higher than that of its single air void. This confirms the previous findings about the change in air-void properties. In general, the method of measuring the shape characteristics of air voids provides a different method to describe the characteristics of air-void formation, growth, connectivity, and expansion. The damage is due to the coalescence of small air voids to form microcracks, which increases the aspect ratio of air voids, thereby reducing the number of air voids and roundness [39]. Based on the observation of a large number of samples, it is determined that the damage controlled by the change of air-void structure occurs in many ways. First of all, these changes may be due to the increase in the existing single air-void size and the propagation of cracks. Secondly, the two separated air voids may merge and expand, forming cracks near the damage. Third, due to deformation, new air voids with crack initiation potential are also formed. This finding is consistent with the two failure types determined by Kim et al. [40].

The damage in the asphalt mixture is the result of microcracks generated in the interfacial transition (ITZ) between asphalt and aggregate and within the binder (cohesive failure).

## 4. Conclusions

This paper aims to quantify the change of air-void characteristics as a method to characterize damage by analyzing the cross-sectional image information of rutting specimens during the test. Firstly, the indoor rutting test was carried out, and the air-void information of the sample interface under different loading times was obtained by digital image technology. Then, the variation law of the internal air-void characteristics of the asphalt mixture with the loading times of the rutting test is systematically studied, and the relationship with the development of rutting is established. Finally, the air-void characteristic parameters are used to quantify the damage in the asphalt mixture and describe the generation and development of damage. According to the research results, the following conclusions can be drawn:

(1) In the process of the rutting test, the change of air-void ratio in the rutting specimen is a dynamic process with the initiation and propagation of the crack. With the increase in cumulative rut depth, the air-void ratio increases first, then decreases to a certain extent, and finally tends to a stable value.

(2) The fractal dimension of air voids increases with the increase in loading times. In other words, with the increase in loading times, the air-void complexity increases. This is consistent with the observation that the section of the specimen expands, and the roughness increases during the test.

(3) In the rutting test, the damage is not only the compaction of voids but also accompanied by the crushing of aggregates and the overflow of asphalt mortar. At the same time, this phenomenon is more obvious for the asphalt mixture with a smaller maximum particle size of aggregate.

(4) The measurement of air-void characteristics and shape characteristics provides a different method to describe the characteristics of air-void formation, growth, connectivity, and expansion of voids in asphalt mixture during a rutting test. After deformation, the total air-void area and average air-void size of the specimen increase, and the total air-void number decreases. Because microcracks are formed in the specimen after rutting damage, the aspect ratio of the air voids increases, and the roundness value decreases.

Therefore, we can use the change of void ratio, void characteristics, and void shape characteristics to quantify the damage type and its evolution process in the rutting formation of the asphalt mixture in future research. Due to the limitations of the technology used in this study, the voids discussed are obtained based on two-dimensional images. It is difficult to distinguish the connection and closure of voids in the three-dimensional case of a mixture, so further research is needed. In addition, in order to determine the main cause of damage to the inner cup of the asphalt mixture, it is necessary to analyze the damage at the contact between aggregate and asphalt and the impact of the contact point between aggregates on the performance.

**Author Contributions:** Conceptualization, K.Z. and L.W.; methodology, L.W. and W.W.; software, K.Z.; resources, H.Y.; data curation, K.Z.; writing—original draft preparation, K.Z.; writing—review and editing, L.W.; supervision, L.W. All authors have read and agreed to the published version of the manuscript.

**Funding:** This research was funded by Beijing Major Science and Technology Projects, grant number (No. Z191100008019002).

**Institutional Review Board Statement:** Not applicable.

**Informed Consent Statement:** Not applicable.

**Data Availability Statement:** Not applicable.

**Conflicts of Interest:** The authors declare no conflict of interest.

## References

1. Lytton, R.L.; Zhang, Y.; Gu, F.; Luo, X. Characteristics of damaged asphalt mixtures in tension and compression. *Int. J. Pavement Eng.* **2018**, *19*, 292–306. [CrossRef]
2. Pratico, F.G.; Vaiana, R.; Moro, A. Dependence of volumetric parameters of hot-mix asphalts on testing methods. *J. Mater. Civ. Eng.* **2014**, *26*, 45–53. [CrossRef]
3. Wróbel, M.; Woszuk, A.; Franus, W. Laboratory methods for assessing the influence of improper asphalt mix compaction on its performance. *Materials* **2020**, *13*, 2476. [CrossRef]
4. Praticò, F.G.; Vaiana, R. A study on volumetric versus surface properties of wearing courses. *Constr. Build. Mater.* **2013**, *38*, 766–775. [CrossRef]
5. Hu, J.; Liu, P.; Wang, D.; Oeser, M. Influence of aggregates' spatial characteristics on air-voids in asphalt mixture. *Road Mater. Pavement Des.* **2018**, *19*, 837–855. [CrossRef]
6. Benzerga, A.A.; Leblond, J.B.; Needleman, A.; Tvergaard, V. Ductile failure modeling. *Int. J. Fract.* **2016**, *201*, 29–80. [CrossRef]
7. Gilabert, F.A.; Garoz, D.; Van Paepegem, W. Macro-and micro-modeling of crack propagation in encapsulation-based self-healing materials: Application of XFEM and cohesive surface techniques. *Mater. Des.* **2017**, *130*, 459–478. [CrossRef]
8. Zheng, L.; Liu, H.; Zuo, Y.; Zhang, Q.; Lin, W.; Qiu, Q.; Liu, Z. Fractal study on the failure evolution of concrete material with single flaw based on DIP technique. *Adv. Mater. Sci. Eng.* **2022**, *2022*, 6077187. [CrossRef]
9. Wang, Y.; Wang, W.; Wang, L. Understanding the relationships between rheology and chemistry of asphalt binders: A review. *Constr. Build. Mater.* **2022**, *329*, 127161. [CrossRef]
10. Wang, L.B.; Frost, J.D.; Voyiadjis, G.Z.; Harman, T.P. Quantification of damage parameters using X-ray tomography images. *Mech. Mater.* **2003**, *35*, 777–790. [CrossRef]
11. Zhang, Y.; Gu, F.; Birgisson, B.; Lytton, R.L. Viscoelasticplastic–fracture modeling of asphalt mixtures under monotonic and repeated loads. *Transp. Res. Rec.* **2017**, *2631*, 20–29. [CrossRef]
12. Zhu, L.; Dang, F.; Xue, Y.; Ding, W.; Zhang, L. Comparative study on the meso-scale damage evolution of concrete under static and dynamic tensile loading using X-ray computed tomography and digital image analysis. *Constr. Build. Mater.* **2020**, *250*, 118848. [CrossRef]
13. Radeef, H.R.; Hassan, N.A.; Mahmud, M.Z.H.; Abidin, A.R.Z.; Ismail, C.R.; Abbas, H.F.; Al-Saffar, Z.H. Characterisation of cracking resistance in modified hot mix asphalt under repeated loading using digital image analysis. *Theor. Appl. Fract. Mech.* **2021**, *116*, 103130. [CrossRef]
14. Xu, G.; Chen, X.; Cai, X.; Yu, Y.; Yang, J. Characterization of Three-Dimensional Internal Structure Evolution in Asphalt Mixtures during Freeze–Thaw Cycles. *Appl. Sci.* **2021**, *11*, 4316. [CrossRef]
15. Yang, B.; Li, H.; Zhang, H.; Xie, N.; Zhou, H. Laboratory investigation on effects of microscopic void characteristics on properties of porous asphalt mixture. *Constr. Build. Mater.* **2019**, *213*, 434–446. [CrossRef]
16. Zhang, Z.; Liu, Q.; Wu, Q.; Xu, H.; Liu, P.; Oeser, M. Damage evolution of asphalt mixture under freeze-thaw cyclic loading from a mechanical perspective. *Int. J. Fatigue* **2021**, *142*, 105923. [CrossRef]
17. Kassem, E.; Masad, E.; Lytton, R.; Chowdhury, A. Influence of air voids on mechanical properties of asphalt mixtures. *Road Mater. Pavement Des.* **2011**, *12*, 493–524. [CrossRef]
18. Xu, H.; Guo, W.; Tan, Y. Internal structure evolution of asphalt mixtures during freeze–thaw cycles. *Mater. Des.* **2015**, *86*, 436–446. [CrossRef]
19. Hassan, N.A.; Airey, G.D.; Hainin, M.R. Characterisation of micro-structural damage in asphalt mixtures using image analysis. *Constr. Build. Mater.* **2014**, *54*, 27–38. [CrossRef]
20. Bhasin, A.; Izadi, A.; Bedgaker, S. Three dimensional distribution of the mastic in asphalt composites. *Constr. Build. Mater.* **2011**, *25*, 4079–4087. [CrossRef]
21. Li, W.; Cao, W.; Ren, X.; Lou, S.; Liu, S.; Zhang, J. Impacts of Aggregate Gradation on the Volumetric Parameters and Rutting Performance of Asphalt Concrete Mixtures. *Materials* **2022**, *15*, 4866. [CrossRef] [PubMed]
22. Ministry of Transport of the People's Republic of China. *Standard Test Methods of Bitumen and Bituminous Mixtures for Highway Engineering*; China Communications Press: Beijing, China, 2011.
23. Sun, P.; Zhang, K.; Han, S.; Liang, Z.; Kong, W.; Zhan, X. Method for the Evaluation of the Homogeneity of Asphalt Mixtures by 2-Dimensional Image Analysis. *Materials* **2022**, *15*, 4265. [CrossRef] [PubMed]
24. Enríquez-León, A.J.; de Souza, T.D.; Aragão, F.T.S.; Braz, D.; Pereira, A.M.B.; Nogueira, L.P. Determination of the air void content of asphalt concrete mixtures using artificial intelligence techniques to segment micro-CT images. *Int. J. Pavement Eng.* **2021**, 1–10. [CrossRef]
25. Ali, H.; Murtaza, G.; Badshah, N. Covariance based image selective segmentation model. *J. Inf. Commun. Technol.-Malays.* **2010**, *4*, 11–19.
26. Wang, L.; Zeng, X.; Yang, H.; Lv, X.; Guo, F.; Shi, Y.; Hanif, A. Investigation and application of fractal theory in cement-based materials: A review. *Fractal Fract.* **2021**, *5*, 247. [CrossRef]

27. Chung, S.Y.; Kim, J.S.; Stephan, D.; Han, T.S. Overview of the use of micro-computed tomography (micro-CT) to investigate the relation between the material characteristics and properties of cement-based materials. *Constr. Build. Mater.* **2019**, *229*, 116843. [CrossRef]
28. Shaheen, M.; Al-Mayah, A.; Tighe, S. A novel method for evaluating hot mix asphalt fatigue damage: X-ray computed tomography. *Constr. Build. Mater.* **2016**, *113*, 121–133. [CrossRef]
29. Ma, X.; Zhou, P.; Jiang, J.; Hu, X. High-temperature failure of porous asphalt mixture under wheel loading based on 2D air void structure analysis. *Constr. Build. Mater.* **2020**, *252*, 119051. [CrossRef]
30. Zhang, C.; Shi, X.; Wang, L.; Yao, Y. Investigation on the Air Permeability and Pore Structure of Concrete Subjected to Carbonation under Compressive Stress. *Materials* **2022**, *15*, 4775. [CrossRef]
31. Wang, Z.; Xiao, J. Evaluation of air void distributions of cement asphalt emulsion mixes using an X-ray computed tomography scanner. *J. Test. Eval.* **2012**, *40*, 273–280. [CrossRef]
32. Jiang, J.; Ni, F.; Dong, Q.; Yao, L.; Ma, X. Investigation of the internal structure change of two-layer asphalt mixtures during the wheel tracking test based on 2D image analysis. *Constr. Build. Mater.* **2019**, *209*, 66–76. [CrossRef]
33. Li, P.; Su, J.; Gao, P.; Wu, X.; Li, J. Analysis of aggregate particle migration properties during compaction process of asphalt mixture. *Constr. Build. Mater.* **2019**, *197*, 42–49. [CrossRef]
34. Wills, J.; Caro, S.; Braham, A. Influence of material heterogeneity in the fracture of asphalt mixtures. *Int. J. Pavement Eng.* **2019**, *20*, 747–760. [CrossRef]
35. Liu, Y.; You, Z.; Dai, Q.; Mills-Beale, J. Review of advances in understanding impacts of mix composition characteristics on asphalt concrete (AC) mechanics. *Int. J. Pavement Eng.* **2011**, *12*, 385–405. [CrossRef]
36. Sefidmazgi, N.R.; Tashman, L.; Bahia, H. Internal structure characterization of asphalt mixtures for rutting performance using imaging analysis. *Road Mater. Pavement Des.* **2012**, *13*, 21–37. [CrossRef]
37. Kim, Y.; Dodbiba, G. A novel method for simultaneous evaluation of particle geometry by using image processing analysis. *Powder Technol.* **2021**, *393*, 60–73. [CrossRef]
38. Coleri, E.; Kayhanian, M.; Harvey, J.T.; Yang, K.; Boone, J.M. Clogging evaluation of open graded friction course pavements tested under rainfall and heavy vehicle simulators. *J. Environ. Manag.* **2013**, *129*, 164–172. [CrossRef]
39. Palvadi, S.; Bhasin, A.; Little, D.N. Method to quantify healing in asphalt composites by continuum damage approach. *Transp. Res. Rec.* **2012**, *2296*, 86–96. [CrossRef]
40. Kim, Y.R.; Aragão, F.T.S. Microstructure modeling of rate-dependent fracture behavior in bituminous paving mixtures. *Finite Elem. Anal. Des.* **2013**, *63*, 23–32. [CrossRef]

Article

# Study on the Extraction of CT Images with Non-Uniform Illumination for the Microstructure of Asphalt Mixture

Lei Zhang [1], Guiping Zheng [1], Kai Zhang [2], Yongfeng Wang [3], Changming Chen [3], Liting Zhao [3], Jiquan Xu [4], Xinqing Liu [4], Liqing Wang [4], Yiqiu Tan [1] and Chao Xing [1,*]

1 School of Transportation Science and Engineering, Harbin Institute of Technology, Harbin 150090, China
2 China State Construction International Holdings Limited, Hong Kong 999077, China
3 CCCC NO. 1 Highway Survey Design & Research Institute Co., Ltd., Xi'an 710065, China
4 Sichuan Gezhouba Batongwan Expressway Co., Ltd., Bazhong 636600, China
* Correspondence: cxing@hit.edu.cn; Tel.: +86-451-86282120

**Abstract:** An adaptive image-processing method for CT images of asphalt mixture is proposed in this paper. Different methods are compared according to the error analysis calculated between the real gradation and 3D reconstruction gradation. As revealed by the test results, the adaptive image-processing method was effective in carrying out different brightness homogenization processes for each image. The Wiener filter with 7 × 7 size filter was able to produce a better noise reduction effect without compromising image sharpness. Among the three methods, the adaptive image-processing method performed best in the accuracy of coarse aggregate recognition, followed by the ring division method and the global threshold segmentation method. The error of the gradation extracted by the adaptive image-processing method was found to be lowest compared with the real gradation. For a variety of engineering applications, the developed method helps to improve the analysis of CT images of asphalt mixtures.

**Keywords:** asphalt mixture; Computed Tomography; image process; microstructure; gradation analysis

## 1. Introduction

As variable multi-phase composite material, asphalt mixture is characterized by significant differences, randomness, and variability among the constituent materials, which contributes to the non-uniqueness of its microstructure. However, the spatial distribution characteristics of each component in the internal structure of asphalt mixture cannot be reflected by the traditional macro-empirical evaluation method, which focuses primarily on the overall features and road performance of asphalt mixture at the macro scale. However, for multi-phase asphalt mixture, the macro mechanical properties depend on its microstructures. In this circumstance, it is necessary to calculate the performance mechanism of asphalt mixture from a meso-level perspective. At present, computed tomography (CT) and digital image-processing technology have been widely applied in micro-scale research on asphalt mixture. It is difficult to properly identify the differential materials in CT images of asphalt mixture, due to factors such as CT imaging mechanism and variations of asphalt-mixture gradation composition. Meanwhile, grayscale values of substances with the same density should theoretically be consistent at different locations, but the scanning energy limitation of industrial CT machines causes them to present varying degrees of lightness and darkness at different distances from the specimen's central axis. Therefore, it remains challenging to accurately identify aggregates, voids, and asphalt mortar. To obtain an accurate extraction of the microstructure of asphalt mixture, it is necessary to further study the image-processing method used for CT images of asphalt mixture.

It was in the early 1990s that research began on the CT image-processing method used for asphalt mixture. At that time, Masad et al. [1] applied CT technology and digital camera technology for the first time to conduct a quantitative analysis of the compaction effect of

the mixture under different compaction modes, which led to success in introducing CT into the research on the microstructure of asphalt mixture, thus opening up a new field of research. In the meantime, research was advanced on the image-processing method applied for composite CT.

CT scanning technology is used to detect the attenuation of X-rays after passing through an object. Then, the attenuation images are outputted to the reconstruction algorithm for completing the reconstruction [2]. Due to the poly-chromatic nature of X-ray energy, however, the attenuation coefficient can cause errors, thus resulting in artifacts within the reconstructed images [3,4]. The methods used to correct artifacts can be categorized into physical filtering methods and software correction methods. Commonly used during scanning, the physical filtering method [5–8] can reduce the artifacts. Meanwhile, however, it also reduces ray intensity, which contributes to SNR (signal noise ratio) reduction. Due to fewer restrictions during application, software correction methods [3,9–11] are also frequently used. For asphalt mixture, however, which is a multiphase material, these methods are more complicated for correcting artifacts due to the lack of prior knowledge about X-ray spectra and material properties.

In recent years, CT scanning technology has been increasingly used for the microstructural characterization of asphalt [12], bituminous mortar [13], and asphalt mixture. Within research on asphalt-mixture CT image processing, the focus has mainly been placed on the methods emloyed to separate aggregate, mortar, and void within the image [14,15]. Currently, commonly used methods include the artificial threshold method [16–18], fuzzy C-means algorithm [19,20], Gaussian mixture model [21], and Otsu algorithm [22–25]. The most widely used is the Otsu algorithm. When Otsu is applied to images with artifacts, however, the segmentation effect is less than satisfactory. To solve this problem, Liu put forward the ring division method [26], which achieved higher segmentation accuracy. Nevertheless, it failed to determine accurately the size of the test piece and its position in the image. Additionally, the method was restricted to reducing the impact of artifacts on image segmentation when faced with the artifacts distributed along the radial non-linear section of the test piece, meaning it was unable completely to address the problem of artifacts.

In order to address the current shortcomings of image-processing methods, this paper proposes a brightening program for asphalt-mixture CT images that can help effectively eliminate the adverse effect of artifacts on image threshold segmentation. The processing effect based on this method was compared with that produced by the ring division method and the global threshold segmentation method, respectively. The research findings of this paper can contribute to more precisely identifying the composition and different materials of asphalt mixtures, accurately ascertaining the microstructure of asphalt mixture, and accomplishing a quantitative analysis of void distribution characteristics, aggregate homogeneity, and microscopic damage characteristics of asphalt mixture. This can provide a thorough understanding of the internal microstructure and properties of asphalt mixture, and improve the relevant test methods and protocols to further improve the performance of pavements. In addition, the method can also be applied to cement concrete, composite materials, and metal materials, so it has promising application potential.

## 2. Materials and Methods

### 2.1. Materials

In this study, andesite produced in Heilongjiang Province and limestone powder produced in Jilin Province were used. The relevant tests were conducted on stone materials, in line with the standards JTGE42-2005 [27] and JTGF40-2004 [28]. According to the test results, all indicators met the requirements. SBS polymer modified asphalt was used in this study, and the properties are shown in Table 1. The lignin fiber content of the SMA asphalt mixture was 0.3%.

Table 1. Properties of SBS modified asphalt.

| Properties | Unit | Test Result | Specification Requirements | Specification |
|---|---|---|---|---|
| Penetration (25 °C, 100 g, 5 s) | 0.1 mm | 66.9 | 60~80 | |
| Ductility (5 °C, 5 cm/min) | cm | 43.3 | ≥30 | JTG F40 |
| Softening point | °C | 66.5 | ≥55 | |

*2.2. Gradation Design*

AC-16, OGFC-16, and SMA-16 test pieces were designed and fabricated according to the standard JTGF40-2004, with three replicates prepared for each sample. The gradation is shown below in Figure 1.

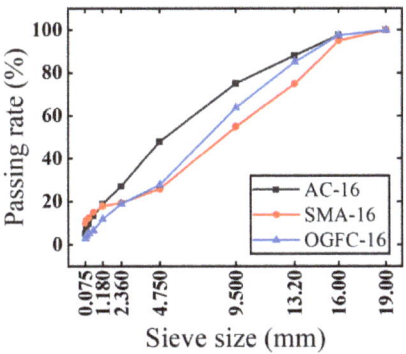

Figure 1. Gradation curves of three kinds of asphalt mixtures.

*2.3. Optimum Asphalt Content and Test Pieces Preparation*

The optimum oil content of the asphalt mixture was determined using the Marshall asphalt mixture design method. The optimum asphalt content of AC-16, SMA-16, and OGFC-16 was 4.5%, 6%, and 3.1%, respectively. The mixtures were compacted 120 times using the rotary compaction method. To obtain more pictures of the test pieces, the height of the test pieces after molding was set to about 150 mm. The volume parameters of the asphalt mixtures are shown in Table 2.

Table 2. Volume parameters of asphalt mixtures.

| Mixtures | Air Voids Content (%) | VMA (%) | VFA (%) |
|---|---|---|---|
| AC-16 | 3.5 | 13.9 | 74.9 |
| SMA-16 | 3.6 | 17.0 | 78.9 |
| OGFC-16 | 19.7 | 28.2 | 30.2 |

*2.4. Scan by Industrial CT*

The CT scanning equipment used for Marshall test pieces was a Phoenix micro focus industrial CT purchased by the School of Transportation Science and Engineering affiliated to Harbin Institute of Technology. The cone-beam scanning industrial CT system and the cone-beam filter back-projection reconstruction algorithm were both applied to reconstruct the faulty images. The scanning voltage/current was 195 kv/95 μA. The critical performance indicators are shown in Table 3. In this study, 150 mm corresponded to 1500 pixels in the image, so the resolution of the asphalt-mixture image recognition was 0.1 mm.

Table 3. Key performance indicators of Phoenix micro-focus industrial CT.

| Maximum Tube Voltage/kV | Maximum Tube Power/kW | Detail Resolution/μm | Minimum Distance from Focus to Test Pieces/mm | Maximum Pixel Resolution (3D)/μm | Geometric Magnification (2D) | Geometric Magnification (3D) |
|---|---|---|---|---|---|---|
| 240 | 320 | 1 | 4.5 | ≤2 | 1.460–180 | 1.46–100 |

For the original model subjected to CT scanning and 3D reconstruction, the cross-sectional images were exported at equal spacing, with the spacing value set to 0.15 mm, and each test piece was capable of exporting a total of 1000 images.

## 3. CT Image Feature Analysis

Taking SMA-16 as an example, there were roughly five peaks shown in the gray histogram of the asphalt-mixture CT image. Under normal conditions, each peak is represented in ascending order of gray scale: background, voids and artifacts, asphalt mortar, and aggregates. In practice, however, the artifacts appeared in the center of the test piece within the image, i.e., the brightness near the center was lower than near the edge, and the brightness along the radius of the test piece was nonlinear, which is largely attributed to ray hardening. Consequently, the gray value of some aggregate shifted to the wave peak of mortar. In case of artifacts being left unprocessed, the aggregate was extracted directly using the global threshold segmentation, causing the aggregate within the artifacts to be recognized as background removal, thus resulting in the loss of aggregate information as shown in Figure 2. This is where the difficulty lies when processing asphalt-mixture CT images.

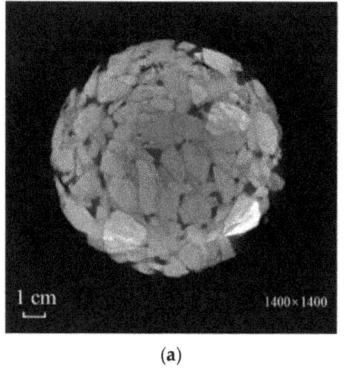
(a)

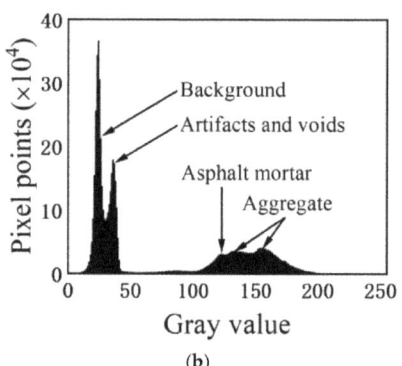
(b)

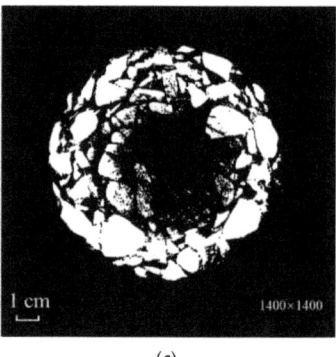
(c)

Figure 2. Characteristics of CT images. (a) Original image; (b) gray histogram; (c) global segmentation.

## 4. Image-Processing Method for Asphalt Mixture

### 4.1. Adaptive CT Image-Processing Method

#### 4.1.1. Obtain the Center and Radius of the Test Piece

It is important to determine the center and radius of the test piece for subsequent image-processing operations. During the actual CT scanning process, however, there is no way to guarantee that the test piece is located at the center of the scanning platform. Thus, the test piece can show different center positions in the images. Therefore, it is necessary to identify the center and radius of the test piece in the image.

Firstly, the matrix of the entire image was extracted into a rectangular coordinate system, with the row and column numbers of the pixel points representing their Y and X coordinates respectively.

Then, the image was segmented using a single threshold. A range of values was set for the radius of the circle surface is, starting from the minimum value, and a circular surface with that radius value was drawn and allowed to move freely over the binarized image. As shown in Figure 3, the maximum number of white pixels it can cover was recorded, and then the radius value was increased by one pixel length. The above steps were repeated until the radius of the circular surface reached the maximum value, or the maximum number of white pixels covered after increasing the radius value by one pixel length remained unchanged. At that time, the iteration process was discontinued.

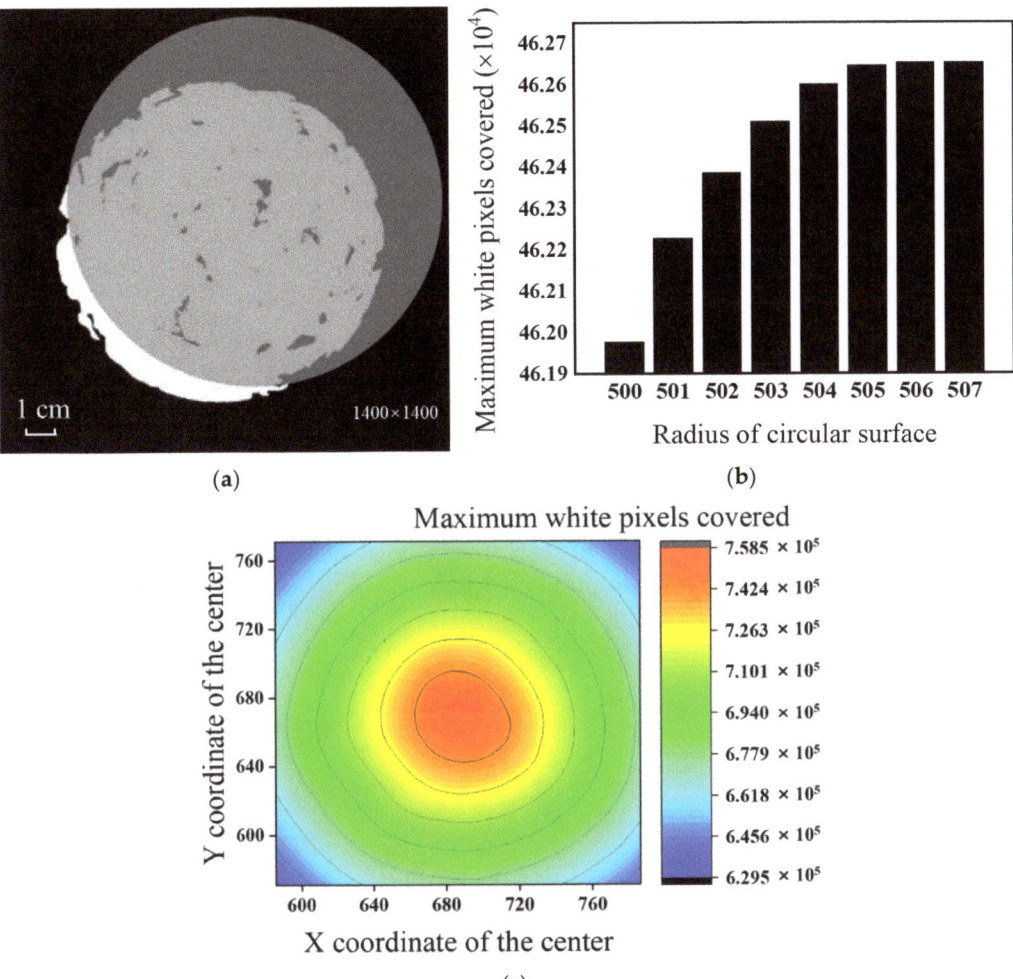

**Figure 3.** Diagram of center and radius's determination. (a) Circular surface covering white pixels; (b) determined radius of the test piece; (c) determined center of the test piece.

Initially, following the method described above, the upper limit on the radius of the circular surface was excessively small. The upper limit was increased and iteration continued. In the second case, the maximum number of white pixels that can be covered by the circular surface with the value of adjacent radius was identical, suggesting that the smaller value was the radius of the test piece. Then, the circular surface was drawn with the radius of the test piece, so that it could be moved within the image. When it covered the maximum number of white pixels, the center of the circular surface overlapped with the center of the test piece, meaning that the center of the test piece could be determined.

Finally, the radius and center of test piece were used for segmenting the image to remove the redundant information while improving the processing efficiency.

4.1.2. Image Brightness Homogenization

Since the distribution of the artifacts was a circular range concentric with the test piece, the brightness adjustment operation was conducted by calculating the distribution of the grayscale of the test piece along the radius direction, with the procedures detailed as follows.

(1) Zero processing of the grayscale of voids. Since the extracted object used in this study was aggregated, it was necessary for mortar and voids to be removed as background. However, mortar cannot be removed due to the presence of artifacts, resulting in the grayscale of mortar and making the aggregates overlap each other. In order to prevent this from impacting the calculation of the statistical distribution of grayscale along the radius, multi-threshold segmentation and image subtraction [22] were first performed to remove the voids, as shown in Figure 4.

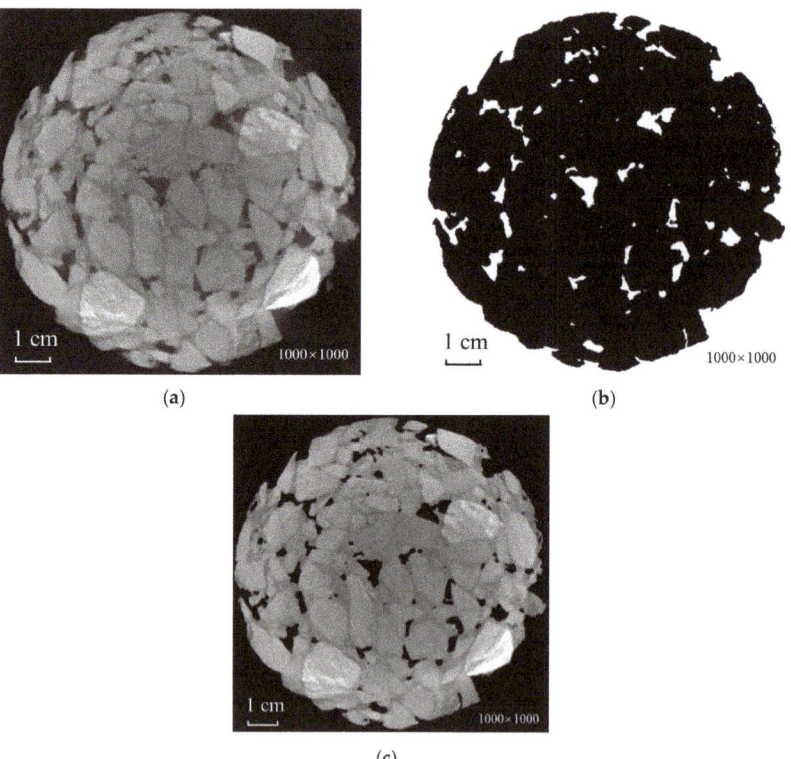

**Figure 4.** Zero processing of grayscale voids. (**a**) Cropped image; (**b**) extraction of voids (white part); (**c**) voids' gray value return to zero.

(2) Calculating the distribution of grayscale along the radial direction. Based on the previously determined center of the test piece, the average gray values of the pixels on the

approximate ring concentric with the test piece were calculated one by one. The expression of each approximate ring is shown in (1), while the approximate ring is shown in Figure 5a.

$$i - 1 < \sqrt{(x - x_0) + (y - y_0)} \leq i \qquad (1)$$

where $x_0$ and $y_0$ represent the horizontal and vertical coordinates of the center of the test piece, respectively; $x$ and $y$ indicate the horizontal and vertical coordinates of the pixel points on the $i$-th approximate ring, respectively; $i$ denotes the outer diameter of the approximate ring, and the value ranges from one pixel length to radius.

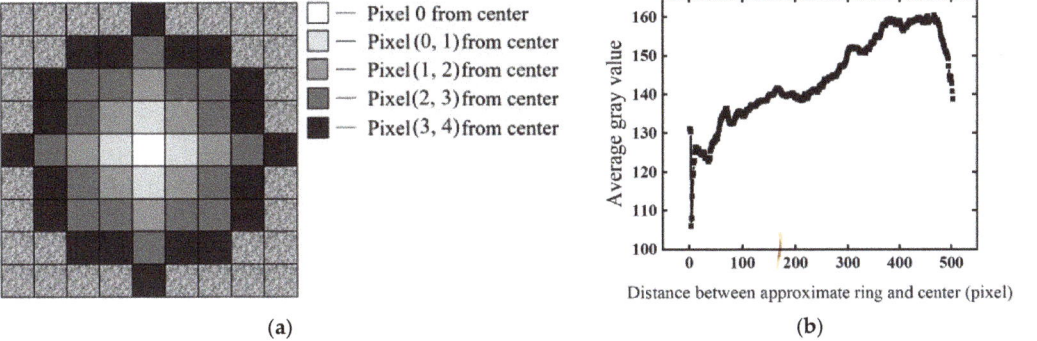

**Figure 5.** Obtained gray distribution along the radial direction. (**a**) Approximate ring concentric with the test piece; (**b**) radial distribution of grayscale.

After determining the approximate ring that different pixels belong to, the average gray value of the pixels on each approximate ring was calculated, as shown in Formula (2).

$$Average_i = Sum_i / Num_i \qquad (2)$$

where $Sum_i$ represents the sum of the pixels' gray value with the gray value of [0, 180] on the $i$-th approximation ring; $Num_i$ indicates the number of pixels with the gray value of [0, 180] on the $i$-th approximation ring; and $Average_i$ denotes the average gray value of pixels with the gray value of [0, 180] on the $i$-th approximation ring.

If the radius of the test piece is 503 pixels in length, 503 average values can be calculated. To avoid affecting the brightening operation, inclusion was limited to pixels with gray values greater than 0 but less than 180 before the average value was calculated. This is because the aggregates with large gray value and voids did not need to be increased in brightness, so they were excluded from the calculation. Figure 5b shows the radial distribution of gray value.

(3) Based on the average gray value of the pixels in each approximate ring, 180 was taken as the target gray value. When the average was no greater than 180, the gray value of the pixels on the $i$-th approximate ring was processed as follows, and the pixels in the whole image whose gray value exceeded 180 were made equal to 180:

$$gray_{x,y} = gray_{x,y} + 180 - Average_i \qquad (3)$$

where $gray_{x,y}$ represents the pixels whose gray value was less than 180 on the $i$-th approximate ring; $x, y$ indicate their horizontal and vertical coordinates; and 180 is the target gray value.

The one-dimensional entropy of an image reflects the average amount of information contained in the image, and provides the aggregation characteristics of gray distribution for the image. The larger the entropy is, the greater the probability of different levels of gray in the image. One-dimensional entropy of the image was introduced to evaluate the brightness homogenization process. The formula is:

$$H = -\sum_0^{255} p_i \times \log_2 p_i \qquad (4)$$

where $H$ represents the image entropy and $p_i$ indicates the probability of the $i$-th gray value appearing in the image, obtainable from the gray histogram.

The results of image entropy calculation before and after processing were between 5.336 and 4.869. The entropy value of images after brightness homogenization was shown to be significantly lower than before processing, suggesting that the gray value of images after processing was more uniform and the processing effect was satisfactory.

The problem of artifacts was effectively reduced after the image had been processed by brightness homogenization as shown in Figure 6. Additionally, the differences in grayscale between the central region and the edge region of the test piece were reduced significantly, and the brightness of the image was more uniform, which facilitated the threshold segmentation of the image.

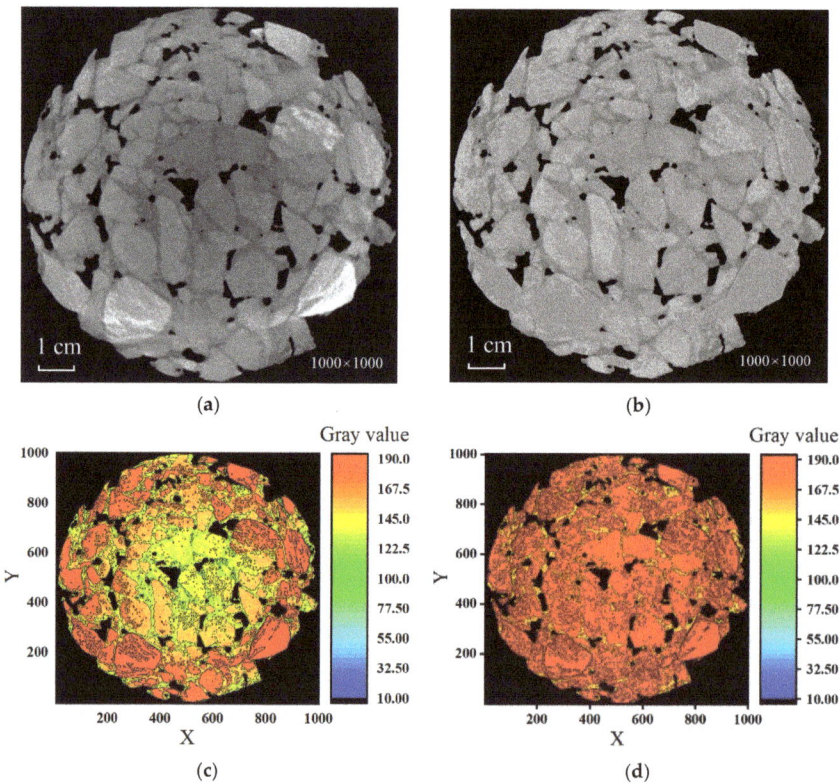

**Figure 6.** Diagram of the effect of luminance homogenization. (**a**) Original image; (**b**) image after brightness homogenization; (**c**) grayscale contour map of original image; (**d**) after brightness homogenization.

4.1.3. Image Filtering and Noise Reduction

The existence of noise makes the threshold segmentation program prone to errors in the recognition of aggregate edge, and can cause adverse effects on the calculation of threshold, as a result of which the threshold segmentation program shows sensitivity to noise. It was thus necessary to filter the image for the purpose of noise reduction. In this study, the adaptive low-pass Wiener filter [29] was applied to filter the grayscale images, with $3 \times 3, 5 \times 5, 7 \times 7, 9 \times 9, 11 \times 11$ selected respectively as filter windows, to estimate the mean and standard deviation of local images, shown in Figure 7.

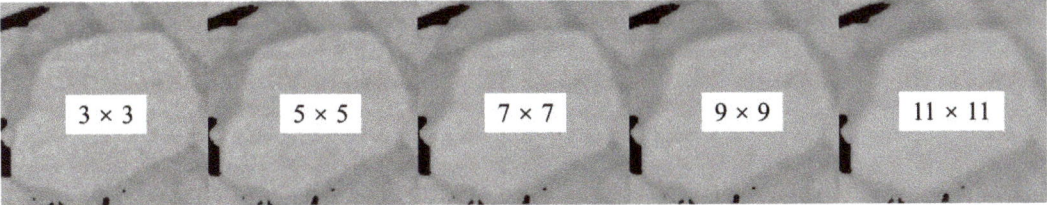

**Figure 7.** Comparison of filtering effects using filters of different sizes.

It can be seen in the figure that the amount of noise around the aggregate particles diminished with the increase of the filter window. Furthermore, the gray value of aggregate became more uniform. Meanwhile, however, the application of a larger filter caused severe loss of image details, thus resulting in deterioration of image clarity and aggregate edge identification. In this study, $7 \times 7$ filter window was applied to reduce the image noise.

The effect of the $7 \times 7$ filter is shown in Figure 8, revealing that the peak of the representative aggregate in the gray histogram narrowed but shifted upward after noise reduction, indicating that the aggregate gray value tended to be uniform and that the effect of noise reduction was evident.

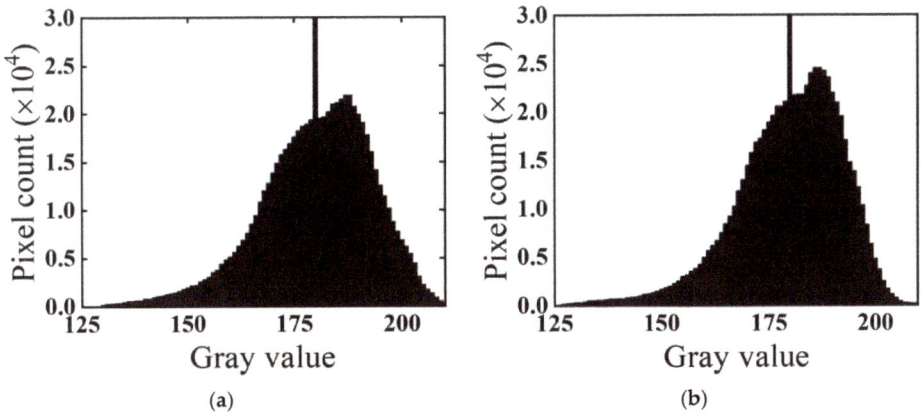

**Figure 8.** Diagram of noise reduction effect. (**a**) Before noise reduction; (**b**) after noise reduction.

4.1.4. Global Otsu Threshold Segmentation

Depending on the grayscale characteristics of the image, the Otsu algorithm divides the image into two parts: background and foreground, with the variance taken as a measurement of the uniformity of grayscale distribution. Thus, the greater the variance between the background and foreground, the more significant are the differences between the two parts of the image. In this situation, the threshold calculated by the Otsu algorithm is considered best when the variance between the two parts is the largest. Therefore, Otsu thresholding based on the whole image was selected to perform binarization. At this stage, the gray value of voids and artifacts in the image was returned to zero, and the image contained as few as three parts, i.e., black background, asphalt mortar, and aggregate. The aim of binarization is to separate the aggregate. Therefore, in this study, Otsu was applied to calculate the double threshold. The smaller threshold was the boundary between the background and the asphalt mortar, while the larger threshold was the boundary between asphalt mortar and aggregate. The higher threshold was used for segmentation, with the gray value of mortar restored to zero. The result is shown in Figure 9.

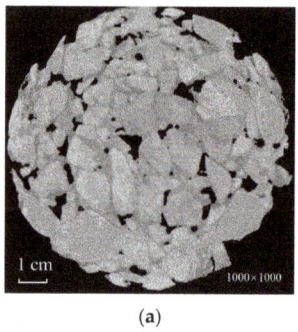 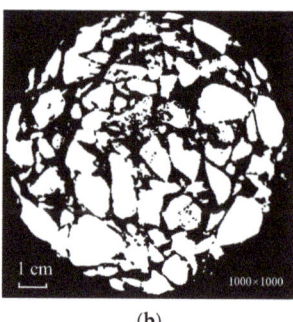

(a)                  (b)

**Figure 9.** Diagram of the adaptive processing method's threshold segmentation. (**a**) Before threshold segmentation; (**b**) after threshold segmentation.

4.1.5. Image Morphological Processing

After thresholding segmentation, the holes require filling and conglutinated aggregates need to be separated. The holes distributed in the image can be classed into two categories. One results from the uneven texture of the aggregate and causes the low grayscale after CT scanning. This can be eliminated by the algorithm, and holes of this kind need to be filled. The other category relates to the proximity between the two adjacent aggregate edges, where the gaps between them become holes after threshold segmentation, and should be retained. This lays a foundation for the subsequent separation of the adhered aggregates by the watershed segmentation algorithm.

In this study, the image was first processed in reverse color, and then eroded [22], as a result of which the adjacent aggregate was separated. In the meantime, the holes inside the aggregates expanded, while remaining closed holes. The closed holes were filled, the holes inside the aggregates were removed, and the image was dilated to restore the size of the aggregates. Finally, the watershed algorithm [17,18,30] was applied to separate further the adhered aggregates. The outcome of the treatment is shown in Figure 10.

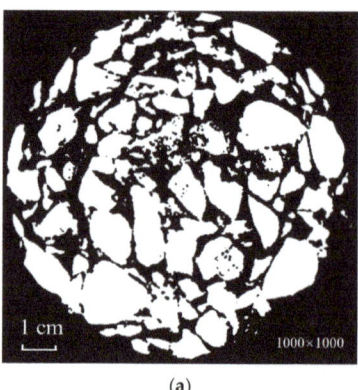 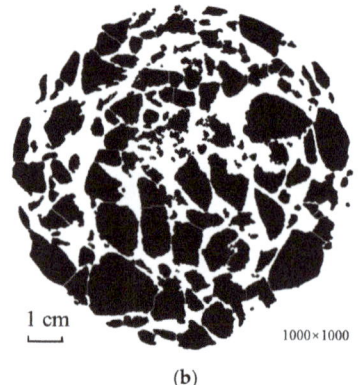

(a)                  (b)

**Figure 10.** Comparison of before and after morphological treatment. (**a**) Before morphological processing; (**b**) after morphological processing.

## 4.2. Ring Division Method

With regard to the CT image processing of asphalt mixture, it has been proposed by some scholars that the mixture can be divided into multiple ring blocks, and that the impact of artifacts on threshold segmentation can be mitigated by threshold segmentation being performed in each part of the image. The final image can be obtained by combining the images after binary processing. The core idea is that the artifacts are distributed in a non-linear way along the radial direction of the test piece. In general, if segmentation based on the global threshold is adopted directly, the closer the gray value of the aggregate to the center of the test piece, the lower is the gray value difference. A more significant gray value difference in the aggregate will contribute to the removal of the aggregate from the center of the test piece. If the image is divided into ring blocks, however, the difference in the aggregate's gray level in each part of the image can be reduced, while the precision of threshold segmentation can be improved.

The asphalt-mixture CT image was split into three rings and one circular surface, in line with the ring division rule, with each part overlapping each other. Each part of the image was filtered and denoised, the threshold was segmented, and the combined image was morphologically processed, as shown in Figures 11 and 12, respectively.

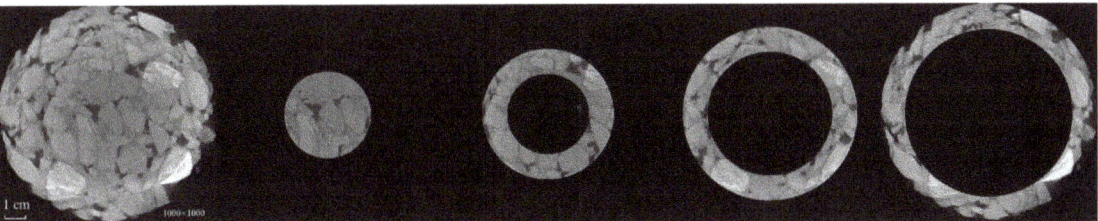

**Figure 11.** Partition of original image.

**Figure 12.** Diagram of the image-processing method based on ring division.

It was found that the CT image-processing method based on ring division was effective for mitigating the impact of artifacts on threshold segmentation, by splitting rings. The loss of aggregate recognition was less severe than with global threshold segmentation, and boundaries were clearly seen when the binary image was combined with a complete image. Compared with the adaptive CT image-processing method, an evident gap was observed in the completeness of the aggregate recognition.

## 5. Grading Verification of 3D Reconstruction Model

To validate the image-processing method, the images processed using an adaptive image-processing method, the ring division method, and a global threshold segmentation method, respectively, were imported into Avizo software for 3D reconstruction [31,32]. The model is shown in Figure 13.

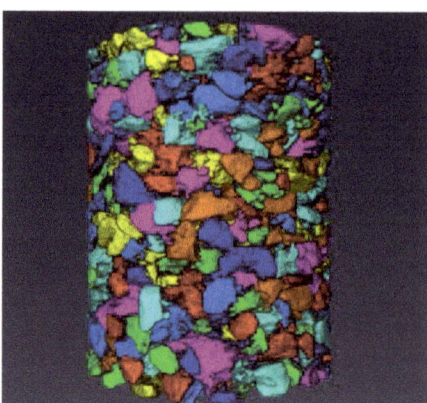

**Figure 13.** Three-dimensional reconstruction model of asphalt mixture.

Applying the extracted volume information for the particles in the model, the particles were regarded as spheres, and virtual sieving was conducted. For the virtual sieving, the software automatically identified the volume of the particles and equate each particle with a sphere of the same volume. The diameter of the sphere was considered to be the diameter of the particle. Because the quality of the CT images was not as high as required, the accuracy of aggregate extraction below 4.75 mm was far from satisfactory. Furthermore, it was difficult for some of the fine aggregate to be separated by the algorithm, due to its close contact and adhesion with coarse aggregates. Its small area caused some fine aggregates to be eroded. Consequently, when the grading curve was drawn, it was suitable only for counting aggregates with a diameter of 4.75 mm or over. The true grading required conversion to a diameter of 4.75 mm or more. The comparison of gradation is shown in Figure 14. When model gradation was evaluated, it was also possible to regard the model gradation as a regression fitting to the true gradation, and to use statistical parameters to evaluate the regression fitting effect.

As shown in Figure 14, it was found that the three-dimensional reconstruction model established from the images processed with the adaptive method was the closest to the real gradation. The error analysis is shown in Table 4. The error of the model using the adaptive method was smaller compared with the method based on ring division, and significantly smaller compared with the method based on global threshold segmentation, suggesting that the adaptive method is more accurate than other methods for the identification and segmentation of aggregates. Thus, it can restore the true gradation of the aggregates to the greatest extent. It was also revealed that there was little difference in accuracy of aggregate recognition between different kinds of asphalt mixtures, which suggests that asphalt mixtures of different kinds have good adaptability to the application of CT images.

**Table 4.** Gradation error analysis.

| Evaluation Parameter | SMA-16 | | | OGFC-16 | | | AC-16 | | |
|---|---|---|---|---|---|---|---|---|---|
| | Adaptive | Ring | Global | Adaptive | Ring | Global | Adaptive | Ring | Global |
| Absolute error | 4.757 | 8.019 | 17.275 | 4.151 | 7.178 | 17.195 | 3.497 | 11.477 | 27.296 |

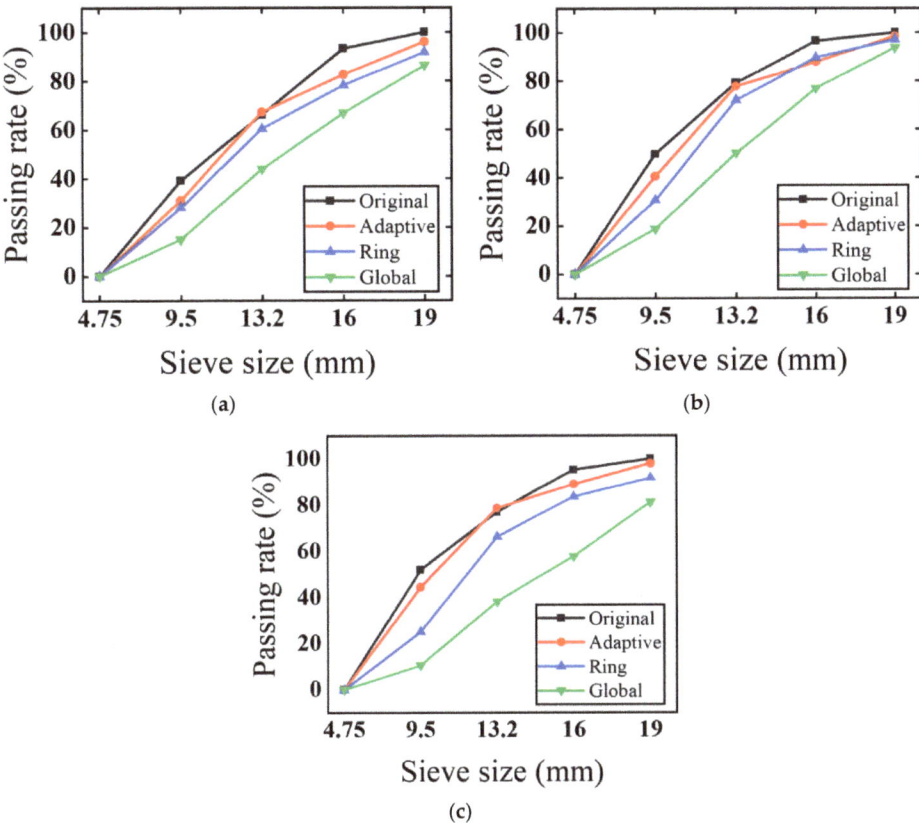

**Figure 14.** Comparison of three test pieces' gradation curves. Adaptive: image processing based on the adaptive image-processing method. Ring: image processing based on ring division. Global: image processing based on global segmentation. (**a**) SMA grading comparison; (**b**) OGFC grading comparison; (**c**) AC grading comparison.

## 6. Conclusions

(1) The adaptive CT image-processing method was based on the recognition of the center and radius of the test piece in the image. After the voids were removed from the image, the gray distribution was counted along the radial direction of the test piece, the grayscale distribution along the radial direction of the test piece was adjusted to homogenize the brightness, and the one-dimensional entropy value of the image was applied to characterize the effect of brightness homogenization. As indicated by the results, the entropy value reduced significantly after the image was brightened. The problem of artifacts in CT images of asphalt mixture has been effectively resolved.

(2) The larger the window of the Wiener filter, the more significant was the noise reduction effect on the CT image. Simultaneously, however, it caused image sharpness to be reduced. The 7 × 7 Wiener filter was able to produce a better effect of noise reduction without compromising image sharpness.

(3) In order to avoid subjectivity caused by the visual observation of the image processing effect, the binary images obtained by the three methods were imported into the 3D reconstruction software to extract the particle information contained in the reconstruction model and to perform virtual sieving. The model's gradation curve of aggregates of 4.75 mm and above was obtained, and the errors between the model gradation

curves and the real gradation curves were characterized by statistical parameters. It was discovered that compared with other methods, model gradation caused the least significant error from the real gradation, the degree of aggregate adhesion was also lower, and the adaptive image-processing method showed a strong adaptability to the processing of CT images for different kinds of mixtures.

**Author Contributions:** L.Z. (Lei Zhang) Writing—review & editing, G.Z., Y.T. Data curation, K.Z., C.X. Writing—original draft, Y.W. Investigation, C.C. Resources, L.Z. (Liting Zhao), X.L. Software, J.X. Validation, X.L. Visualization, L.W. Methodology, Y.T. Supervision, C.X. Funding Acquisition. All authors have read and agreed to the published version of the manuscript.

**Funding:** This paper was supported by the National Key Research and Development Program of China (No. 2018YFB1600203), National Natural Science Foundation of China (No. 51908168), Postdoctoral Science Foundation of China (No. 2019M651192 and No. 2020T130150), and Heilongjiang Postdoctoral Fund (No. LBH-Z19163).

**Institutional Review Board Statement:** Not applicable.

**Informed Consent Statement:** Not applicable.

**Data Availability Statement:** All the data available in main text.

**Conflicts of Interest:** The authors declare that they have no known competing financial interests or personal relationships that could have appeared to influence the work reported in this paper.

## References

1. Masad, E.; Muhunthan, B.; Shashidhar, N.; Harman, T. Internal Structure Characterization of Asphalt Concrete Using Image Analysis. *J. Comput. Civ. Eng.* **1999**, *13*, 88–95. [CrossRef]
2. De Chiffre, L.; Carmignato, S.; Kruth, J.-P.; Schmitt, R.; Weckenmann, A. Industrial applications of computed tomography. *Cirp Ann. Manuf. Technol.* **2014**, *63*, 655–677. [CrossRef]
3. Gao, H.; Zhang, L.; Chen, Z.; Xing, Y.; Li, S. Beam Hardening Correction for Middle-Energy Industrial Computerized Tomography. *IEEE Trans. Nucl. Sci.* **2006**, *53*, 2796–2807. [CrossRef]
4. Hanna, R.D.; Ketcham, R. X-ray Computed Tomography of Planetary Materials: A Primer and Review of Recent Studies. *Geochemistry* **2017**, *77*, 547–572. [CrossRef]
5. Lifton, J.; Mcbride, J. The Application of Beam Hardening Correction for Industrial X-Ray Computed Tomography. In Proceedings of the 5th International Symposium on NDT in Aerospace, Singapore, 13–15 November 2013.
6. Jennings, R.J. A Method for Comparing Beam-Hardening Filter Materials for Diagnostic Radiology. *Med. Phys.* **1988**, *15*, 588–599. [CrossRef] [PubMed]
7. Wang, M. *Industrial Tomography: Systems and Applications*; Elsevier: Amsterdam, The Netherlands, 2015.
8. Chen, S.; Xi, X.; Li, L.; Luo, L.; Han, Y.; Wang, J.; Yan, B. A Filter Design Method for Beam Hardening Correction in Middle-Energy X-Ray Computed Tomography. In Proceedings of the 8th International Conference on Digital Image Processing (ICDIP 2016), Chengu, China, 20–22 May 2016. [CrossRef]
9. Hammersberg, P.; Mångård, M. Correction for Beam Hardening Artefacts in Computerised Tomography. *J. X-ray Sci. Technol.* **1998**, *8*, 75–93.
10. Yan, C.H.; Whalen, R.; Beaupre, G.; Yen, S.; Napel, S. Reconstruction Algorithm for Polychromatic CT Imaging: Application to Beam Hardening Correction. *IEEE Trans. Med. Imaging* **2002**, *19*, 1–11. [CrossRef]
11. Yang, F.; Zhang, D.; Zhang, H.; Huang, K. Cupping Artifacts Correction for Polychromatic X-Ray Cone-Beam Computed Tomography Based on Projection Compensation and Hardening Behavior. *Biomed. Signal Process. Control* **2020**, *57*, 101823. [CrossRef]
12. Hasheminejad, N.; Pipintakos, G.; Vuye, C.; De Kerf, T.; Ghalandari, T.; Blom, J.; Bergh, W.V.D. Utilizing deep learning and advanced image processing techniques to investigate the microstructure of a waxy bitumen. *Constr. Build. Mater.* **2020**, *313*, 125481. [CrossRef]
13. Margaritis, A.; Pipintakos, G.; Varveri, A.; Jacobs, G.; Hasheminejad, N.; Blom, J.; Bergh, W.V.D. Towards an enhanced fatigue evaluation of bituminous mortars. *Constr. Build. Mater.* **2021**, *275*, 121578. [CrossRef]
14. Chen, L.; Wang, Y.H. Improved Image Unevenness Reduction and Thresholding Methods for Effective Asphalt X-Ray CT Image Segmentation. *J. Comput. Civ. Eng.* **2017**, *31*, 04017002. [CrossRef]
15. Zhang, S.W.; Zhang, X.N.; Wu, Z.Y.; Shi, L.W. Research on Asphalt Mixture Injury Digital Image Based on Enhancement and Segmentation Processing Technology. *Appl. Mech. Mater.* **2013**, *470*, 832–837. [CrossRef]
16. Adhikari, S.; You, Z. Investigating the Sensitivity of Aggregate Size within Sand Mastic by Modeling The microstructure of an Asphalt Mixture. *J. Mater. Civ. Eng.* **2011**, *5*, 580–586. [CrossRef]

17. Hassan, N.A.; Airey, G.D.; Khan, R.; Collop, A.C. Nondestructive Characterisation of The Effect of Asphalt Mixture Compaction on Aggregate Orientation and Segregation Using X-ray Computed Tomography. *Int. J. Pavement Res. Technol.* **2012**, *5*, 84–92. [CrossRef]
18. Shi, L.; Wang, D.; Jin, C.; Li, B.; Liang, H. Measurement of Coarse Aggregates Movement Characteristics within Asphalt Mixture Using Digital Image Processing Methods. *Measurement* **2020**, *163*, 107948. [CrossRef]
19. Bezdek, J.C.; Ehrlich, R.; Full, W.E. FCM: The Fuzzy C-Means Clustering Algorithm. *Comput. Geosci.* **1984**, *10*, 191–203. [CrossRef]
20. Wei, J.J.; Li, H.B.; Wan, C. X-Ray CT Image Segmentation of Asphalt Concrete Based on Fuzzy C-Means. *Appl. Mech. Mater.* **2012**, *170–173*, 3444–3448. [CrossRef]
21. Zeng, L.W.; Zhang, S.X.; Zhang, X.N. The Research on Aggregate Microstructure Uniformity Image Processing of Asphalt Mixture Based on Computer Scanning Technology. *Adv. Mater. Res.* **2013**, *831*, 393–400. [CrossRef]
22. Tan, Y.; Liang, Z.; Xu, H.; Xing, C. Internal deformation monitoring of granular material using intelligent aggregate. *Autom. Constr.* **2022**, *139*, 104265. [CrossRef]
23. Tan, Y.; Liang, Z.; Xu, H.; Xing, C. Research on Rutting Deformation Monitoring Method Based on Intelligent Aggregate. *IEEE Trans. Intell. Transp. Syst.* **2022**, 3175060. [CrossRef]
24. Yao, Q.; Qi, S.; Li, H.; Yang, X.; Li, H. Infrared Image-Based Identification Method for The Gradation of Rock Grains Using Heating Characteristics. *Constr. Build. Mater.* **2020**, *264*, 120216. [CrossRef]
25. Bruno, L.; Parla, G.; Celauro, C. Image Analysis for Detecting Aggregate Gradation in Asphalt Mixture from Planar Images. *Constr. Build. Mater.* **2012**, *28*, 21–30. [CrossRef]
26. Liu, J.H.; Li, Z. Image Segmentation and Its Effect of Asphalt Mixtures Using Computed Tomography Images Method. *J. Chongqing Jiaotong Univ.* **2011**, *30*, 1335.
27. *JTGE42-2005*; Test Methods of Aggregate for Highway Engineering. People's Communications Press: Beijing, China, 2005.
28. *JTGF40-2004*; Technical Specification for Construction of Highway Asphalt Pavements. People's Communications Press: Beijing, China, 2004.
29. Kumar, S.; Kumar, P.; Gupta, M.; Nagawat, A.K. Performance Comparison of Median and Wiener Filter in Image De-Noising. *Int. J. Comput. Appl.* **2010**, *12*, 24–28. [CrossRef]
30. Campbell, A.I.; Murray, P.; Yakushina, E.; Marshall, S.; Ion, W. New Methods for Automatic Quantification of Microstructural Features Using Digital Image Processing. *Mater. Des.* **2018**, *141*, 395–406. [CrossRef]
31. Li, Y.; Chi, Y.; Han, S.; Zhao, C.; Miao, Y. Pore-throat Structure Characterization of Carbon Fiber Reinforced Resin Matrix Composites: Employing Micro-CT and Avizo Technique. *PLoS ONE* **2021**, *16*, e0257640. [CrossRef] [PubMed]
32. Xu, S.; Chen, H.; Yang, Y.; Gao, K. Fatigue Damage of Aluminum Alloy Spot Welded Joint Based on Defects Reconstruction. *J. Eng. Mater. Technol.* **2019**, *142*, 021001. [CrossRef]

Article

# Improved Procedure for the 3D Reconstruction of Asphalt Concrete Mesostructures Considering the Similarity of Aggregate Phase Geometry between Adjacent CT Slices

Chao Wang [1], Hui Xu [1], Yan Zhang [2], Yiren Sun [1,*], Weiying Wang [3,*] and Jingyun Chen [1]

1 School of Transportation and Logistics, Dalian University of Technology, Dalian 116024, China
2 City Institute, Dalian University of Technology, Dalian 116600, China
3 College of Transportation Engineering, Tongji University, Shanghai 201804, China
* Correspondence: sunyiren@dlut.edu.cn (Y.S.); wywang@tongji.edu.cn (W.W.)

**Abstract:** Existing image segmentation algorithms used for the computed tomography (CT) images of asphalt concrete mostly ignore the similarity of aggregate phase geometry between adjacent CT slices, thus increasing the variability in the aggregate phase pixel values between adjacent slices and leading to a large number of model defects, e.g., interconnected aggregates, flaky aggregates, and incomplete aggregates. The developed mesostructural models with these defects pose a challenge to following simulation operations. To address this issue, an improved procedure for the 3D reconstruction of asphalt concrete mesostructures considering the similarity of aggregate phase geometry between adjacent slices was developed, which includes two adjacent-slice pixel-value-correction algorithms, a multi-directional multiple-correction method, and an image pixel interpolation process. First, the bilinear interpolation algorithm was employed to improve the pixel density of 2D CT images and the average filtering algorithm was used to reduce the noise of the CT images. Subsequently, the OTSU method was employed to separate the asphalt mortar matrix phase from the aggregate phase, and the marker-based watershed segmentation method was used to separate the interconnected aggregates. Finally, the adjacent-slice pixel-value-correction algorithm was used to recover the similarity of aggregate phase geometry between adjacent CT slices, and the multi-directional multiple-correction method was used to further enhance the geometric similarity. The results show that the developed 3D reconstruction procedure removes most of the model defects in the 3D mesostructural model of asphalt concrete, thus realistically maintaining the 3D spatial distribution features and contour characteristics.

**Keywords:** asphalt concrete; 3D reconstruction; digital image processing (DIP); computed tomography (CT); mesostructure

## 1. Introduction

In traditional pavement analysis and design, asphalt concrete is often considered a homogeneous material and its mechanical properties can be determined using relevant laboratory tests [1,2]. However, asphalt concrete is essentially a heterogeneous material, which actually contains the aggregate phase, asphalt mortar matrix phase, and air void phase at the mesoscale [3,4], and thus its overall mechanical behavior is closely associated with the interaction between various components in the mesostructure.

Numerical simulation approaches, such as discrete element (DE) and finite element (FE) methods, have been applied to investigate the mesomechanical behavior of asphalt concrete; however, a reliable mesostructural model of asphalt concrete needs to be realistically established before using these numerical methods to deeply explain the mechanical behavior and damage evolution mechanism of asphalt concrete at the mesoscale. Currently, the methods for establishing asphalt concrete mesostructures are classified into two major categories, namely the random aggregate generation method [5–8] and the CT-based digital image processing (DIP) technique [9–11]. Compared with the random aggregate generation method, the CT-based DIP technology can better reflect the real mesostructural morphological characteristics and spatial distribution of aggregates. Two types of CT-based DIP approaches, i.e., the 2D and 3D methods, are involved. The 2D methods are now more widely used in the asphalt paving community due to their simplicity in simulation. However, the 2D methods cannot reasonably explain the mechanical behavior because aggregates in asphalt concrete actually interlock with each other in a 3D fashion. In the common procedure for 3D CT image processing, filtering algorithms are first employed to reduce the noise of the 2D CT slice grayscale images. Then, image segmentation algorithms are adopted to segment the aggregate phase and the asphalt mortar matrix phase in 2D grayscale images to obtain 2D binary images. Finally, the processed 2D binary images of asphalt concrete are reconstructed into a 3D mesostructural voxel model along the CT acquisition direction. In this procedure, image segmentation is a key step that aims to facilitate the development of a geometrically reasonable 3D model by separating the aggregate phase from the asphalt mortar phase in 2D slices and eliminating interconnected aggregates in 2D slices. In the subsequent simulation, using these geometrically reasonable 3D numerical models will help improve the modeling accuracy and ensure computational convergence.

Among image segmentation techniques, the manual threshold selection (TH) [12] is the most basic one. The efficacy of this method is limited by the skill proficiency of the operator and it cannot guarantee the reproducibility of the image processing results. Therefore, it is not suitable for the high-volume image processing task required for 3D reconstruction [13]. OTSU is an automatic threshold determination method [14]. This method first assumes a threshold value in the grayscale range of image pixels, then classifies the pixels in the grayscale image into target and background categories according to the assumed threshold value, and finally uses the variance of the grayscale values of the two categories as the judgment index to obtain the best threshold value for this image by exhaustively searching for the maximum value of the variance [15,16]. Gong et al. [17] developed a 2D OTSU method. This method improves the segmentation quality of low signal-to-noise ratio images by simultaneously considering the gray values of image pixels and their 2D coordinate information [18]. Bhandari et al [19] developed the 3D-OTSU method. This method simultaneously considers the pixel gray value, neighborhood pixel gray median, and neighborhood pixel gray average to improve the segmentation quality of the image with a low signal-to-noise ratio and low contrast [20].

Although both the TH and OTSU methods can segment the aggregate phase and asphalt mortar matrix phase in the 2D CT slice images of asphalt concrete, a large number of model defects can be found in the obtained 3D model, such as interconnected aggregates, flaky aggregates, and incomplete aggregates, after 3D reconstruction is performed on the CT slice sets that have undergone the 2D image segmentation. Actually, there are huge differences in aggregate morphology for 2D slices and 3D models. The main reason for this discrepancy is that existing studies mostly ignore the connection of the pixels characterizing the aggregate phase between the adjacent slices in the procedure of obtaining optimal 2D binary image segmentation results, which inevitably increases the geometric differences in the aggregate phases between adjacent slices, and the accumulation of such geometric differences eventually leads to a large number of model defects in the constructed 3D voxel models. These model defects can adversely impact the convergence of simulation operations and even the correctness of simulation results. Due to the presence of numerous aggregates in the asphalt concrete mesostructural model, these defects are difficult to remove manually using software.

In the set of asphalt concrete CT slices, the 2D geometry and location information of the same aggregate possesses a natural similarity between adjacent slices. From a pixel perspective, this connection manifests itself as a continuity of pixel information at the same coordinates between adjacent slices. In order to overcome the above-mentioned model defects, this study developed a new procedure for the 3D reconstruction of asphalt concrete mesostructures considering the similarity of aggregate phase geometry between adjacent slices, which includes two adjacent-slice pixel-value-correction algorithms, a multi-directional multiple-correction method, and an image pixel density increase process.

## 2. Methodology

### 2.1. Preparation of Asphalt Concrete Specimens

In this study, asphalt concrete was considered as a two-phase particulate composite consisting of coarse aggregates (greater than 2.36 mm) and an asphalt mortar matrix containing the remaining fine aggregates, mineral fillers, and air voids.

To demonstrate the effectiveness of the developed procedure for the 3D reconstruction of asphalt concrete mesostructures, asphalt concrete specimens with a nominal maximum aggregate size (NMAS) of 13.2 mm were prepared. The asphalt used was PG 58–22 binder. The asphalt mix with an asphalt content of 3.88% was compacted using a Superpave gyratory compactor (Pine, Grove City, PA, USA). Finally, the compacted asphalt concrete specimens were cut into cylindrical specimens with a diameter of 65 mm and a height of 75 mm with an air void content of 4 ± 1%. Detailed information on the aggregate gradations of the asphalt concrete and asphalt mortar can be found elsewhere [21].

### 2.2. CT Image Acquisition

To obtain the mesostructural information of asphalt concrete, CT scanning was performed on cylindrical asphalt concrete specimens. The CT scanning equipment employed was the Germany Diondo d2 universal (Diondo, Hattingen, Germany) micro nano focus CT system with a 270 kV and 72 µA X-ray source. A set of 2D slice images (stack) representing the asphalt concrete mesostructure was reconstructed using the software XMReconstructor coupled to the micro-CT system. The stack displayed different grayscales in its 2D slice images. The slice spacing was 0.1 mm, with a resolution of 78 µm/pixel, and a total of 750 slices were displayed at 3.94 MB per slice.

The complete sample information of the asphalt concrete specimen consisted of 750 8-bit slices of 1507 pixels × 914 pixels. In each CT slice, pixels of different grayscale intensity levels between 0 (black) and 255 (white) represent different components. Higher-density components tend to yield higher grayscale intensities. In this way, the distribution of materials in the sample can be well identified based on the grayscale intensity values.

### 2.3. Bilinear Interpolation Algorithm

To avoid the jagged distortion of the aggregate contours in CT slices during 2D image segmentation, the bilinear interpolation algorithm was used, which can increase the pixel density of an image without changing the image size. The interpolation process is shown in Figure 1.

Assuming that the pixel number of the image after pixel filling is $k^2$ times the pixel number of the image before pixel filling:

$$k = \frac{q+1}{2} \quad (1)$$

$$\eta_n^{enh}(i+u, j+v) = (1-\frac{u}{q})(1-\frac{v}{q})\eta_n(i,j) + (1-\frac{u}{q})\frac{v}{q}\eta_n(i,j+1) + \frac{u}{q}(1-\frac{v}{q})\eta_n(i+1,j) + \frac{u}{q}\frac{v}{q}\eta_n(i+1,j+1) \quad (2)$$

where $\eta_n$ is the grayscale value of the pixel point in the $n$th slice before pixel interpolation; $\eta_n^{enh}$ is the grayscale value of the new pixel point in the $n$th slice after pixel interpolation; $q+1$ is the number of pixels in horizontal or vertical orientation after pixel interpolation; $i$ is the x-axis pixel coordinate of the corresponding known pixel point; $j$ is the y-axis pixel coordinate of the known pixel point; $u$ is the x-axis coordinate of the new pixel point relative to the surrounding four known pixel points; and $v$ is the y-axis coordinate of the new pixel point relative to the surrounding four known pixel points.

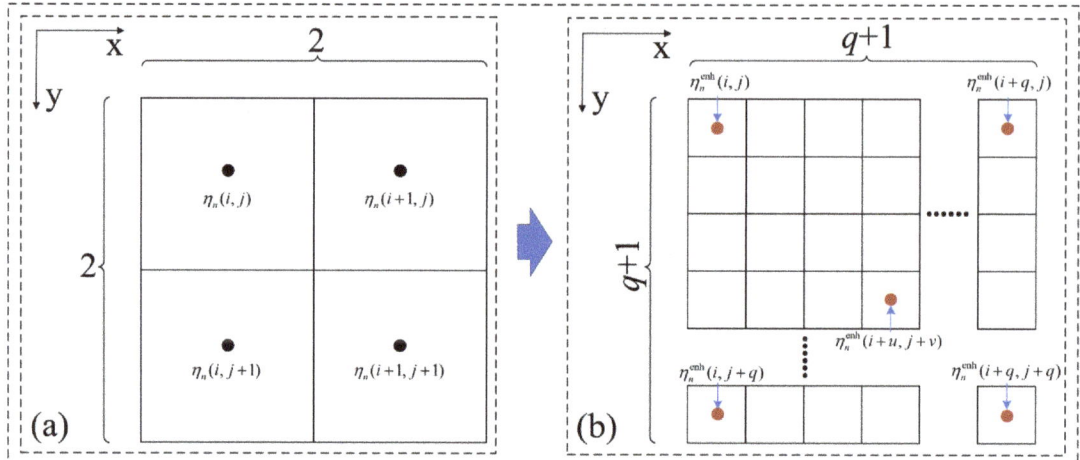

**Figure 1.** Schematic diagram of the bilinear interpolation algorithm: (**a**) before interpolation; (**b**) after interpolation.

### 2.4. Two-Dimensional Image Segmentation Algorithm

In this study, the OTSU method was used to separate the aggregate phase from the asphalt mortar matrix phase, and then the marker-based watershed splitting method was used to split the interconnected aggregates.

The OTSU method classifies the pixels in the grayscale image into two categories, target and background, according to the assumed threshold values, as shown in Figure 2. It first uses the variance of the grayscale values of the two categories as the judgment index to obtain the best threshold value for this figure by an exhaustive search and then transforms the grayscale image into a binary image containing the two categories of target (aggregate phase) and background (asphalt mortar matrix phase) according to the best threshold value.

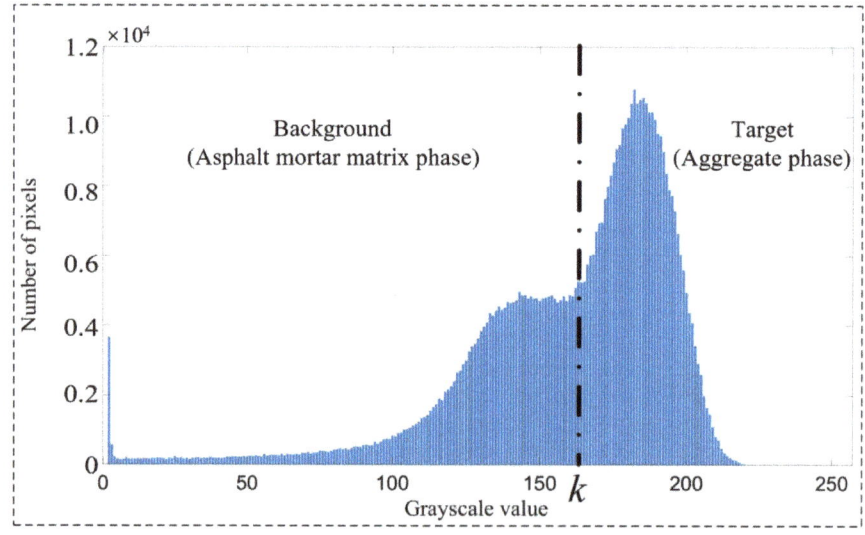

**Figure 2.** Schematic diagram of the OTSU method.

The grayscale values of CT grayscale images range from 0 to 255, with a total of 256 grayscale levels. It is assumed that the number of pixels corresponding to the grayscale value $i$ is $N_i$; the optimal threshold is $k$; the grayscale value less than $k$ is the background class; and the grayscale value greater than or equal to $k$ is the target class. The interclass variance can be expressed as:

$$\delta^2(k) = \omega_{\text{back}}[\mu - \mu_{\text{back}}(k)]^2 + \omega_{\text{targ}}[\mu - \mu_{\text{targ}}(k)]^2 \tag{3}$$

$$\mu = \sum_{i=0}^{255} iP_i \tag{4}$$

$$\mu_{\text{back}}(k) = \frac{\sum_{i=0}^{k-1} iP_i}{\sum_{i=0}^{k-1} P_i} \tag{5}$$

$$\mu_{\text{targ}}(k) = \frac{\sum_{i=k}^{255} iP_i}{\sum_{i=k}^{255} P_i} \tag{6}$$

$$\omega_{\text{back}}(k) = \sum_{i=0}^{k-1} P_i \tag{7}$$

$$\omega_{\text{targ}}(k) = \sum_{i=k}^{255} P_i \tag{8}$$

$$P_i = \frac{N_i}{\sum_{i=0}^{255} N_i} \tag{9}$$

where $\delta^2(k)$ is the corresponding interclass variance when the optimal threshold is assumed to be $k$; $\mu$ is the mean gray value of the whole CT slice; $\mu_{\text{back}}(k)$ is the mean gray value of the corresponding background class pixels when the optimal threshold is assumed to be $k$; $\mu_{\text{targ}}(k)$ is the mean gray value of the corresponding target class pixels when the optimal threshold is assumed to be $k$; $\omega_{\text{back}}(k)$ is the ratio of the total number of background class pixels to the total number of CT slice pixels when the optimal threshold is assumed to be $k$; $\omega_{\text{targ}}(k)$ is the ratio of the total number of target class pixels to the total number of CT slice pixels when the optimal threshold is assumed to be $k$; and $P_i$ is the ratio of pixels with gray value $i$ to the total number of CT slice pixels.

At the maximum value $\max[\delta^2(k)]$, the optimal threshold value $k$ is determined for this slice. At this point, the CT grayscale image will be transformed into a binary image containing only white (aggregate phase) and black (asphalt mortar phase):

$$\varphi(x,y) = \begin{cases} 1 & i(x,y) \geq k \\ 0 & i(x,y) < k \end{cases} \tag{10}$$

where $\varphi(x,y)$ is the binary value of the corresponding coordinate of the binary image after segmentation; $i(x,y)$ is the gray value of the corresponding coordinate of the grayscale image before segmentation; $x$ is the horizontal coordinate corresponding to the pixel in the CT slice image; and $y$ is the vertical coordinate corresponding to the pixel in the CT slice image.

The marker-based watershed segmentation method was applied to segment the interconnected aggregates. By means of Matlab programming, the binary image was first subjected to limit erosion operations to obtain the internal markers of the aggregates. Then, the middle point of the aggregate-to-aggregate contact was calculated as the external markers using the distance function, and the 0-value pixels were used to connect the external markers at different locations as watershed ridges to obtain the watershed ridge segmentation layer. Finally, the watershed ridge segmentation layer was superimposed on the corresponding binary image to separate the inner markers and complete the segmentation of interconnected aggregates.

*2.5. Developed Adjacent-Slice Pixel-Value-Correction Algorithms*

From the pixel perspective, the pixel value distribution of the same aggregate in the CT slice set of asphalt concrete specimens has a natural similarity between adjacent CT slices due to the continuity of geometry. Based on this continuity of pixel values in adjacent layers in the CT slice set, two adjacent-slice pixel-value-correction algorithms are proposed in this paper.

Refer to the direction perpendicular to the CT slice plane as the CT acquisition direction, and define this direction as the Z-axis. Take the set of CT binary images that have experienced 2D image segmentation as the initial CT slice set, whose pixel information in each CT slice can be expressed as:

$$\varphi_n(x,y) = \begin{cases} 0 \\ 1 \end{cases} \tag{11}$$

where $n$ is the serial number of the CT slice in the acquisition direction (Z-axis) (the value range is 1–750); $x$ is the coordinate of the horizontal corresponding to the pixel in the CT slice; $y$ is the coordinate of the vertical corresponding to the pixel in the CT slice.

Depending on the processing purpose, this paper provides two adjacent-slice pixel-correction algorithms. It should be noted that all slices in the initial slice set need to have a uniform image size and pixel density.

Algorithm 1: First, take the intersection of adjacent-slice aggregate phase pixels in the initial slice set. Then, process the intersection of adjacent-slice aggregate phase pixels by morphological expansion operations. Finally, record the processed results in a new slice.

Algorithm 2: Firstly, take the union of adjacent-slice aggregate phase pixels of the initial slice set. Then, process the union of adjacent-slice aggregate phase pixels by morphological erosion operations. Finally, record the processed results in a new slice.

Apply Algorithm 1 or Algorithm 2 to all adjacent slices in the initial slice set to obtain a new slice set with one round of correction completed. Algorithm 1 is suitable for removing model defects in the 3D mesostructural model of asphalt concrete that cannot be well handled in the 2D image segmentation process, e.g., interconnected aggregates, flaky aggregates, and incomplete aggregates. Algorithm 2 is suitable for repairing the loss of pixel information in the outer contours of aggregates during processing. A detailed description of the two adjacent slice pixel correction algorithms is given as follows.

2.5.1. Algorithm 1: First Take the Intersection and Then Perform Morphological Expansion (IMEX)

The intersection of the nth slice $\varphi_n$ and $(n+1)$th slice $\varphi_{n+1}$ in the initial slice set is taken, and the result of the operation is recorded in a new slice $\varphi_n^I$. The pixel values of this slice can be expressed as:

$$\varphi_n^I(x,y) = [\varphi_n(x,y)] \cap [\varphi_{n+1}(x,y)] \tag{12}$$

Figure 3 illustrates the computational process of the IMEX algorithm from the pixel level. It can be found that after performing the intersection-taking operation, the overlapping region of the aggregate phases in the adjacent slices is recorded into the new slice $\varphi_n^I$, while the non-overlapping region is deleted. Therefore, the new slice $\varphi_n^I$ will lose a small amount of pixel information in the aggregate phase. In order to ensure the original volume fraction of aggregates in asphalt concrete without significant changes, it is necessary to take morphological expansion operations for the aggregate phase of the new slice, and the processed result is denoted as $\varphi_n^{\text{IMEX}}$. The 2D four-connected domain kernel function (0 1 0, 1 1 1, 0 1 0) was chosen as the morphological operator in this study.

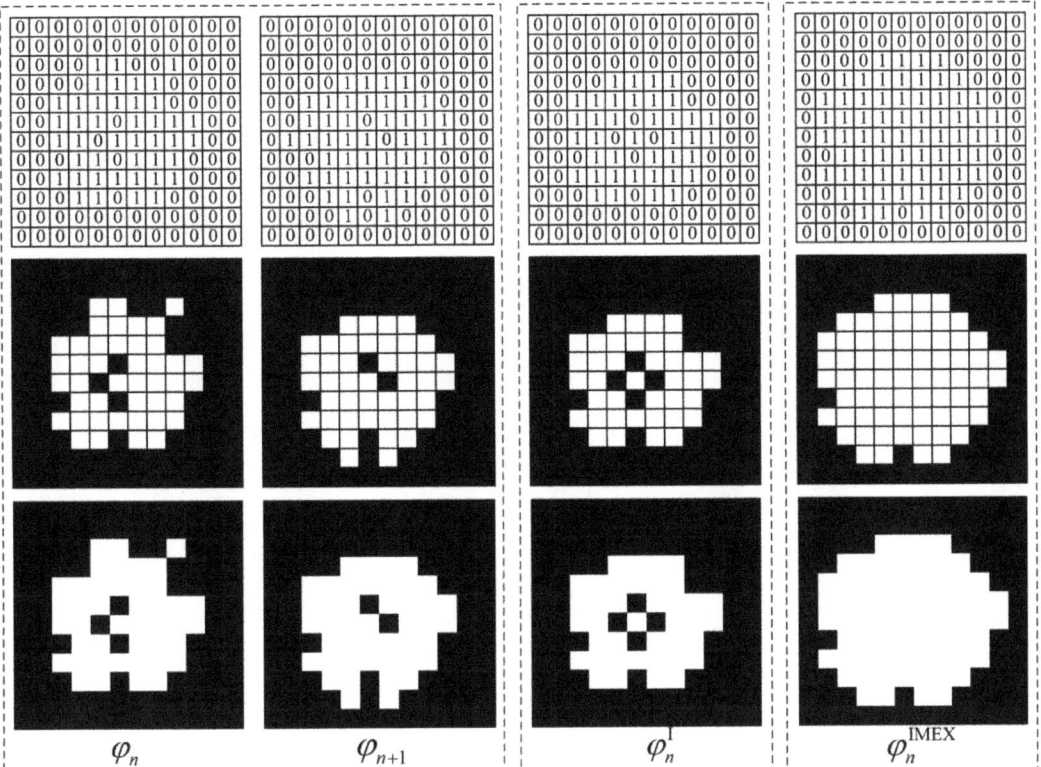

**Figure 3.** Schematic diagram of Algorithm 1 (IMEX).

2.5.2. Algorithm 2: First Take the Union and Then Perform Morphological Erosion (UMER)

The union of the $n$th slice $\varphi_n$ and $(n + 1)$th slice $\varphi_{n+1}$ in the initial slice set is taken, and the result of the operation is recorded in a new slice $\varphi_n^{\text{U}}$. The pixel values of this slice can be expressed as:

$$\varphi_n^{\text{U}}(x,y) = [\varphi_n(x,y)] \cup [\varphi_{n+1}(x,y)] \qquad (13)$$

Figure 4 illustrates the computational process of the UMER algorithm from the pixel level. It can be found that after performing the union-taking operation, the overlapping region and non-overlapping region of the aggregate phases in the adjacent slices are recorded into the new slice $\varphi_n^{\text{U}}$. Therefore, the new slice $\varphi_n^{\text{U}}$ will add a small amount of pixel information from the aggregate phase. To ensure the original volume fraction of the aggregates in the asphalt concrete and reduce the interconnected aggregates, it is necessary to take morphological erosion operations for the aggregate phase of the new slice. The processed result is denoted as $\varphi_n^{\text{UMER}}$.

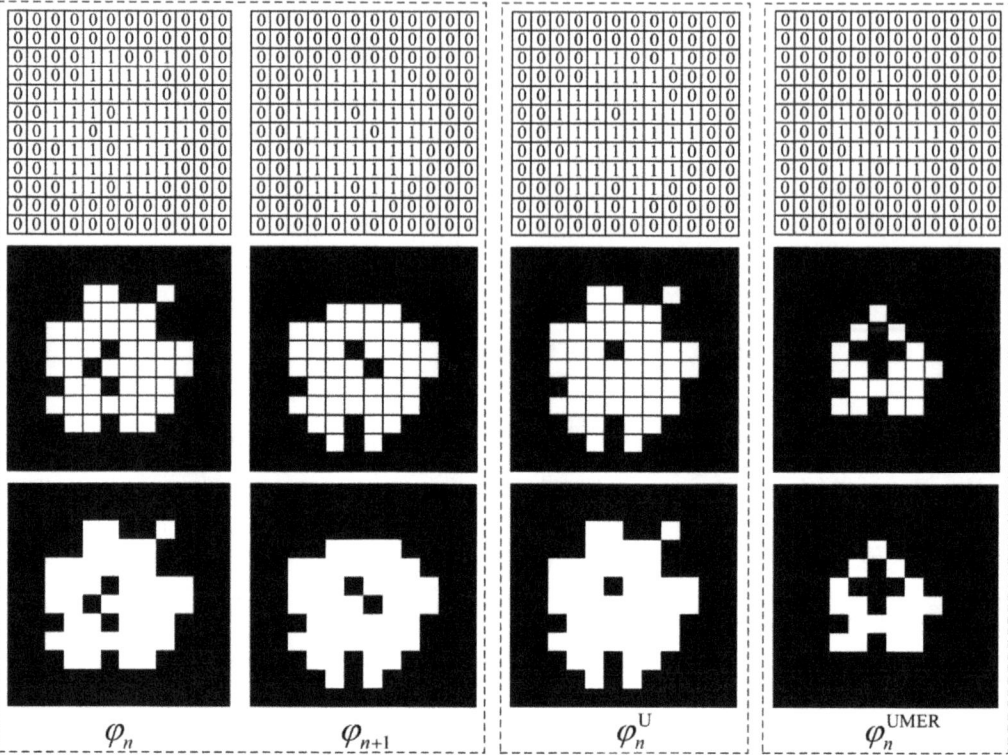

**Figure 4.** Schematic diagram of Algorithm 2 (UMER).

2.5.3. Multiple Correction

As described above, two adjacent-slice pixel-value-correction algorithms are proposed. After one round of correction processing, the aggregate phase in the new slice ($\varphi_n^{IMEX}$ or $\varphi_n^{UMER}$) integrates the aggregate phase morphological features of the $n$th slice $\varphi_n$ and the $(n + 1)$th slice $\varphi_{n+1}$ in the initial slice set. Taking the IMEX algorithm as an example, the multiple-correction process is shown in Figure 5. After two rounds of correction processing, the aggregate phase in the new slice $\varphi_n^{IMEX\_2}$ integrates the aggregate phase morphological features of $\varphi_n$, $\varphi_{n+1}$, and $\varphi_{n+2}$ in the initial slice set. Therefore, it can be expected that as the number of multiple corrections ($i$) increases, the new slice $\varphi_n^{IMEX\_i}$ obtained will integrate more information from the initial slice set and the recovery effect of the similarity of aggregate phase geometry between adjacent CT slices will be enhanced. However, 1-value pixels that distribute interruptedly along the acquisition direction in the slice set can become linear aggregates due to overcorrection. A limit should be set on the number of multiple corrections to avoid the appearance of linear aggregates. This limit is related to the slice spacing of the slice set, and the limit of multiple corrections is 10 in this paper.

It should be noted that after each round of correction processing, the pixel value of the slice at the end of the previous slice set is lost. The slice spacing chosen in this paper is 0.1 mm, i.e., the slice lost in each round of the correction process corresponds to the voxel value information of 0.1 mm thickness at the edge of the model. This model voxel loss is extremely small and thus negligible under the premise of limiting the number of multiple corrections.

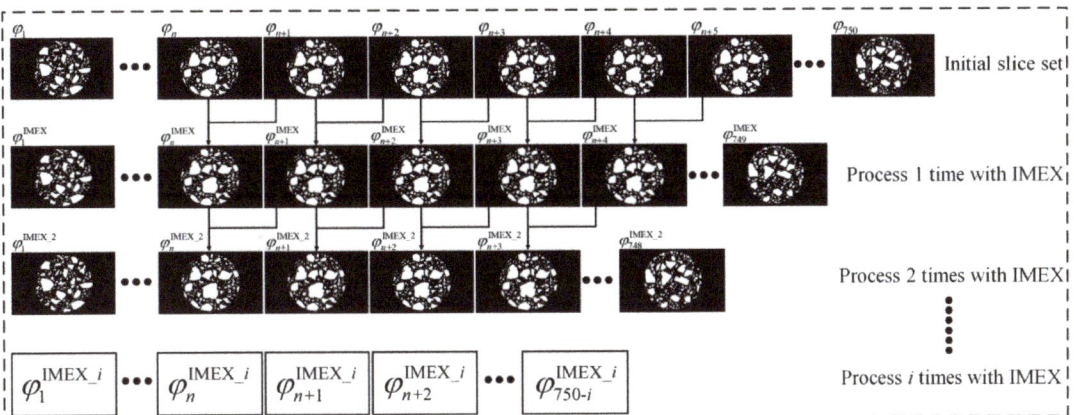

**Figure 5.** Schematic diagram of the multiple-correction method.

### 2.5.4. Multi-Directional Multiple Correction

Based on the Matlab toolbox, the binary slice set of the asphalt concrete can be obtained from any other direction by using the numerical model re-slicing algorithm and the 3D voxel reconstruction algorithm to simulate the CT nondestructive scanning process.

As shown in Figure 6, multi-directional multiple correction is a combination of single-directional multiple-correction processes in multiple new directions. The single-directional multiple-correction process is to first re-slice the asphalt concrete 3D voxel model along a new direction to obtain a new slice set. Then, a multiple-correction process, as described above, is applied to the new slice set to recover and strengthen the similarity of aggregate phase geometry between the adjacent new slices. Finally, to facilitate the implementation of the multiple corrections in the next new direction, the corrected new slice set is reconstructed in 3D to obtain the new asphalt concrete mesostructural model.

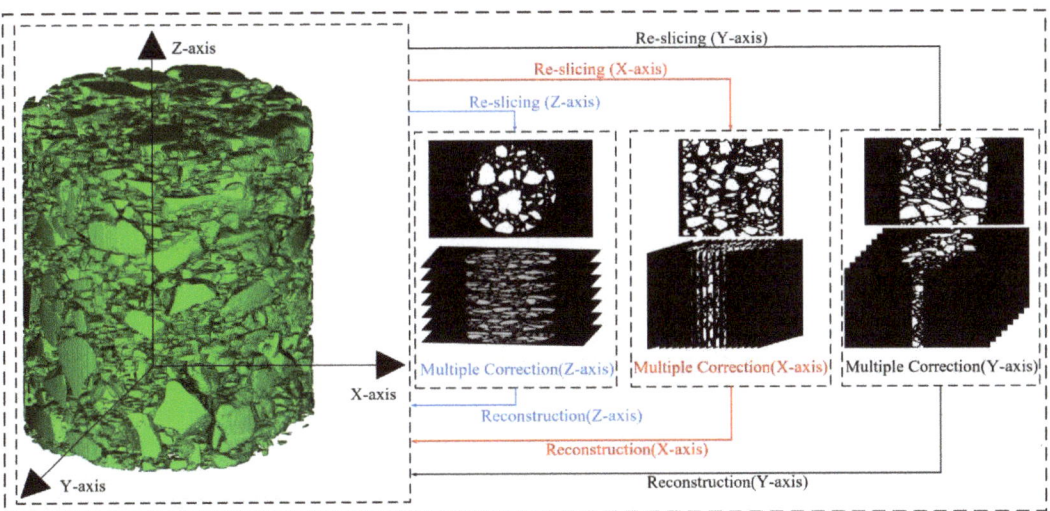

**Figure 6.** Schematic diagram of the multi-directional multiple-correction method.

The numerical model re-slicing algorithm converts the voxel values $\psi(x,y,z)$ with 3D coordinate information to the pixel values $\varphi_x(y,z)$, $\varphi_y(x,z)$ or $\varphi_z(x,y)$ with 2D coordinate information along a certain direction and then arranges the pixel values by coordinates to generate a 2D slice set. Conversely, the 3D voxel reconstruction algorithm converts the pixel values $\varphi_x(y,z)$, $\varphi_y(x,z)$, or $\varphi_z(x,y)$ along the acquisition direction to obtain the voxel values, $\psi(x,y,z)$, and then arranges the voxel values by coordinates to generate a 3D model. The parameter relationships are as follows:

$$\psi(x,y,z) = \begin{cases} \varphi_x(y,z) \\ \varphi_y(x,z) \\ \varphi_z(x,y) \end{cases} \tag{14}$$

where $\psi$ is a voxel value with 3D coordinate information; $\varphi_x$ is a pixel value with 2D coordinate information in a slice perpendicular to the x-axis; $\varphi_y$ is a pixel value with 2D coordinate information in a slice perpendicular to the y-axis; and $\varphi_z$ is a pixel value with 2D coordinate information in a slice perpendicular to the z-axis.

### 2.6. Developed Procedure for the 3D Reconstruction of Asphalt Concrete Mesostructure

The bilinear interpolation algorithm introduced in Section 2.3 was employed to improve the pixel density of 2D CT images and the average filtering algorithm was used to reduce the noise of CT images. Subsequently, the OTSU method described in Section 2.4 was employed to separate the asphalt mortar matrix phase from the aggregate phase, and the watershed segmentation method based on markers was used to separate the interconnected aggregates. Finally, the IMEX and the UMER proposed in Sections 2.5.1 and 2.5.2 was employed to recover the similarity of aggregate phase geometry between adjacent slices, and the recovery effect of geometric similarity was enhanced through the multi-directional multiple-correction method developed in Section 2.5.4.

Figure 7 shows the details of the improved procedure for 3D reconstruction. It includes CT image preprocessing (image pixel interpolation and image filtering), 2D image segmentation (OTSU image segmentation, watershed segmentation, and image pixel density restoration), and multi-directional multiple correction (re-slicing, multiple correction, and voxel reconstruction).

The workstation configuration for performing the 3D reconstruction process included an Intel®Core TM i7-9700 3.0GHz processor (Dell, Lundrock, TX, USA), 64GB RAM (DDR4), and operating system version Windows 10 19044.2006 (Professional Edition), and the processing software version is Matlab 2021.

According to the new process of 3D reconstruction given in Section 2.6, this study provides the processing results of each stage of 3D reconstruction, as shown in Figure 8.

The CT image pre-processing process is shown in Figure 8b–d. First, the CT slice set image information was read and the CT slice set was uniformly grayed out to obtain the CT grayscale slice set in order to avoid image format transformation due to file dumping.

Then, the pixel density of the CT grayscale slice set was increased based on the bilinear interpolation algorithm to prevent the contour distortion of the aggregate phase caused by image filtering and 2D image segmentation.

Finally, there was a loss of light intensity when the X-ray passes through the model because of the power limitation of CT. The overall grayscale of the CT grayscale image shows a distribution trend of low center and high surroundings (Figure 9a), which will reduce the ability of the numerical algorithm to recognize pixels representing different materials in CT grayscale slices. In this study, the average filtering algorithm was selected to filter the CT grayscale slices with high pixel density, which enhanced the global contrast of the pixel grayscale values between the asphalt mortar matrix phase and the aggregate phase in the CT grayscale slice (Figure 9c) and weakened the noise impact to improve the image segmentation quality of the OTSU algorithm for low-contrast images.

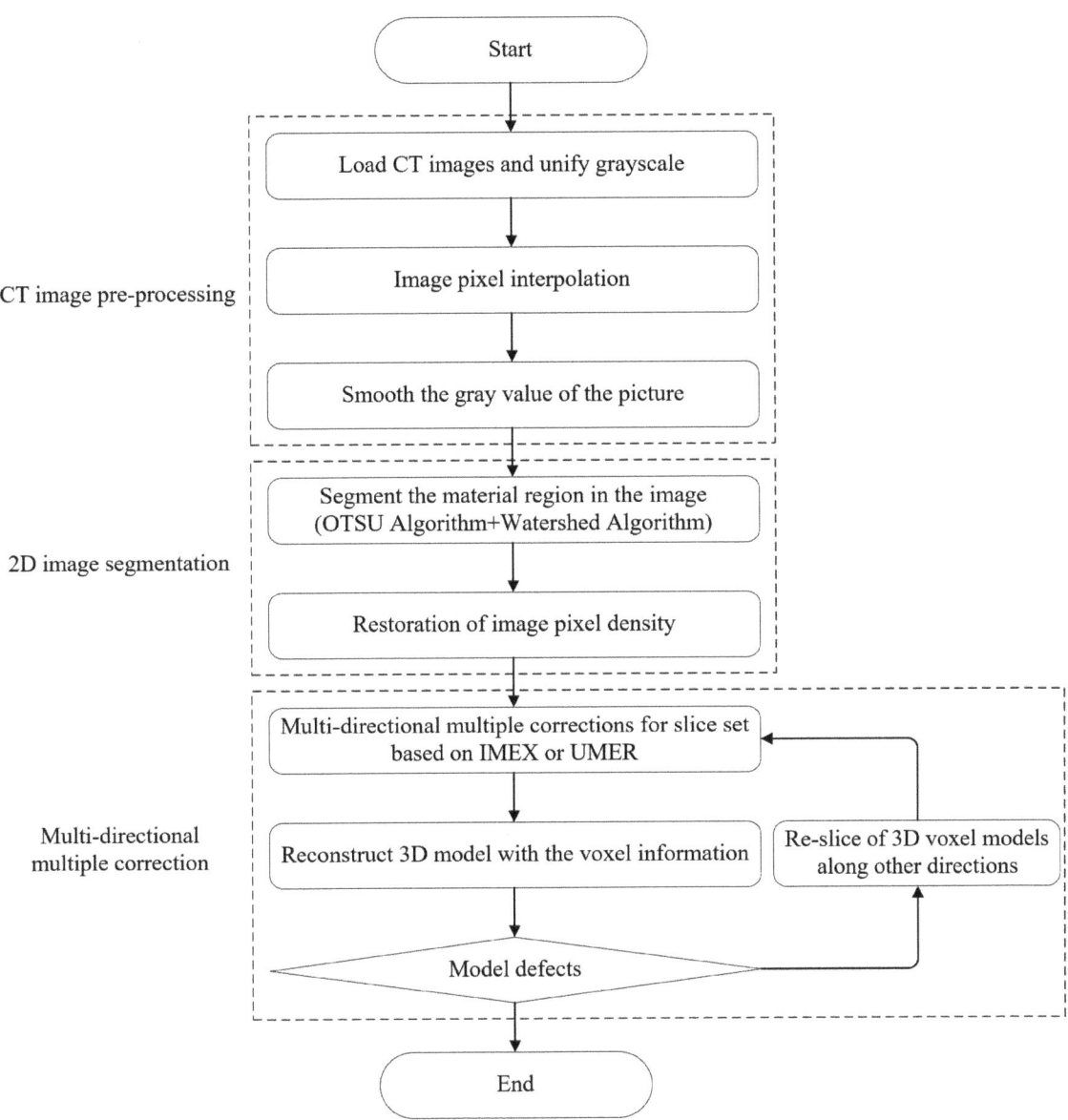

**Figure 7.** Flow chart of the improved procedure for the 3D reconstruction of an asphalt concrete mesostructure considering the similarity of aggregate phase geometry between adjacent slices.

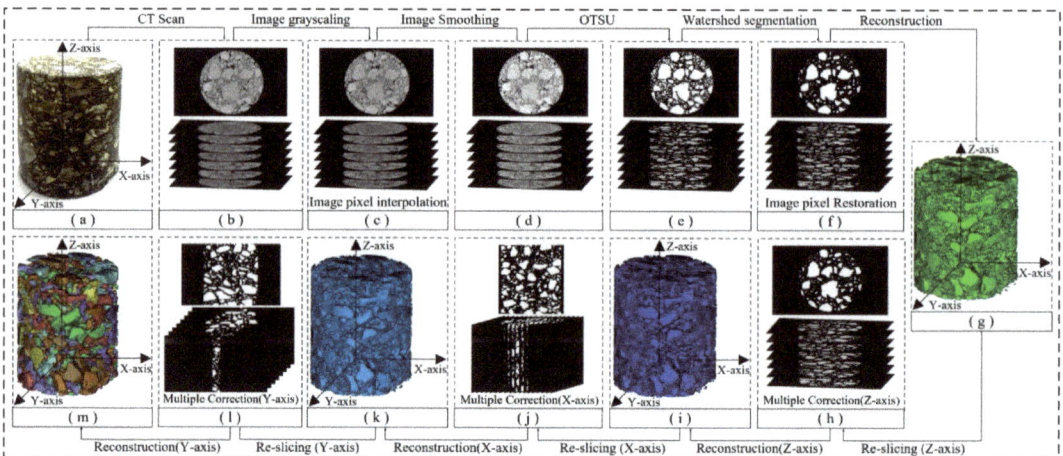

**Figure 8.** Processing results of each stage of the improved procedure for 3D reconstruction: (**a**) asphalt concrete specimen; (**b**) original CT images of asphalt concrete specimen; (**c**) CT images after grayscaling and pixel interpolation; (**d**) CT images after smoothing; (**e**) CT images after the OTSU segmentation; (**f**) CT images after the watershed segmentation and pixel restoration; (**g**) Completing the 3D voxel reconstruction; (**h**) Completing the Z-axis multiple correction; (**i**) Completing the 3D voxel reconstruction along the Z-axis; (**j**) Completing the X-axis multiple correction; (**k**) Completing the 3D voxel reconstruction along the X-axis; (**l**) Completing the Y-axis multiple correction; (**m**) Completing the 3D voxel reconstruction along the Y-axis.

The 2D image segmentation process is shown in Figure 8d–f. First, the optimal threshold of each slice of the CT grayscale slice set was calculated based on the OTSU method's exhaustive calculation, and the grayscale image was segmented into a binary image containing only the asphalt mortar matrix phase and the aggregate phase according to the optimal threshold (Figure 9d).

Then, the segmentation line layer was obtained by using the watershed segmentation method based on markers, and it was overlaid on the corresponding CT binary slice to complete the segmentation of the interconnected aggregates. At this time, a large number of randomly scattered 0-value or 1-value pixel blocks appeared in the CT binary slice, which mainly included fine aggregates with diameters less than 2.36 mm and erroneous data under the influence of noise could be removed using the image morphology method. The processing results are shown in Figure 9e.

Finally, to ensure that the voxel model obtained from the subsequent 3D reconstruction is consistent with the dimensions of the original specimen, the pixel density of the CT binary slice set was restored based on the inverse operation of the bilinear interpolation algorithm after the 2D image segmentation was completed (Figure 9f).

The multi-directional multiple-correction process of the asphalt concrete 3D voxel model is shown in Figure 8g–m. Based on the multi-directional multiple-correction method proposed in Section 2.5.4, the 3D voxel model of the asphalt concrete with model defects was corrected along the Z-axis, X-axis, and Y-axis successively.

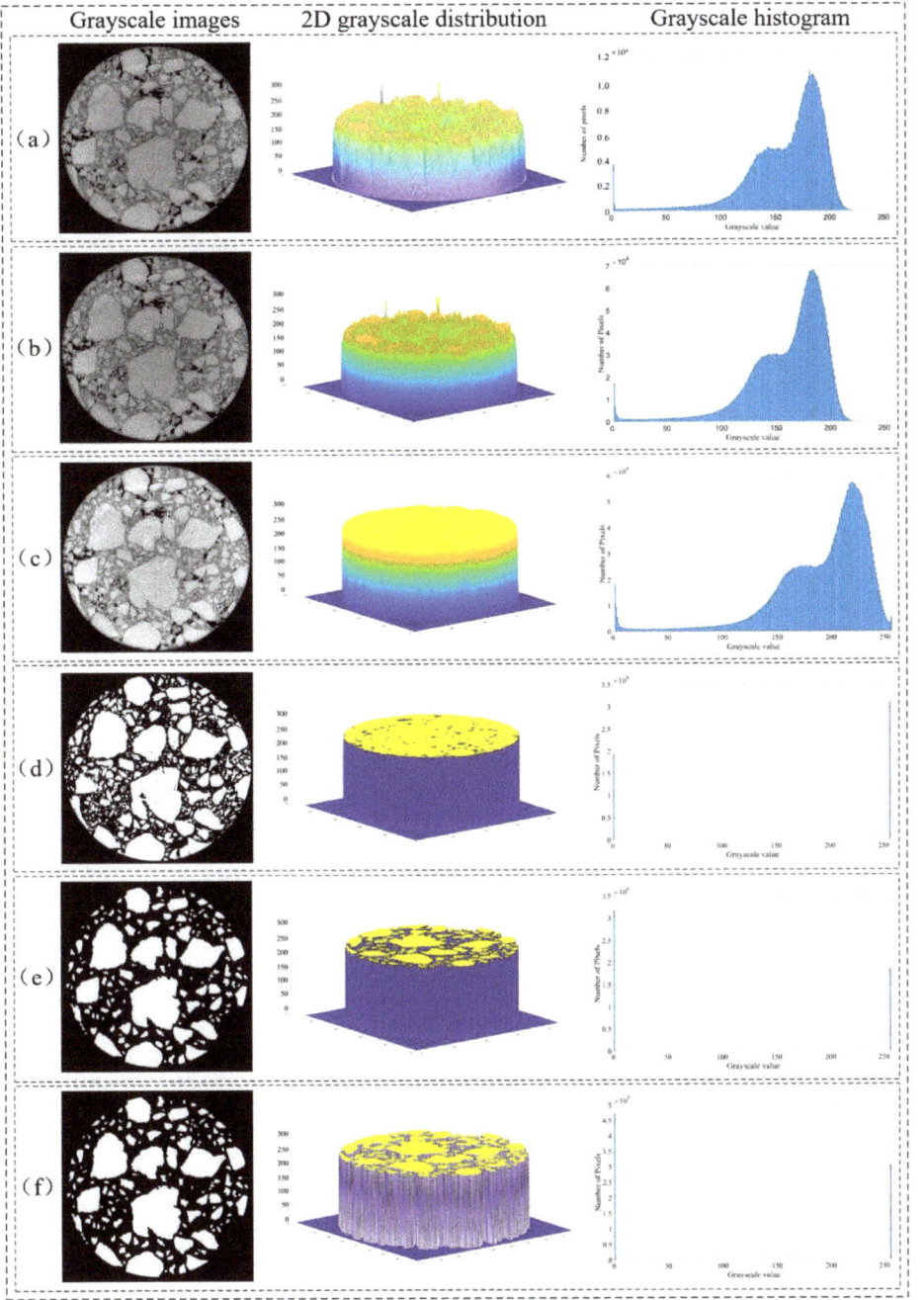

**Figure 9.** Changes in the grayscale of CT slice images during CT image preprocessing and 2D image segmentation stages: (**a**) after uniform grayscale; (**b**) after increasing image pixel density; (**c**) after the filtering process; (**d**) after OTSU segmentation; (**e**) after watershed segmentation; (**f**) after restoring image pixel density.

## 3. Results and Discussion

Figure 10 shows the 3D models at each stage under the improved procedure for 3D reconstruction. These models are discussed in detail in this section. Table 1 shows the main defects in the 3D model at each stage.

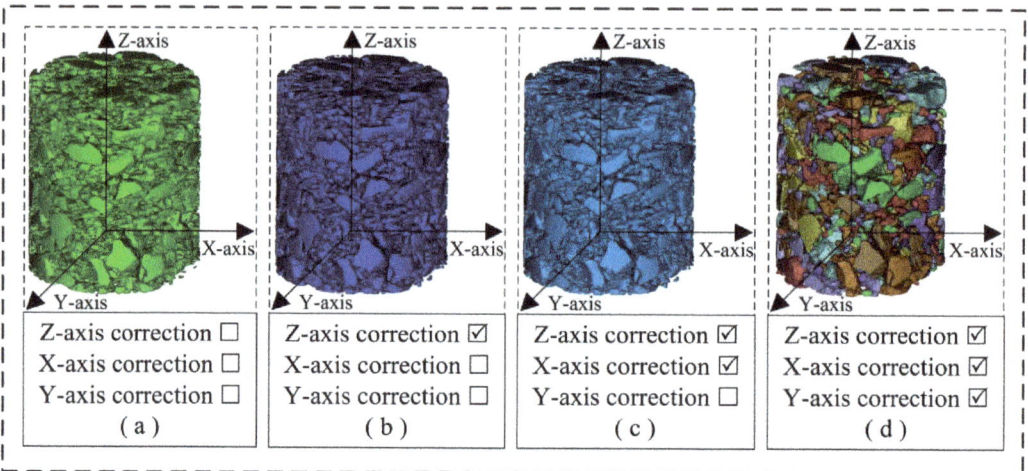

**Figure 10.** Three-dimensional voxel models at each stage of the improved procedure for 3D reconstruction: (**a**) Model A: obtained from 3D reconstruction of the CT slice set after 2D image segmentation; (**b**) Model B: Z-axis multiple correction completed on the basis of Model A; (**c**) Model C: X-axis multiple correction completed on the basis of Model B; (**d**) Model D: Y-axis multiple correction completed on the basis of Model C.

**Table 1.** The main defects in the 3D model at each stage.

| Model | Main Model Defects |
| --- | --- |
| A | Interconnected aggregates<br>Incomplete aggregates<br>Flaky aggregates |
| B | Interconnected aggregates<br>Incomplete aggregates<br>Linear aggregates |
| C | Interconnected aggregates<br>Incomplete aggregates |
| D | Interconnected aggregates (few) |

### 3.1. Two-Dimensional Slice Analysis

Using the numerical model re-slicing algorithm can obtain sections in any direction of the 3D model, which will facilitate the observation of the mesostructure of the asphalt mixture model.

As shown in Figure 11, the 3D voxel model (Figure 10a) was re-sliced to obtain its vertical sections (XZ section and YZ section). It can be found that the morphology of the aggregate phase in the vertical section differs greatly from that in the horizontal section (XY section). In the asphalt concrete voxel model obtained by the 3D reconstruction of the CT slice set that has completed 2D image segmentation, there actually exists a large number of flaky aggregates distributed perpendicular to the Z-axis (CT acquisition direction), and the problems of interconnected aggregates and incomplete aggregates are very serious.

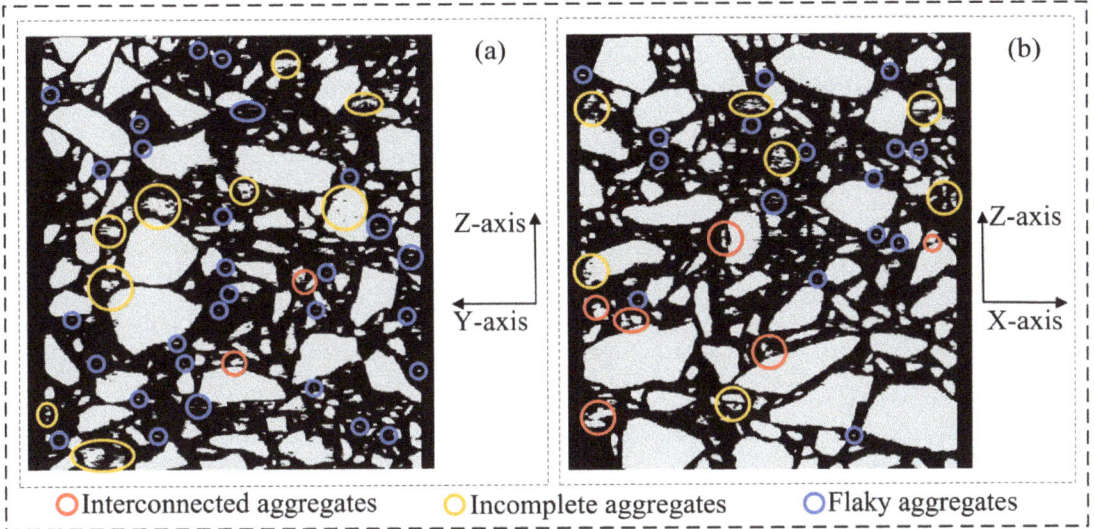

**Figure 11.** Asphalt concrete 3D voxel model sections (corresponding to Figure 10a): (**a**) YZ section; (**b**) XZ section.

As shown in Figure 12, the 3D voxel model (Figure 10b) with the Z-axis multiple correction completed was re-sliced. In its vertical sections (XZ section and YZ section), it can be found that the Z-axis multiple correction is very effective in removing the flaky aggregates distributed perpendicular to the Z-axis, and the problems of incomplete aggregates and interconnected aggregates have been solved to some extent.

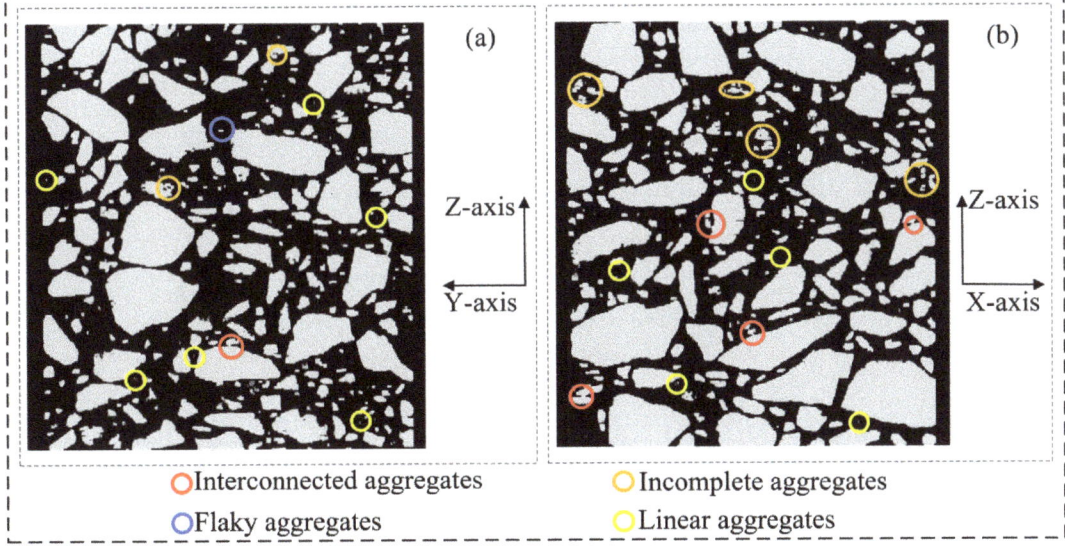

**Figure 12.** Three-dimensional asphalt concrete voxel model sections (corresponding to Figure 10b): (**a**) YZ section; (**b**) XZ section.

However, due to the cumulative effect of the multiple-correction algorithm in the correction direction, when the number of multiple corrections is high, 1-value pixels that distribute interruptedly along the acquisition direction in the slice set become linear aggregates distributed along the correction direction. It is verified that this linear aggregate will be automatically eliminated during subsequent corrections in other directions.

As shown in Figure 13, the model before Z-axis multiple correction (Figure 10a) and the model after Z-axis multiple correction (Figure 10b) are re-sliced. In the XY sections of the two models, it can be found that although the sliced images processed by 2D image segmentation have good aggregate separation characteristics, there are actually a large number of model defects such as flaky aggregates, incomplete aggregates, and interconnected aggregates in the processed binary images due to the neglect of the similarity of aggregate phase geometry between adjacent slices. More importantly, the geometry and distribution of these model defects in 3D space are random in nature and cannot be directly confirmed from XY sections by visual comparison or 2D parametric characterization methods.

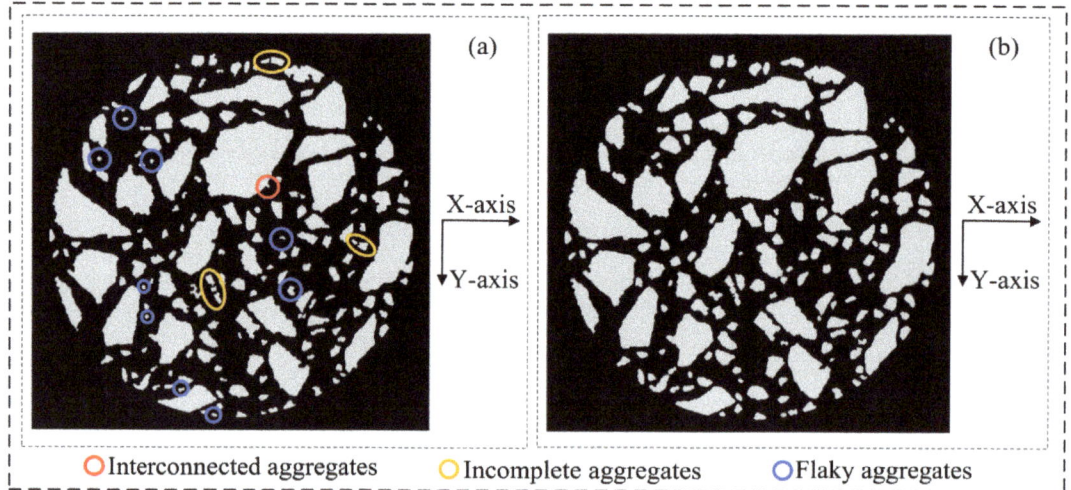

**Figure 13.** Asphalt concrete 3D voxel model sections (XY section): (**a**) before Z-axis multiple correction (corresponding to Figure 10a); (**b**) after Z-axis multiple correction (corresponding to Figure 10b).

As shown in Figure 14, in the XZ section and YZ section of the 3D voxel model (Figure 10c) that has completed X-axis multiple correction, it can be found that the linear aggregates parallel to the Z-axis distribution in the 3D voxel model (Figure 10b) have largely disappeared after the X-axis correction, but there is still a small number of interconnected aggregates and incomplete aggregates.

As shown in Figure 15, in the XZ section and YZ section of the 3D voxel model (Figure 10d) that has completed the Y-axis multiple correction, it can be found that the model defects in the asphalt concrete voxel model have been well resolved at this time.

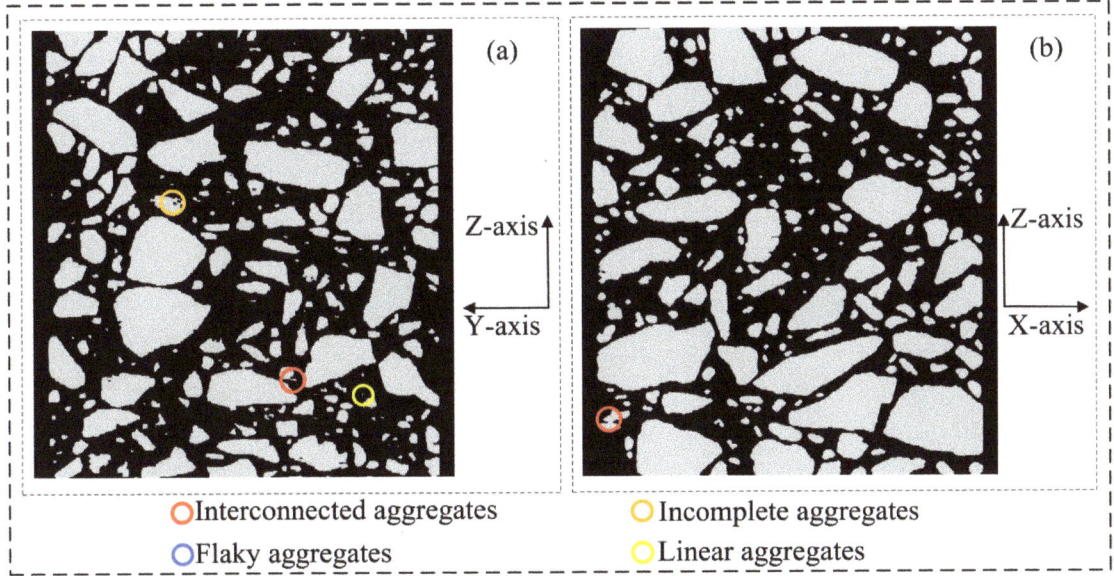

**Figure 14.** Asphalt concrete 3D voxel model sections (corresponding to Figure 10c): (**a**) YZ section; (**b**) XZ section.

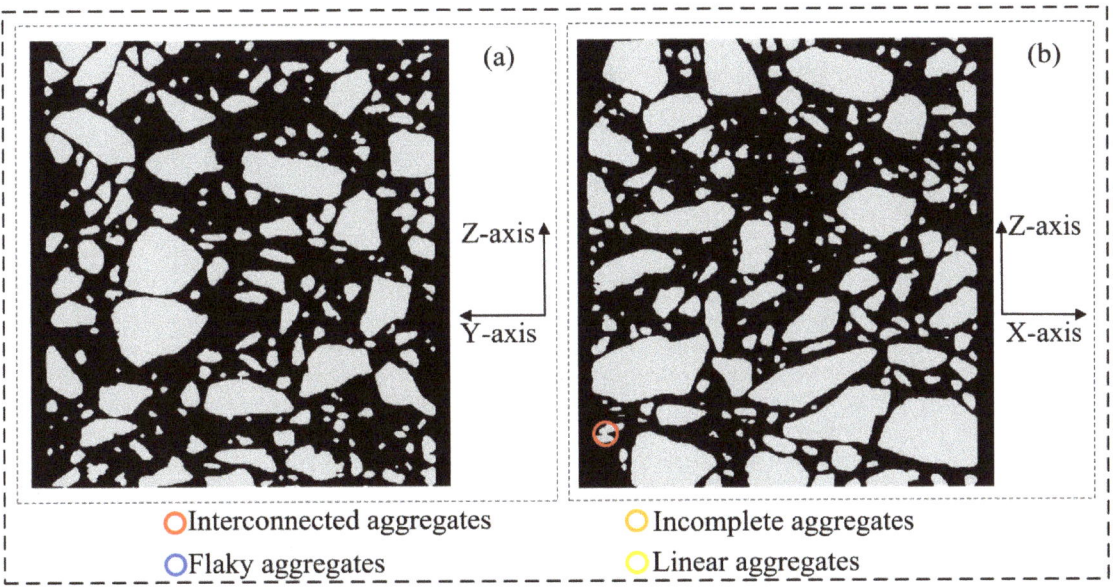

**Figure 15.** Asphalt concrete 3D voxel model sections (corresponding to Figure 10d): (**a**) YZ section; (**b**) XZ section.

## 3.2. Three-Dimensional Model Analysis

In the 3D voxel model, the aggregate volume can be determined based on the number of voxels contained in a single aggregate particle. The aggregates in the uncorrected 3D voxel model (Figure 10a) and the 3D voxel model that has completed multi-directional multiple correction (Figure 10d) are arranged in order of volume. The interconnected aggregates are multiple independent aggregates that are very close to each other and thus incorrectly identified as a single particle during image processing. They are generally larger in volume than real aggregates. To facilitate the presentation, large-volume aggregates (mainly interconnected aggregates) and small-volume aggregates (mainly flaky aggregates) are extracted from both the corrected and uncorrected voxel models. The extraction results are shown in Figure 16.

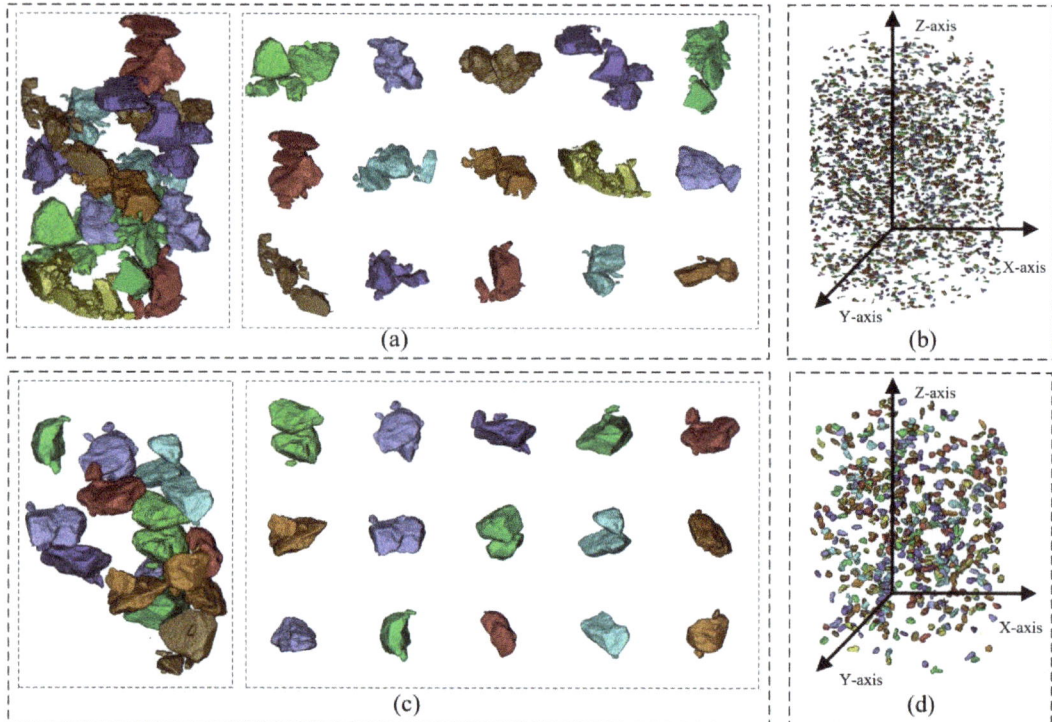

**Figure 16.** Aggregate extraction results for the 3D voxel model: (**a**) large volume aggregates without considering the similarity of aggregate phase geometry between adjacent slices (Figure 10a); (**b**) small volume aggregates without considering the similarity of aggregate phase geometry between adjacent slices (Figure 10a); (**c**) large volume aggregates considering the similarity of aggregate phase geometry between adjacent slices (Figure 10d); (**d**) small volume aggregates considering the similarity of aggregate phase geometry between adjacent slices (Figure 10d).

It can be found that in large-volume aggregates (Figure 16a,c), the 3D aggregates constructed by the procedure without considering the similarity of aggregate phase geometry between adjacent slices restore the 3D spatial distribution characteristics and contour characteristics of the real aggregates to a certain extent, but the model defect of interconnected aggregates is very serious and often appears along with flaky aggregates and incomplete aggregates, forming more complex comprehensive model defects. These comprehensive model defects will greatly increase the workload and difficulty of manual processing. In contrast, the aggregates constructed by the improved procedure for the 3D reconstruction of asphalt concrete mesostructures considering the similarity of aggregate phase geometry between adjacent slices have good 3D spatial distribution characteristics and contour characteristics, while the model defects are greatly reduced and the remaining small amount of interconnected aggregates can be easily solved by hand.

In small-volume aggregates (Figure 16b,d). A large number of independently distributed flaky aggregates exist in the fine aggregates constructed without considering the similarity of aggregate phase geometry between adjacent slices, and they are mostly distributed in the whole asphalt concrete model at an angle perpendicular to the Z-axis. In this paper, after the multiple correction of the 3D voxel model along Z-axis, the flaky aggregates are effectively eliminated and the real 3D spatial distribution information of a large number of small aggregates in the asphalt concrete is preserved, which is finally recovered to the small aggregates with real 3D contour characteristics after the X-axis multiple correction and Y-axis multiple correction.

Figure 17 shows the details of the concave contour of the 3D aggregate constructed by the improved procedure for 3D reconstruction and the 2D slice image corresponding to this contour. It can be found that the partially closed holes in the aggregate particles in the 2D slice result from slicing the 3D aggregate outer contour depression at a specific angle. Based on the proposed IMEX and UMER algorithms, these 2D closed holes were preserved by using the developed multi-directional multiple-correction method. The final 3D aggregate model obtained preserves the concave characteristics of the actual aggregate outer contour.

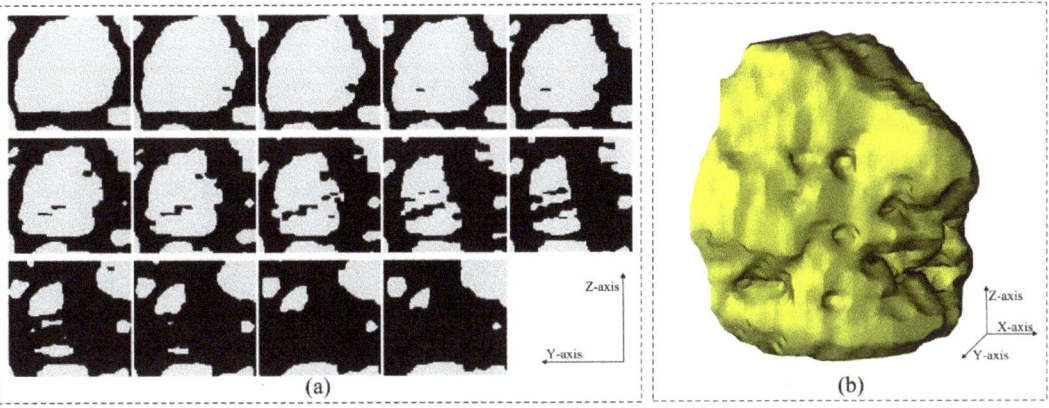

**Figure 17.** Schematic diagram of the concave contour of the aggregate: (**a**) 2D slice; (**b**) 3D model (Figure 10d).

The geometric similarity between adjacent slices is naturally present and would not disappear due to the change in aggregate particle size or in material properties. Therefore, the developed procedure for 3D reconstruction is also effective for other asphalt concretes with different aggregate sizes, or other heterogeneous materials with significant density differences.

However, it should be noted that the greater the spacing between adjacent CT slices, the weaker the geometric similarity of adjacent CT slices. The developed procedure for 3D reconstruction requires that the spacing between adjacent CT slices should be less than or equal to 0.3 mm.

## 4. Summary and Conclusions

This paper develops an improved procedure for the 3D reconstruction of asphalt concrete mesostructures considering the similarity of aggregate phase geometry between adjacent slices, which includes two adjacent-slice pixel-value-correction algorithms, a multi-directional multiple-correction method, and an image pixel interpolation process. In the 3D reconstruction procedure, we used numerical algorithms to reproduce the natural geometric continuity of the asphalt concrete in order to eliminate model defects. The adjacent-slice pixel-value-correction algorithm was used to recover the similarity of aggregate phase geometry between adjacent CT slices, and the suggested multi-directional multiple-correction method was employed to further enhance the geometric similarity. The image pixel interpolation was applied to increase the image pixel density. Based on the analysis and results of this study, the following conclusions were drawn:

- Using the developed procedure for 3D reconstruction can efficiently eliminate the vast majority of model defects in the asphalt concrete mesostructural model.
- By means of the proposed adjacent slice pixel correction algorithms, the multiple corrections implemented along the CT acquisition direction can effectively remove the model defects (interconnected aggregates, incomplete aggregates, and flaky aggregates) distributed perpendicular to the CT acquisition direction in the 3D voxel model.
- Based on the proposed adjacent slice pixel correction algorithms, the closed holes in the 2D slice corresponding to the concave features of the 3D aggregate can be preserved by using the developed multi-directional multiple-correction method. The resulting 3D aggregate model will have the concave characteristics of the actual aggregate outer contour.
- The multi-directional multiple-correction method can more accurately evaluate the image segmentation effect of 2D slice images by acquiring 3D model slices in different directions.
- The image pixel interpolation process can increase the pixel density of the image to avoid the distortion of aggregate contours during 2D image segmentation.

Although this study demonstrates the effectiveness of the developed procedure for 3D reconstruction in removing model defects from the 3D mesostructure of asphalt concrete, it still cannot completely avoid manual processing; that is, the final modeling results are still influenced by subjective factors from the operator. In future studies, more effective algorithms should be designed to achieve the fully automated modeling of the 3D mesostructure of asphalt concrete, and the validity of the method will be verified using different asphalt concretes.

**Author Contributions:** Methodology, C.W. and Y.S.; Investigation, C.W., H.X., Y.Z., W.W. and J.C.; Writing—original draft, C.W.; Writing—review & editing, Y.S.; Supervision, Y.S. All authors have read and agreed to the published version of the manuscript.

**Funding:** This study was sponsored by the National Natural Science Foundation of China (51808098 and 51878122), the Natural Science Foundation of Liaoning Province (2022-MS-140), and Fundamental Research Funds for the Central Universities (DUT22JC22). The support is gratefully acknowledged.

**Institutional Review Board Statement:** Not applicable.

**Informed Consent Statement:** Not applicable.

**Data Availability Statement:** Data sharing is not applicable to this article.

**Conflicts of Interest:** The authors declare no conflict of interest.

## References

1. Al-Mosawe, H.; Thom, N.; Airey, G.; Albayati, A. Linear viscous approach to predict rut depth in asphalt mixtures. *Constr. Build. Mater.* **2018**, *169*, 775–793. [CrossRef]
2. PLi, P.; Liu, J.; Zhao, S. Implementation of stress-dependent resilient modulus of asphalt-treated base for flexible pavement design. *Int. J. Pavement Eng.* **2018**, *19*, 439–446.
3. Sadd, M.H.; Dai, Q.L.; Parameswaran, V.; Shukla, A. Microstructural simulation of asphalt materials: Modeling and experimental studies. *J. Mater. Civ. Eng.* **2004**, *16*, 107–115. [CrossRef]
4. Guddati, M.N.; Feng, Z.; Kim, Y. Toward a micromechanics-based procedure to characterize fatigue performance of asphalt concrete, Bituminous Paving Mixtures 2002: Materials and Construction. *Transp. Res. Board Natl. Res. Counc.* **2002**, *1789*, 121–128. [CrossRef]

5. Castillo, D.; Caro, S.; Darabi, M.; Masad, E. Influence of aggregate morphology on the mechanical performance of asphalt mixtures. *Road Mater. Pavement Des.* **2018**, *19*, 972–991. [CrossRef]
6. Yin, A.; Yang, X.; Yang, Z. 2D and 3D fracture modeling of asphalt mixture with randomly distributed aggregates and embedded cohesive cracks. *Procedia IUTAM* **2012**, *6*, 114–122. [CrossRef]
7. Zhou, B.; Wang, J. Random generation of natural sand assembly using micro x-ray tomography and spherical harmonics. *Geotech. Lett.* **2015**, *5*, 6–11. [CrossRef]
8. Chung, S.-Y.; Kim, J.-S.; Stephan, D.; Han, T.-S. Overview of the use of micro-computed tomography (micro-CT) to investigate the relation between the material characteristics and properties of cement-based materials. *Constr. Build. Mater.* **2019**, *229*, 13. [CrossRef]
9. Masad, E.; Muhunthan, B.; Shashidhar, N.; Harman, T. Internal structure characterization of asphalt concrete using image analysis. *J. Comput. Civ. Eng.* **1999**, *32*, 88–95. [CrossRef]
10. du Plessis, A.; Boshoff, W.P. A review of X-ray computed tomography of concrete and asphalt construction materials. *Constr. Build. Mater.* **2019**, *199*, 637–651. [CrossRef]
11. Sun, Y.; Zhang, Z.; Gong, H.; Zhou, C.; Chen, J.; Huang, B. 3D Multiscale Modeling of Asphalt Pavement Responses under Coupled Temperature-Stress Fields. *J. Eng. Mech.* **2022**, *148*, 15. [CrossRef]
12. Klimczak, M.; Jaworska, I.; Tekieli, M. 2D Digital Reconstruction of Asphalt Concrete Microstructure for Numerical Modeling Purposes. *Materials* **2022**, *15*, 18. [CrossRef] [PubMed]
13. Enríquez-León, A.J.; de Souza, T.D.; Aragão, F.T.S.; Braz, D.; Pereira, A.M.B.; Nogueira, L.P. Determination of the air void content of asphalt concrete mixtures using artificial intelligence techniques to segment micro-CT images. *Int. J. Pavement Eng.* **2021**, *10*, 3973–3982. [CrossRef]
14. Otsu, N. Threshold Selection Method from Gray-Level Histograms. *IEEE Trans. Syst. Man Cybern.* **1979**, *9*, 62–66. [CrossRef]
15. AlSaeed, D.H.; Bouridane, A.; ElZaart, A.; Sammouda, R. Two modified Otsu image segmentation methods based on Lognormal and Gamma distribution models. In Proceedings of the 2012 International Conference on Information Technology and e-Services (ICITeS), Sousse, Tunisia, 24–26 March 2012; p. 5.
16. HCai, H.; Yang, Z.; Cao, X.; Xia, W.; Xu, X. A New Iterative Triclass Thresholding Technique in Image Segmentation. *IEEE Trans. Image Process.* **2014**, *23*, 1038–1046.
17. Gong, J.; Li, L.; Chen, W. Fast recursive algorithms for two-dimensional thresholding. *Pattern Recognit.* **1998**, *31*, 295–300. [CrossRef]
18. Zhao, G.Z.; Zhu, G.X.; Zeng, Y.R.; Zhang, T.X.; Xu, H.Z. Infrared image segmentation with 2D OTSU method based on particle swarm optimization. In Proceedings of the 5th International Symposium on Multispectral Image Processing and Pattern Recognition, Wuhan, China, 15 November 2007.
19. Bhandari, A.K.; Ghosh, A.; Kumar, I.V. A local contrast fusion based 3D Otsu algorithm for multilevel image segmentation. *IEEE-CAA J. Autom. Sin.* **2020**, *7*, 200–213. [CrossRef]
20. Bhandari, A.K.; Kumar, I.V. A context sensitive energy thresholding based 3D Otsu function for image segmentation using human learning optimization. *Appl. Soft. Comput.* **2019**, *82*, 35. [CrossRef]
21. Du, C.; Sun, Y.; Chen, J.; Zhou, C.; Liu, P.; Wang, D.; Oeser, M. Coupled Thermomechanical Damage Behavior Analysis of Asphalt Pavements Using a 2D Mesostructure-Based Finite-Element Method. *J. Transp. Eng. Part B-Pavements* **2021**, *147*, 13. [CrossRef]

**Disclaimer/Publisher's Note:** The statements, opinions and data contained in all publications are solely those of the individual author(s) and contributor(s) and not of MDPI and/or the editor(s). MDPI and/or the editor(s) disclaim responsibility for any injury to people or property resulting from any ideas, methods, instructions or products referred to in the content.

Article

# Identify the Micro-Parameters for Optimized Discrete Element Models of Granular Materials in Two Dimensions Using Hexagonal Close-Packed Structures

Xiaodong Zhou [1,2], Dongzhao Jin [2], Dongdong Ge [2,3], Siyu Chen [2,4] and Zhanping You [2,*]

1. Rizhao City Transportation Bureau, Rizhao 276800, China; xzhou3@mtu.edu
2. Department of Civil, Environmental, and Geospatial Engineering, Michigan Technological University, 1400 Townsend Drive, Houghton, MI 49931-1295, USA; dge1@csust.edu.cn (D.G.); siychen@mtu.edu (S.C.)
3. National Engineering Research Center of Highway Maintenance Technology, Changsha University of Science & Technology, Changsha 410114, China
4. School of Transportation, Southeast University, Nanjing 211189, China
* Correspondence: zyou@mtu.edu

Citation: Zhou, X.; Jin, D.; Ge, D.; Chen, S.; You, Z. Identify the Micro-Parameters for Optimized Discrete Element Models of Granular Materials in Two Dimensions Using Hexagonal Close-Packed Structures. *Materials* 2023, *16*, 3073. https://doi.org/10.3390/ma16083073

Academic Editors: Eddie Koenders and Danuta Barnat-Hunek

Received: 12 February 2023
Revised: 9 April 2023
Accepted: 10 April 2023
Published: 13 April 2023

Copyright: © 2023 by the authors. Licensee MDPI, Basel, Switzerland. This article is an open access article distributed under the terms and conditions of the Creative Commons Attribution (CC BY) license (https://creativecommons.org/licenses/by/4.0/).

**Abstract:** The widely used simple cubic-centered (SCC) model structure has limitations in handling diagonal loading and accurately representing Poisson's ratio. Therefore, the objective of this study is to develop a set of modeling procedures for granular material discrete element models (DEM) with high efficiency, low cost, reliable accuracy, and wide application. The new modeling procedures use coarse aggregate templates from an aggregate database to improve simulation accuracy and use geometry information from the random generation method to create virtual specimens. The hexagonal close-packed (HCP) structure, which has advantages in simulating shear failure and Poisson's ratio, was employed instead of the SCC structure. The corresponding mechanical calculation for contact micro-parameters was then derived and verified through simple stiffness/bond tests and complete indirect tensile (IDT) tests of a set of asphalt mixture specimens. The results showed that (1) a new set of modeling procedures using the hexagonal close-packed (HCP) structure was proposed and was proved to be effective, (2) micro-parameters of the DEM models were transit form material macro-parameters based on a set of equations that were derived based on basic configuration and mechanism of discrete element theories, and (3) that the results from IDT tests prove that the new approach to determining model micro-parameters based on mechanical calculation is reliable. This new approach may enable a wider and deeper application of the HCP structure DEM models in the research of granular material.

**Keywords:** granular material; discrete element method; modeling theories; hexagonal close-packed structure; asphalt mixture

## 1. Introduction

The origin of the discrete element method (DEM) can be traced back to the late 1970s, when it was developed by Cundall and Strack to address the complexities associated with granular materials, owing to their inherently discrete nature [1]. Then, the DEM was introduced into the modeling of the asphalt mixture. Buttlar and You utilized it to model and examine the workings and efficiency of asphalt materials [2,3]. Another study simulated the viscoelastic behavior of asphalt mixture using micromechanical parameters obtained from a dynamic shear rheometer in the Simple Performance Test (SPT) [4]. A 3D microstructure-based Discrete Element Method (DEM) model was created by combining multiple 2D models and then was used to calculate the stress-strain behavior during repeated loading conditions [5]. The results of a laboratory test indicated that the 3D model had better agreement with the test results than the associated 2D models [6]. The contact models form the foundational mechanism in the DEM, and its micro-parameter determination is a critical part. Researchers developed several approaches to relate the micro-parameters with the macro-parameters of the asphalt mixture. For the compacted asphalt mixture at room temperature, the dynamic modulus test was used to determine the viscoelastic parameters of the Burgers model [7]. The dynamic modulus could be used to reflect the stress and strain response by specific load directly. The dynamic modulus test is conducted at temperatures of −10 °C, 10 °C, 21 °C, 37 °C, and 54 °C at loading frequencies of 0.1 Hz, 0.5 Hz, 1 Hz, 5 Hz, 10 Hz, and 25 Hz at each temperature and is specified in AASHTO T342. The simulation results were in agreement with the results obtained from laboratory tests. The creep test was used to determine the viscoelastic parameters in models that were based on microstructure [8]. The dynamic shear rheometer test [9] and the constant strain rate uniaxial compression test were conducted to calculate the time-dependent contact stiffness of the Burgers model [10,11]. Similar contact models and parameter calculations were used in the prediction of the mechanical properties of asphalt [12]. The internal forces configuration of the asphalt mixture was evaluated through the use of established DEM models. The parameter determination for the samples in the compaction process is more difficult than the compacted samples due to the high flowability of asphalt at high temperatures. Chen, Huang et al. proposed an indirect approach to predict the viscoelastic parameters at high temperatures [13]. In their study, serval dynamic modulus tests were performed at low temperatures, and the nonlinear regression analysis was used to obtain the mater curve of the asphalt mixture. Then, the viscoelastic parameters at high temperatures were predicted through the asphalt mixture master curve [14,15].

Another crucial aspect of the DEM model is the use of modeling techniques. You and Buttlar were pioneers in utilizing image processing to construct 2D model structures. They utilized grayscale images obtained through optical scanning to establish microfabric DEM models [2,3]. The simple performance test employed similar models to forecast the dynamic modulus and phase angles of the asphalt mixture [4]. The concept of constructing a 3D model by stacking 2D models was developed based on the 2D DEM models [5]. Then, the development of 3D DEM models utilizing ball/clump elements as basic building blocks was carried out. Ball elements are widely adopted in 3D DEM models for simulating the performance of asphalt mixture due to its simplicity and clear visual aid. For example, the uniaxial compression test was simulated by ball-based DEM models [10]. The modified model was demonstrated as having the capability of simulating creep tests. For the purpose of simplifying the DEM models for asphalt mixtures, researchers typically treat the mixture as a two-phase material composed of coarse aggregates and asphalt mastic [16]. The ball elements were also utilized to represent the asphalt mastic, which is composed of fine aggregate, fines, and asphalt. The ball-based models have an obvious disadvantage due to their inability to represent the irregular shapes of aggregates. New modeling approaches were developed to model the aggregates with more realistic shapes. A proposal was made to utilize randomly generated irregular particles to visualize and simulate the micro-scale properties of the asphalt mixture under mechanical loading [17]. The investigation into the effect of aggregate shape on the diffusivity of asphalt mastic utilized random packing models of ellipsoidal and convex polyhedral particles [18]. Additionally, researchers have utilized realistic aggregate shapes to improve the accuracy of asphalt mixture simulations. Techniques such as X-ray CT and image processing were employed to generate DEM models featuring realistic aggregate shapes [19,20]. A more precise method, the individual aggregate reconstruction technology, was proposed to establish DEM models for asphalt mixture [21,22]. Fracture behavior in asphalt concrete laboratory specimens is able to bridge a vital link in the design of asphalt concrete paving mixtures and pavement structures. A two-dimensional particle flow software package (PFC-2D) was used to study the complex crack behavior observed in asphalt concrete fracture tests [23]. A computer simulation using the discrete element method (DEM) is presented in order to understand and visualize how crushing initiates and develops inside a simulated pavement structure [24,25]. Yu et al. [26] studied the effect of aggregate size distribution and angularity distribution on dynamic modulus using a 3D discrete element method (DEM).

The above research primarily focused on the properties of the compacted asphalt mixture. The compaction process of asphalt mixture is characterized by frequent and intense material movement and changes in contact force. As a result, the contact models and modeling techniques for this process are distinct from those used for compacted asphalt mixtures. There are limited studies that focus on the compaction of asphalt mixture. Wang et al. compared the fundamental mechanics of asphalt compaction using both FEM and DEM and emphasized that DEM can simulate aggregate translation and rotation [27]. The DEM models have also been shown to provide valuable theoretical support for intelligent compaction. Chen and Huang et al. utilized the Burgers model to simulate the compaction of asphalt mixture using DEM [28]. In a subsequent study, they simulated gyratory compaction, vibration compaction, and kneading compaction using an open-source code [13]. Gong et al. established shape-based DEM models to simulate Superpave gyratory compactor (SGC) tests, which simulate the field compaction process of asphalt mixture, and reported agreement between the results of laboratory compaction tests and simulation results [29,30]. This research introduced realistic aggregate shapes into the DEM simulation. However, the established models had limitations on the total number of elements, compaction dynamics, and parameters determination.

Yu Liu et al. [7] introduced a set of theoretical calculations for DEM models using the Burgers contact model and cubic-centered cubic (SCC) ball array structures. This theoretical calculation approach has been proved reliable in predicting the dynamic modulus of asphalt mixture. In order to extend the use of DEM to a variety of performance tests, the hexagonal close-packed (HCP) ball array was used as the basic model structure. It is clear that the HCP structure is more complicated than the SCC structure and has a different mechanical structure. Therefore, a new set of theoretical calculations is needed.

This study aims to establish a hexagonal close-packed discrete element model for granular material with the ability to transit diagonal loading and performance based on Poisson's ratio. To achieve this objective, first, new generation procedures of the HCP ball array for the 2D DEM models were proposed; second, a theatrical approach that was used to transition from material macro-properties to contact micro-parameters values in the 2D DEM models were derived based on the basic configuration and mechanism of discrete element theories; and third, the contact stiffness, bond strength, and an-isotropic properties were discussed and verified by comparing IDT results between designed 2D DEM models and laboratory tests.

## 2. Model and Methods

The research methodology of this study is shown in Figure 1, where the generation method of the new HCP model is introduced in the modeling procedures of the hexagonal close-packed generation method section. The process involves several steps, starting with the generation of nonoverlap clumps obtained through scanning aggregates of varying grain sizes with a 3D scanner, followed by grain-size expansion of clumps where the clump sizes are progressively increased to reach their intended dimensions. Next, hexagonal close-packed (HCP) balls are generated, and these balls are then grouped based on clump geometry, where the classification of an HCP ball as either coarse aggregate or asphalt mastic depends on whether its center position falls within a clump. Finally, the installation of contact properties is carried out using the contact-bond model, which has been shown to effectively simulate the fracture behavior of asphalt mixtures in previous studies. The corresponding mechanical calculation for contact micro-parameters was then derived and verified through simple stiffness/bond tests in the 2D DEM model and verified with theoretical values. Finally, an indirect tensile (IDT) test in the 3D DEM modeling generated by the HCP model structure and laboratory test results is compared.

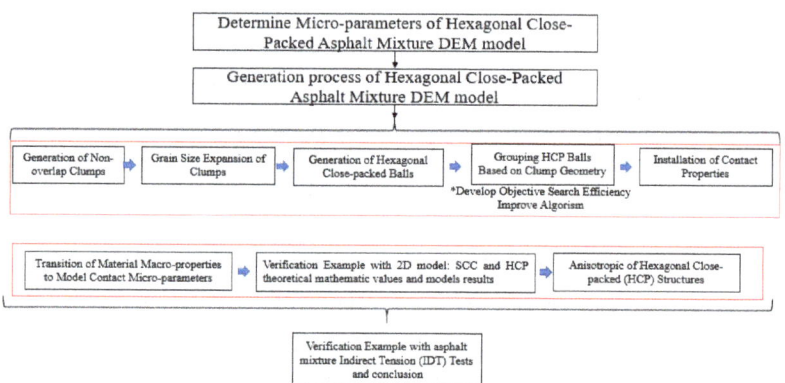

**Figure 1.** Research methodology in this research.

In previous studies, the simple cubic-centered (SCC) ball array was widely used as the basic model structure, as shown in Figure 2a. This kind of model is proved to be effective in the simulation and in the prediction of the dynamic modulus of the asphalt mixture. The SCC structure can transit loading in vertical and horizontal directions efficiently. However, it lacks the ability to transit diagonal loading and performance based on Poisson's ratio. To make up for this disadvantage, the hexagonal close-packed (HCP) ball array (see Figure 2b) was used.

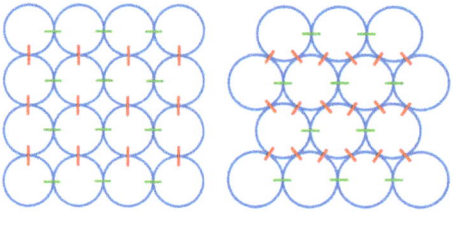

(a) Simple cubic-centered    (b) hexagonal close-packed (HCP)

**Figure 2.** Two different types of the model structure used in DEM modelling (2D view).

This study aims to carry out a reliable approach to determine the micro-parameters in the DEM models. Based on the basic configuration and mechanism of discrete element theories, the transition from material macro-properties to contact micro-parameters was derived. The contact stiffness, bond strength, and an-isotropic properties were discussed and verified by designed DEM models.

*2.1. Contact Stiffness (kn) without Bonding*

2.1.1. Case of SCC

The basic mechanical unit of SCC can be described as a single contact with two balls (Figure 3). This unit can be treated as a single-spring system. The contact force ($F$) and stress ($\sigma$) can be expressed as Equations (1) and (2).

$$F = k_n \cdot \delta \tag{1}$$

$$\sigma = E \cdot \varepsilon \tag{2}$$

where, $k_n$ is stiffness of the contact, $\delta$ is the displacement at the contact, $E$ is the material modulus, and $\varepsilon$ is strain.

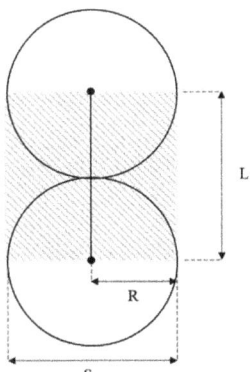

**Figure 3.** Basic Mechanical Unit of SCC Ball Array.

In regard to unit dimensions, the stress and strain at contact can also be expressed as:

$$\sigma = \frac{F}{S} \tag{3}$$

$$\varepsilon = \frac{\delta}{L} \tag{4}$$

where $S$ is the area of the contact plane and $L$ is the length of contact.

Submit Equations (2)–(4) into Equation (1)

$$E = \frac{L}{S} k_n, \ L = 2R, \ S = 2R \tag{5}$$

where $R$ is the radius of SCC balls. One important point to mention is that the third dimension L is hidden in the calculation of S.

Eventually, the material modulus ($E$) of an SCC array can be expressed by contact stiffness ($k_n$):

$$E = k_n \tag{6}$$

2.1.2. Case of HCP

The basic unit of HCP is the combination of three closed contact balls, as seen in Figure 4. This unit can be treated as a simple truss system. The contact force $F$ and displacement $\delta$ are the combinations of the vertical portion of $F'$ and $\delta'$:

$$F = 2 \cdot F' \cdot \cos\theta \tag{7}$$

$$F' = k_n \cdot \delta' \tag{8}$$

$$\delta' = \delta \cdot \cos\theta \tag{9}$$

As the angle $\theta$ of the truss equals 30 degrees, the contact force $F$ can be expressed as:

$$F = 2\cos^2\theta \cdot k_n \cdot \delta = \frac{3}{2} k_n \cdot \delta \tag{10}$$

The length of the truss system is calculated as the vertical portion of the connection between the two balls:

$$L = 2\cos\theta R, \ S = 2R \tag{11}$$

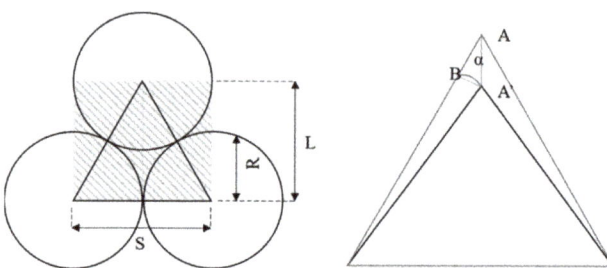

**Figure 4.** Basic Mechanical Unit of HCP Ball Array.

Eventually, the material modulus (*E*) of a unidimensional hex array can be related to contact stiffness ($k_n$) as:

$$E = \frac{3\sqrt{3}}{4}k_n \qquad (12)$$

This section may be divided by subheadings. It should provide a concise and precise description of the experimental results and their interpretation as well as the experimental conclusions that can be drawn.

2.1.3. Validation Example

A set of DEM models was used to verify the reliability of Equations (6) and (12). The ratio of model height versus model width was set as 2.0. Due to hardware and computational power limitations, the validation model dimension is constrained to a limited size, which is worth noting. Consider the impact of model sizes, as shown in Figure 5, where four groups of models were tested with scales ranging from 5 × 10, 10 × 20, and 20 × 40 to 40 × 80.

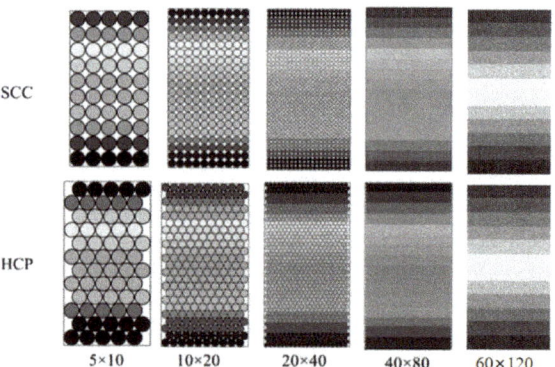

**Figure 5.** Stiffness Validation Models of SCC and HCP Ball Arrays.

The contact stiffness was set as $1 \times 10^5$ for all the tested models. All boundaries were rigid and confined. Vertical displacements of 1% model height per second were applied on the top plane, according to Equations (6) and (22). The theoretical material moduli should be 100 kPa and 129.8 kPa. The obtained material moduli from DEM model are shown in Figure 6.

The model scale has a significant impact on the obtained material moduli from the DEM model. HCP was more sensitive to the model scale than was SCC. The obtained material moduli of the SCC group from the DEM model were close to the 100 kPa theoretical value. The HCP group reached 96.22% of the theoretical value (124.9/129.8 kPa) when using the 60 × 120 configuration.

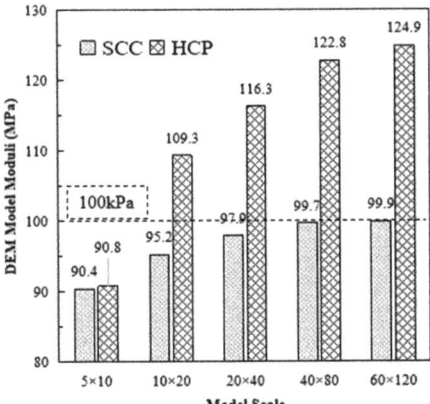

**Figure 6.** DEM Model Moduli of SCC and HCP Stiffness Validation Models without Boding.

*2.2. Contact Stiffness (kn) with Bonding*

2.2.1. Case of SCC

The bonding condition makes no difference to the SCC ball arrays, since the bonding plane is perpendicular to the vertical direction.

2.2.2. Case of HCP

The bonding condition makes no difference to the SCC ball arrays, since the bonding plane is perpendicular to the vertical direction. The bonding plane in the HCP ball arrays has an angle of θ degrees in the horizontal direction (Figure 7). Thus, the shear force at the bonding plane contributes a vertical component when applied to vertical loading.

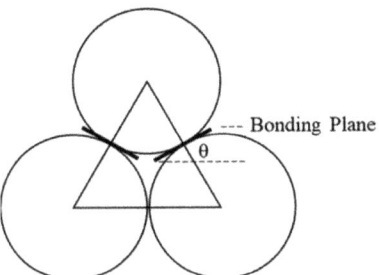

**Figure 7.** Bonding Plane in Basic Mechanical Unit of HCP Ball Array.

The shear force at contact plane can be expressed as:

$$F_s = \begin{cases} k_s \delta_s, & before\ slip \\ \mu F_n, & slip \end{cases} \quad (13)$$

$$\delta_n = \delta cos\theta, \quad \delta_s = \delta sin\theta \quad (14)$$

where $k_s$ is the shear stiffness at bonding plane, $\delta_s$ is the displacement in shear direction, $\mu$ is the friction coefficient, and $F_n$ is the normal contact force at bonding plane.

Then, the total force in vertical direction equals

$$F = F_n cos\theta + F_s sin\theta \quad (15)$$

Submit Equations (13) and (14) into (15):

$$F = 2(k_n cos^2\theta + k_s sin^2\theta)\delta \tag{16}$$

The material modulus before slip then equals

$$E = \frac{\frac{F}{S}}{\frac{\delta}{L}} = \frac{2(k_n cos^2\theta + k_s sin^2\theta)\delta/2R}{\frac{\delta}{2Rcos\theta}} = \frac{3\sqrt{3}}{4}k_n + \frac{\sqrt{3}}{4}k_s \tag{17}$$

2.2.3. Case of HCP

The configuration of the validation example is the same as the models used in the previous section. Contact stiffness ($k_n$) without bonding $k_n$ and $k_s$ were set as $1 \times 10^5$. The theoretical value according to Equation (17) was 173.2 kPa. The simulation results are shown in Figure 8.

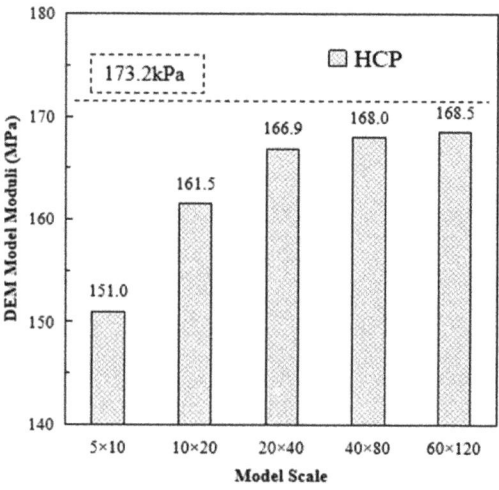

Figure 8. Obtained Moduli of HCP Stiffness validation Models with Bonding.

Model scale also has significant impacts on the material moduli of DEM models. As the model scale increased, the obtained material moduli from the DEM model were closer to the theoretical value. When using the $60 \times 120$ configuration, the obtained material moduli from the DEM model reached 97.29% (168.5/173.2) of the theoretical value. Considering the hardware calculation efficiency and the scale effects influence, a $40 \times 80$ configuration was used in this study.

2.3. Contact Bond Strength

2.3.1. Case of SCC

The tensile bond strength $T_F$ can be expressed as:

$$T_F = T_\sigma \cdot S \tag{18}$$

In the case of the unidimensional ball array, refer to Equation (5):

$$T_\sigma = \frac{T_F}{2Rt}, \quad t = 1 \tag{19}$$

### 2.3.2. Case of HCP

The tensile bond strength $T_F$ can be expressed as:

$$T_F = 2T_\sigma \cos\theta = \sqrt{3}T_\sigma \tag{20}$$

$$T_\sigma = \frac{\sqrt{3}T_F}{2Rt}, \; t = 1 \tag{21}$$

The relationship of contact stiffness in the contact interface is:

$$\frac{1}{k_n} = \frac{1}{k_{n1}} + \frac{1}{k_{n2}}. \tag{22}$$

where $k_n$ is the aggregate-mastic interface normal contact stiffness, $k_{n1}$ is the normal contact stiffness of aggregate, and $k_{n2}$ is the normal contact stiffness of mastic; $k_s$ used the same method.

### 2.3.3. Validation Example

Setting $T_F = 0.5$ N and $R = 0.01$ m, according to Equations (19) and (21), the theoretical values of SCC and HCP models are 25 Pa and 43.3 Pa, respectively. The obtained value from DEM model are 24.7 Pa (98.8% of theoretical value) and 38.37 Pa (88.6% of theoretical value), respectively.

### 2.4. An-Isotropic of Hexagonal Close-Packed (HCP) Structures

The vertical direction and horizontal direction of the HCP ball array are different. The an-isotropic properties can be written as:

$$E = \frac{\sin\varphi}{\sin\theta}|i|k_n + \left(\cos\varphi - \frac{\sin\varphi}{\sin\theta}\cos\theta\right)|j|k_s$$

in which, $\phi$ is among 0–30 degree, $\theta$ equals to 30 degrees.

In the case of material modulus, the modulus constant in the horizontal direction is $|i| = \frac{2\sqrt{3}}{3}$, and in the vertical direction it is $|j| = \frac{3\sqrt{3}}{4}$. Then, the an-isotropic properties of the material are plotted in Figure 9.

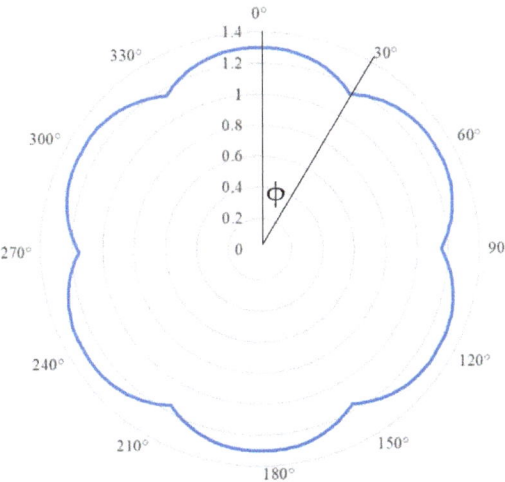

**Figure 9.** Material Moduli An-isotropic of Hexagonal Close-Packed (HCP) Structures.

## 3. Modeling Procedures of Modified Random Generation Method with Realistic Coarse Aggregate Shapes

To enhance the efficiency of the model, a modified random generation method with realistic coarse aggregate shapes was introduced in this study. The new method employed the cross-section of 3D models as the 2D model geometries rather than the directed generation of 2D models. The new modeling procedures are described in the following steps:

- Generation of Nonoverlap Clumps

The clump geometries were obtained through scanning aggregates of varying grain sizes using a 3D scanner. The methods for generating clumps and determining grain size have been described in prior studies [21,31]. The clump grain sizes were determined based on the mixture design. The clumps were generated within a 100 × 63 mm cylinder container, with each clump being generated at 70% of its intended grain size to ensure successful generation. The coarse aggregates of grain sizes G2, G3, and G4 were generated in succession, as depicted in Figure 10(1). It is worth mentioning that the coarse aggregate can also be directly introduced via a compacted model through the compaction process.

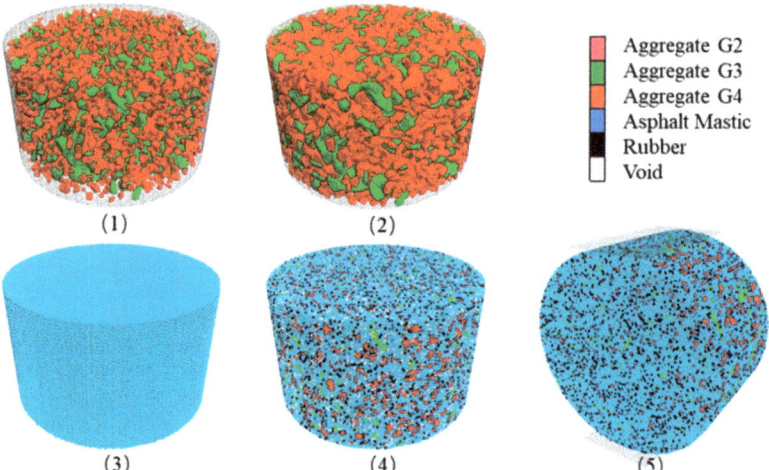

**Figure 10.** Model setup process of the rubber modified HMA: (**1**) clumps generation process; (**2**) diameter expansion procedure of clumps; (**3**) HCP balls generation process; (**4**) setup of different group of HCP balls; (**5**) indirect tensile test process.

- Grain-Size Expansion of Clumps

The clump grain sizes were increased until they reached their target dimensions, as depicted in Figure 10(2). The expansion procedure involved several iterations to prevent excessive overlapping in a single expansion. In the example shown, the clump grain sizes were expanded 10 times with an expansion factor of approximately 1.03631121 for each step, calculated as (1/0.7) (1/10). To minimize overlap between clumps, the model was run until the maximum overlap ratio dropped below 0.1%. While the overlap ratio could be calculated by iterating through the entire clump set, this method would add unnecessary computational strain to the computer. As an alternative, the maximum overlap ratio could be estimated by monitoring the leading contact force. To limit clump movement and enhance efficiency, a high damping ratio of 0.7 was assigned to all clumps.

- Generation of Hexagonal Close-Packed (HCP) Balls

The two lattice structures that result in the highest density for equal-diameter ball arrangements are the cubic-centered cubic (SCC) and the hexagonal close-packed (HCP). In this study, the hexagonal arrangement was chosen. For ease of ball labeling, the balls were generated within a cubic space, and then any balls outside the cylindrical container boundary were removed, as illustrated in Figure 10(3).

- Grouping HCP Balls Based on Clump Geometry (Objective Search Efficiency Improved Algorism)

The classification of an HCP ball into either coarse aggregate or asphalt mastic depends on whether its center position falls within a clump. The number of HCP balls representing rubber particles and voids was determined based on the mixture design. These two groups of HCP balls were then randomly selected from within the mastic. The final grouping results are displayed in Figure 10(4). The most time-consuming step in this section is the objective search of overlap detection, which determines the group properties of HCP balls. Thus, here we proposed an improved objective algorism. The original algorism needs to loop the ball list and clump list (pebble list) from beginning to end, as shown in Figure 11. The required steps for a model with 88,489 balls and 142,266 pebbles are 12.6 billion.

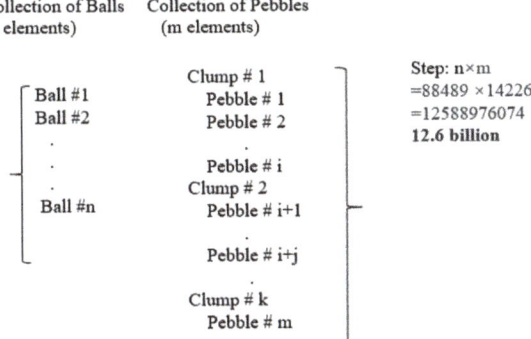

Figure 11. Original Objective Search Algorism.

The improved algorism decreases the calculation steps by narrowing down the search area, as shown in Figure 12.

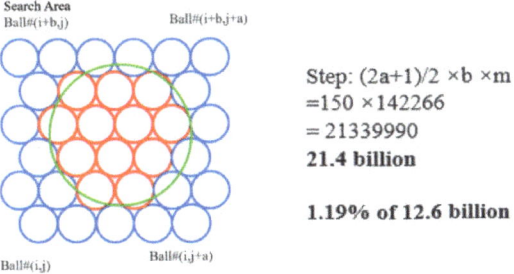

Figure 12. Illustration of the Objective Search Algorism: Clump pebble (Green Circle), Search Area (Blue Circle), Aggregate Cluster (Red Circle).

- First, find the location and diameter of the current pebble.
- Second, calculate the extended coverage area.
- Third, determine if the ball is within the pebble area.

By estimation, the improved objective algorism requires 21.4 million steps to finish the calculation, which saves about 98.9% of calculation time.

- Installation of Contact Properties

The contact-bond model was selected because it has been demonstrated to effectively simulate the fracture behavior of asphalt mixtures. Although nearly all aggregates and rubber particles are covered by asphalt, there is bond strength between directly connected aggregates and rubber particles. Furthermore, the linear contact model was designated as the default model for all subsequent contacts (following fracture), and the contact properties would be derived from the parent particles.

## 4. Validation Example with Indirect Tension (IDT) Tests

The IDT test is an effective method for evaluating the low-temperature cracking performance of asphalt mixture [32]. This section designed a group of indirect tension (IDT) tests in laboratory to verify the reliability of the proposed mechanical parameters transition.

### 4.1. Mixture Design and DEM models

Three mixture designs were selected; see Table 1. The IDT test setup is shown in Figure 13. The test speed is 50 mm/s, and the load and displacements during the test are recorded and compared with the DEM model.

**Table 1.** Mixture Design for Parameter Validation Tests.

| Sieve Size (mm) | Passing (%) | | |
| --- | --- | --- | --- |
| | Mix#1 | Mix#2 | Mix#3 |
| 19 | 100 | 100 | 100 |
| 12.5 | 100 | 100 | 94 |
| 9.5 | 100 | 97 | 86 |
| 4.75 | 94 | 75 | 71 |
| 2.36 | 69 | 54 | 54 |
| 1.18 | 46 | 36 | 38 |
| 0.6 | 32 | 25 | 26 |
| 0.3 | 20 | 15 | 16 |
| 0.15 | 13 | 7 | 8 |
| 0.075 | 8.5 | 4.8 | 4.4 |
| Asphalt content (%) | 7 | 5.8 | 5.4 |

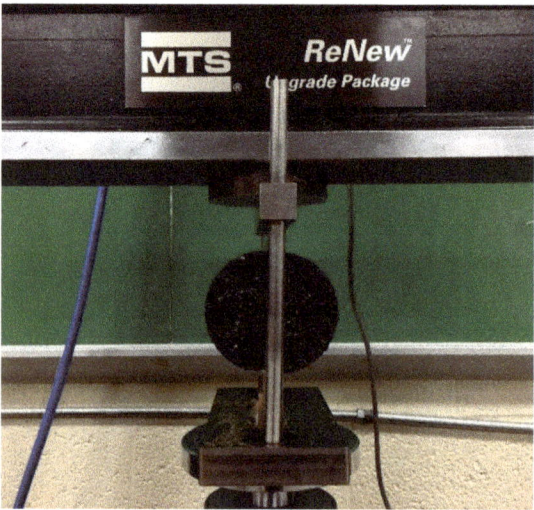

**Figure 13.** Indirect tensile strength test of asphalt mixture.

The models with three mixture designs are shown in Figure 14.

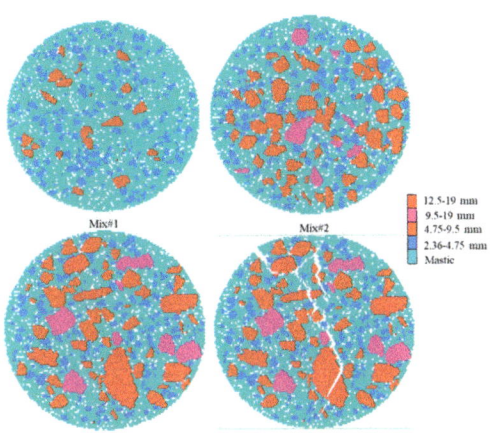

**Figure 14.** DEM Models for Parameter Validation Tests.

*4.2. Calculation of Model Micro-Parameters*

The micro-parameters were calculated based on the equations derived from the previous section, and the contact model parameters were calculated based on materials' macro-properties [7,33–35], as shown in Table 2.

**Table 2.** Martial Properties and Micro-parameter.

|  |  | Aggregate | Mastic | Aggregate-Mastic Interface |
|---|---|---|---|---|
| Elastic moduli, $E$, Pa | $E = \frac{3\sqrt{3}}{4}k_n + \frac{\sqrt{3}}{4}k_s$ | 20 GPa | 300 MPa | - |
| Poisson's ratio | $v = \frac{E}{2G} - 1$ | 0.2 | 0.5 | - |
| Tensile strength, $\sigma$, Pa | $\sigma = \frac{\sqrt{3}}{2}\frac{T_f}{R}$ | 15.27 MPa | 7.04 MPa | 6.33 MPa |
| Shear strength, $\tau$, Pa | $\tau = \frac{2S_f}{R}$ | 30.54 MPa | 13.45 MPa | 12.10 MPa |
| Stiffness ratio, $k^*$ | $k^* = \frac{k_n}{k_s} = 2(v+1)$ | 2.4 | 3.0 | 2.33 |
| Normal stiffness, $k_n$, N/m | $k_n = \frac{4\sqrt{3}E}{3(3+k^*)}t$ | $7.23 \times 10^8$ | $7.11 \times 10^6$ | $1.41 \times 10^7$ |
| Shear stiffness, $k_s$, N/m | $k_s = \frac{k_n}{k^*}$ | $3.01 \times 10^8$ | $3.05 \times 10^6$ | $6.04 \times 10^6$ |
| Tensile bond break force, $T_f$, N | $T_f = \frac{2\sqrt{3}R\sigma}{3}t$ | 529.04 | 243.82 | 219.44 |
| Shear bond break force, $S_f$, N | $S_f = \frac{R\tau}{2}t$ | 458.16 | 201.69 | 181.52 |
| Friction coefficient |  | 1.07 | 0.58 | 0.58 |

*4.3. IDT Results and Discussion*

The IDT results from laboratory tests and DEM simulation are shown in Figure 15. As shown in Figure 15a, a total of 9 laboratory specimens belonging to 3 groups were tested. For Mix1, Mix2, and Mix3, the average peak forces were 6.25 kN, 10.12 kN, and 10.81 kN, respectively. In general, the mixture type has the largest coarse aggregate grain size (Mxi3) and presented the highest peak force (tensile strength). Accordingly, Mix3 showed the steepest increasing rate (material moduli) and minimum displacement at the force peak (ultimate strain). After specimen failure, the decrease curves were relatively gentle compared to the increase curve. In this test, the standard deviations caused by the differences in specimens and test errors were relatively large but were still in a reasonable range.

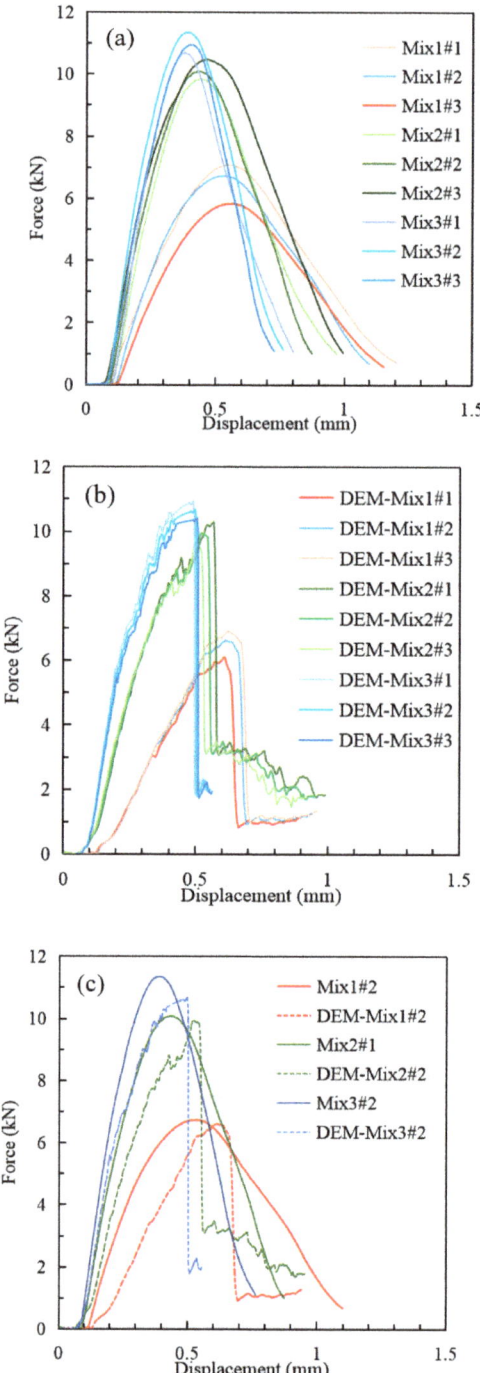

**Figure 15.** IDT Results from Laboratory Tests and DEM simulation. (**a**) Laboratory Tests. (**b**) DEM Simulation. (**c**) Comparison between Laboratory Tests and DEM Simulation.

Figure 15b shows the results of the DEM simulation for Mix1, Mix2, and Mix3. The average peak forces were 6.21 kN, 9.96 kN, and 10.56 kN, respectively. The results were close to that of laboratory tests, with relative errors ranging from 0.64% to 2.31%. The standard deviations of IDT results were at the same level as the comparison group. With zigzag data point curves, the results were much "rougher" than that of the laboratory control group. This is caused by the limited model scale. Actually, in the authors' other studies there are models with millions of elements (less than 10,000 elements were used in this study) that could present more smooth curves, especially in 2D. The authors' intention is to showcase the ability and reliability of their models by utilizing limited scales. After specimen failure, sharp drops were observed. There were two major reasons. First, the limited model scale led to large jumps at each of the failures, and second, the 2D models had less freedom than reality in which cracks could develop in lateral directions.

To compare the results, one force/displacement curve (whose test value is in the middle) for each mixture type was chosen, as shown in Figure 15c. The peak value of DEM simulations causes more displacement than the in lab because the initial loading stage of the DEM needs a process to "compact" the model into a denser status to achieve better loading transfer efficiency. However, prior to specimen failure, the peaking value and other parts of loading curves have good consistency with lab results. Considering that all the parameters are based on theoretical calculation without adjustment and with limitations on minimum element size, the DEM simulation results are reasonable, and the parameter calculation is reliable.

Compared to other models that utilize the discrete element method, which often exhibit a relative error exceeding 10%, the error in the results of this study is relatively small. It is noteworthy that the parameters in this study are derived using formulas rather than iteratively fitting them based on simulation results, as is commonly practiced in general studies. Additionally, this study employs a minimum-cost two-dimensional model with limited dimensions and scale, and the calculation of contact parameters is based on laboratory experiments that may have some degree of error fluctuation. Given these constraints, simulating loading curves of three different graded mixtures with consistent trends is still a challenging task, despite slight differences in the curves.

The crux of this paper lies in utilizing a theoretical calculation method to derive contact parameters for the discrete element method when simulating particulate matter instead of relying on iterative back-calculation fitting, as in typical research. Starting from a theoretical level, this method is more logical, resource-efficient, and reproducible.

## 5. Conclusions

In this study, a new approach for modeling procedures and determining parameters was proposed to enhance the integration of discrete element models (DEMs) in asphalt simulation. The following conclusions were drawn:

(1) A new approach for modeling procedures using the hexagonal close-packed (HCP) structure was proposed. This method, which employs realistic coarse aggregate morphology from 3D scanning, was demonstrated to be effective and can help save time and resources by reducing the need for laboratory samples. An objective search-efficiency improvement algorism is developed in this process.

(2) Micro-parameters of the DEM models were transformed from material macro-parameters using a set of equations that were derived based on the basic configuration and mechanism of discrete element theories. The effectiveness of the DEM models in simulating the indirect tensile strength for asphalt mixtures was demonstrated. The results were close to that of laboratory tests, with relative errors ranging from 0.64% to 2.31%.

(3) The key contribution of this research is the use of a reliable approach for determining model micro-parameters through mechanical calculation instead of a radical and inefficient iteration method for model parameter fitting. This new approach has the potential to expand and deepen the application of HCP structure DEM models in granular material research.

**Author Contributions:** Conceptualization, X.Z.; Methodology, X.Z. and Z.Y.; Software, Z.Y.; Investigation, X.Z., D.J., D.G. and S.C.; Resources, Z.Y.; Data curation, X.Z. and D.J.; Writing—original draft, X.Z. and D.J.; Writing—review & editing, D.J., S.C. and Z.Y.; Visualization, D.G.; Supervision, Z.Y. All authors have read and agreed to the published version of the manuscript.

**Funding:** This research received no external funding.

**Institutional Review Board Statement:** Not applicable.

**Informed Consent Statement:** Not applicable.

**Data Availability Statement:** The datasets generated during analyzed during the current study are available from the corresponding author on reasonable request.

**Conflicts of Interest:** The authors declare no conflict of interest.

## References

1. Cundall, P.A.; Strack, O.D. A discrete numerical model for granular assemblies. *Geotechnique* **1979**, *29*, 47–65. [CrossRef]
2. Buttlar, W.; You, Z. Discrete element modeling of asphalt concrete: Microfabric approach. *Transp. Res. Rec. J. Transp. Res. Board* **2001**, *1757*, 111–118. [CrossRef]
3. You, Z.; Buttlar, W.G. Discrete Element Modeling to Predict the Modulus of Asphalt Concrete Mixtures. *J. Mater. Civ. Eng.* **2004**, *16*, 140–146. [CrossRef]
4. Abbas, A.; Masad, E.; Papagiannakis, T.; Harman, T. Micromechanical modeling of the viscoelastic behavior of asphalt mixtures using the discrete-element method. *Int. J. Geomech.* **2007**, *7*, 131–139. [CrossRef]
5. You, Z.; Adhikari, S.; Dai, Q. Three-Dimensional Discrete Element Models for Asphalt Mixtures. *J. Eng. Mech.* **2008**, *134*, 1053–1063. [CrossRef]
6. Wang, C.; Zhou, X.; Liu, P.; Lu, G.; Wang, H.; Oeser, M. Study on pre-compaction of pavement graded gravels via imaging technologies, artificial intelligent and numerical simulations. *Constr. Build. Mater.* **2022**, *345*, 128380. [CrossRef]
7. Liu, Y.; Dai, Q.; You, Z. Viscoelastic Model for Discrete Element Simulation of Asphalt Mixtures. *J. Eng. Mech.* **2009**, *135*, 324–333. [CrossRef]
8. You, Z.; Liu, Y.; Dai, Q. Three-Dimensional Microstructural-Based Discrete Element Viscoelastic Modeling of Creep Compliance Tests for Asphalt Mixtures. *J. Mater. Civ. Eng.* **2011**, *23*, 79–87. [CrossRef]
9. Jin, D.; Ge, D.; Zhou, X.; You, Z. Asphalt Mixture with Scrap Tire Rubber and Nylon Fiber from Waste Tires: Laboratory Performance and Preliminary ME Design Analysis. *Buildings* **2022**, *12*, 160. [CrossRef]
10. Cai, W.; McDowell, G.R.; Airey, G.D. Discrete element visco-elastic modelling of a realistic graded asphalt mixture. *Soils Found.* **2014**, *54*, 12–22. [CrossRef]
11. Zhou, X.; Chen, S.; Jin, D.; You, Z. Discrete Element Simulation of the Internal Structures of Asphalt Mixtures with High Content of Tire Rubber. In *Advances in Transportation Geotechnics IV: Proceedings of the 4th International Conference on Transportation Geotechnics Volume 1*; Springer: Berlin/Heidelberg, Germany, 2022.
12. Dondi, G.; Vignali, V.; Pettinari, M.; Mazzotta, F.; Simone, A.; Sangiorgi, C. Modeling the DSR complex shear modulus of asphalt binder using 3D discrete element approach. *Constr. Build. Mater.* **2014**, *54* (Suppl. C), 236–246. [CrossRef]
13. Chen, J.; Huang, B.; Shu, X.; Hu, C. DEM Simulation of Laboratory Compaction of Asphalt Mixtures Using Open Source Code. *J. Mater. Civ. Eng.* **2015**, *27*, 04014130. [CrossRef]
14. Jin, D.; Meyer, T.K.; Chen, S.; Boateng, K.A.; Pearce, J.M.; You, Z. Evaluation of lab performance of stamp sand and acrylonitrile styrene acrylate waste composites without asphalt as road surface materials. *Constr. Build. Mater.* **2022**, *338*, 127569. [CrossRef]
15. Jin, D.; Ge, D.; Chen, S.; Che, T.; Liu, H.; Malburg, L.; You, Z. Cold in-place recycling asphalt mixtures: Laboratory performance and preliminary ME design analysis. *Materials* **2021**, *14*, 2036. [CrossRef] [PubMed]
16. Guan, Y.; Guan, H. Algorithms for modelling 3D flexible pavements and simulation of vibration cutting by the DEM. *Int. J. Pavement Eng.* **2019**, *20*, 1127–1139. [CrossRef]
17. Liu, Y.; You, Z. Visualization and Simulation of Asphalt Concrete with Randomly Generated Three-Dimensional Models. *J. Comput. Civ. Eng.* **2009**, *23*, 340–347. [CrossRef]
18. Liu, L.; Shen, D.; Chen, H.; Xu, W. Aggregate shape effect on the diffusivity of mortar: A 3D numerical investigation by random packing models of ellipsoidal particles and of convex polyhedral particles. *Comput. Struct.* **2014**, *144* (Suppl. C), 40–51. [CrossRef]
19. Peng, Y.; Harvey, J.T.; Sun, L.-J. Three-Dimensional Discrete-Element Modeling of Aggregate Homogeneity Influence on Indirect Tensile Strength of Asphalt Mixtures. *J. Mater. Civ. Eng.* **2017**, *29*, 04017211. [CrossRef]
20. Dan, H.-C.; Zhang, Z.; Chen, J.Q.; Wang, H. Numerical Simulation of an Indirect Tensile Test for Asphalt Mixtures Using Discrete Element Method Software. *J. Mater. Civ. Eng.* **2018**, *30*, 04018067. [CrossRef]
21. Liu, Y.; Zhou, X.; You, Z.; Yao, S.; Gong, F.; Wang, H. Discrete element modeling of realistic particle shapes in stone-based mixtures through MATLAB-based imaging process. *Constr. Build. Mater.* **2017**, *143* (Suppl. C), 169–178. [CrossRef]
22. Zhou, X.; Chen, S.; Ge, D.; Jin, D.; You, Z. Investigation of asphalt mixture internal structure consistency in accelerated discrete element models. *Constr. Build. Mater.* **2020**, *244*, 118272. [CrossRef]

23. Kim, H.; Buttlar, W.G. Micromechanical fracture modeling of asphalt mixture using the discrete element method. In *Advances in Pavement Engineering*; American Society of Civil Engineers: Reston, VA, USA, 2005; pp. 1–15.
24. Vallejo, L.E.; Lobo-Guerrero, S.; Hammer, K. Degradation of a granular base under a flexible pavement: DEM simulation. *Int. J. Geomech.* **2006**, *6*, 435–439. [CrossRef]
25. Kim, H.; Wagoner, M.P.; Buttlar, W.G. Simulation of fracture behavior in asphalt concrete using a heterogeneous cohesive zone discrete element model. *J. Mater. Civ. Eng.* **2008**, *20*, 552–563. [CrossRef]
26. Yu, H.; Shen, S. Impact of aggregate packing on dynamic modulus of hot mix asphalt mixtures using three-dimensional discrete element method. *Constr. Build. Mater.* **2012**, *26*, 302–309. [CrossRef]
27. Wang, L.; Zhang, B.; Wang, D.; Yue, Z. Fundamental Mechanics of Asphalt Compaction through FEM and DEM Modeling. In *Analysis of Asphalt Pavement Materials and Systems*; American Society of Civil Engineers: Reston, VA, USA, 2007.
28. Chen, J.; Huang, B.; Chen, F.; Shu, X. Application of discrete element method to Superpave gyratory compaction. *Road Mater. Pavement Des.* **2012**, *13*, 480–500. [CrossRef]
29. Gong, F.; Liu, Y.; Zhou, X.; You, Z. Lab assessment and discrete element modeling of asphalt mixture during compaction with elongated and flat coarse aggregates. *Constr. Build. Mater.* **2018**, *182*, 573–579. [CrossRef]
30. Gong, F.; Zhou, X.; You, Z.; Liu, Y.; Chen, S. Using discrete element models to track movement of coarse aggregates during compaction of asphalt mixture. *Constr. Build. Mater.* **2018**, *189*, 338–351. [CrossRef]
31. Liu, Y.; Zhou, X.; You, Z.; Ma, B.; Gong, F. Determining Aggregate Grain Size Using Discrete-Element Models of Sieve Analysis. *Int. J. Geomech.* **2019**, *19*, 04019014. [CrossRef]
32. Jin, D.; Wang, J.; You, L.; Ge, D.; Liu, C.; Liu, H.; You, Z. Waste cathode-ray-tube glass powder modified asphalt materials: Preparation and characterization. *J. Clean. Prod.* **2021**, *314*, 127949. [CrossRef]
33. Khattak, M.J.; Khattab, A.; Rizvi, H.R.; Das, S.; Bhuyan, M.R. Imaged-based discrete element modeling of hot mix asphalt mixtures. *Mater. Struct.* **2015**, *48*, 2417–2430. [CrossRef]
34. Yang, X.; Dai, Q.; You, Z.; Wang, Z. Integrated experimental-numerical approach for estimating asphalt mixture induction healing level through discrete element modeling of a single-edge notched beam test. *J. Mater. Civ. Eng.* **2015**, *27*, 04014259. [CrossRef]
35. Feng, H.; Pettinari, M.; Stang, H. Study of normal and shear material properties for viscoelastic model of asphalt mixture by discrete element method. *Constr. Build. Mater.* **2015**, *98*, 366–375. [CrossRef]

**Disclaimer/Publisher's Note:** The statements, opinions and data contained in all publications are solely those of the individual author(s) and contributor(s) and not of MDPI and/or the editor(s). MDPI and/or the editor(s) disclaim responsibility for any injury to people or property resulting from any ideas, methods, instructions or products referred to in the content.

Article

# Optimization of Embedded Sensor Packaging Used in Rollpave Pavement Based on Test and Simulation

Zhoujing Ye [1], Yanxia Cai [2,3,4], Chang Liu [1], Kaiji Lu [2,3,4], Dylan G. Ildefonzo [5] and Linbing Wang [5,6,*]

1. National Center for Materials Service Safety, University of Science and Technology Beijing, Beijing 100083, China; yezhoujing@ustb.edu.cn (Z.Y.); g20199178@xs.ustb.edu.cn (C.L.)
2. Beijing Zhonglu Gaoke Highway Technology Co., Ltd., Beijing 100088, China; 13311530926@163.com (Y.C.); kj.lu@rioh.cn (K.L.)
3. Research and Development Center of Transport Industry of New Materials, Technologies Application for Highway Construction and Maintenance, Beijing 100088, China
4. Research Institute of Highway Ministry of Transport, Beijing 100088, China
5. Center for Smart and Green Civil Systems, Virginia Tech, Blacksburg, VA 24060, USA; ildefonzo@vt.edu
6. Joint USTB-Virginia Tech Lab on Multifunctional Materials, University of Science and Technology Beijing, Beijing 100083, China
* Correspondence: wangl@vt.edu

**Abstract:** Rollpave pavement, as a rollable prefabricated asphalt pavement technology, can effectively reduce the overall road closure time required for pavement construction and maintenance. Sensors can be integrated into Rollpave pavement, thereby avoiding sensor damage that may otherwise result from high temperatures and compactive forces during the rolling process, as well as pavement structural damage resulting from cutting and drilling. However, the embedment of sensors into Rollpave pavement still presents certain challenges, namely poor interfacial synergy between the embedded sensor and the asphalt mixture. To solve this problem, three-point bending tests and dynamic response FEM simulations were used to optimize the embedded sensor's packaging. The influence of sensor embedment on Rollpave pavement under different working conditions was analyzed. Results of these analyses show that low temperature and the epoxy resin negatively affect the bending performance of specimens, and that packaging with cylindrical shape, flat design, and consisting of a material with modulus similar to that of the asphalt mixture should be preferred. This study is conducive to improve the intellectual level and service life of road infrastructure.

**Keywords:** embedded sensor; Rollpave pavement; three-point bending test; dynamic response; finite element analysis; packaging optimization

## 1. Introduction

With the increase of urban traffic volume, road construction and maintenance activities are more likely to cause serious traffic congestion and increased delays for the traveling public. In an effort to shorten road construction and maintenance time, Rollpave pavement, a new construction technology, was first proposed in the Netherlands [1]. Rollpave is a rollable pavement technology that is prefabricated in a manufacturing facility. During manufacturing, the pavement's structural layers are fabricated, with specific functions, and then rolled up onto special reels for transport. Upon arrival at the construction site, the prefabricated pavement is unrolled like a carpet, allowing for rapid paving of the pavement layer [2,3]. Use of Rollpave technology can significantly improve the construction speed of asphalt pavement. Rollpave pavement has shown excellent potential as a pavement maintenance and repair technology [4,5].

The current research on Rollpave pavement mainly focuses on curling construction technology, new flexible pavement materials, and pavement structure functional integration. By studying the addition of new materials such as polyurethane, Wang et al. created Rollpave pavement with improved noise-reducing and anti-skid properties and improved the overall noise-reduction performance of this paving technology [6,7]. Steinauer explored the addition of energy harvesting and storage technologies for thermal, light, and piezoelectric energy in Rollpave pavement [8]. Guo et al. used high-elasticity asphalt to improve Rollpave pavement's anti-fatigue and -permanent-deformation performance and developed supporting rolling equipment [9]. Dong et al. proposed a new type of modified asphalt for Rollpave mixtures [10]. Feng et al. proposed a "prefabricated asphalt pavement" for use in areas that cannot support construction requiring large-scale machinery [11]. Dai conducted research on the structure, performance, construction technology, and noise reduction capability of Rollpave pavement [12].

Rollpave pavement shows great technical advantages, especially in the ease of sensor embedment, a unique advantage for enabling the development of intelligent roadways. Prefabrication allows for more accurate placement of sensors within the pavement material, reduces the risk of sensor damage caused by high temperatures and forces from rolling and compaction, and removes the need for cutting and drilling of the pavement for sensor installation. When embedded with sensors, Rollpave pavement will have sensing functions for monitoring traffic information and evaluating the performance of the pavement structure. However, sensor embedment can adversely affect the performance of asphalt pavement structures, due to poor interfacial coupling between the sensor and asphalt mixture. Additionally, the evaluation metrics and experimental schemes for sensor coupling with the asphalt mixture interface are not presently well-defined [13,14]. In light of this fact, the evaluation of the influence of sensor embedment on Rollpave pavement and optimization of the sensor embedment scheme constitute topics of great research significance.

In this paper, a three-point bending test is conducted on asphalt beam specimens with embedded sensors, and the effect of sensor embedment on the bending performance of Rollpave asphalt mixture beam specimens is evaluated. A computational simulation of the dynamic response of Rollpave pavement with embedded sensors under moving vehicle loads is created. The results of this simulation are used to analyze the time-dependent characteristics of the maximum sensor–mixture interfacial stress under various operating conditions. Taken together, this paper illustrates the influence of sensor embedment on Rollpave pavement, with the aim of providing a reference for sensor packaging optimization. Improvement of sensor integration in Rollpave pavements will pave the way for future intelligent roadways with prolonged service lives.

## 2. Three-Point Bending Test by the Embedded Sensor

As Rollpave pavement requires prefabrication off-site, storage and transportation must be considered during manufacturing. Presently, the preferred method to accommodate storage and transportation is to form a thin, prefabricated asphalt layer and roll it into a cylindrical shape. However, in this form, the pavement structure will experience tensile and compressive stresses. To ensure structural performance during storage, it is necessary to evaluate the bending performance of the Rollpave pavement, as well as the impact of sensor embedment. The three-point bending test is suitable for determining the mechanical properties of the bending failure of asphalt mixtures at specified temperatures and loading rates. Therefore, the three-point bending test is performed on the asphalt mixture beam specimens, in order to evaluate the effect of the embedded sensor on the bending performance of Rollpave pavement.

*2.1. Test Conditions*

Several tests were conducted to analyze the effects of different factors on the bending performance of asphalt mixture beam specimens. Testing conditions are summarized in Table 1. The influence factors (independent variables) analyzed include sensor embedment, sensor size, sensor shape, sensor packaging materials, ambient temperature, and bonding materials.

Table 1. Designed test conditions.

| No. | Packaging Material | Sensor Shape | Sensor Size | Bonding Material | Temperature | Influence Factor |
|---|---|---|---|---|---|---|
| LA0 | Without embedding sensor | / | / | No adhesive | 10 °C | Sensor embedment |
| LA1 | Stainless steel | Cylinder | Φ 40 mm H 15 mm | No adhesive | 10 °C | Control group |
| LA2 | Stainless steel | Cylinder | Φ 40 mm H 15 mm | Epoxy | 10 °C | Bonding materials |
| LA3 | Stainless steel | Cylinder | Φ 40 mm H 15 mm | No adhesive | −10 °C | Temperature |
| LA4 | Stainless steel | Cuboid | L 40 mm M 40 mm H 15 mm | No adhesive | 10 °C | Shape |
| LA5 | Cast nylon | Cylinder | Φ 40 mm H 15 mm | No adhesive | 10 °C | Packaging material |
| LA6 | Stainless steel | Cylinder | Φ 50 mm H 15 mm | No adhesive | 10 °C | Sensor size (diameter) |
| LA7 | Stainless steel | Cylinder | Φ 30 mm H 15 mm | No adhesive | 10 °C | |
| LA8 | Stainless steel | Cylinder | Φ 40 mm H 20 mm | No adhesive | 10 °C | Sensor size (thickness) |
| LA9 | Stainless steel | Cylinder | Φ 40 mm H 10 mm | No adhesive | 10 °C | |

## 2.2. Specimen Preparation

### 2.2.1. Preparation of Asphalt Mixture

In this experiment, the asphalt mixture used to prepare the Rollpave pavement specimens has an oil-stone ratio of 7%, basalt as coarse aggregate, machine-made sand as fine aggregate, and bitumen penetration grade 70 as asphalt binder. The specific gradation of the mixture used is shown in Table 2.

Table 2. Gradation of composite aggregate.

| | Cumulative Percent Passing Each Sieve (%) | | | | | | | | |
|---|---|---|---|---|---|---|---|---|---|
| Aperture sizes (mm) | 13.2 | 9.5 | 4.75 | 2.36 | 1.18 | 0.6 | 0.3 | 0.15 | 0.075 |
| Gradation | 100 | 93 | 47 | 30 | 20 | 17 | 15 | 12 | 10 |

For the preparation of the asphalt mixture especially used for Rollpave pavement, high-viscosity additive (HVA) with 12% base asphalt, as shown in Figure 1, is added to mixed and stirred well. HVA can improve the asphalt viscosity and increase inter-aggregate bond strength, resulting in improved resistance to rutting and fatigue, low temperature crack resistance, and overall asphalt mixture stability.

**Figure 1.** High-viscosity additive.

2.2.2. Sensor Packaging

Different sensor packages were designed and fabricated for the experiments. The different sensor packages used are shown in Figure 2. Packaging materials are composed of either 304 stainless steel or cast nylon. Package shapes are either cylindrical or cuboid, and sensor sizes cover different diameters and thicknesses. The dimensions of the packaging are mainly determined by both the size of the sensor chip and limitations of packaging processing technology. The packaging, as designed, can be used to protect built-in sensor chips, including MEMS accelerometers, temperature sensors, humidity sensors, vibration sensors, pressure sensors, and displacement sensors [15,16].

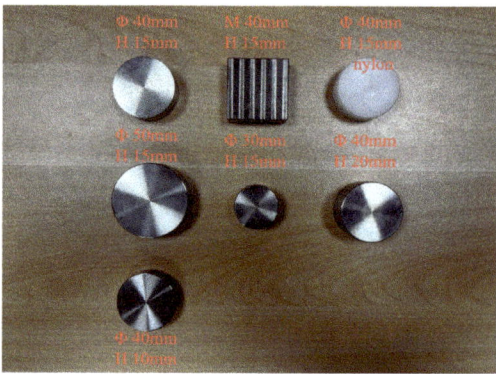

**Figure 2.** Different sensor packages.

2.2.3. Fabrication of Beam Specimen

During asphalt mixing, rutting plate specimens with embedded sensors are fabricated. The specific process for the fabrication of rutting plates is as follows: (1) the mold is preheated and the bottom and sides of the mold are brushed with oil; (2) paper is installed in the mold; (3) the asphalt mixture is evenly placed into the mold with a small shovel, and it is ensured that the center is slightly higher than the sides; (4) after placing the sensor in the designated position, the remaining asphalt mixture is added, and then a small, preheated hammer is used to tamp and level the mixture from the edges to the center. The rutting plate specimens are then cut into beam specimens for the three-point bending tests, as shown in Figures 3 and 4.

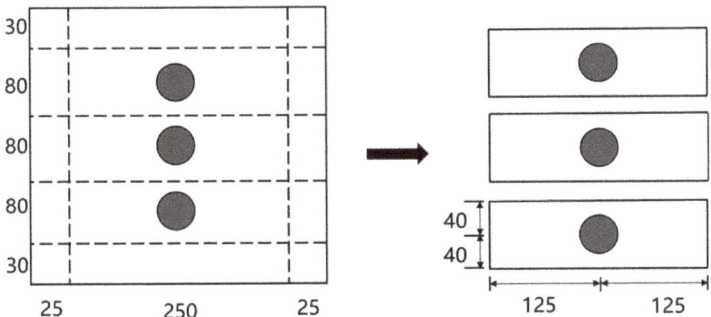

**Figure 3.** Dimension of beam specimen (unit: mm).

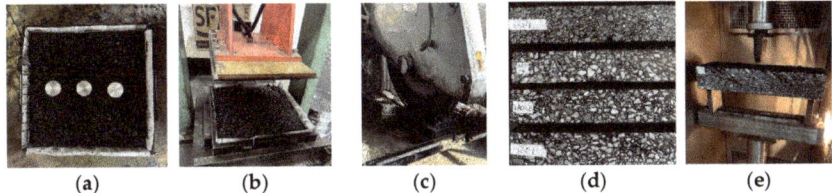

**Figure 4.** Beam specimen preparation. (**a**) Sensor positioning; (**b**) rutting plate specimens; (**c**) cutting of specimens; (**d**) beam specimens marking; (**e**) a three-point bending test.

In consideration of the boundary effect, the sensor is embedded in the center of the rutting plate specimen (sensors are embedded 1.25 cm away from the specimen surface) to maintain a certain distance from the specimen boundary [17]. According to the *Standard Test Methods of Bitumen and Bituminous Mixtures for Highway Engineering (JTG E20-2011)*, the rutting plate specimens are cut into beam specimens, with dimensions of 250 × 80 × 40 mm (length × width × thickness). The beam specimens are then placed in a temperature-controlled chamber, where they are kept at the set temperature for no less than four hours, until the internal temperature reaches ±0.5 °C of the test temperature. Once the test temperature is reached, the test specimen is taken out and placed on the supports of the three-point bending test machine. The distance between the fulcrum points was determined to be 200 ± 0.5 mm, so that the upper pressure head and the lower pressure head were kept parallel, and the two sides were equidistant. The hydraulic press then applies a concentrated load at the span center, at a loading rate of 50 mm/min, until specimen fracture occurs, at which point the specimen is considered to have failed. Three beam specimens were made for each working condition. Based on the average load and displacement of the three specimens, the load–midspan deflection curve for each case is obtained.

### 2.3. Evaluation Index of Bending Performance

As an index to measure the critical failure of specimens, flexural tensile strength and strain have opposite trends, making it challenging to quantitatively evaluate the bending performance by using both. Therefore, the bending strain energy is chosen as the evaluation metric [18]. The bending strain energy refers to the area enclosed by the stress–strain curve, and the x-axis before the stress reaches the peak value; this concept is illustrated in Figure 5. The unit of strain energy is $KJ/m^3$. The bending strain energy characterizes the energy absorbed by the specimen before failure. The greater the bending strain energy, the greater the energy required to fail the specimen.

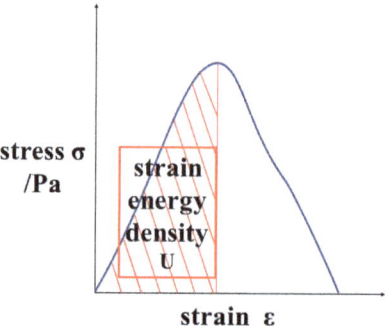

**Figure 5.** Bending strain energy.

According to the load–displacement curve and Equations (1) and (2), the stress–strain curve is calculated. In addition, flexural tensile strength $R_B$, beam bottom maximum flexural tensile strain $\varepsilon_B$ at bending failure, and bending stiffness modulus $S_B$ can be calculated according to Equations (3)–(5):

$$\sigma_B = \frac{3FL}{2bh^2} \tag{1}$$

$$\varepsilon_B = \frac{6hd}{L^2} \tag{2}$$

$$R_B = \frac{3P_B L}{2bh^2} \tag{3}$$

$$S_B = \frac{R_B}{\varepsilon_B} \tag{4}$$

$$U = \int \sigma d\varepsilon \tag{5}$$

where $F$ is the load; $P_B$ is the maximum load when the specimen fails; $h$ is the height of the specimen; $B$ is the width of the specimen; $d$ is the midspan deflection when the specimen is loaded; $L$ is the span length of the specimen; $U$ is the bending strain energy; and all other variables are as designated above.

In consideration of the curling process of Rollpave pavement, the beam bottom maximum flexural tensile strain is selected as the evaluation index. The greater the beam bottom maximum flexural tensile strain of specimens before loading failure, the easier it is for the Rollpave pavement crimping process to be achieved. The bending stiffness modulus (Equation (4)) represents the ability of the beam specimens to resist bending deformation within elastic limit. The higher the value of bending stiffness modulus, the less effective the beam specimens are at resisting bending deformation within the elastic limit.

### 2.4. Influence Analysis of Bending Properties

The load–midspan deflection curves of the specimens under various test conditions are obtained by the recorder attached to the three-point bending test frame and shown in Figure 6.

The values of the evaluation indices were calculated according to Equations (1) through (5), with testing condition LA1 set as the control group marked with yellow columns. The results are shown in Figure 7.

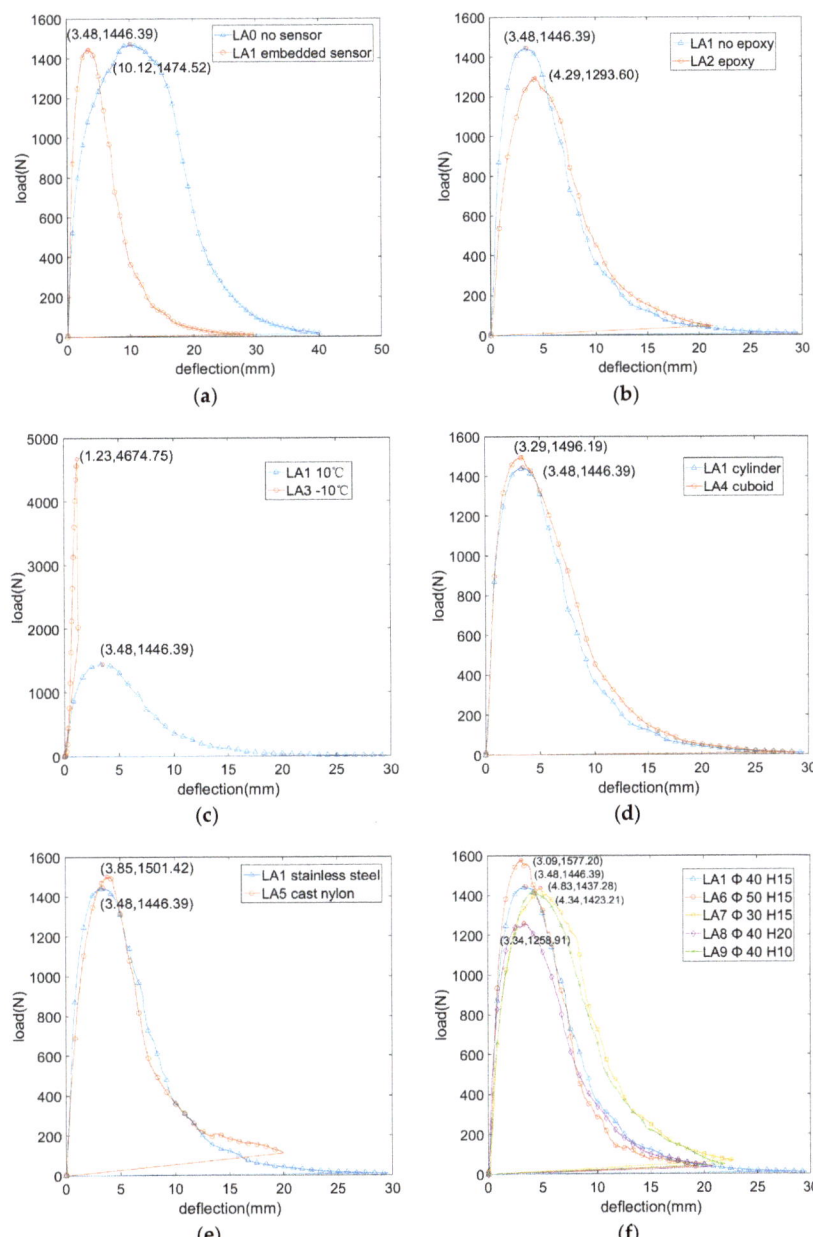

**Figure 6.** The load–midspan deflection curve under various test conditions. (**a**) Influence of embedded sensor; (**b**) influence of epoxy coating; (**c**) influence of temperature; (**d**) influence of sensor packaging shape; (**e**) influence of packaging material; (**f**) influence of sensor packaging size.

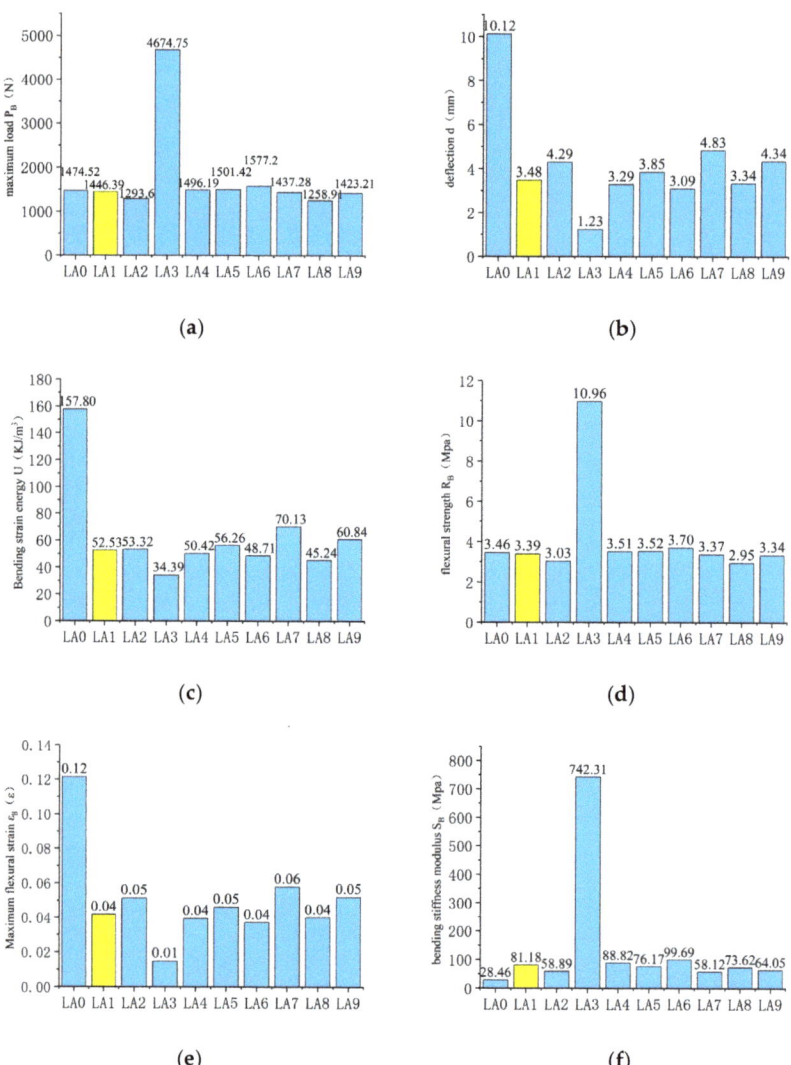

**Figure 7.** Evaluation index values for bending performance. (**a**) Maximum load $P_B$ at specimen failure; (**b**) mid-span deflection $d$; (**c**) bending strain energy $U$; (**d**) flexural tensile strength $R_B$; (**e**) beam bottom maximum flexural tensile strain $\varepsilon_B$. (**f**) bending stiffness modulus $S_B$.

2.4.1. Influence of Embedded Sensor

The influence of the embedded sensors on the bending performance of the beam specimens is analyzed by comparing test conditions LA0 and LA1. According to Figure 6a, it is shown that, with the increase of midspan deflection, specimens containing an embedded sensor experience a loading rate increase that is significantly higher than that of the specimen without an embedded sensor. The specimen containing an embedded sensor reached failure load in a shorter amount of time than did the specimen without an embedded sensor. According to Figure 6a, the maximum loads of the two tests are only slightly different, with maximum load rates of 1474.52 and 1446.39 N for testing conditions LA0 and LA1, respectively. Despite the small difference in maximum loading, the midspan deflections experience differed greatly between the two specimens, with deflection of 10.12 and 3.48 mm for testing conditions LA0 and LA1, respectively. Sensor embedment is, thus, observed to decrease midspan deflection at beam specimen failure load by 65.58%. By comparing the evaluation index values of test conditions LA0 and LA1 from Figure 7c, the bending strain energy of beam specimens decreased by 66.71%, from 157.8 to 52.53 KJ/m$^3$ after embedding the sensor. According to Figure 7e, it was observed that the embedment of sensors makes beam specimens more prone to damage. There is little difference between flexural tensile strength for testing conditions LA0 and LA1. The beam bottom maximum flexural tensile strains for testing conditions LA0 and LA1 are 0.12 and 0.04 ε, respectively. According to Figure 7f, the bending stiffness modulus of condition LA1 is 81.18 MPa, greater than 28.46 MPa of condition LA0. It is, therefore, observed that the ability of the specimen with an embedded sensor to resist bending deformation is diminished, compared to the specimen without an embedded sensor.

The sensor is embedded in the midspan of the beam specimen, and it is known that the stiffness of the sensors is greater than that of the asphalt mixture, making it more difficult to produce local deformation of the sensor. Therefore, under the same midspan deflection, the deformation of specimens with a sensor embedded is greater than that of specimens without a sensor embedded. There exists a stress concentration at the interface between the sensor and asphalt mixture, which will lead to the compression peeling of the sensor surrounding the asphalt mixture, as shown in Figure 8.

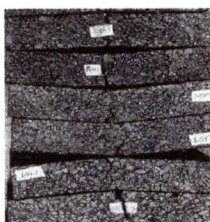 

**Figure 8.** Specimen failure after the three-point bending test.

### 2.4.2. Influence of Epoxy Coating

The influence of epoxy resin bonding on the bending performance of beam specimens is analyzed by comparing test conditions LA1 and LA2. According to Figure 6b, the failure load of the beam specimens with epoxy resin (test condition LA1) is 1446.39 N, while the failure load of beam specimens (testing condition LA2) is 1293.60 N; such a small difference in failure load is deemed insignificant. The load increase rates between conditions LA1 and LA2 are similar, and the corresponding deflections are 3.48 and 4.29 mm for LA1 and LA2, respectively. According to Figure 7c, conditions LA1 and LA2 have similar bending strain energy values, 52.53 and 52.32 KJ/m$^3$, respectively. According to Figure 7d,e, the values of flexural tensile strength and beam bottom maximum flexural tensile strain are similar for both test conditions. According to Figure 7f, the bending stiffness modulus of LA1 is 81.18 MPa, which is greater than the 58.89 MPa bending stiffness modulus of LA2. It is, thus, observed that epoxy resin as a binder can indeed enhance the bending performance of specimens, but the effects are not significant. It is additionally understood that the adhesion between the asphalt mixture and sensor is good at high temperatures.

### 2.4.3. Influence of Temperature

By comparing testing conditions LA1 and LA3, the effect of temperature on the beam specimens' bending performance is analyzed. Figure 6c shows that temperature has a significant impact on the bending performance of beam specimens. When the loading temperature is −10 and 10 °C, the beam specimens failure loads are 1446.39 N and 4674.75 N, respectively. The corresponding midspan deflections are 3.48 and 1.23 mm for conditions

LA1 and LA3, respectively. Figure 7 shows that the bending strain energies of LA1 and LA3 are 52.53 and 34.39 KJ/m$^3$, respectively. Moreover, there are significant differences in flexural tensile strength, beam bottom maximum flexural tensile strain, and bending stiffness modulus.

At low temperatures, asphalt mixtures can become hard and brittle. As the bending stiffness modulus increases, the resistance to bending deformation within the elastic limit decreases significantly. Therefore, the load at failure is larger, and the corresponding deflection is smaller. For bending strain energy, the energy required for specimens to fail at low temperature is lower, meaning that the specimens are more likely to fail at a low temperature than they are at normal temperature. Therefore, low-temperature fabrication of Rollpave pavement with embedded sensors should be avoided.

2.4.4. Influence of Sensor Packaging Shape

The influence of sensor shape (cylinder or cuboid) on the bending performance of beam specimens is analyzed by comparing testing conditions LA1 and LA4. According to Figure 6d, the failure load and deflection of beam specimens embedded with cylindrical and cuboid sensors are similar, and the corresponding midspan deflections are 3.48 and 3.29 mm for conditions LA1 and LA4, respectively. According to Figure 7c, cylindrical sensor packaging topology can increase the bending strain energy required for specimens to break to a small extent, but no significant difference is observed between the two specimens. No significant differences in the values of flexural tensile strength and beam bottom maximum flexural tensile are observed between testing conditions LA1 and LA4. According to Figure 7f, the bending stiffness modulus for conditions LA1 and LA4 are 81.18 and 88.82 MPa, respectively. Sensor packaging shape is observed to have little influence on the bending performance of beam specimens. To a small extent, the cylindrical topology of the sensor packaging can slightly improve the bending performance of specimens, but this does not constitute a significant improvement.

2.4.5. Influence of Packaging Material

The influence of sensor materials (stainless steel or cast nylon) on beam specimen bending performance is analyzed by comparing testing conditions LA1 and LA5. Figure 6e shows that the failure loads of the beam specimens with embedded sensors in stainless steel packaging (condition LA1) is 1446.4 N, while the failure load of beam specimens embedded with sensors in cast nylon packaging is 1501.4 N. Midspan deflections are 3.48 and 3.85 mm for conditions LA1 and LA5, respectively. According to Figure 7, the bending strain energy of the specimen containing an embedded sensor packaged in cast nylon is 56.26 KJ/m$^3$, while the bending strain energy of the specimen containing an embedded sensor packaged in stainless steel is 52.53 KJ/m$^3$. Flexural tensile strength, beam bottom maximum flexural tensile strain, and bending stiffness modulus are all similar between the two specimens tested. The bending stiffness modulus of condition LA1 is 81.18 MPa, which is greater than 76.17 MPa, which is observed for condition LA5. The sensor material is, thus, thought to have little influence on the bending performance of beam specimens, though the use of cast nylon sensor packaging can improve the bending performance of specimens to some extent. It is hypothesized that improvements seen in specimens with cast nylon sensor packaging are owed to the fact that the elastic modulus of nylon is closer to that of the road surface, as well as the fact that the nylon surface is rougher than the surface of the stainless steel, which can create better adhesion at the sensor–asphalt interface.

2.4.6. Influence of Sensor Packaging Size

By comparing the LA1, LA6, LA7, LA8, and LA9 test conditions, the impact of the sensor size is analyzed in two ways. First, the impact of the sensor diameter is analyzed by selecting three beam specimens containing sensors with packaging of the same thickness but different diameters. In this case, beam specimens LA1, LA6, and LA7 are selected for comparison, as they all have sensors of 15 mm thickness. Figure 6f shows that the midspan deflection at failure for specimens containing embedded sensors with diameters of 30, 40, and 50 mm (corresponding to specimens LA7, LA1, and LA6) are 4.83, 3.48, and 3.09 mm, respectively. Next, the impact of sensor thickness is analyzed by selecting three beam specimens containing sensors with packaging of the same diameter but different thicknesses. In this case, beam specimens LA1, LA8, and LA9 are selected for comparison as they all have sensors of 40 mm diameters. The midspan deflection at failure for beam specimens with sensor thicknesses of 10, 15, and 20 mm (corresponding to specimens LA9, LA1, and LA8) are 4.34, 3.48, and 3.34 mm, respectively. According to Figure 7, based on the values of bending strain energy and bending stiffness moduli across testing conditions, it is observed that specimens containing sensor packaging of both smaller diameter and thickness have higher strain energy at failure. That is, smaller embedded sensor packaging results in stronger resistance to deformation in the beam specimen. It is, therefore, necessary to reduce the sensor package size as much as possible, in order to improve the bending performance of the pavement.

## 3. Dynamic Response Simulation of Rollpave Pavement

In the three-point bending test, the loading force is fundamentally different from that of a road surface under vehicle load [19]. The loading force of three-point bending test is concentrated force, while the loading force of vehicle is a uniform load. In addition, the specimen size of three-point bending test is also different from the real road surface, which may lead to differences in experimental results. To further explore the influence of embedded sensors on Rollpave pavement, a simulation of the dynamic response of Rollpave pavement under vehicle moving loading is carried out.

### 3.1. Model Establishment

#### 3.1.1. Pavement Structure and Materials

Referring to the pavement structure of the National Center for Materials Service Safety (NCMS) test road, The elastic modulus, Poisson's ratio, and damping ratio were derived from *the Specifications for Design of Highway Asphalt Pavement (JTG D50-2017)* [20]. The pavement structure and material parameters are shown in Table 3, in which, the upper layer is Rollpave pavement.

Table 3. Pavement structure and material parameters.

| No. | Structure | Thickness [cm] | Elastic Modulus [Mpa] | Poisson's Ratio | Density [kg/m$^3$] | Damping Ratio |
|---|---|---|---|---|---|---|
| 1 | Upper layer Rollpave pavement | 4 | 1100 | 0.35 | 2500 | 0.05 |
| 2 | Middle layer AC-20 | 6 | 1300 | 0.35 | 2400 | 0.05 |
| 3 | Lower layer AC-25 | 8 | 1200 | 0.35 | 2400 | 0.05 |

#### 3.1.2. Model Size and Mesh Generation

A 3D finite element model (FEM) of Rollpave pavement with embedded sensors is established using Abaqus software, as illustrated in Figure 9. The road model is of dimensions 1.0 × 1.0 × 0.18 m (length × width × depth). The driving direction is taken to be the same as that of the positive $x$-axis, and the vertical direction is taken to be the same as that of the positive $z$-axis. The cylindrical sensor of dimensions 40 (diameter) and 15 mm (height) is centered on the Rollpave pavement. The top of the sensor is 1.25 cm from the road surface, the bottom is 1.25 cm from the bottom of the Rollpave pavement, and the sides are 0.48 m from the road boundary, reducing the influence of the model boundary on the sensor stress analysis.

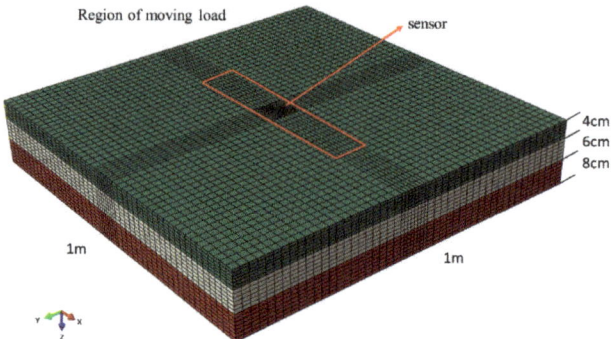

**Figure 9.** A 3D model of Rollpave pavement with an embedded sensor.

The road model was idealized using linear hexahedral elements of type C3D8R. The length and width of elements were set as 2 × 2 cm. The meshes of the loading area were refined, and the model was meshed into 54,986 elements.

3.1.3. Constrain Condition and Moving Load

The three direction movements and rotations were restrained at the bottom of the model, and the normal directions were restrained, corresponding to the four sides of the model. Uniaxial vehicle loads are idealized as moving uniform rectangular loads and imparted on the model, using the Abaqus subroutine DLOAD module [21]. The loading area was set to be in the middle of the model. The length of the moving load area was set to 0.5 m, and the width to 0.216 m. Three moving load working conditions with magnitudes of 0.7, 1.4, and 2.1 MPa were considered. Further, working conditions with loading speeds of 10 and 20 m/s was considered. The most unfavorable working condition was selected as the loading condition of the subsequent simulation.

*3.2. Error Analysis of Mesh Generation*

Due to the difference in the mesh generation between cylindrical and cuboid sensor shapes, the element shape and position at the interface are also different, which may result in errors in subsequent stress analyses. An error analysis of the mesh generation is performed during subsequent stress analyses to correct for errors arising from differences in mesh generation. Figure 10 shows the mesh generation results for the cuboid and cylindrical sensors. The cylindrical sensor is 40 mm in diameter and 15 mm in height, and the cuboid sensor is 40 mm in length and width and 15 mm in height.

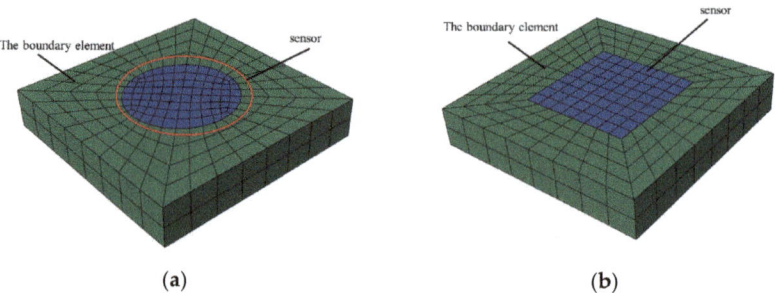

**Figure 10.** Mesh generation of sensors with different shapes. (**a**) Embedded cylinder sensor; (**b**) embedded cuboid sensor.

After the meshes are generated for the different sensor packaging shapes, other variables are controlled. By setting the sensor material properties (the pavement materials), boundary conditions (merge connection with pavement elements), and moving loads (0.7 MPa, 20 m/s) to the same working conditions, the maximum stress–time curves of the elements at the interface under different mesh generations are compared, as illustrated in Figure 11.

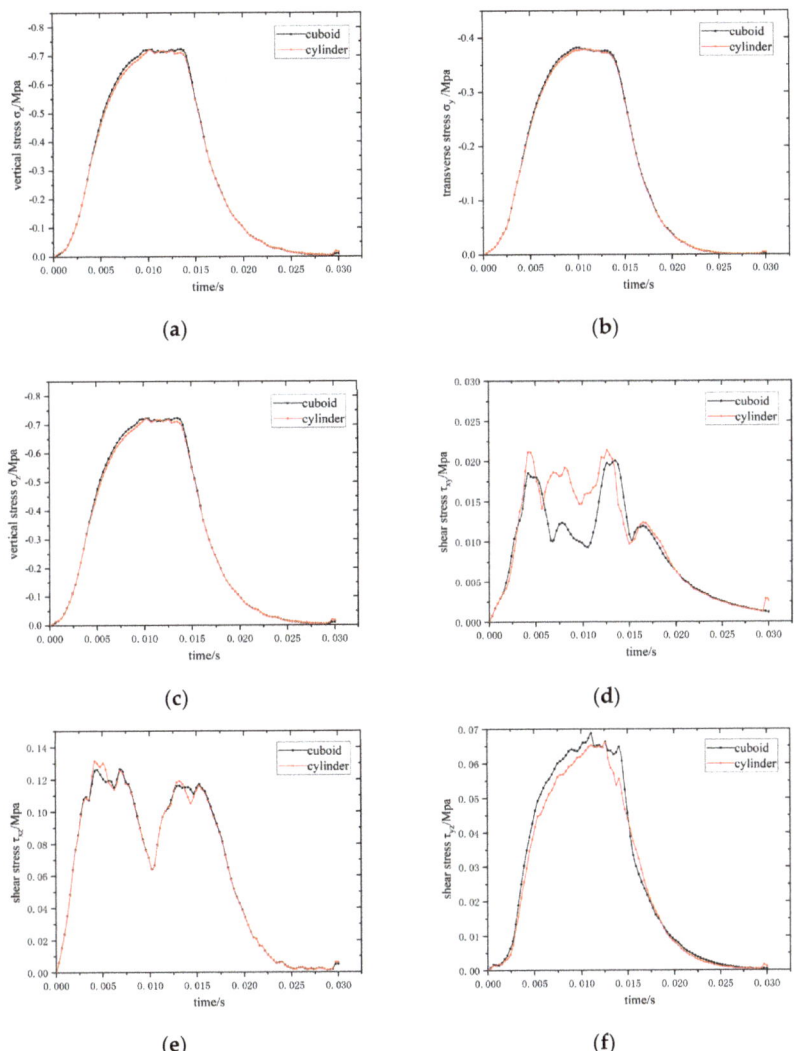

**Figure 11.** Maximum stress–time curves of elements at the interface. (**a**) Longitudinal stress ($\sigma_x$); (**b**) transverse stress ($\sigma_y$); (**c**) vertical stress ($\sigma_z$); (**d**) shear stress ($\tau_{xy}$); (**e**) shear stress ($\tau_{xz}$); (**f**) shear stress ($\tau_{yz}$).

No significant difference is observed between the maximum stresses of elements at the material interface during moving load between the two mesh types. The existing two types of mesh generation can, thus, be used to analyze the stress. No large error, caused by differences in the meshes, are observed.

Under vehicle load, the maximum normal stress in the pavement structure is vertical stress, $\sigma_z$, and the maximum shear stress is $\tau_{xz}$ (shear stress along the axis of the driving direction). Therefore, vertical stress, $\sigma_z$, and shear stress, $\tau_{xz}$, are selected as the parameters for evaluating the stress concentration at the interface between the sensor and pavement material. Greater values of the selected parameters correspond to higher stress concentrations at this interfacial boundary.

### 3.3. Sensor Packaging Optimization

The mechanical properties of sensor packaging materials are different than those of the pavement materials. The maximum vertical stress, $\sigma_z$, and shear stress, $\tau_{xz}$, in all elements at the interface are used to evaluate the synergistic performance between the embedded sensor and asphalt mixtures and are used to generate the maximum stress–time curves. Additionally, sensor packaging optimization is carried out for different driving conditions and sensor packaging shapes, materials, and sizes, as shown in Table 4.

**Table 4.** Multiple working conditions.

| Set Conditions | Operating Conditions | Test Purpose |
|---|---|---|
| Shape: cylinder<br>Size: φ40 * 15 mm<br>Materials: stainless steel | Load magnitude: 0.7 Mpa, 1.4 Mpa, 2.1 Mpa<br>Speed: 20 m/s, 10 m/s | Compare different loading conditions |
| Materials: stainless steel<br>Load magnitude: 2.1 Mpa<br>Speed: 10 m/s | Cuboid: 40 * 40 * 15 mm<br>Cylinder: φ40 * 15 mm | Compare different sensor shapes |
| Shape: cylinder<br>Size: φ40 * 15 mm<br>Load magnitude: 2.1 Mpa<br>Speed: 10 m/s | Stainless steel<br>Cast nylon | Compare different sensor materials |
| Materials: stainless steel<br>Shape: cylinder<br>Load magnitude: 2.1 Mpa<br>Speed: 10 m/s | Change in thickness<br>10/15/20/25/30 * φ40 mm<br>Change in diameter<br>Φ20/30/40/50/60 * 15 mm | Compare different sensor sizes |

#### 3.3.1. Comparison of Driving Conditions

The most unfavorable working conditions for the synergistic performance between the pavement structure and embedded sensor are analyzed under various driving conditions. Driving conditions include combinations of vehicle speed and load magnitudes and are composed of vehicle speeds of 10 and 20 m/s, with load magnitudes of 0.7, 1.4, and 2.1 MPa. Figure 12 shows the maximum stress–time curves of elements at the interface under different driving conditions.

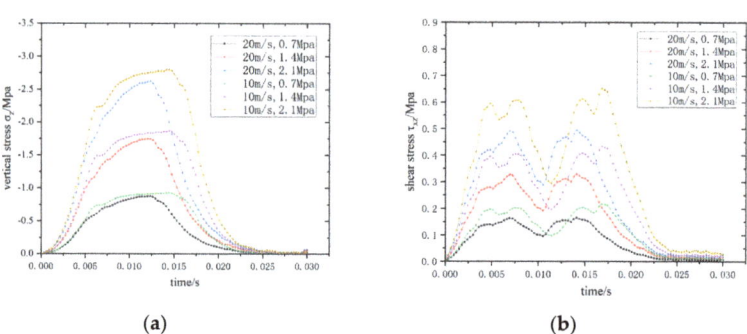

**Figure 12.** Maximum stress–time curve of elements at the interface under different driving conditions. (a) $\sigma_z$-time; (b) $\tau_{xz}$-time.

The maximum vertical stress and maximum shear stress of the elements at the interface increase, both as the amplitude of the moving load increases and velocity of the moving load decreases. The combination of a speed of 10 m/s and moving load amplitude of 2.1 MPa represent the most unfavorable condition. Accordingly, this combination is adopted for the subsequent optimization analysis.

3.3.2. Comparison of Sensor Shapes

Analysis of the impact of sensor packaging shape on mechanical performance is completed by comparing the cuboid sensor packaging with the cylindrical sensor packaging. In both cases, stainless steel is selected as the packaging material. The driving condition is selected as previously described. Figure 13 shows the maximum stress–time curve of elements at the interface under different sensor shapes.

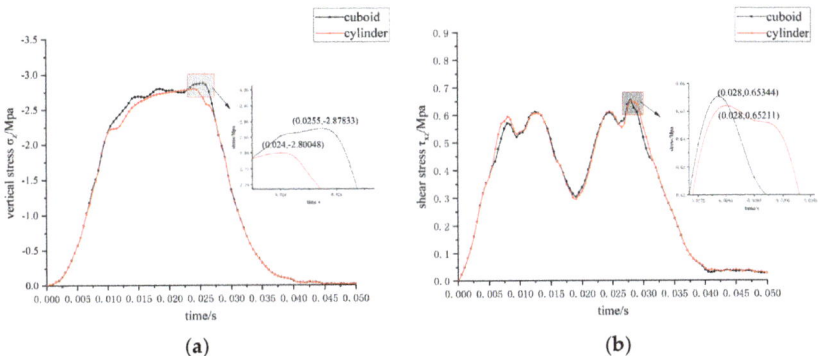

**Figure 13.** Maximum stress–time curve of elements at the interface under different sensor shapes. (a) $\sigma_z$-time; (b) $\tau_{xz}$-time.

Since the sensor is small, relative to the size of the road model, a difference in sensor packaging shape has little effect on the results of the stress analysis. Compared to the cuboid sensor packaging, the cylindrical sensor packaging performs only slightly better at low speed and heavy load. The maximum vertical stress of the cylindrical sensor packaging is reduced by 2.7%, compared to that of the cuboid sensor packaging, and the maximum shear stress is reduced by 0.2%. Therefore, the cylindrical packaging is selected as the optimum sensor packaging shape.

3.3.3. Comparison of Sensor Materials

An analysis of sensor packaging materials was completed. Packaging materials consider the sensor include stainless steel and cast nylon, the material parameters of which are shown in Table 5. The maximum stress–time curve of elements at the interface, under different sensor materials, is shown in Figure 14.

**Table 5.** Material parameters of different sensor packaging materials.

| Material | Elastic Modulus (Mpa) | Poisson's Ratio | Density (kg/m$^3$) |
|---|---|---|---|
| Cast nylon | 2500 | 0.35 | 930 |
| Stainless steel | 200,000 | 0.3 | 8000 |

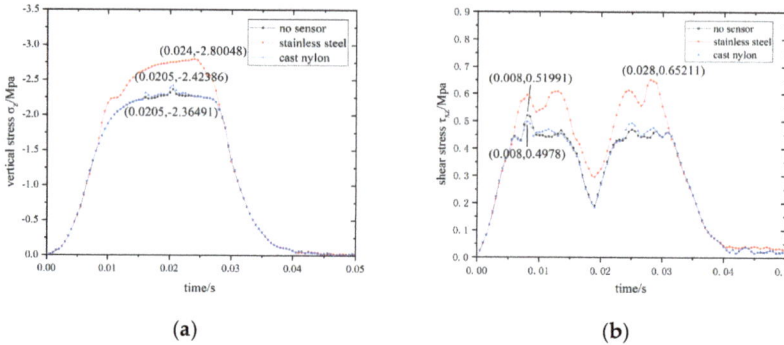

**Figure 14.** Maximum stress–time curve of elements at the interface under different sensor materials. (a) $\sigma_z$-time; (b) $\tau_{xz}$-time.

When cast nylon is used as the sensor packaging material, both vertical and shear stress are closest to the pavement without an embedded sensor. Maximum vertical stress under the cast nylon packaging is reduced by 13.45%, and the maximum shear stress is reduced by 23.66%, when compared to the stainless steel sensor packaging. Therefore, cast nylon is observed to be the more optimum of the two sensor packaging materials.

### 3.3.4. Comparison of Sensor Sizes

An analysis on the effect of sensor packaging size on stress concentration in Rollpave pavement was completed. First, the effects of different sensor thicknesses are compared and analyzed by holding the sensor packaging diameter constant at 40 mm. Second, the effects of varying sensor packaging diameters are compared and analyzed by holding sensor thickness constant at 15 mm. The results of these analyses are shown in Figure 15.

With the sensor diameter held constant, both the maximum vertical stress and shear stress decrease with decreasing sensor thickness. It, thus, follows that the smaller the sensor thickness, the less stress concentration. However, the maximum vertical stress decreases as the sensor diameter increases, and there is no obvious correlation between the maximum shear stress and sensor diameter.

For the above simulation, it should be noted that the connection between the road and sensor elements at the interface is set to "Merge" in the ABAQUS program; thus, the nodal displacements at the interface are always consistent. However, in reality, debonding may occur at the interface between the sensor and asphalt mixture.

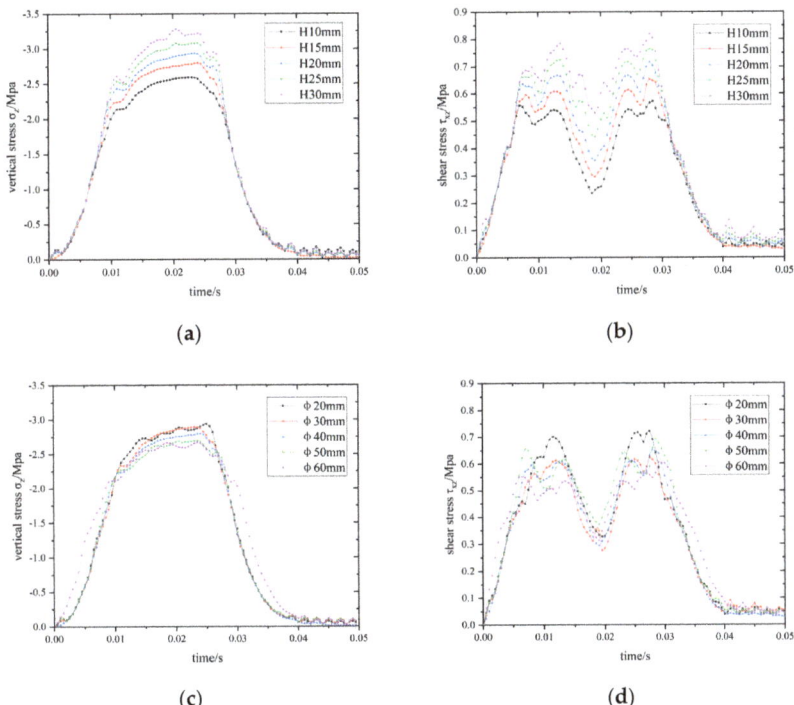

**Figure 15.** Maximum stress–time curve of elements at the interface under different sensor sizes. (**a**) $\sigma_z$-time curve, sensor thicknesses; (**b**) $\tau_{xz}$-time curve, sensor thicknesses; (**c**) $\sigma_z$-time curve, sensor diameters; (**d**) $\tau_{xz}$-time curve, sensor diameters.

## 4. Conclusions

In this paper, the influence of embedded sensors in Rollpave pavement are analyzed via three-point bending test and dynamic response simulation, and the embedded sensor packaging used in Rollpave pavement are optimized. The conclusions are as follows:

(1) Bending strain energy, the amount of energy absorbed by the specimens before destruction, was found to be an effective parameter for the characterization of the bending performance of beam specimens containing an embedded sensor. Under the action of vehicle loading, stress concentrations will appear at the interface between the sensor and asphalt mixture. The maximum normal stress is vertical stress, $\sigma_z$, and the maximum shear stress is in the x-z plane, $\tau_{xz}$; these parameters can be used to evaluate the stress concentration level at the sensor-pavement interface.

(2) The results of the three-point bending test show that the embedment of sensors significantly reduced beam specimens' ability to resist bending deformation. At low temperatures, the failure load of the specimens increases, and the deflection decreases, making the specimens more prone to failure than they would be at normal temperature. Use of epoxy resin as an adhesive does not effectively enhance the bending performance of the specimens. Unexpectedly, it was found that, at high temperatures, the asphalt mixture bonded well to the sensor, even without the addition of an adhesive. Use of the cylindrical sensor packaging shape with cast nylon encapsulation, combined with the reduction of sensor size, was shown to increase the bending performance of the specimens.

(3) The simulation results show that the most unfavorable vehicle loading conditions for roadway structural health are the combination of low speed of travel and a heavy moving load. It was found that the selection of sensor packaging materials (e.g., nylon)

with modulus, similar to the pavement materials, are better choices. Analysis of sensor size revealed that smaller and more circular sensor packaging exhibits better performance. Further, sensor thickness was observed to have a greater effect on bending performance than sensor diameter. For optimum performance, the use of a flat sensor packaging design and minimization of sensor packaging height are both recommended.

(4) In the future, drawing and shear tests should be carried out to evaluate the effect of the sensor surface texture on the bonding performance and better define the failure mechanism of the sensor–asphalt mixture interface. This study optimizes the embedded sensor packaging used in Rollpave pavement. The interfacial synergy between the embedded sensor and asphalt mixture can be improved, which is helpful to prolong the service life of intelligent roadways.

**Author Contributions:** Methodology, Z.Y. and L.W.; investigation, Y.C. and C.L.; formal analysis, Z.Y., Y.C. and C.L.; writing-original draft preparation, Z.Y., C.L. and L.W.; writing—review and editing, D.G.I., Z.Y. and L.W.; resource support, Y.C., K.L. and L.W. All authors have read and agreed to the published version of the manuscript.

**Funding:** This research was supported by science and technology innovation fund project of the Research Institute of Highway Ministry of Transport (2019-C537), Fundamental Research Funds for the Central University (FRF-TP-19-050A1, FRF-BD-19-001A, FRF-MP-19-014) and the Beijing Major Science and Technology Projects (Z191100008019002).

**Institutional Review Board Statement:** Not applicable.

**Informed Consent Statement:** Not applicable.

**Data Availability Statement:** Data is contained within the article.

**Conflicts of Interest:** The authors declare no conflict of interest.

## References

1. Houben, L.J.M.; van der Kooij, J.; Naus, R.W.M.; Bhairo, P.D. APT Testing of modular pavement structure 'Rollpave' and comparison with conventional asphalt motorway structures. In Proceedings of the 2nd International Conference on Accelerated Pavement Testing, Minneapolis, MS, USA, 26–29 September 2004; pp. 1–24.
2. Molenaar, J.M.M.; Montfort, J. *Super Stille Deklaag Rubber Rollpave (in Dutch)*; DRHEI 3A-58; Road and Hydraulic Engineering Institute: Delft, The Netherlands, 2011.
3. Dommelen, A.E.; Kooij, J.V.D.; Houben, L.J.M.; Molenaar, A.A.A. LinTrack APT research supports accelerated implementation of innovative pavement concepts in the Netherlands. In Proceedings of the 2nd International Conference on Accelerated Pavement Testing, Minneapolis, MS, USA, 26–29 September 2004; pp. 1–19.
4. Mao, X.; Zhou, K. Rollpave Pavement Construction Technology and Its Development Trend. *West. China Commun. Sci. Technol.* **2017**, *3*, 10–15. [CrossRef]
5. Ingram, L.S.; Herbold, K.D.; Rasmussen, R.O.; Baker, T.E.; Brumfield, J.W.; Felag, M.E.; Ferragut, T.R.; Grogg, M.G.; Lineman, L.R. *Superior Materials, Advanced Test Methods and Specifications in Europe*; Federal Highway Administration U.S. Department of Transportation: Washington, WA, USA, 2004.
6. Wang, D.; Schacht, A.; Chen, X.; Oeser, M.; Steinauer, B. Feasibility study on the innovative construction method of a 'prefabricated and rollable road'. *BAUTECHNIK* **2013**, *90*, 614–621. [CrossRef]
7. Wang, D.; Schacht, A.; Chen, X.; Liu, P.; Oeser, M.; Steinauer, B. Innovative Treatment to Winter Distresses Using a Prefabricated Rollable Pavement Based on a Textile-Reinforced Concrete. *J. Perform. Constr. Facil.* **2016**, *30*, 4014008. [CrossRef]
8. Yu, H.; Ma, T.; Wang, D.; Wang, Z.; Lv, S. Review on China's pavement engineering research 2020. *China J. Highw. Transp.* **2020**, *33*, 1–66. [CrossRef]
9. Guo, Y. The Asphalt of Carpet of Mix Design Performance and Evaluation. M.D. Thesis, Chongqing Jiaotong University, Chongqing, China, June 2015.
10. Dong, Y.; Liu, Q.; Cao, D.; Zhang, Y. Development and application of modified asphalt dedicated to Rollpave. *J. Highw. Transp. Res. Dev.* **2015**, *32*, 12–17.
11. Feng, Z. Mechanical Behavior of Separable Precast Airfield Pavement Applied to the Desert Highway. M.D. Thesis, South China University of Technology, Guangzhou, China, April 2016.
12. Dai, S. Study on the Performance of Curling Prefabricated Noise Reduction Pavement. M.D. Thesis, Xinjiang University, Urumqi, China, May 2019.
13. Ye, Z.; Xiong, H.; Wang, L. Collecting comprehensive traffic information using pavement vibration monitoring data. *Comput. Civ. Infrastruct. Eng.* **2019**, *35*, 134–149. [CrossRef]

14. Hou, Y.; Li, Q.; Zhang, C.; Lu, G.; Ye, Z.; Chen, Y.; Wang, L.; Cao, D. The State-of-the-Art Review on Applications of Intrusive Sensing, Image Processing Techniques, and Machine Learning Methods in Pavement Monitoring and Analysis. *Engineering* **2021**, *7*, 845–856. [CrossRef]
15. Fedele, R.; Praticò, F.G.; Pellicano, G. The prediction of road cracks through acoustic signature: Extended finite element modeling and experiments. *J. Test. Eval.* **2019**, *49*, 20190209. [CrossRef]
16. Barriera, M.; Pouget, S.; Lebental, B.; Van Rompu, J. In Situ Pavement Monitoring: A Review. *Infrastructures* **2020**, *5*, 18. [CrossRef]
17. Dan, H.C.; Yang, D.; Zhao, L.H.; Wang, S.P.; Zhang, Z. Meso-scale study on compaction characteristics of asphalt mixtures in Superpave gyratory compaction using SmartRock sensors. *Constr. Build. Mater.* **2020**, *262*, 120874. [CrossRef]
18. Dong, Y. Study on Material Properties and Construction Technologies of Rollable Prefabricated Asphalt Pavement. Ph.D. Thesis, Southeast University, Nanjing, China, January 2015.
19. Huang, X.; Zheng, J.; Feng, D. *Road Subgrade and Pavement Engineering*, 6th ed.; China Communications Press: Beijing, China, 2019; pp. 201–205.
20. Liao, G.; Huang, X. *The Application of Abaqus Finite Element Software in Road Engineering*, 2rd ed.; Southeast University Press: Nanjing, China, 2019; pp. 162–196.
21. Liu, P.; Xing, Q.; Wang, D.; Oeser, M. Application of Dynamic Analysis in Semi-Analytical Finite Element Method. *Materials* **2017**, *10*, 1010. [CrossRef] [PubMed]

*Article*

# Influence of Basalt Fibers on the Crack Resistance of Asphalt Mixtures and Mechanism Analysis

**Bangwei Wu [1], Weijie Meng [1], Ji Xia [1] and Peng Xiao [1,2,\*]**

[1] College of Architectural Science and Engineering, Yangzhou University, Yangzhou 225127, China; wubw@yzu.edu.cn (B.W.); MX120190454@yzu.edu.cn (W.M.); MZ120200952@yzu.edu.cn (J.X.)
[2] Research Center for Basalt Fiber Composite Construction Materials, Yangzhou University, Yangzhou 225127, China
\* Correspondence: pengxiao@yzu.edu.cn; Tel.: +86-0514-8797-9418

**Abstract:** The paper aims to investigate the influence of basalt fiber (BF) on the crack resistance of the asphalt mixture and conduct a mechanical analysis. First, two typical asphalt mixtures, namely AC-13 and SMA-13, were designed. The impact of BF on the mixture design results was analyzed. Then, several macroscopic tests, namely the four-point bending test, indirect tensile test, and semicircular bending test (SCB), were conducted to assess the effect of BF on the cracking resistance of asphalt mixtures. Finally, the influence of BF on the cracking resistance of asphalt mixtures was analyzed based on an environmental scanning electron microscope (ESEM) observation. The results show that: (1) BF increases the optimal asphalt content of AC13 and decreases the optimal asphalt content of SMA-13, which is caused by the different asphalt-absorption capacity of BF and lignin fiber (LF). (2) BF enhances both the fatigue crack resistance and temperature crack resistance of asphalt mixtures. The enhancement on the SMA-13 is more significant, indicating that the enhancement of BF on asphalt mixtures is related to the type of aggregate gradation. (3) BFs in the asphalt mixture lap each other to form a spatial network structure. Such structure can effectively improve the crack resistance of the mixture by dispersing the load stress and preventing the flow of asphalt mastic. The study results provide an effective method to design crack-resistant asphalt mixtures.

**Keywords:** basalt fiber; asphalt mixture; crack resistance; environmental scanning electron microscope

**Citation:** Wu, B.; Meng, W.; Xia, J.; Xiao, P. Influence of Basalt Fibers on the Crack Resistance of Asphalt Mixtures and Mechanism Analysis. *Materials* **2022**, *15*, 744. https://doi.org/10.3390/ma15030744

Academic Editor: Simon Hesp

Received: 20 December 2021
Accepted: 17 January 2022
Published: 19 January 2022

**Publisher's Note:** MDPI stays neutral with regard to jurisdictional claims in published maps and institutional affiliations.

**Copyright:** © 2022 by the authors. Licensee MDPI, Basel, Switzerland. This article is an open access article distributed under the terms and conditions of the Creative Commons Attribution (CC BY) license (https://creativecommons.org/licenses/by/4.0/).

## 1. Introduction

Complex climates and increasing traffic loads have a severe negative impact on the cracking resistance of asphalt pavements [1]. Engineers, therefore, have tried to improve the performance of pavements with external admixtures [2,3]. Fibers and polymers are the two main admixtures for asphalt mixtures. Fibers have already been applied for asphalt mixtures for more than 50 years [4]. After the 1990s, with the use of stone mastic asphalt (SMA) in pavements, more fibers were adopted in asphalt mixtures [3], such as synthetic fibers [5], kenaf fiber [6], glass fiber [7], etc. Scholars studied the influence of fibers on the asphalt mixture performance, and concluded that fibers can effectively improve mixture performances [8,9]. In recent years, basalt fiber (BF), a green and high-performance material, has received increasing attention. BF is made of natural basalt stone. It is made by quickly drawing the molten basalt lava at 1400–1500 °C. No other chemical additives are used in the production process, and there is no wastewater, gas, or slag discharge [10,11]. Therefore, BF is an environment-friendly fiber. Many researchers have tried to use basalt fibers in asphalt mixtures. Morova investigated the usability of basalt fibers in hot mix asphalt mixtures. Based on the Marshall stability test results, he found that the best fiber ratio was 0.5% by the mass of asphalt mixtures [12]. Celauro used basalt fibers for urban asphalt pavement. He argued that basalt fibers proved to satisfy specific needs in road technique, and fibers were likely to slightly increase the micro-texture of asphalt mixtures [13]. Zhang used a three-dimensional fiber distribution model to investigate the effect of BF distribution on the flexural–tensile rheological performance of asphalt mortar. He argued that the flexural–tensile rheological value under the horizontal-oriented fiber is minimum [14]. Katharine et al. [15] investigated the microstructure of basalt fibers and the fibers' asphalt mastic at different magnifications with scanning electron microscopy. The finding shows that the fibers are randomly distributed in the asphalt mixtures.

The crack resistance of asphalt mixtures has been considered a significant factor in the design of asphalt pavements. According to the report from the National Cooperative Highway Research Program (NCHRP) 9–57, cracks in asphalt mixtures were divided into four categories: temperature-related cracks, load-related cracks, fatigue cracks, and reflection cracks [16]. Many methods have been developed to determine the cracking resistance of asphalt mixtures. Zhou suggested an IDEAL cracking test for asphalt mixture design. He believed this method was sensitive to RAR content, asphalt content, aggregate gradation, and other variables [17]. Turipan used many indicators based on semi-circular bending (SCB) tests to evaluate factors affecting the cracking resistance of asphalt mixtures. He argued that the Flexibility Index (FI) was more suitable than peak load to judge the cracking property of asphalt mixes [18]. Yan compared SCB-IFIT, un-notched SCB-IFIT, and IDEAL-CT for measuring the cracking resistance of asphalt mixtures. He found that IDEAL-CT has a high correlation with SCB-IFIT but with a lower variation in results [19]. Poulikakos [20] compared and analyzed the four-point bending (FPB) and cantilever beam two-point bending fatigue test methods and found significant differences between the two results. He suggested four-point bending to evaluate the fatigue performance of asphalt mixtures. Hasan [21] compared the difference between four-point bending fatigue and semi-circular bending tests and found that four-point bending fatigue is better for distinguishing coarse and fine mixes. There are many more studies on the cracking performance testing of asphalt mixtures [22–24], and it is difficult to cover them all for the length of this article. In summary, it can be found that it is hard to evaluate different cracking using a single test method, because different cracking mechanisms cause different types of cracks. For example, Islam argued that the current fatigue model in the Mechanistic-Empirical (ME) Design Guide was incomplete due to the fact that the temperature-induced fatigue damage was not considered [25]. In his study of the cracking properties of fiber-reinforced polymer matrix composites, Budiman pointed out that the addition of fibers may alter the way the matrix cracks [26]. Thus, for the fiber-reinforced asphalt mixtures, multiple test methods are needed to fully evaluate their cracking properties.

The incorporation of fibers into asphalt mixtures has gained more and more attention. Many researchers argued that fibers help to improve the cracking resistance of asphalt mixtures. On the other hand, cracking resistance is rather critical when it comes to the design of asphalt mixtures [27]. As a new fiber, BF's effect on the cracking resistance of asphalt mixtures has not been explored enough. The suitability of BF for use with different graded asphalt mixtures has also not yet been studied. Thus, to further evaluate the effect of BF on the cracking resistance of asphalt mixtures, this paper designed two typical asphalt mixtures and used three tests to determine the cracking performance of the asphalt mixtures. An environmental scanning electron microscopy (ESEM) was also used to analyze the enhancement mechanism of BF. The conclusions of this study help to improve the crack resistance of asphalt mixtures and understand the reinforcement mechanism of fiber asphalt mixtures.

## 2. Materials and Asphalt Mixture Design

### 2.1. Materials

#### 2.1.1. Asphalt

SBS-modified asphalt is widely used in China for asphalt pavements. The asphalt used in the research is SBS-modified asphalt, and the main technical properties of SBS asphalt are shown in Table 1. Its technical properties met the JTG F40-2004 Technical Specification for Construction of Highway Asphalt Pavements requirements.

**Table 1.** Results of SBS asphalt properties.

| Index | Specification Requirements | Value | Test Method |
|---|---|---|---|
| Penetration at 25 °C/0.1 mm | 60~80 | 71 | ASTM D 5 |
| Softening point/°C | ≮55 | 64 | ASTM D 2398 |
| Ductility at 5 °C/cm | ≮30 | 48 | ASTM D 113 |
| Viscosity at 135 °C/Pa.s | ≯3 | 1.8 | ASTM D 4402 |
| Elastic recovery at 25 °C/% | ≮65 | 76 | ASTM D 6084 |

2.1.2. Fibers

Basalt fiber (BF) and lignin fiber (LF), two types of fibers, were used in this experiment. The appearance of both fibers is as shown in Figure 1, and their basic properties are shown in Table 2. LF was only used for SMA-13. BF was used for both SMA-13 and AC-13.

(a)  (b)

**Figure 1.** The appearance of the two fibers: (a) BF and (b) LF.

**Table 2.** Properties of modifiers.

| Characteristics | Types BF | LF |
|---|---|---|
| Color | Golden brown | Gray |
| Form | Smooth | Loose flocculent |
| Single fiber diameter/μm | 13~16 | ≈13 |
| Length/mm | 6 | 0.8 |
| Density/(g·cm$^{-3}$) | 2.715 | 1.295 |
| Breaking strength/Mpa | ≥2000 | <300 |
| Melting point/°C | 1600 | 230 |

2.1.3. Mineral Aggregates

Two types of natural stone, limestone and basalt, were used for asphalt mixture design. The aggregates were within four size ranges, namely 10–15, 5–10, 3–5, and 0–3 mm, respectively. The basic properties of the aggregates are shown in Table 3.

**Table 3.** Results of aggregate properties.

| Aggregate Size (mm) | 10–15 mm | | 5–10 mm | | 3–5 mm | | 0–3 mm | |
|---|---|---|---|---|---|---|---|---|
| | Limestone | Basalt | Limestone | Basalt | Limestone | Basalt | Limestone | Basalt |
| Bulk relative density | 2.753 | 2.831 | 2.746 | 2.807 | 2.721 | 2.886 | 2.635 | 2.895 |
| Apparent relative gravity | 2.776 | 2.931 | 2.778 | 2.936 | 2.768 | 2.927 | 2.695 | 2.967 |
| Water absorption (%) | 0.30 | 0.12 | 0.42 | 0.16 | 0.62 | 0.48 | 0.84 | 0.82 |

## 2.2. Asphalt Mixture Composition Design

AC-13 and SMA-13 were designed with the Marshall method for this study. The two mixtures are the most widely used in China. The nominal maximum particle size of aggregates is 13.2 mm. The aggregate gradations of AC-13 and SMA-13 are shown in Figure 2.

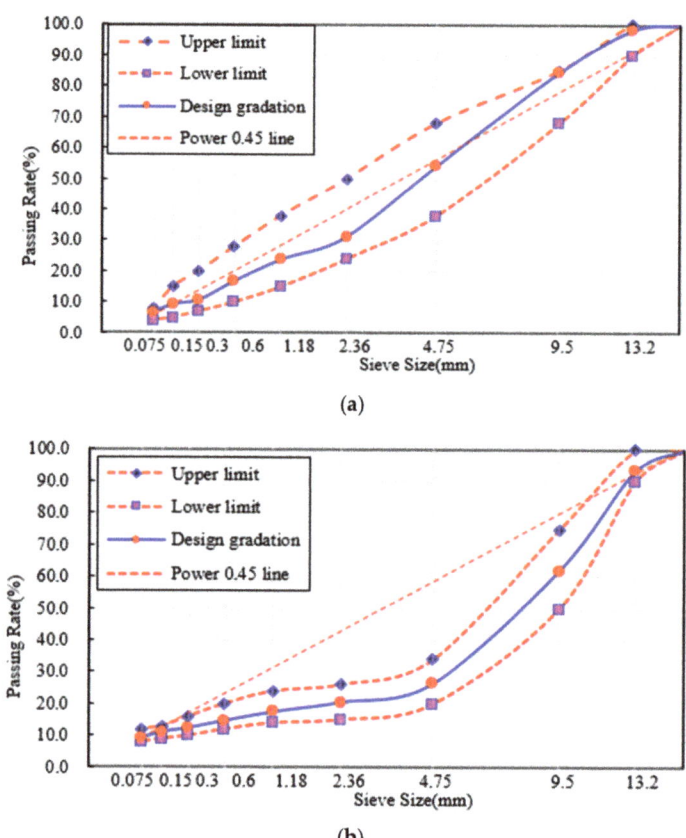

Figure 2. Aggregate gradation: (a) AC-13 and (b) SMA 13.

Four asphalt mixtures, namely SBS AC-13, SBS+BF AC-13, SBS+LF SMA-13, and SBS+LF+BF SMA-13, were prepared. The BF constitutes 0.4% of the total asphalt mixture mass in AC-13 (BF 0.35% of the volume of AC-13). In SBS+LF+BF SMA-13, BF and LF account for 0.3% and 0.1% of the total asphalt mixture mass, respectively (BF 0.29% of the volume of SMA-13, LF 0.2% of the volume of SMA-13). In the control group of SBS+LF SMA-13, LF accounts for 0.3% of the whole asphalt mixture mass (LF 0.61% of the volume of SMA-13). Such fiber contents are based on the Chinese specification T/CHTS 10016–2019 Technical Guideline for Construction of Asphalt Pavement with Basalt Fiber.

For the fabrication of fiber asphalt mixtures, a "dry mix" process was used. The fibers were first blended with the aggregate for 90 s. Then, the asphalt was added to ensure the uniform dispersion of fibers in the asphalt mixture. The volume parameters of the four asphalt mixtures are shown in Table 4.

Table 4. Volume parameters of AC-13 and SMA-13.

| Items | Optimal Asphalt Content (OAC) (%) | | Voids Volume (VV) (%) | | Voids in the Mineral Aggregate (VMA) (%) | | Voids Filled with Asphalt (VFA) (%) | |
|---|---|---|---|---|---|---|---|---|
| | AC-13 | SMA-13 | AC-13 | SMA-13 | AC-13 | SMA-13 | AC-13 | SMA-13 |
| Without BF | >4.7 | >5.8 | >4.1 | >3.8 | >14.2 | >17.2 | >71.1 | >77.9 |
| With BF | 4.9 | 5.5 | 4.1 | 3.9 | 14.3 | 16.7 | 71.3 | 76.6 |
| Specification | - | - | 3~6 | 3–4 | ≮14.0 | ≮16.5 | 65~75 | 75–85 |

As shown in Table 4, BF increases the optimal asphalt content of AC-13 and decreases that of SMA-13. BF has a specific asphalt absorption capacity, so the addition of basalt fibers to AC-13 causes an increase in optimal asphalt content. However, the BF has a lower asphalt absorption ability than LF, and the amount of LF in SBS+LF SMA-13 is more than that in SBS+LF+BF SMA-13, causing the optimal asphalt content of the latter to be lower.

## 3. Test Methods

### 3.1. Four-Point Bending Fatigue Test

The four-point bending fatigue test was carried out in line with the AASHTO T 321 specification [28]. A formed asphalt mixture slab (size 4000 × 3000 × 80 mm) was cut into beams of size 380 × 63.5 × 50 mm. Four beams can be formed per asphalt mixture slab. This specimen was tested in the UTM (universal testing machines) software operating system with the test temperature of 15 ± 0.5 °C and the specimen preloaded at the target stress level for 50 cycles. The strain levels for this study were 450, 650, and 850 µε. The test index represents the number of fatigue actions with accumulated dissipated energy.

### 3.2. Indirect Tensile Test

Indirect tensile tests were carried out according to AASHTO T322 [29]. The specimens were formed with the standard Marshall compaction method. The specimens were tested in a UTM-25 test machine at a temperature of 15 ± 0.5 °C. The test procedure will be terminated when the specimen reaches a vertical deformation value of 10 mm. Tensile strength and toughness index ($T_I$) are used as test indicators. $T_I$ is calculated according to Equation (1), where $A_d$ is the area of the lower curve corresponding to $d/d_p$ after dimensionless processing, $A_p$ is the area of the lower curve corresponding to the peak after dimensionless processing, $d$ is any deformation value greater than $d_p$, and $d_p$ is the deformation value corresponding to the peak load:

$$T_I = \frac{A_d - A_p}{d/d_p - 1} \quad (1)$$

### 3.3. Semi-Circular Bending Test (SCB)

The SCB test was conducted in strict accordance with the AASHTO TP105 [30]. This test was conducted to determine the crack expansion performance of the asphalt mixture at a room temperature of 25 °C with a vertical displacement of 50 mm/min applied. It requires a pre-cut joint of a certain length at the bottom of the specimen in advance. The main evaluation indicators for this test are the fracture energy ($G_f$) and the flexibility index (FI). $G_f$ is the fracture energy, and it is calculated according to Equation (2), where $W_f$ is the integral of the load-displacement curve, and $Area_{lig}$ is the ligament area and the thickness of the specimen ($t$ = 50 mm). The FI was adopted to reflect the crack propagation rate, and it was calculated using Equation (3), where |m| is the absolute value of the slope at the inflection point after the peak of the loading value. The *FI* value is negatively correlated with the crack propagation rate. Four specimens were used to determine the *FI*.

$$G_f = \frac{W_f}{Area_{lig}} \times 10^6 \qquad (2)$$

$$FI = \frac{G_f}{|m|} \times 0.01 \qquad (3)$$

### 3.4. Environmental Scanning Electron Microscope (ESEM)

To study the distribution of basalt fibers in the asphalt mixture and the bond between the fibers and the asphalt mastic, the ESEM was used. The microstructural features of the damaged surface of samples that were taken from indirect tensile fractures were observed with ESEM to analyze the nature of the mechanical contribution of BF to the asphalt mixture.

The main flow chart for this study is shown in Figure 3.

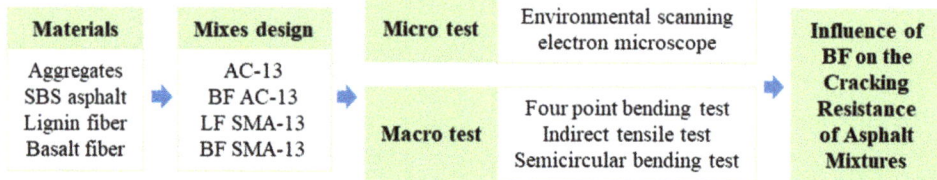

**Figure 3.** Flow chart of this study.

## 4. Results and Discussion

### 4.1. Four-Point Bending Fatigue Test Results

The fatigue life and cumulative dissipated energy results of the four-point bending fatigue test are shown in Figures 4 and 5. The result is the average of five parallel measurements.

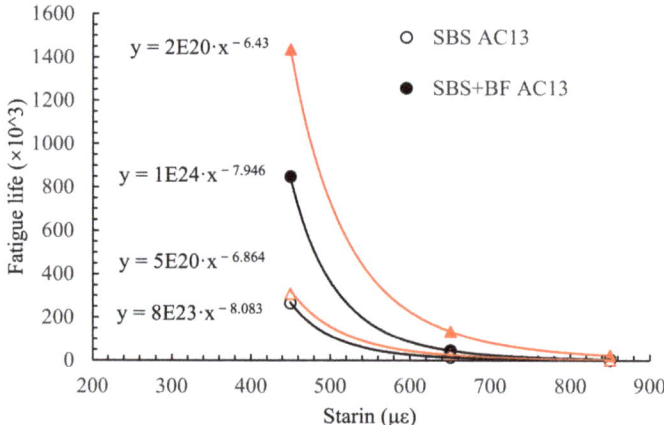

**Figure 4.** Fatigue life of the asphalt mixtures.

From Figure 4, the fatigue life of the four asphalt mixtures declines with the increasing levels of the control strain. A higher strain level causes more distress in asphalt mixtures, resulting in a lower fatigue life. Moreover, the addition of BF significantly increases the fatigue life of mixtures; for example, the fatigue life of AC-13 with and without BF is $264.7 \times 10^3$ and $848.88 \times 10^3$, where the latter is 3.2 times more than the former. Fibers can disperse the stress and reduce the stress concentration phenomenon in the asphalt mixture, resulting in a higher fatigue life. Another point that can be observed from Figure 4 is that the improvement of BF on fatigue life is about three times in AC and four times in SMA, indicating that the enhancement of BF is more significant in SMA mixtures. SMA is a skeleton-dense mixture, and AC is a suspension-dense mixture. Their different aggregate structures cause different synergistic effects with BF, leading to different enhancements in fatigue life.

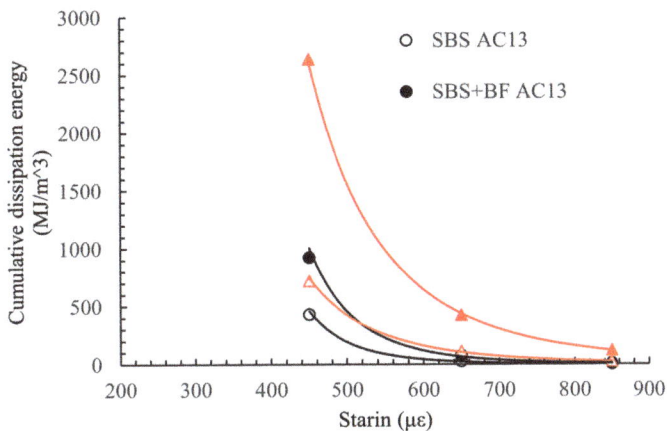

**Figure 5.** Cumulative dissipated energy results for the four-point bending fatigue test.

Moreover, the fatigue life–strain relationship is fitted according to the classical equation, where $N$ is the fatigue life, $\varepsilon$ is the strain, and $A$ and $m$ are fitted parameters related to the material properties. The smaller the $m$, the more sensitive the fatigue life is to the strain level. The fitting results are shown in Figure 4. It can be seen that the fitted parameter $m$ becomes smaller after BF is introduced to the mixtures. The value of $m$ is 8.083 and 6.864 for AC-13 and SMA-13 without BF, and $m$ is 7.946 and 6.430 after BF is introduced. This phenomenon indicates that BF reduces the sensitivity of the asphalt mixtures' fatigue life to strain, which helps to improve the fatigue life of asphalt mixtures at high strain levels, allowing the asphalt mixtures to withstand heavier traffic loads.

In terms of energy dissipation, the overall trend in cumulative dissipation energy for the four asphalt mixtures is consistent. The cumulative dissipation energy decreases with the increasing levels of the control strain. The cumulative dissipation energy of asphalt mixtures also increases significantly with the addition of BF, increasing by 1.8 times in AC-13 and about 3.0 times in SMA-13. However, the trend of variation decreases. That is, the cumulative dissipation energy shows a decreasing trend with increasing strain control levels. However, it is clear that the dissipation energy increases somewhat with the addition of BF, indicating that the dissipation energy is not all caused by damage to the asphalt mixture itself, which indicates that the addition of BF reduces the energy consumed by damage to the mixture and improves the performance of the mixture by forming a three-dimensional mesh structure with uniform dispersion. The comparison of the trends of cumulative dissipation energy before and after the addition of BF shows that the latter is smaller than the former, implying that BF reduces the energy consumed due to damage to the mixture and thus prolongs the fatigue life of the mixture. Moreover, Figure 5 shows that the trend of the decreasing dissipation rate curves of the asphalt mixtures became slower with a longer stable decreasing phase after the addition of BF. The energy consumption per load cycle is higher with BF for the same micro-strain conditions, indicating that the asphalt mixtures with BF are more ductile and relatively challenging to deform.

Taken together, both AC-13 and SMA-13 show a significant improvement in fatigue life and cumulative dissipation energy after the introduction of BF. In particular, BF is more effective within SMA-13. The fatigue life of SMA-13 and AC-13 decreases with the increasing strain control levels, regardless of whether BF is added or not, which is in accordance with some other researchers' findings [31,32].

### 4.2. Indirect Tensile Test Results

The results of the indirect tensile tests are shown in Figures 6 and 7. As can be seen from Figure 6, the addition of BF to AC-13 and SMA-13 increases the maximum load value. The deformation curve of the asphalt mixture after the peak value tends to level off and decreases at a slower rate after the addition of BF, which indicates that the addition of BF improves the deformation resistance of the asphalt mixture. Fibers limit the development of cracks in the asphalt mixture, resulting in the asphalt mixture exhibiting a higher toughness. The reason for this fact is that the BF is probably distributed in the asphalt mixture to form a local spatial network structure, which transmits stress well and dissipates it, and effectively hinders the relative slip between the particles [33].

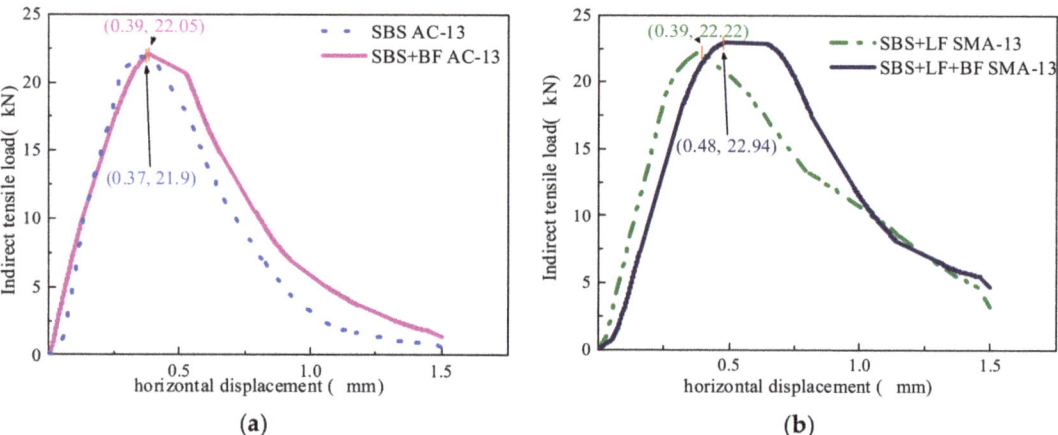

**Figure 6.** Horizontal displacement results of the indirect tensile test: (**a**) AC-13 and (**b**) SMA 13.

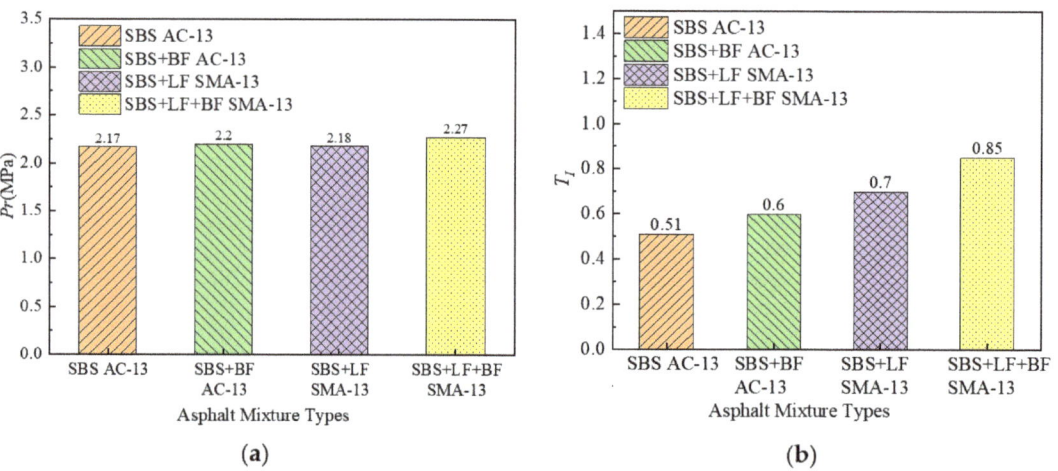

**Figure 7.** Indirect tensile test results: (**a**) tensile strength and (**b**) Resilience Index.

As can be seen from Figure 7, the increase in tensile strength of the asphalt mixture after the addition of fibers is not significant, and it does not appear to be ideal to judge the improvement of the BF on the cracking resistance of the asphalt mixture from the perspective of tensile strength. The tensile strength of mixtures with BF is only 2–4% higher than that without BF. However, the TI of the asphalt mixture increased substantially after the addition of BF. TI reflects the toughness of asphalt mixtures. The higher the TI, the better the toughness of the asphalt mixtures. A higher TI after the addition of BF indicates that BF can effectively improve the cracking resistance of the mixture. After the crack initiation, the BF in the asphalt mixture bears the more external load and reduces the load borne by the asphalt mixture matrix, retarding the development of the cracks. With the addition of BF, the TI of AC-13 and SMA-13 improved by 17.65% and 21.43%, respectively, indicating that the improvement of BF on asphalt mixture toughness is related to the mixture type, and the improvement of BF on SMA toughness is more remarkable than that on AC. This phenomenon is consistent with the fatigue test results. As previously discussed, basalt fibers can be stressed together with the asphalt mixture, which on the one hand allows the asphalt mixture to withstand greater external loads. On the other hand, the fibers retard the development of cracks in the asphalt mixture, allowing the asphalt mixture to exhibit higher cracking resistance.

### 4.3. Semicircular Bending Test Results

The results of the SCB test are shown in Figure 8. The fracture energy, Gf, is significantly higher for the BF mixtures than the asphalt mixtures without BF. AC-13 with BF improved by approximately 36% compared to AC-13 without BF, and SMA-13 with BF increased by 38% compared to SMA-13 without BF. Gf is the fracture energy during the asphalt mixture cracking process. According to the fracture mechanics theory, the cracking of a material is a process of continuous energy consumption. The higher the fracture energy, the less likely the material is to crack. From the test results in Figure 8, it can be seen that basalt fibers significantly increase the Gf of the asphalt mixture, making the asphalt mixture less likely to crack. It is because basalt fibers form an inter-lap spatial network in the asphalt mixture, allowing the basalt fibers to share the external load with the asphalt mixture.

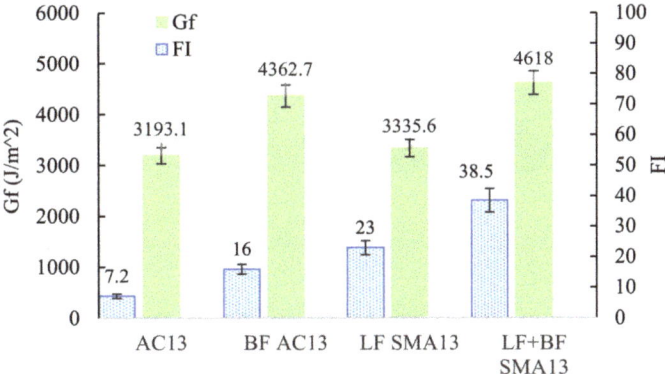

**Figure 8.** SCB test results.

The variation pattern of FI is consistent with that of Gf. The FI with BF is significantly better than the FI without BF. The FI values of the AC-13 and SMA-13 increase by as much as two times and 67% with the introduction of BF. The cracking of the asphalt mixture is divided into two stages: the first stage is the sprouting of cracks, and the second stage is the expansion of cracks [34]. The FI characterizes the expansion ability of cracks. The experimental results showed that basalt fibers retarded the development of cracks, making microscopic cracks within the asphalt mixture less likely to develop into macroscopic cracks.

As mentioned before, the BF acted as reinforcement during the cracking of asphalt mixtures. BF can form a network of inter-lap spaces to bear the load, which improves the stress concentration phenomenon inside the asphalt mixture and retards the crack development, resulting in improved cracking performance of asphalt mixtures.

### 4.4. Microscopic Morphology of Fiber Asphalt Mixture Sliced

The damaged surface microstructure of the basalt fiber asphalt mixtures was observed with ESEM, and the results are shown in Figure 9. Figure 9a,b are the microscopic images of AC-13 and SMA-13 without BF at a magnification of 50×. Figure 9c,d are at the microscopic images of BF AC-13 at a magnification of 50× and 200×, respectively.

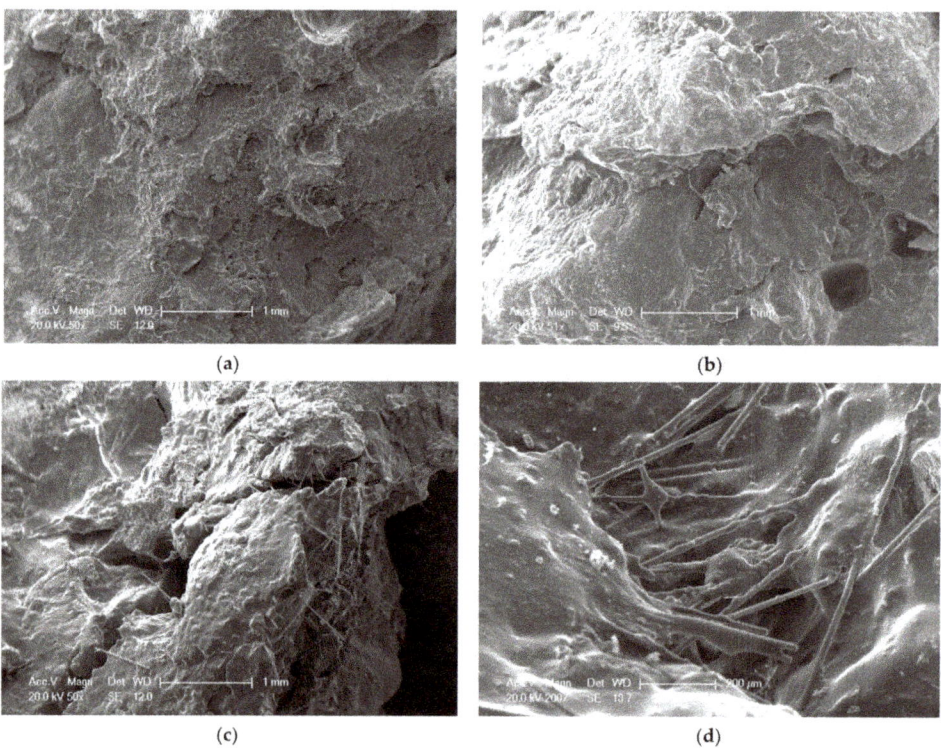

**Figure 9.** Microscopic images of asphalt mixtures: (**a**) SBS AC-13 (50×), (**b**) SBS+LF SMA-13 (50×), (**c**) SBS+BF AC-13 (50×), and (**d**) SBS+BF AC-13 (200×).

As shown in Figure 9a, the asphalt mixture without BF has a relatively flat section with a certain degree of peeling. Moreover, by comparing Figure 9a,b, it can be seen that BF has no significant effect on the microscopic image of the asphalt mixture. In Figure 9b, the microscopic images of SMA-13 with LF, LF is barely visible in the mixture, indicating that LF only plays a role in absorbing the extra free asphalt in the asphalt mixture without significantly improving the mechanical properties of the asphalt mixture [35]. In contrast, BF can be easily observed in Figure 9c, and a clearer supporting microscopic image is shown in Figure 9d. According to the composite theory, adding fiber is an effective method to improve the strength and toughness of materials [36]. The microscopic pictures of BF asphalt mixtures show that BFs are coated with asphalt, indicating that the adhesion between BFs and asphalt is good, which is conducive to the cooperative work of the fiber and asphalt mixture. Moreover, the section of the asphalt mixture with BF shows that numerous BFs are randomly dispersed in the asphalt mixture in three dimensions and lapped together to form a spatial network. BF's strength and modulus are relatively higher than asphalt mixtures, so BFs bear more loads than asphalt mixtures. This fiber network limits the movement of asphalt and aggregate and increases the asphalt mixture strength, making the asphalt mixtures less prone to cracking. At the same time, once the asphalt mixture sprouts cracks, this fiber network can improve the toughness of asphalt mixtures by delaying the development of cracks. Thus, BFs effectively improve the crack resistance of the asphalt mixture. Furthermore, the fatigue test results imply that the effect of the BF network within asphalt mixtures is strain dependent, and BF improves the fatigue life more at a higher strain level. As discussed previously, the modulus of BF is relatively higher than that of asphalt mixtures; thus, when a fibrous asphalt mix is stressed, the asphalt mix matrix deforms first and transfers the force to the fiber network. The higher the strain on the asphalt matrix, the greater the force shared by the fiber network and the more pronounced the strengthening effect of the fiber network.

Figure 10 provides a comprehensive demonstration of the cracking resistance of BF in different asphalt mixtures. As shown in Figure 10, there is a clear difference in the BF improvement on the crack resistance of different types of asphalt mixtures. From all aspects of the analysis, SMA-13 has the best crack resistance, partly due to the reinforcing effect of basalt fibers and partly because the higher amount of asphalt in SMA facilitates the crack resistance of the asphalt mixture. From Figure 10, it can be observed that BFs provide a greater enhancement to the performance of SMA-13. Therefore, besides well-known factors such as BF content, the type of asphalt mixture and the fit of the BF also exert a vital influence, which should be paid attention to by engineers when selecting the type of asphalt mixture.

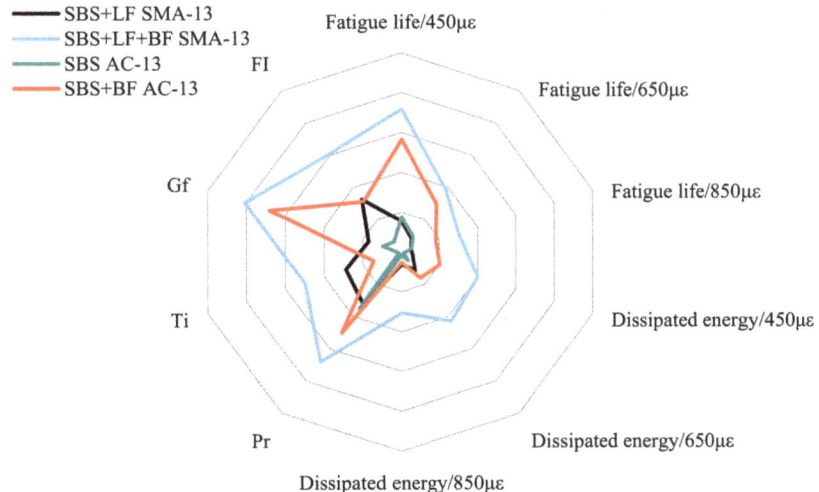

**Figure 10.** Crack resistance of different asphalt mixtures.

## 5. Conclusions

In this study, the influence of BF on the crack resistance of asphalt mixtures was evaluated. Based on the previous analysis and discussion, the following conclusions can be drawn:

(1) The addition of BF significantly increased the fatigue life and the accumulated dissipation energy of the mixture, by up to 3~4 times in AC-13 and by about 4~5 times in SMA-13, resulting in asphalt mixtures being less prone to fatigue cracking.

(2) After adding BF, the indirect tensile strength of the asphalt mixture slightly increased. The TI substantially increased. BF significantly increased the toughness and improved the anti-cracking properties of the mixture.
(3) BF improved both the FI and Gf in the SCB test, indicating that BF can delay the development of cracks.
(4) BFs were randomly dispersed in the asphalt mixture and lapped together to form a spatial network. This fiber network made asphalt mixtures less prone to cracking and delayed the development of cracks.
(5) The reinforcing effect of BF is related to aggregate gradation and the strain level. The reinforcing effect of BF was more pronounced for SMA13. Moreover, the higher the strain in the asphalt mix matrix, the greater the reinforcing effect of BF.

In the future work, more evaluation of the interaction between BF and the asphalt mixture is needed to further clarify the reinforcement mechanism of BF.

**Author Contributions:** Conceptualization, B.W. and P.X.; methodology, B.W.; software, J.X.; writing—original draft preparation, W.M.; writing—review and editing, B.W.; visualization, J.X.; supervision, P.X.; project administration, P.X.; funding acquisition, B.W. and W.M. All authors have read and agreed to the published version of the manuscript.

**Funding:** This research was funded by the National Natural Science Foundation of China, grant number 52008365, and the Postgraduate Research & Practice Innovation Program of Jiangsu Province, Grant Number YKYCX20_010.

**Institutional Review Board Statement:** Not applicable.

**Informed Consent Statement:** Not applicable.

**Data Availability Statement:** The data presented in this study are available from the corresponding author upon request.

**Conflicts of Interest:** The authors declare no conflict of interest.

## References

1. Elkashef, M.; Williams, R.C.; Cochran, E. Investigation of fatigue and thermal cracking behavior of rejuvenated reclaimed asphalt pavement binders and mixtures. *Int. J. Fatigue* **2018**, *108*, 90–95. [CrossRef]
2. Shanbara, H.K.; Ruddock, F.; Atherton, W. A laboratory study of high-performance cold mix asphalt mixtures reinforced with natural and synthetic fibers. *Constr. Build. Mater.* **2018**, *172*, 166–175. [CrossRef]
3. Abtabi, S.M.; Sheikhzadeh, M. Fiber reinforced asphalt-concrete—A review. *Constr. Build. Mater.* **2010**, *24*, 871–877. [CrossRef]
4. Wu, B.; Wu, X.; Xiao, P.; Chen, C.; Xia, J.; Lou, K. Evaluation of the long-term performances of SMA-13 containing different fibers. *Appl. Sci.* **2021**, *11*, 5145. [CrossRef]
5. Kim, M.-J.; Kim, S.; Yoo, D.-Y.; Shin, H.-O. Enhancing mechanical properties of asphalt concrete using synthetic fibers. *Constr. Build. Mater.* **2018**, *178*, 233–243. [CrossRef]
6. Pirmohammad, S.; Shokorlou, Y.M.; Amani, B. Laboratory investigations on fracture toughness of asphalt concretes reinforced with carbon and kenaf fibers. *Eng. Fract. Mech.* **2020**, *226*, 106875. [CrossRef]
7. Fakhri, M.; Hosseini, S.A. Laboratory evaluation of rutting and moisture damage resistance of glass fiber modified warm mix asphalt incorporating high RAP proportion. *Constr. Build. Mater.* **2017**, *134*, 626–640. [CrossRef]
8. Bonica, C.; Toraldo, E.; Andena, L.; Marano, C.; Mariani, E. The effects of fibers on the performance of bituminous mastics for road pavements. *Compos. Part B Eng.* **2016**, *95*, 76–81. [CrossRef]
9. Pirmohammad, S.; Shokorlou, Y.M.; Amani, B. Influence of natural fibers (kenaf and goat wool) on mixed mode I/II fracture strength of asphalt mixtures. *Constr. Build. Mater.* **2020**, *239*, 117850. [CrossRef]
10. Fiore, V.; Scalici, T.; Di Bella, G.; Valenza, A. A review on basalt fiber and its composites. *Compos. Part B Eng.* **2015**, *74*, 74–94. [CrossRef]
11. Slebi-Acevedo, C.J.; Lastra-González, P.; Pascual-Muñoz, P.; Castro-Fresno, D. Mechanical performance of fibers in hot mix asphalt: A review. *Constr. Build. Mater.* **2019**, *200*, 756–769. [CrossRef]
12. Morova, N. Investigation of usability of basalt fibers in hot mix asphalt concrete. *Constr. Build. Mater.* **2013**, *47*, 175–180. [CrossRef]
13. Celauro, C.; Praticò, F. Asphalt mixtures modified with basalt fibres for surface courses. *Constr. Build. Mater.* **2018**, *170*, 245–253. [CrossRef]
14. Zhang, X.; Gu, X.; Lv, J. Effect of basalt fiber distribution on the flexural–tensile rheological performance of asphalt mortar. *Constr. Build. Mater.* **2018**, *179*, 307–314. [CrossRef]

15. Kathari, P.M.; Sandra, A.K.; Sravana, P. Experimental investigation on the performance of asphalt binders reinforced with basalt fibers. *Innov. Infrastruct. Solut.* **2018**, *3*, 76. [CrossRef]
16. Zhou, F.; Newcomb, D. *Experimental Design for Field Validation of Laboratory Tests to Assess Cracking Resistance of Asphalt Mixtures*; Draft final report of NCHRP 9-57; Texas A&M Transportation Institute: College Station, TX, USA, 2016.
17. Zhou, F.; Im, S.; Sun, L.; Scullion, T. Development of an IDEAL cracking test for asphalt mix design and QC/QA. *Road Mater. Pavement Des.* **2017**, *18* (Suppl. S4), 405–427. [CrossRef]
18. Mandal, T.; Ling, C.; Chaturabong, P.; Bahia, H.U. Evaluation of analysis methods of the semi-circular bend (SCB) test results for measuring cracking resistance of asphalt mixtures. *Int. J. Pavement Res. Technol.* **2019**, *12*, 456–463. [CrossRef]
19. Yan, C.; Zhang, Y.; Bahia, H.U. Comparison between SCB-IFIT, un-notched SCB-IFIT and IDEAL-CT for measuring cracking resistance of asphalt mixtures. *Constr. Build. Mater.* **2020**, *252*, 119060. [CrossRef]
20. Poulikakos, L.D.; Pittet, M.; Dumont, A.-G.; Partl, M.N. Comparison of the two point bending and four point bending test methods for aged asphalt concrete field samples. *Mater. Struct.* **2015**, *48*, 2901–2913. [CrossRef]
21. Hasan, M.; Ahmad, M.; Hasan, A.; Faisal, H.M.; Tarefder, R.A. Laboratory performance evaluation of fine and coarse-graded asphalt concrete mix. *J. Mater. Civ. Eng.* **2019**, *31*, 04019259. [CrossRef]
22. Daniel, J.S.; Corrigan, M.; Jacques, C.; Nemati, R.; Dave, E.V.; Congalton, A. Comparison of asphalt mixture specimen fabrication methods and binder tests for cracking evaluation of field mixtures. *Road Mater. Pavement Des.* **2019**, *20*, 1059–1075. [CrossRef]
23. Kaseer, F.; Yin, F.; Arámbula-Mercado, E.; Martin, A.E.; Daniel, J.S.; Salari, S. Development of an index to evaluate the cracking potential of asphalt mixtures using the semi-circular bending test. *Constr. Build. Mater.* **2018**, *167*, 286–298. [CrossRef]
24. Ozer, H.; Al-Qadi, I.L.; Singhvi, P.; Bausano, J.; Carvalho, R.; Li, X.; Gibson, N. Prediction of pavement fatigue cracking at an accelerated testing section using asphalt mixture performance tests. *Int. J. Pavement Eng.* **2018**, *19*, 264–278. [CrossRef]
25. Islam, M. Thermal Fatigue Damage of Asphalt Pavement. Ph.D. Thesis, University of New Mexico, Albuquerque, NM, USA, 2015.
26. Budiman, B.A.; Adziman, F.; Sambegoro, P.L.; Nurprasetio, I.P.; Ilhamsyah, R.; Aziz, M. The role of interfacial rigidity to crack propagation path in fiber reinforced polymer composite. *Fibers Polym.* **2018**, *19*, 1980–1988. [CrossRef]
27. Im, S.; Karki, P.; Zhou, F. Development of new mix design method for asphalt mixtures containing RAP and rejuvenators. *Constr. Build. Mater.* **2016**, *115*, 727–734. [CrossRef]
28. AASHTO T321. *Standard Method of test for Determining the Fatigue Life of Compacted Asphalt Mixtures Subjected to Repeated Flexural Bending*; AASHTO: Washington, DC, USA, 2017.
29. AASHTO T322. *Standard Method of Test for Determining the Creep Compliance and Strength of Hot Mix Asphalt (HMA) Using the Indirect Tensile Test Device*; AASHTO: Washington, DC, USA, 2011.
30. AASHTO TP105. *Standard Method of Test for Determining the Fracture Energy of Asphalt Mixtures Using the Semicircular Bend Geometry (SCB)*; AASHTO: Washington, DC, USA, 2015.
31. Liu, Q.; Schlangen, E.; Van de Ven, M.; Van Bochove, G.; Van Montfort, J. Evaluation of the induction healing effect of porous asphalt concrete through four point bending fatigue test. *Constr. Build. Mater.* **2012**, *29*, 403–409. [CrossRef]
32. Lopez-Montero, T.; Miro, R.; Botella, R.; Pérez-Jiménez, F.E. Obtaining the fatigue laws of bituminous mixtures from a strain sweep test: Effect of temperature and aging. *Int. J. Fatigue* **2017**, *100*, 195–205. [CrossRef]
33. Ziari, H.; Aliha, M.R.M.; Moniri, A.; Saghafi, Y. Crack resistance of hot mix asphalt containing different percentages of reclaimed asphalt pavement and glass fiber. *Constr. Build. Mater.* **2020**, *230*, 117015. [CrossRef]
34. Birgisson, B.; Montepara, A.; Romeo, E.; Roncella, R.; Napier, J.; Tebaldi, G. Determination and prediction of crack patterns in hot mix asphalt (HMA) mixtures. *Eng. Fract. Mech.* **2008**, *75*, 664–673. [CrossRef]
35. Sadeghian, M.; Namin, M.L.; Goli, H. Evaluation of the fatigue failure and recovery of SMA mixtures with cellulose fiber and with SBS modifier. *Constr. Build. Mater.* **2019**, *226*, 818–826. [CrossRef]
36. Christensen, R.M.; Zywicz, E. A three-dimensional constitutive theory for fiber composite laminated media. *J. Appl. Mech.* **1990**, *57*, 948–955. [CrossRef]

*Article*

# Laboratory Study on the Stability of Large-Size Graded Crushed Stone under Cyclic Rotating Axial Compression

Bo Tan [1,2], Tao Yang [1], Heying Qin [1,2,*] and Qi Liu [1]

[1] College of Civil and Architecture Engineering, Guilin University of Technology, Guilin 541004, China; 2000015@glut.edu.cn (B.T.); 1020180360@glut.edu.cn (T.Y.); 15264714363@163.com (Q.L.)
[2] Guangxi Key Laboratory of New Energy and Building Energy Saving, Guilin 541004, China
* Correspondence: qinheyinglcx@163.com

**Citation:** Tan, B.; Yang, T.; Qin, H.; Liu, Q. Laboratory Study on the Stability of Large-Size Graded Crushed Stone under Cyclic Rotating Axial Compression. *Materials* **2021**, *14*, 1584. https://doi.org/10.3390/ma14071584

Academic Editor: Christophe Petit

Received: 9 February 2021
Accepted: 20 March 2021
Published: 24 March 2021

**Publisher's Note:** MDPI stays neutral with regard to jurisdictional claims in published maps and institutional affiliations.

**Copyright:** © 2021 by the authors. Licensee MDPI, Basel, Switzerland. This article is an open access article distributed under the terms and conditions of the Creative Commons Attribution (CC BY) license (https://creativecommons.org/licenses/by/4.0/).

**Abstract:** In this paper, the stability of large-size graded crushed stone used for road base or cushioning under repeated load is investigated. Using an in-house developed device, large-size crushed stone mix was compacted and molded by the vibration and rotary compaction method. Cyclic rotating axial compression was applied, and the shakedown theory was used to study the cumulative deformation of the large-size crushed stone specimens. The effects of gradation parameters on the cumulative strain and stability behavior were analyzed, and the critical stability and failure loads were determined according to the shakedown theory. The test results indicate that there are three obvious instability behavior stages of large-size graded crushed stone under cyclic rotating axial compression: elastic stability, plastic creep, and incremental plastic failure. Large-size graded crushed stone has a higher critical stability load stiffness than conventional-size graded crushed stone. The critical shakedown load of the specimen is mainly affected by the skeleton structure performance, and the critical failure load by the properties of the crushed stone material. Increasing the content and compactness of large-size crushed stone in the specimen can improve the stiffness and stability performance, and to achieve improvements, the content of large-size crushed stone should be controlled between 22% and 26%. The critical shakedown load increases with the increase in the California bearing ratio (CBR) value, while, on the other hand, the CBR value has little relationship with the critical failure load.

**Keywords:** large-size graded crushed stone; cyclic rotating axial compression; shakedown theory; cumulative axial strain; long-term stability; critical load

## 1. Introduction

Large-size graded crushed stone, which is a typical elastic–plastic granular material [1–6], finds its applications as road base, highway cushion, and as airstrips. Its nominal maximum particle size is generally between 25 and 63 mm. The large-size graded crushed stone used in engineering applications has better stiffness, load bearing capacity, and pressure stability than the conventional-size graded crushed stone [7–10]. The deformation of road base caused by long-term vehicular cyclic loads is generally divided into elastic and plastic deformation. Elastic deformation has little fluctuation and can be recovered, while plastic deformation will gradually accumulate and can seriously affect the long-term stability of the road structure. However, there are only limited applications of large-size graded crushed stone in inland and foreign projects. The reason is that the performance and stability of large-size graded crushed stone are not fully understood.

Existing experimental studies on the stability of road base were mainly focused on conventional-size graded crushed stone or other granular materials. Cyclic loading was applied to the specimens by a repeated load triaxial (RLT) or material testing system (MTS) tester manufactured by the American MTS company to observe their deformations, and the stability was examined [11,12]. In recent years, numerous researchers have advanced understanding by adopting the shakedown theory to analyze the deformation behavior of granular materials under cyclic loading. Werkmeister [13,14] analyzed the stability of granular materials under cyclic loading using RLT tests and evaluated the deformation behavior of granular materials by shakedown theory. It was noted that the deformations of granular materials under cyclic loading can be divided into three stable behaviors: stable plastic behavior, plastic creep behavior, and incremental failure behavior, and the corresponding classification criteria and analysis methods were proposed. Xiao [15] performed both constant and variable confining pressure tests on the granular materials by using an advanced cyclic loading RLT device. It was found that the dynamic stress states can model the moving wheel loads effectively and shakedown theory can accurately describe the deformation behavior of granular materials. Several studies [16–20] also indicated that the critical shakedown load of granular materials under cyclic loading increased linearly with the increase in material yield stress. Moreover, the stability behavior of granular materials under cyclic loading is closely related to the confining pressure, dynamic stress amplitude, and fine aggregate content.

The cyclic loading was exerted by a traditional RLT tester designed for testing granular materials under axial and confining pressures and an MTS tester used in previous studies to test the deformation behavior of granular materials under cyclic loading. The actual compaction at a road construction site is usually due to the simultaneous action of axial pressure and shear stress [12,15]. Therefore, the combined effect of axial pressure and shear stress should be considered when studying the deformation behavior of granular materials under cyclic loading.

In this paper, adopting the rotation compression used in the design method of asphalt mixtures such as Superpave, cyclic rotating axial compression tests were conducted to study the deformations of the large-sized graded crushed stone under cyclic loading using a novel testing device developed in-house to achieve the simultaneous action of axial pressure and shear stress. The stability behavior was then evaluated by the shakedown theory. The step-by-step filling method [21] and the I method [22] were applied to design the gradation of large-size crushed stone mix, and the influences of gradation parameters on the mechanical properties and stability of the mix were analyzed. A CBR (California bearing ratio) test was carried out to investigate the relationship between the CBR values and the stability and critical shakedown load. This research provides a theoretical basis for the optimal design of large-sized graded crushed stone for engineering applications.

## 2. Shakedown Theory
### 2.1. Principles of Shakedown Theory

The shakedown theory, also known as the structural stability theory, was initially developed to study the deformation characteristics of metallic structures with clear elastic–plastic behavior and subject to the joint action of temperature and loads [12]. Subsequently, the theory was adopted by Sharp et al. for studying pavement material structure [23,24]. Based on the numerous previous investigations and reviews, it is considered that materials or structures generally exhibit three types of unstable behaviors, i.e., elastic stability, plastic creep, and incremental plastic failure, and two critical loads, i.e., critical shakedown load and critical failure load [25,26].

It is generally believed that when the stresses in the road material or structure are lower than a certain value, the strain increase rate becomes smaller with the increase in the number of load cycles, and the strain tends to stabilize. The corresponding stress is generally defined as the critical shakedown load, which refers to a threshold value of the loading stress when the material reaches the elastic shakedown state. If the cyclic loading is smaller than the critical shakedown load, the response of the structure is elastic. The final plastic deformations tend to be a stable and structural failure due to excessive accumulation of plasticity will not occur. The instability of the specimen under this stress level is called the shakedown state [14]. When the stresses in the road material are greater than a certain value, the stable state is destroyed, the strain rate shows no sign of decreasing, may even increase, and accumulates rapidly. The stress when this happens is defined as the critical failure load. The pavement material or structure will be destroyed when the load is greater than the critical failure load, and the specimen will find itself in the state of incremental plastic failure. When the load is greater than the critical shakedown load but less than the critical failure load, the specimen is in the plastic creep state [12], the strain of the specimen in this state increases slowly with the increase in the number of load cycles, and the strain rate stays small.

The graded crushed stone mixture has a clear skeleton structure and obvious cumulative deformation behavior when subjected to cyclic loading. In practical engineering applications, cumulative deformation will occur when the road is subjected to excessive cyclic loading, and the internal structure will be destroyed when the accumulated deformation becomes excessive. This will cause cracking between the road base and surface layer, which will affect the overall stability of the road structure [27]. Therefore, to study the stability of the large-sized graded crushed stone mixture, it is necessary to analyze its critical shakedown load, critical failure load, and deformation relationships. The optimal gradation design method suitable for the large-sized crushed stone can be developed by analyzing the relationship between the critical loads, deformations, and gradation type of crushed stone.

*2.2. Stability Test and Behavior Evaluation*

There are two main testing methods for studying the structural stability of granular materials: RLT test and MTS test. Both methods can perform triaxial cyclic loading tests, while the MTS testing machine can also perform axial cyclic loading tests. Using the two experimental methods, researchers have achieved significant scientific progress. In reference [28], the state of a granular material subjected to cyclic loads was divided into three ranges: range A (plastic shakedown), range B (plastic creep), and range C (incremental collapse). The critical shakedown load is at the transition point between ranges A and B, while the critical failure load is at the transition point between ranges B and C. The loading process was divided into two stages: post-compression and secondary cyclic compression. The stability behavior curve is shown in Figure 1.

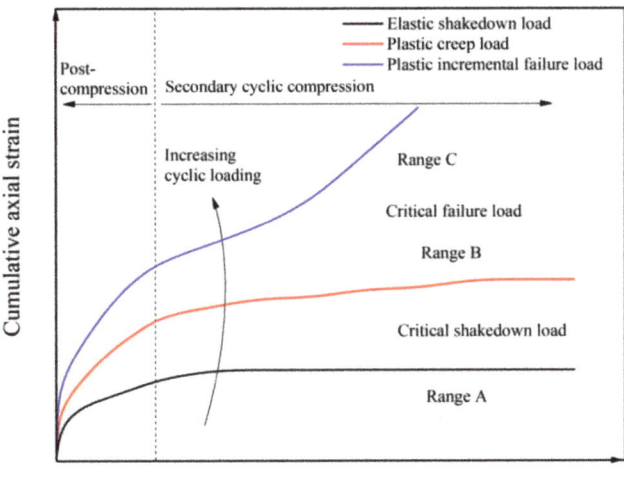

**Figure 1.** Deformation behavior of granular material under cyclic loading.

Using RLT tests and analyzing a large number of studies, Werkmeister [13,14] formulated a standard method for evaluating the stability behavior of granular materials under cyclic loading. The standard method is based on the cumulative axial strain generated by 3000 to 5000 load cycles. The change rate of cumulative axial strain between two loading times can be calculated, and then the specimen state can be evaluated according to that rate. The specific evaluation criteria are as follows:

Range A: $\Delta\varepsilon_{5000} - \Delta\varepsilon_{3000} < 4.5 \times 10^{-5}$, the material is in the elastic stability state.
Range B: $4.5 \times 10^5 < \Delta\varepsilon_{5000} - \Delta\varepsilon_{3000} < 4.5 \times 10^4$, the material is in the plastic creep state.
Range C: $\Delta\varepsilon_{5000} - \Delta\varepsilon_{3000} > 4.5 \times 10^4$, the material is in the incremental plastic failure state.

In recent years, Chen [28] improved the standard method and verified it using the research data of Tao [29], Werkmeister [14], Perez [30], and Gu [12]. The rationality of the method was also proved. The standard method is based on the creep formula proposed by Yin [31] to describe granular materials:

$$\Delta\varepsilon = \frac{\psi'_0 \log[(t+t_0)/t_0]}{1 + (\psi'_0/\Delta\varepsilon_1) \log(t+t_0)}, \qquad (1)$$

where $\Delta\varepsilon$ represents the cumulative creep strain, $t$ refers to the number of load cycles during creep, and $\Psi'_0$, $t_0$, and $\Delta\varepsilon_1$ are constants.

In this study, $t$ and $t_0$ in Equation (1) are replaced by the number of cycles $N$ and $N_0$ (Figure 2), and the constants are replaced by $n_s$ and $m_s$, The following formula results:

$$\Delta\varepsilon = \frac{\log[(N_S + N_0)]}{n_s + m_s \times \log[(N_S + N_0)/N_0]}. \qquad (2)$$

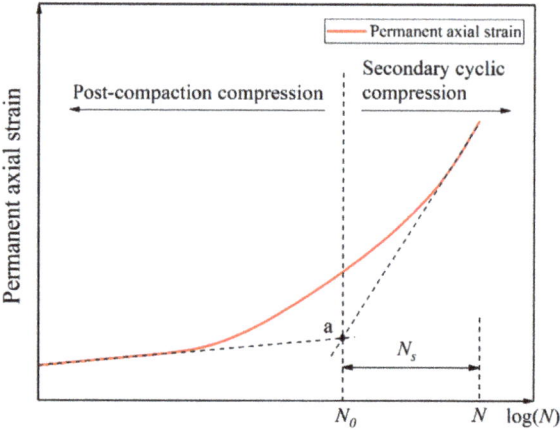

**Figure 2.** Selection of numbers of rotating axial compression cycles, reprinted with permission from [28]. Copyright 2019 Elsevier.

In Equation (2), $1/m_s$ represents the cumulative axial strain of the specimen when $N$ tends to infinity, and $1/n_s$ represents the slope between $\Delta\varepsilon$ and the strain curve (Figure 3).

In Figure 2, a represents the turning point on the permanent axial strain curve, $N_0$ represents the cycle number of the turning point on the permanent axial strain curve, $N$ represents the total cycle number in a test, $N_S$ is equal to $N$ minus $N_0$.

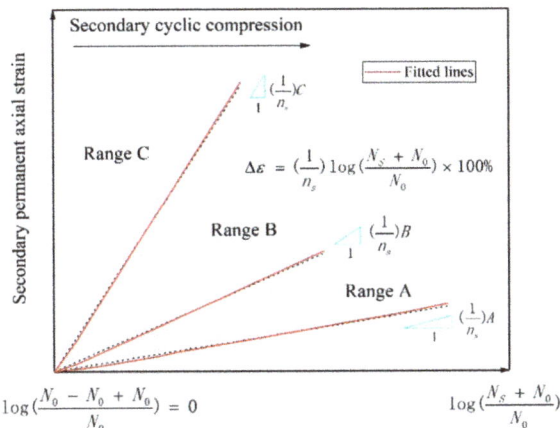

**Figure 3.** Relationship between cumulative axial strain and rotation number, reprinted with permission from [28]. Copyright 2019 Elsevier.

Further, the following relationship can be obtained:

$$1/n_s = \frac{\Delta \varepsilon}{\log[(N_S + N_0)/N_0] \times 100\%} \quad (3)$$

and $1/n_s$ can be used to represent the rate of cumulative axial strain change and to evaluate the stability behavior of granular materials under cyclic rotating axial compression.

Based on a summary and analysis of the existing research data, Chen [28] obtained the specific numerical relationship between $1/n_s$ and stability behavior, as follows:

Range A: $1/n_s \leq 0.1$, the specimen is in the elastic stability state.
Range B: $0.1 < 1/n_s \leq 0.434$, the specimen is in the plastic creep state.
Range C: $0.434 < 1/n_s$, the specimen is in the incremental plastic failure state.

Based on this method, in this research, the evaluation rules for stability behavior of the graded crushed stone specimen under cyclic rotating axial compression are proposed, and the critical loads of the specimen that indicate when the plastic creep and failure occur are obtained.

## 3. Raw Materials and Specimen Preparation

### 3.1. Raw Materials

1. Fine aggregate: crushed granite crushed stone particles with particle size below 4.75 mm used in the Guigang Expressway in Guangxi.
2. Coarse aggregate: crushed granite crushed stone particles with particle size from 4.75 to 53 mm used in the Guigang Expressway in Guangxi. The properties of the crushed stone materials were tested according to specification [32], and these properties are shown in Table 1.

**Table 1.** Technical specifications of aggregates.

| Aggregate (mm) | Crushing Value (%) | Apparent Density (g/cm³) | Dry Density (g/cm³) | Gross Relative Volumetric Density | Water Absorption Rate (%) | Needle Flake Content (%) |
|---|---|---|---|---|---|---|
| 19–53 | 21.57 | 2.729 | 2.672 | 2.646 | 1.24 | 7.30 |
| 9.5–19 | 14.08 | 2.753 | 2.691 | 2.662 | 1.32 | 6.58 |
| 4.75–9.5 | | 2.735 | 2.702 | 2.684 | 0.70 | 6.89 |
| 0–4.75 | | 2.720 | 2.705 | 2.694 | 0.92 | |
| Specification requirements | ≤28 | Measured data | Measured data | Measured data | Measured data | ≤15 |

## 3.2. Specimen Preparation Method

The specimen molding device used in this study for the crushed stone mixture preparation was an in-house developed road material vibration rotary compaction device shown in Figure 4. As shown in Figure 4, the device uses air pressure to protect the pressurized parts, the rotation and vibration motors of the device are placed under the console, and the pressure and displacement sensors are set on the right side of the indenter. The device performs vibration, rotation, and static compaction, and the three functions can be applied individually or in combination. The specific performance parameters of the device are as follows: vibration frequency of 3000 times/min, vibration amplitude of 0.6 mm, rotation rate of 5 rpm, and static pressure of 100–700 kPa. The device can accomplish a variety of different specimen preparation methods. In order to better simulate the rolling mechanism of road rollers in the process of road construction, a preparation method for the graded crushed stone mixture specimen using vibration and rotary compaction was proposed.

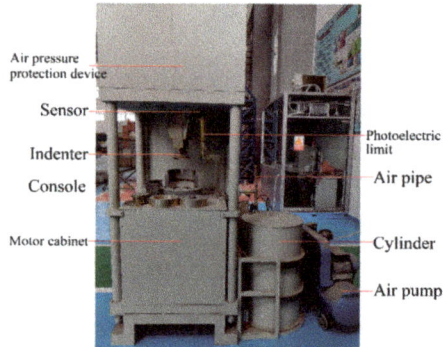

**Figure 4.** Road material vibration rotary compaction device.

The specific specimen preparation steps were as follows:
1. Take 5000 g of crushed stone mixture according to intended gradation.
2. Add water to the 5000 g crushed stone mixture according to the best moisture content of 4.1% and mix evenly. Seal with plastic and maintain for 12 h.
3. Place the crushed stone mixture in two layers and compact each layer for 4 min.

As shown in Figure 5, the inner diameter and height of the specimen tube were 150 and 230 mm, respectively. The height of the prepared specimens was usually between 120 and 125 mm, and the compacted density of crushed stone was between 2.3 and 2.45 g/cm$^3$, which was close to the value for practical engineering applications.

**Figure 5.** Crushed stone mixture specimen after preparation.

## 4. Experimental Program

### 4.1. Loading Method of Cyclic Rotating Axial Compression

The road material vibration rotary compaction device can apply a cyclic rotating axial compression load by performing rotation and axial compression concurrently. The device and its working principles are shown in Figure 6. First, the parameters that control the operation of the device are entered using a computer connected to the device using an optical fiber cable. Then, the indenter of the device begins to drop and apply pressure when it touches the top surface of the crushed stone mixture. When the pressure reaches the set value, the rotation and vibration motor of the device starts to work. Finally, experimental data are obtained by the operating software of the device. The combined effect of axial stress and shear stress was achieved by the simultaneous application of rotation and compression actions, with the axial pressure controlled by a computer during the loading process.

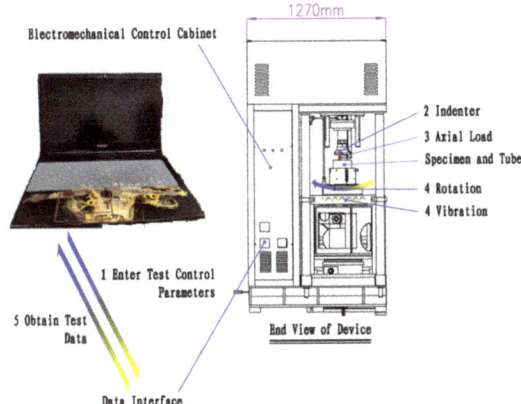

**Figure 6.** Schematic diagram of instrument loading system.

The indenter of the road material vibration rotary compaction device can automatically identify the load during the test; thus, after setting the load, the device can automatically adjust the load according to the measured load values to ensure the load is stable and remains in the set range of about 20% of the set value. An example of the loading process and load waveform during testing is shown in Figure 7a.

As shown in Figure 7a, the loading process included three stages: pressurization, during which the axial load increased up to the set value for the first time; stabilization, when the axial load attained the set value and paused for a short term; and finally, the stage when the axial load fluctuated. Figure 7b shows the changing load values during the loading process. Figure 8 shows the axial displacement of a specimen during loading in the road material vibration rotary compaction device.

In this study, a road material vibration rotary compaction device was used to apply cyclic rotating axial pressure on large-size graded crushed stone specimens. According to the authors of [28], the experiments can be terminated when the deformation trends of specimens become clear under repeated loading, therefore, 100 load rotations were adopted (20 min). In order to test the shakedown behavior and calculate the critical load of large-size crushed stone specimens under cyclic loading, experimental displacement results need to be obtained from specimens subjected to different load levels [28], therefore, the axial load was divided into seven levels. The cumulative axial strain of the graded crushed stone specimens under the different levels of cyclic rotating axial pressure was collected by a computer.

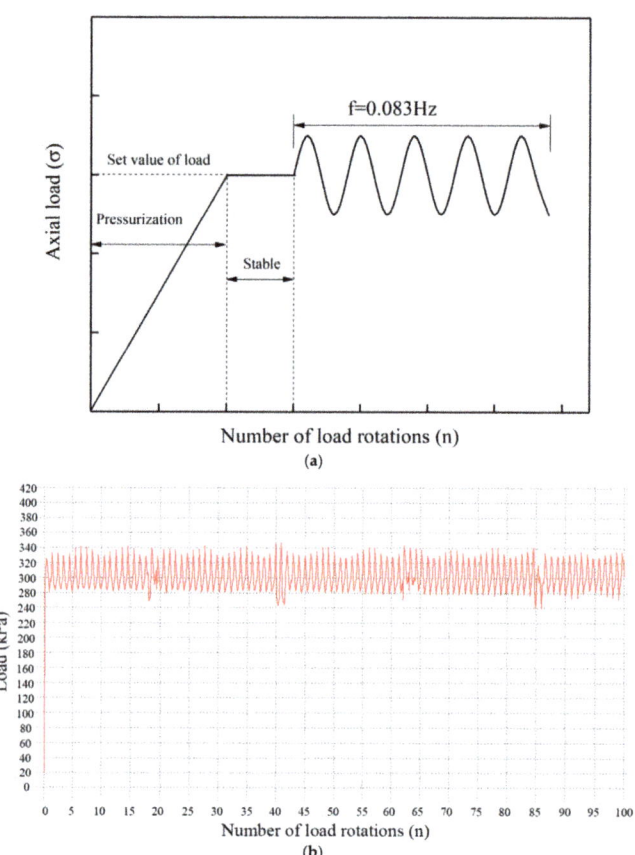

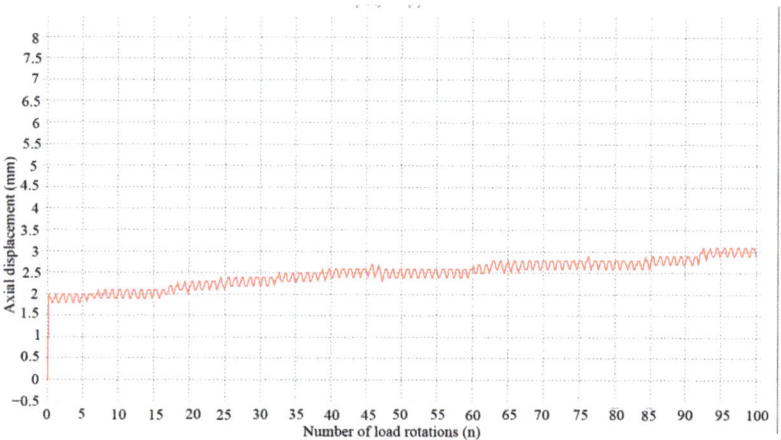

**Figure 7.** (**a**) Loading process and test load waveform; (**b**) real-time load during cyclic rotating axial compression.

**Figure 8.** Real-time displacement under cyclic rotating axial compression.

The shakedown theory was used to evaluate the deformation behavior and calculate the critical shakedown and critical failure loads of the graded crushed stone—first, according to Equation (3) to calculate values of $1/n_s$ and evaluate the shakedown behavior of specimens under different loads based on evaluation critical and second, according to Equations (4) and (5) to calculate the critical load of different gradation crushed stone specimens under cyclic loading. Finally, the influence of different factors on the shakedown behavior and the stability of the specimen under cyclic loading were analyzed.

*4.2. Influence of Gradation on Stability*

The step-by-step filling method [21] was used to design the three-level coarse aggregate ratio, and the I method [22] to design the fine aggregate ratio. Then five groups with three samples per group of large-size graded crushed stone gradations to be subjected to a single load level were obtained by mixing coarse and fine aggregates, these type of gradations are named as DG-1 to 5 (Design Gradation) in Table 2. The median values of the upper and lower limits of the design ranges were calculated according to the gradation design ranges recommended by the specification [32]. The median value of the specification was taken as the gradation of the conventional-size crushed stone mixture, and this type of gradation is named as SG-1 (Standard Gradation) in Table 2. The gradation graph of SG-1 and DG-1 to 5 are shows in Figure 9. In this research, three specimens of each gradation type of crushed stone were used for testing the deformation behavior under each load level, and the average displacement values of three specimens under every five cycles of cyclic loading were calculated for the subsequent analysis. Hence, this study used a total of 126 specimens to test the deformation behavior of crushed stone mixtures under cyclic loading. Finally, the stability of large-size graded crushed stone and conventional-size graded crushed stone under cyclic rotating axial compression was analyzed and compared.

Table 2. Screening pass rate of each gradation.

| Gradation Group | Mass Percentage Passing through Screen Hole (mm)/% | | | | | | | | |
|---|---|---|---|---|---|---|---|---|---|
| | 53 | 37.5 | 31.5 | 26.5 | 19 | 9.5 | 4.75 | 1.18 | 0.6 | 0.075 |
| DG-1 | 100 | 97.05 | 92.87 | 73.57 | 51.93 | 48.00 | 28.92 | 14.13 | 9.98 | 3.43 |
| DG-2 | 100 | 96.89 | 92.47 | 72.12 | 49.29 | 44.96 | 22.35 | 9.41 | 6.18 | 1.70 |
| DG-3 | 100 | 97.15 | 93.11 | 74.48 | 53.58 | 49.62 | 28.92 | 14.13 | 9.98 | 3.43 |
| DG-4 | 100 | 97.47 | 93.87 | 77.30 | 58.71 | 55.19 | 36.77 | 20.63 | 15.59 | 6.58 |
| DG-5 | 100 | 97.15 | 93.10 | 74.23 | 52.09 | 47.98 | 28.92 | 14.13 | 9.98 | 3.43 |
| SG-1 | 100 | 100 | 100 | 95.00 | 82.50 | 55.00 | 38.00 | 22.50 | 16.50 | 5.00 |

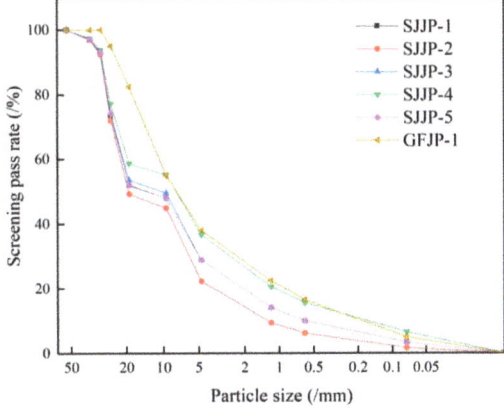

Figure 9. Gradation graph for all gradations.

## 4.3. Influence of Skeleton Structure Performance on Long-Term Stability Based on CBR Values

The CBR values of six groups of graded crushed stone were measured. This study used five specimens of each crushed stone gradation type for testing the CBR values, and the average CBR values of five specimens were calculated for the subsequent analysis. Hence, this study obtained 30 CBR values for crushed stone. The CBR value is an important parameter influencing the local load bearing capacity and skeleton structure properties of granular materials. By analyzing the relationship between the CBR value and the cumulative axial strain and critical load of the specimen under cyclic rotating axial compression, the influence of the skeleton structure performance on the specimen stability was studied.

## 5. Analysis of Test Results

### 5.1. Deformation Relationship for Large-Size Graded Crushed Stone under Cyclic Rotating Axial Compression

The deformation trends of graded crushed stone became clear after 100 cycles of compression, and the experiments were terminated at 100 load cycles (20 min). The relationship between the cumulative axial strain and the number of load rotations for the large-size graded crushed stone after 100 cycles of cyclic axial compression with different load levels is shown in Figure 10, and the cumulative axial strain rate curve of all types of specimens is shown in Figure 11.

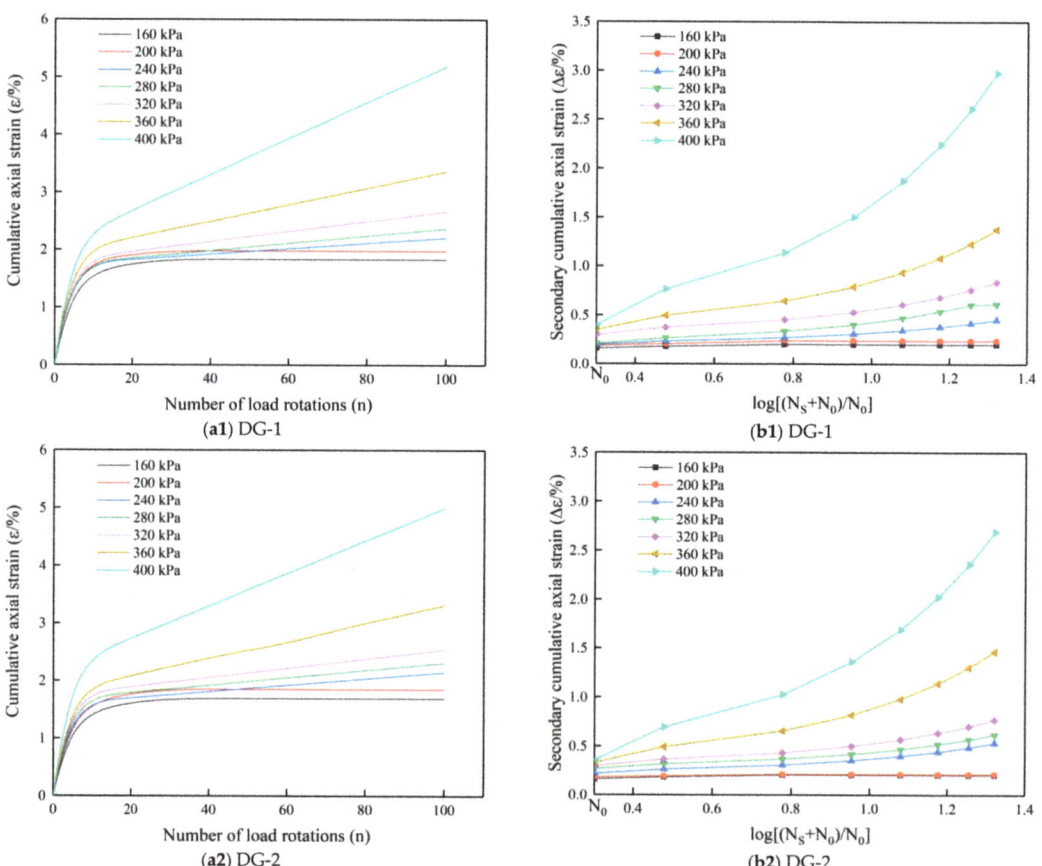

**Figure 10.** *Cont.*

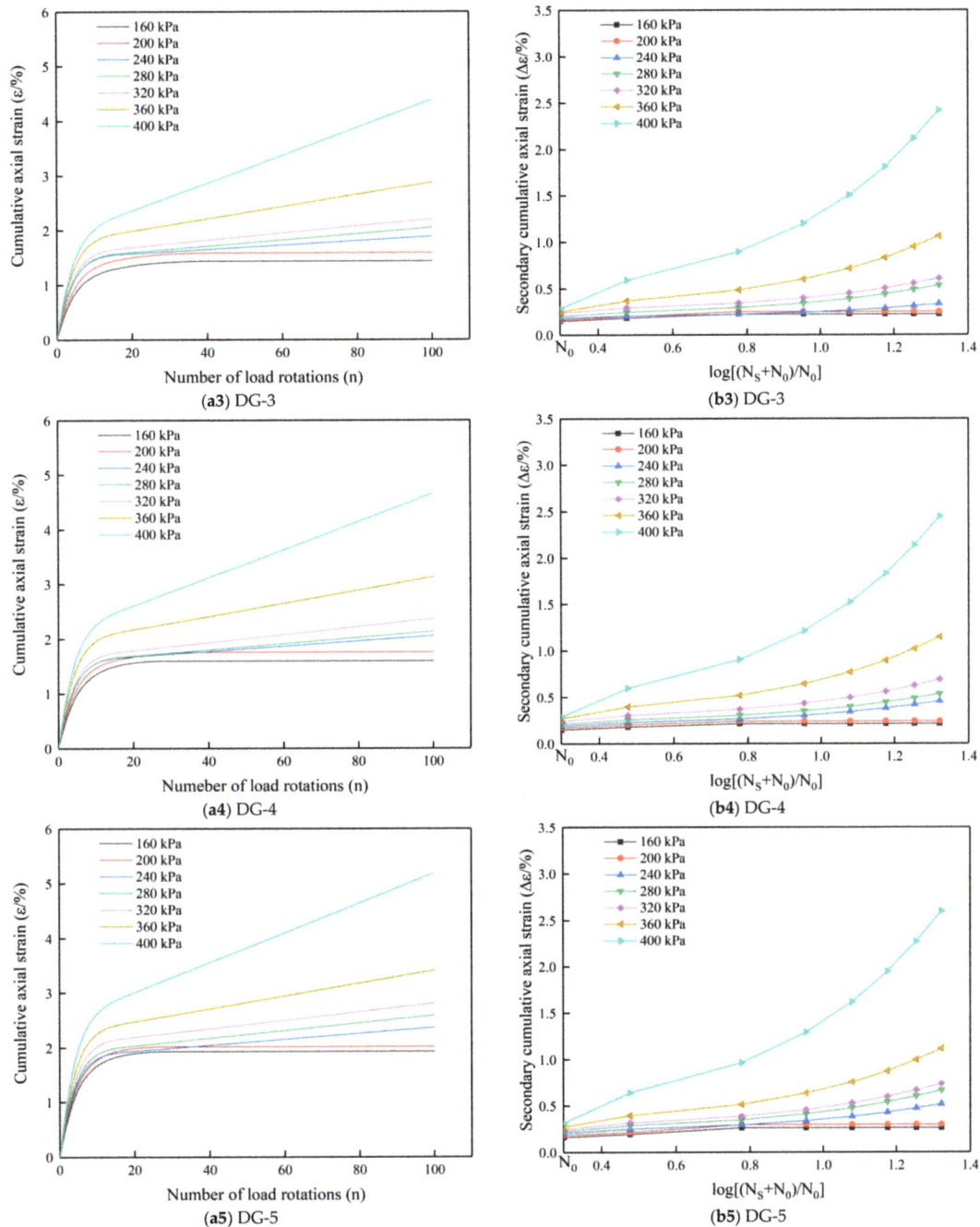

**Figure 10.** (**a1–a5**) Cumulative axial strain of large-size graded crushed stone of five gradation types under different loads; (**b1–b5**) secondary cumulative axial strain versus $\log[(N_S + N_0)/N_0]$ of large-size graded crushed stone of five gradation types under different loads.

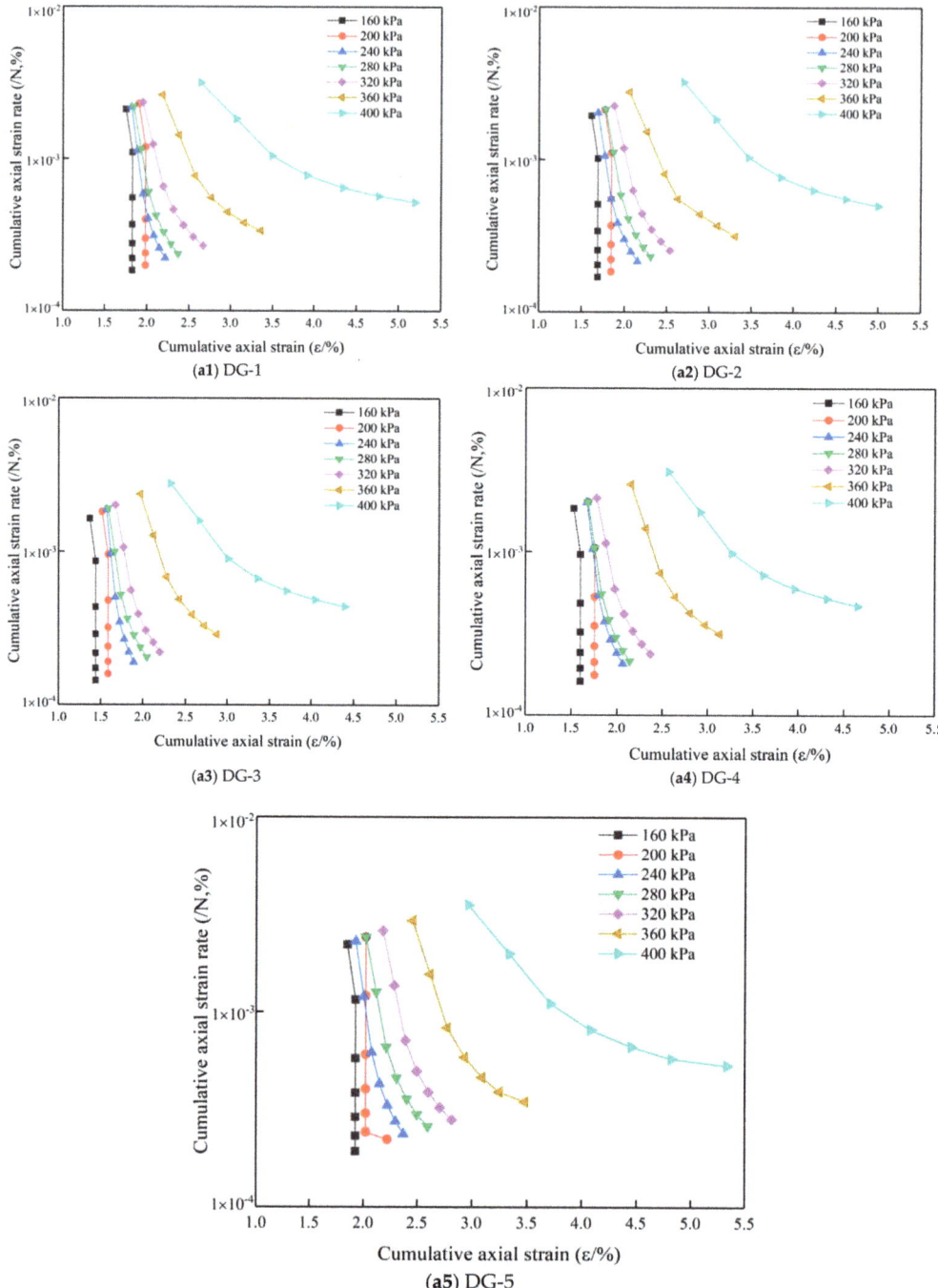

**Figure 11.** Axial strain change rate. (**a1**) DG-1; (**a2**) DG-2; (**a3**) DG-3; (**a4**) DG-4; (**a5**) DG-5.

As shown in Figure 10a, the test results indicate that the deformation relationships for the large-size graded crushed stone specimens under cyclic rotating axial compression differ at different load levels. When the load was 160 and 200 kPa, the cumulative axial strain of the specimens did not increase after the elastic deformation stage, and the specimens were essentially in the stable elastic state. However, when the load was between 240 and 320 kPa, the specimens first went through the rapid strain accumulation stage of the elastic deformation stage, but subsequently, the cumulative axial strain increased only slightly with the increase of the number of load rotations. Then, when the load reached 360 kPa the cumulative axial strain of the specimens still increased rapidly after the elastic deformation stage. Finally, when the load was 400 kPa the cumulative axial strain change rate became very high, and it is assumed that the skeleton structure of the specimens had been destroyed, which clearly signifies the incremental plastic failure state.

Figure 10(a1–a5) shows that the deformation behavior of graded crushed stone under cyclic loading can be divided into two distinct stages. In the first stage ($0 < N < N_0$), the specimens mainly underwent elastic deformations due to post-compaction by cyclic loading; in the second stage ($N_0 < N$), the specimens mainly showed plastic deformations under the secondary cyclic compression. Figure 10(b1–b5) shows the relationship between secondary cumulative axial strain and $\log[(N_S + N_0)/N_0]$, and demonstrates that the secondary cumulative axial strain was proportional to $\log[(N_S + N_0)/N_0]$ when $N_0 < N$.

Figure 11 shows the variation of the axial strain rate of the specimens under cyclic loading. The test results show that the strain rate of the specimens decreased sharply with the increase in the number of load rotations under different levels of cyclic axial compression load, and with the increase of the load level, the cumulative axial strain and change rate of the axial strain also increased.

In reference [22], the stability of graded crushed stone used in highway base under cyclic loading was explored experimentally. The cyclic loading used in this study simulated wheel rolling, and the specific deformation relationship for the graded crushed stone (SG-1) under cyclic loading is shown in Figure 12a. Figure 12b shows the deformation relationship for the conventional-size graded crushed stone (SG-1) under cyclic rotating axial compression. The deformation trends in the two diagrams are essentially the same. Under cyclic loading, the graded crushed stone first underwent elastic deformations and then tended to stabilize or continue to accumulate plastic deformations.

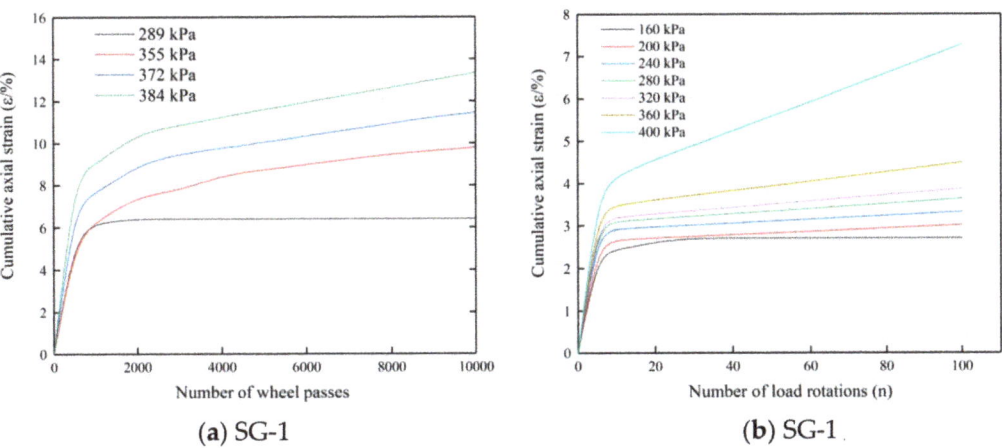

**Figure 12.** (a) Deformation relationship for graded crushed stone mixture under cycling wheel load; (b) deformation relationship for graded crushed stone mixture under cyclic rotating axial compression.

The axial strain of graded crushed stone under a cycling wheel rolling load was larger than that under a cyclic rotating axial pressure, because the former is an unconfined test, while the latter is a confined test in a steel tube. Based on the comparative analysis of the test data in [19,33], the final cumulative axial strain of granular materials under cyclic loading was between 1% and 15%. Therefore, the cyclic rotational axial compression method proposed in this paper is valid for testing the stability of graded crushed stone and can correctly capture the deformation relationship for the graded crushed stone under cyclic loading.

### 5.2. Influence of Gradation on Shakedown Stability

#### 5.2.1. Stability Analysis of Specimens with Different Gradations

The shakedown theory is used to evaluate the deformation relationship for graded crushed stone specimens under different load levels. According to Equation (3), to calculate values of $1/n_s$ ($N_0 = 5$, $N = 100$, and $N_S = N − N_0$), the calculation results as listed in Table 3. The specific evaluation method evaluates the deformation behavior of graded crushed stone specimens under cyclic rotating axial compression by calculating the slope of the strain curve, $1/n_s$, as mentioned in Section 2.2. The evaluation criteria are as follows:

Range A: $1/n_s \leq 0.1$, the specimen is in the elastic stability state.
Range B: $0.1 < 1/n_s \leq 0.434$, the specimen is in the plastic creep state.
Range C: $0.434 < 1/n_s$, the specimen is in the incremental plastic failure state.

**Table 3.** Values of $1/n_s$ value and shakedown behavior of graded crushed stone under different loads.

| Gradation Group | $1/n_s$ (160 kPa) | Behavior | $1/n_s$ (200 kPa) | Behavior | $1/n_s$ (240 kPa) | Behavior | $1/n_s$ (280 kPa) | Behavior | $1/n_s$ (320 kPa) | Behavior | $1/n_s$ (360 kPa) | Behavior | $1/n_s$ (400 kPa) | Behavior |
|---|---|---|---|---|---|---|---|---|---|---|---|---|---|---|
| DG-1 | 0.058 | A | 0.058 | A | 0.173 | B | 0.289 | B | 0.289 | B | 0.635 | C | 1.847 | C |
| DG-2 | 0.058 | A | 0.058 | A | 0.175 | B | 0.291 | B | 0.349 | B | 0.640 | C | 1.803 | C |
| DG-3 | 0.057 | A | 0.057 | A | 0.115 | B | 0.229 | B | 0.286 | B | 0.573 | C | 1.604 | C |
| DG-4 | 0.058 | A | 0.058 | A | 0.115 | B | 0.173 | B | 0.289 | B | 0.635 | C | 1.617 | C |
| DG-5 | 0.056 | A | 0.084 | A | 0.168 | B | 0.280 | B | 0.336 | B | 0.672 | C | 1.793 | C |
| SG-1 | 0.059 | A | 0.117 | B | 0.117 | B | 0.235 | B | 0.352 | B | 0.645 | C | 2.462 | C |

The results of the calculations and evaluation of shakedown behavior are shown in Table 3. According to the relationship between the $1/n_s$ values and load level, the specific values of critical loads were calculated by interpolation. The formula for calculating the critical shakedown load is as follows:

$$F_a = \frac{0.1 - \left(\frac{1}{n_{s1}}\right)}{\left(\frac{1}{n_{s2}}\right) - \left(\frac{1}{n_{s1}}\right)} \times 40 + F_{a1}, \quad (4)$$

where $F_a$ represents the critical shakedown load, $1/n_{s1}$ indicates that the value of the shakedown behavior parameter, $1/n_s$, is less than 0.1 but close to 0.1, $1/n_{s2}$ indicates that the value of $1/n_s$ is greater than 0.1 but close to 0.1, and $F_{a1}$ represents the test load value corresponding to $1/n_{s1}$.

The formula for critical failure load is as follows:

$$F_p = \frac{0.434 - \left(\frac{1}{n_{s3}}\right)}{\left(\frac{1}{n_{s4}}\right) - \left(\frac{1}{n_{s3}}\right)} \times 40 + F_{p1}, \quad (5)$$

where $F_p$ represents the critical failure load, $1/n_{s3}$ indicates that the value of the shakedown behavior parameter, $1/n_s$, is less than 0.434 but close to 0.434, $1/n_{s4}$ indicates that the value of $1/n_s$ is greater than 0.434 but close to 0.434, and $F_{p1}$ represents the test load value corresponding to $1/n_{s3}$.

The calculation results are shown in Table 4 and demonstrate that the critical shakedown load of the conventional-size graded crushed stone specimens is more than 10% lower than that of the large-size graded crushed stone specimens, while there is little difference in the critical failure load between the two types of specimens. When the specimens are subjected to a lower cyclic rotating axial compression, the main source of deformation is the rearrangement of particles caused by particles slipping inside the graded crushed stone mixture. When the cyclic rotating axial load increases to the value of the critical failure load, the large-size crushed stones will be broken, and the relative slip between the particles will form the overall plastic deformation.

**Table 4.** Critical loads of six gradation groups of graded crushed stone.

| Gradation Group | Critical Shakedown Load (kPa) | Critical Failure Load (kPa) |
|---|---|---|
| DG-1 | 214.609 | 336.763 |
| DG-2 | 214.359 | 331.684 |
| DG-3 | 229.655 | 340.627 |
| DG-4 | 229.474 | 336.763 |
| DG-5 | 207.619 | 331.667 |
| SG-1 | 188.276 | 331.195 |

Previous studies [34–36] indicate that the deformation caused by the overall rearrangement of the particles generated by the slip in graded crushed stone mixtures was mainly related to their density, gradation characteristics, and voids. Therefore, specimens with a higher density and stronger skeleton structure exhibited smaller slip deformations under the same load. It can be observed that the skeleton structure performance in SJJP-3 and SJJP-4 is better than in the other specimens. In the graded crushed stone specimen, the relative slip deformation between the particles caused by the crushing of large-size crushed stone is mainly related to the crushing and compaction condition, which does not differ significantly under the same load [37,38].

5.2.2. Effect of Content and Density of Large-Size Crushed Stone on Critical Load

The particle size of large-size crushed stone is larger than 26.5 mm. The relationship between the large-size crushed stone content, the critical shakedown load, and the specimen density is shown in Figure 12. The data of large-size crushed stone content and density are shown in Table 5, and the effect of large-size crushed stone content on the long-term stability is analyzed.

**Table 5.** Contents and densities of large-size crushed stone specimens.

| Gradation Group | Content of Large-Size Crushed Stone with Size over 26.5 mm (%) | Density (g/cm$^3$) |
|---|---|---|
| DG-1 | 26.43 | 2.336 |
| DG-2 | 27.88 | 2.350 |
| DG-3 | 25.52 | 2.373 |
| DG-4 | 22.70 | 2.371 |
| DG-5 | 25.77 | 2.321 |
| SG-1 | 5.00 | 2.343 |

As shown in Figure 13, the critical shakedown load of the specimen is not only related to the large-size crushed stone content but also related to the density of the specimen; thus, a joint analysis can better describe this relationship. The lowest content of large-size crushed stone in the conventional-size graded crushed stone (GFJP-1) was 5%, much lower than that in the other five groups. The critical shakedown load was also the lowest, as shown in Figure 13a. The content of large-size crushed stone in the five groups of large-size graded crushed stone specimens (SJJP-1 to 5) ranged between 22% and 28%, and the relationship between the large-size crushed stone content and its critical shakedown load was not clear. According to the relationship between the density specimens and the critical shakedown load in Figure 13b, the critical shakedown load increased with the increase in the density. The large-size crushed stone content of SJJP-3 was 25.52% and that of SJJP-4 was 22.70%. The density of these two groups was the highest among the six groups of specimens, and the critical shakedown loads were larger than those of the other groups of specimens. Therefore, for the crushed stone mixture specimen, when the content of large-size crushed stone was between 22% and 28%, and the critical shakedown load increased with the increase in density.

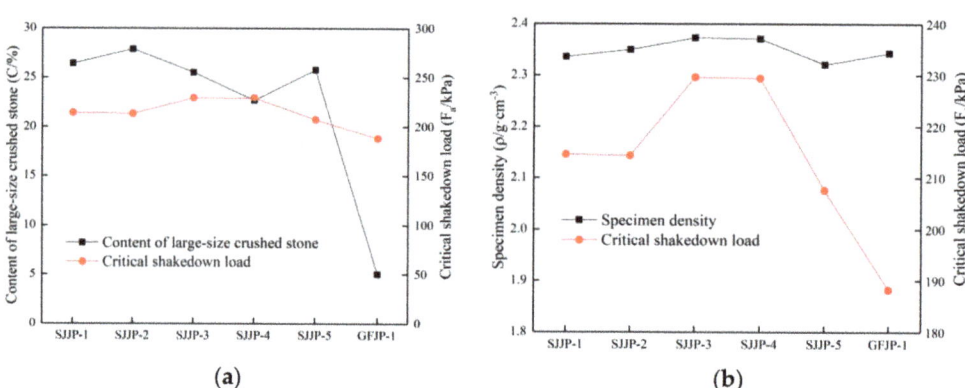

**Figure 13.** (a) Relationship between critical shakedown load and content of large-size crushed stone; (b) relationship between critical shakedown load and specimen density.

The relationships between the content of large-size crushed stone, the critical failure load, and the specimen density are shown in Figure 14. It is noted that the critical failure load of the specimens under cyclic loading was not strongly related to the content and density of large particle-size crushed stone. The reason is that the deformation of large-size graded crushed stone is mainly related to the stone crushing condition under a high level of cyclic rotating axial compression as mentioned in Section 4.2, and the stone crushing condition under the same load was almost the same.

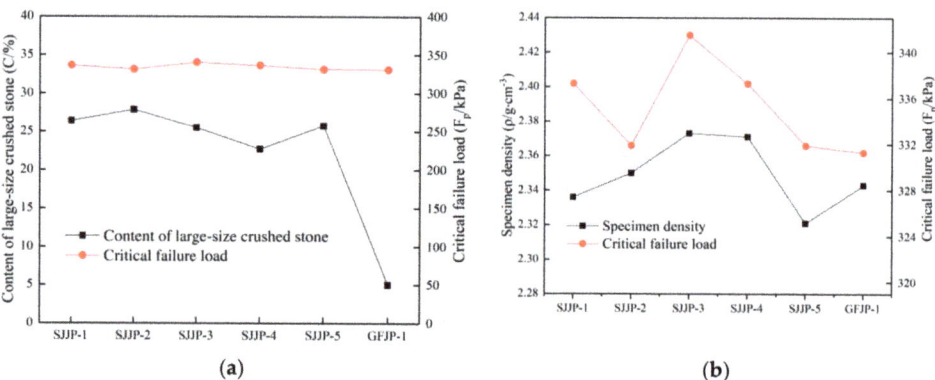

**Figure 14.** (a) Relationship between critical failure load and content of large-size crushed stone; (b) relationship between critical failure load and specimen density.

5.2.3. Effect of Content and Density of Large-Size Crushed Stone on Cumulative Axial Strain

The relationship between the content and density of large-size crushed stone and the cumulative axial strain under the 280 kPa load is shown in Figure 15. It is noted that the large-size crushed stone content of GFJP-1 was 5%, which is lower than those of other specimens, and the cumulative axial strain was the highest. The specimens with a lower large-size crushed stone content had low stiffness. In the five groups of the large-size graded crushed stone specimens (SJJP-1 to 5), the content of large-size crushed stone was between 22% and 28%. The cumulative axial strain of specimens was not closely related to the large-size crushed stone content but rather to the density of specimen. The cumulative axial strain decreased with the increase in specimen density. The large-size crushed stone content of SJJP-3 was 25.52%, while in SJJP-4 it was 22.70%. The densities of the two specimens were similar and higher than those of the other four specimens, but the cumulative axial strains of the two specimens were smaller than those of the other specimens. It can be seen that when there was little difference in the content of large-size crushed stone, the main factor affecting the specimen stiffness was the density of the specimen. Therefore, when the content of large-size crushed stone was between 22% and 28%, the cumulative axial strain decreased with the increase in the specimen density.

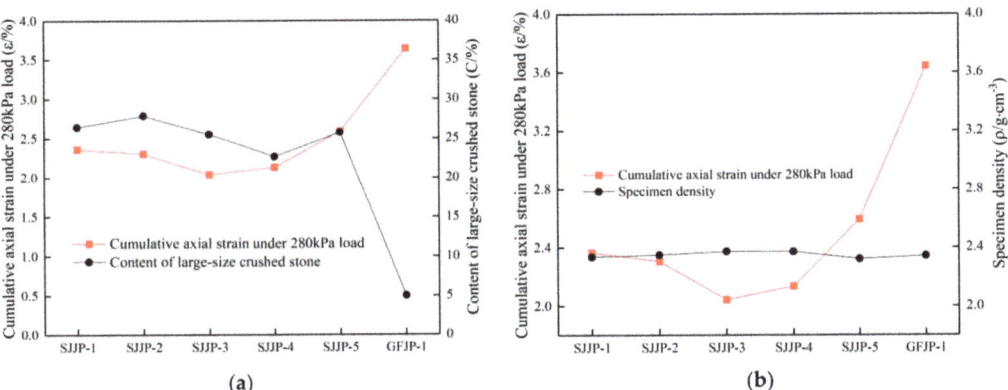

**Figure 15.** (a) Relationship between cumulative axial strain and content of large-size crushed stone content under load of 280 kPa; (b) relationship between cumulative axial strain and specimen density under load of 280 kPa.

Based on the above analysis, it can be concluded that the large-size graded crushed stone has better stiffness and load bearing capacity than the conventional-size graded crushed stone. When the content of the large-size crushed stone is appropriate, the higher the specimen density is, the larger the stiffness and load bearing capacity is. In the six groups of graded crushed stone specimens, the stiffness and load bearing capacity of SJJP-3 and SJJP-4 were better than those of the other specimens, thus the best gradation design range of large-size crushed stone is between those of SJJP-3 and SJJP-4.

5.2.4. Regression Analysis of the Joint Effect of Content and Density of Large-Size Crushed Stone

The joint effect of the large-size crushed stone content and density on the cumulative axial strain and critical shakedown load of DG-1 to 5 samples under a cyclic rotating axial compression of 280 kPa were analyzed. The coupling factor was used to quantify the joint effect of large-size crushed stone content and density, and the specific coupling factor formula as follows:

$$V = \frac{\rho - 2.4693}{-0.0464 \times C} \times 10, \quad (6)$$

where $V$ is the coupling factor, $\rho$ is the specimen density (g/cm$^3$), and $C$ is the content of large-size crushed stone (%). The coupling factors of large-size crushed stone content and density of DG-1 to 5 were calculated, and the linear regression analysis was carried out between the coupling factors and the cumulative axial strain and critical shakedown load under a cyclic rotating axial compression of 280 kPa. The results are shown in Figure 16. It was found that there is a good linear relationship between the large-size crushed stone content and density coupling factor of DG-1 and the cumulative axial strain and critical shakedown load under a cyclic rotating axial compression of 280 kPa. The analysis demonstrated that the joint effect of large-size crushed stone's content and density on its mechanical properties and anti-deformation ability is strong.

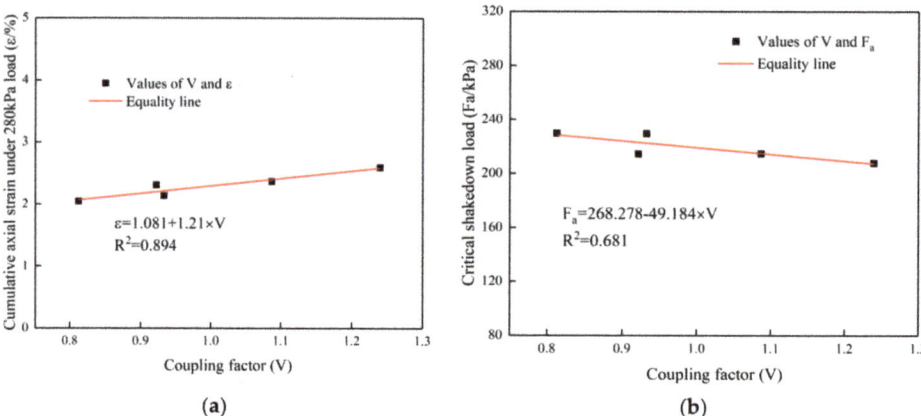

**Figure 16.** (a) Regression analysis between coupling factor and cumulative axial strain under a cyclic rotating axial compression of 280 kPa; (b) regression analysis between coupling factor and critical shakedown load.

### 5.3. Relationship between CBR Values, Strain, and Critical Load

#### 5.3.1. Relationship between CBR Values and Cumulative Axial Strain of Specimen

The curves in Figure 17 show that the cumulative axial strain of the specimens with low CBR values is larger than that of other specimens under the same level load, which indicates that the specimens with weaker skeleton structure have weaker stiffness and therefore develop larger deformation under the same load level. Moreover, the CBR values of the conventional-size graded crushed stone are smaller than those of the large-size graded crushed stone, and the cumulative axial strains of the conventional-size graded crushed stone are also larger than those of the large-size graded crushed stone. Thus, the large-size graded crushed stone has better skeleton structure properties than the conventional-size graded crushed stone.

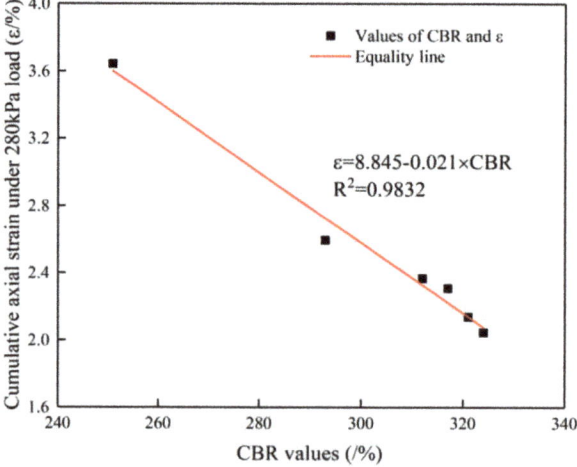

**Figure 17.** Relationship between CBR values and cumulative axial strain under a 280 kPa load.

### 5.3.2. Relationship between CBR Values and Critical Load

The curves in Figure 18a show that the CBR values are closely related to the specimen's critical shakedown load under cyclic loading. The critical shakedown load increases with an increase in the CBR value of the specimen, which indicates that the skeleton structure of large-size graded crushed stone strongly influences its stiffness, and the stronger the skeleton structure is, the larger the stiffness is. The correlation between the CBR values and the critical failure load of the specimen in Figure 18b is not as clear as that in Figure 18a, because there is little difference in the stone crushing condition under the same loads.

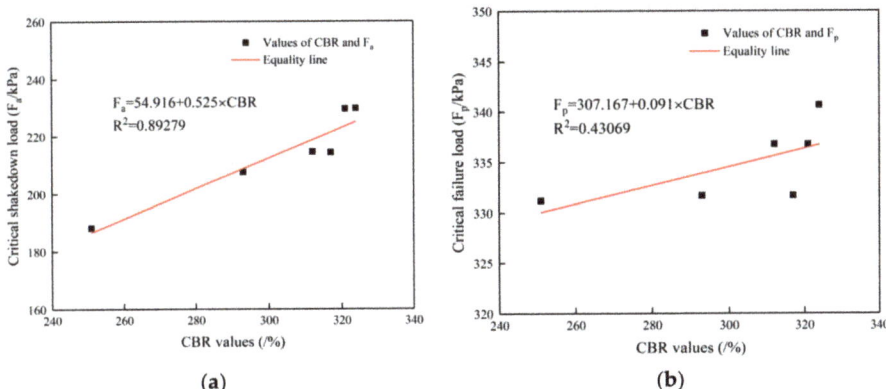

**Figure 18.** (a) Relationship between CBR values and critical shakedown load; (b) relationship between CBR values and critical failure load.

From the above analyses of the test results, it can be concluded that there is a strong correlation between the CBR values of the large-size graded crushed stone and its critical shakedown load under cyclic loading. The critical shakedown load increases with the increase in the CBR value, while there is no strong correlation between the CBR value and the critical failure load. The reason is that the failure resistance of the specimen is mainly related to the properties of the crushed stone but not to the skeleton structure performance.

## 6. Conclusions

This research investigated the deformation behavior of granular materials under cyclic loading using the traditional RLT tester and the MTS tester. The deformation behavior of large-size graded crushed stone was tested under the simultaneous action of axial pressure and shear stress using the road material vibration rotary compaction device. Different gradations of the selected materials were tested in order to evaluate the effect of the content of large-size crushed stone and density. This study also introduced a new method to calculate the critical load of granular materials under cyclic loading based on shakedown theory. Some of the main conclusions can be summarized as follows:

1. In the first part of the research, the cumulative axial strain of 126 crushed stone specimens under cyclic loading was analyzed. The cumulative axial strain of large-size graded crushed stone specimen will eventually become stable and no longer increase under a cyclic rotating axial compression of 200 kPa. When the load is between 240 and 320 kPa, the cumulative axial strain of large-size graded crushed stone will still increase slowly after elastic deformations, but the strain rate will decrease rapidly. When the load is increased to 360 kPa, the cumulative axial strain of the large-size graded crushed stone increases rapidly during the entire process of cyclic loading, and the strain rate is maintained at a large value. It is observed that the deformation relationship and strain values of the specimens under cyclic rotating axial compression are similar to those in existing literature. Therefore, it is feasible to

apply cyclic rotating axial compression to study the long-term stability of large-size graded crushed stone used in road base.

2. In the second part, the critical load of graded crushed stone specimens under cyclic loading was calculated based on the shakedown theory and the new calculation method. The maximum critical shakedown load of large-size graded crushed stone was 229.655 kPa and the lowest was 207.619 kPa, respectively. The critical shakedown load of the conventional-size graded crushed stone was 188.276 kPa, making the critical shakedown load of the large-size graded crushed stone at least 10% higher than that of the conventional-size graded crushed stone. This indicates that the large-size graded crushed stone can be used in engineering applications, and it deforms less. The analysis clearly showed that the content of the large-size crushed stone, density, and skeleton structure performance influence the critical shakedown load. On the contrary, there is no significant difference between the critical failure load of the large-size graded crushed stone and that of the conventional-size graded crushed stone, with both being between 330 and 345 kPa.

3. In the third part, the relationship between the cumulative axial strain and gradation of graded crushed stone was analyzed. When the content of large-size graded crushed stone with a size over 26.5 mm was between 22% and 28%, the higher the density was, and the larger the stiffness, the better the skeleton structure performance and the load bearing capacity were. Therefore, the content of the large-size crushed stone in the large-size graded crushed stone mixture should be controlled between 22% and 26%, and the density should be as high as possible.

4. In addition, the skeleton structure performance of crushed stone influences the stability behavior of crushed stone specimens under cyclic loading was observed. The CBR values of 30 specimens and the deformation data of 126 specimens under cyclic loading test showed that there is a linear relationship between CBR values and cumulative axial strain and critical shakedown load of crushed stone specimens. The linear correlation coefficient between CBR values and critical shakedown load was 0.89279 and cumulative axial strain of specimens under a 280 kPa load cyclic loading was 0.9832, respectively.

5. In short, this study proposed the method is using the vibration and rotary compaction method for preparation of specimens of crushed stone and testing their stability under cyclic rotating axial compression in the road material vibration rotatory compaction device. The method achieved simultaneous testing under axial pressure and shear stress using the device. The test data of 126 crushed stone specimens show that this is a simple and practical method for testing the stability of granular materials under cyclic loading, and it enables analyzing the correlations between deformation behavior and gradation parameters. Therefore, this method could be used for investigating the stability of granular materials under a cycling load.

**Author Contributions:** Conceptualization, B.T. and T.Y.; methodology, T.Y.; validation, H.Q., T.Y. and Q.L.; formal analysis, T.Y.; investigation, B.T.; resources, B.T.; data curation, T.Y. and Q.L.; writing—original draft preparation, T.Y.; writing—review and editing, T.Y. and H.Q.; supervision, B.T.; project administration, B.T.; funding acquisition, B.T. All authors have read and agreed to the published version of the manuscript.

**Funding:** This study was funded by the National Natural Science Foundation of China, grant number 51768015.

**Institutional Review Board Statement:** Not applicable.

**Informed Consent Statement:** Not applicable.

**Data Availability Statement:** Data sharing is not applicable to this article.

**Conflicts of Interest:** The authors declare no conflict of interest.

## References

1. Lekarp, F.; Dawson, A. Modelling permanent deformation behaviour of unbound granular materials. *Constr. Build. Mater.* **1998**, *12*, 9–18. [CrossRef]
2. Rahman, M.S.; Erlingsson, S. A model for predicting permanent deformation of unbound granular materials. *Road Mater. Pavement Des.* **2015**, *16*, 653–673. [CrossRef]
3. Rahman, M.S.; Erlingsson, S. Predicting permanent deformation behaviour of unbound granular materials. *Int. J. Pavement Eng.* **2015**, *16*, 587–601. [CrossRef]
4. Lekarp, F.; Isacsson, U.; Dawson, A. State of the art I:Resilient response of unbound aggregates. *J. Transp. Eng.* **2000**, *126*, 66–75. [CrossRef]
5. Lekarp, F.; Isacsson, U.; Dawson, A. State of the art II:Permanent strain response of unbound aggregates. *J. Transp. Eng.* **2000**, *126*, 76–83. [CrossRef]
6. Chen, J.F.; Xu, M.; Song, E.X.; Cao, G.X. Large Scale Triaxial Testing on Mechanical Properties of Broken Limestone Under Various Stress Paths. *Eng. Mech.* **2012**, *29*, 195–201. [CrossRef]
7. Fu, Q.L. Research on Composition Design Parameters and Method of Open-Graded Large Stone Asphalt Mixes. Ph.D. Thesis, Chang'an University of Technology, Xi'an, China, 2011. [CrossRef]
8. Fu, Q.L.; Wei, J.G.; Zhou, X.Z. Evaluation of low temperature performance of open-graded large stone asphalt mixes. *J. Build. Mater.* **2020**, *23*, 896–903. [CrossRef]
9. Dave, E.V.; Buttlar, W.G. Thermal reflective cracking of asphalt concrete overlays. *Int. J. Pavement Eng.* **2010**, *11*, 477–488. [CrossRef]
10. Li, C.; Wang, F.; Deng, X.; Li, Y.; Zhao, S. Testing and prediction of the strength development of recycled-aggregate concrete with large particle natural aggregate. *Materials* **2019**, *12*, 1891. [CrossRef]
11. Gu, F.; Zhang, Y.; Luo, X.; Sahin, H.; Lytton, R.L. Characterization and prediction of permanent deformation properties of unbound granular materials for Pavement ME Design. *Constr. Build. Mater.* **2017**, *155*, 584–592. [CrossRef]
12. Zhang, J.Q. Study on Shakedown Behavior of Granular Material Used in Pavement and Shakedown Analysis of Pavement Structures. Ph.D. Thesis, South China University of Technology, Guangzhou, China, 2012.
13. Werkmeister, S. Shakedown analysis of unbound granular materials using accelerated pavements test results from New Zealand's CAPTIF facility. In Proceedings of the Geoshanghai International Conference, Shanghai, China, 6–8 June 2006; pp. 220–228.
14. Werkmeister, S. Permanent Deformation Behaviour of Unbound Granular Materials in Pavement Constructions. Ph.D. Thesis, Dresden University of Technology, Dresden, Germany, 2003.
15. Xiao, Y.; Zheng, K.; Chen, L.; Mao, J. Shakedown analysis of cyclic plastic deformation characteristics of unbound granular materials under moving wheel loads. *Constr. Build. Mater.* **2018**, *167*, 457–472. [CrossRef]
16. Wichtmann, T.; Rondón, H.A.; Niemunis, A.; Triantafyllidis, T.; Lizcano, A. Prediction of permanent deformations in pavementsusing a high-cycle accumulation model. *J. Geotech. Geoenviron. Eng.* **2010**, *136*, 728–740. [CrossRef]
17. Sas, W.; Głuchowski, A.; Gabryś, K.; Soból, E.; Szymański, A. Deformation Behavior of Recycled Concrete Aggregate during Cyclic and Dynamic Loading Laboratory Tests. *Materials* **2016**, *9*, 780. [CrossRef] [PubMed]
18. Jiang, H.; Bian, X.; Jiang, J.; Chen, Y. Dynamic performance of high-speed railway formation with the rise of water table. *Eng. Geol.* **2016**, *206*, 18–32. [CrossRef]
19. Chen, Q.; Abu-Farsakh, M.; Voyiadjis, G.Z.; Souci, G. Shakedown analysis of geogrid-reinforced granular base material. *J. Mater. Civ. Eng.* **2013**, *337*, 337–346. [CrossRef]
20. Abu-Farsakh, M.Y.; Chen, Q. Evaluation of geogrid base reinforcement in flexible pavement using cyclic plate load testing. *Int. J. Pavement Eng.* **2011**, *12*, 275–288. [CrossRef]
21. Chen, Z.D.; Yuan, W.J.; Gao, C.H. Research on design method of multilevel dense built gradation. *China J. Highw. Transp.* **2006**, *19*, 32–37. [CrossRef]
22. Peng, B. Gradation design method based on method of I change. *J. Wuhan Univ. Technol.* **2005**, *29*, 751–754. [CrossRef]
23. Sharp, R.W.; Booker, J.R. Shakedown of Pavements under Moving Surface Loads. *J. Transp. Eng.* **1984**, *110*, 1–14. [CrossRef]
24. Sharp, R.W. Shakedown Analysis and the Design of Pavement under Moving Surface Load. Ph.D. Thesis, University of Sydney, Sydney, Australia, 1983.
25. Dawson, A.R.; Mundy, M.J.; Huhtala, M. European Research into Granular Material for Pavement Bases and Subbases. *Transp. Res. Rec.* **2000**, *1721*, 91–99. [CrossRef]
26. Zhou, Y.M.; Tan, Z.M.; Liu, S.W.; Niu, K.M. Analysis of Near-corner Stresses in Concrete Pavement Structure. *Eng. Mech.* **2010**, *27*, 105–110. [CrossRef] [PubMed]
27. Werkmeister, S.; Dawson, A.R.; Wellner, F. Permanent Deformation Behavior of Granular Materials and the Shakedown Concept. *Road Mater. Pavement Des.* **2005**, *6*, 31–51. [CrossRef]
28. Chen, W.B.; Feng, W.Q.; Yin, J.H.; Borana, L.; Chen, R.P. Characterization of permanent axial strain of granular materials subjected to cyclic loading based on shakedown theory. *Constr. Build. Mater.* **2019**, *198*, 751–761. [CrossRef]
29. Tao, M.; Mohammad, L.N.; Nazzal, M.D.; Zhang, Z.; Wu, Z. Application of Shakedown Theory in Characterizing Traditional and Recycled Pavement Base Materials. *J. Transp. Eng.* **2010**, *136*, 214–222. [CrossRef]
30. Pérez, I.; Medina, L.; Romana, M.G. Permanent deformation models for a granular material used in road pavements. *Constr. Build. Mater.* **2006**, *20*, 790–800. [CrossRef]

31. Yin, J.H. Non-linear creep of soils in oedometer tests. *Geotechnique* **1999**, *49*, 699–707. [CrossRef]
32. Test Methods of Aggregate for Highway Engineering. In *JTGE42-2005, Occupation Standard of the People's Republic of China*; China Communications Press: Beijing, China, 2005.
33. Xiao, J.; Zhang, D.; Wei, K.; Luo, Z. Shakedown behaviors of railway ballast under cyclic loading. *Constr. Build. Mater.* **2017**, *155*, 1206–1214. [CrossRef]
34. Ge, H.; Sha, A.; Han, Z.; Xiong, X. Three-dimensional Characterization of Morphology and Abrasion Decay Laws for Coarse Aggregates. *Constr. Build. Mater.* **2018**, *188*, 56–67. [CrossRef]
35. Ardah, A.; Chen, Q.M.; Abu-Farsakh, M.Y. Evaluating the Performance of Very Weak Subgrade Soils Treated/stablized with Cementitious Materials for Sustainable Pavements. *Transp. Geotech.* **2017**, *11*, 107–119. [CrossRef]
36. Pouranian, M.R.; Haddock, J.E. Determination of voids in the mineral aggregate and aggregate skeleton characteristics of asphalt mixtures using a linear-mixture packing model. *Constr. Build. Mater.* **2018**, *188*, 292–304. [CrossRef]
37. Zhang, J.W.; Wang, H.L.; Chen, S.J.; Li, Y.L. Bearing Deformation Characteristics of Large-Size Broken Rock. Journal of China Coal Society. *J. China Coal Soc.* **2018**, *43*, 1000–1007. [CrossRef]
38. Li, J.; Yu, J.; Xie, J.; Ye, Q. Performance Degradation of Large-Sized Asphalt Mixture Specimen under Heavy Load and Its Affecting Factors Using Multifunctional Pavement Material Tester. *Materials* **2019**, *12*, 3814. [CrossRef] [PubMed]

Article

# Interlayer Shear Characteristics of Bridge Deck Pavement through Experimental and Numerical Analysis

Weidong Chen [1], Bing Hui [2] and Ali Rahman [3,4,*]

1. Chang'an-Dublin International College of Transportation, Chang'an University, Xi'an 710064, China
2. School of Highway, Chang'an University, Xi'an 710064, China
3. School of Civil Engineering, Southwest Jiaotong University, Chengdu 610031, China
4. Key Laboratory for Highway Engineering of Sichuan Province, Southwest Jiaotong University, Chengdu 610031, China
* Correspondence: arahman@swjtu.edu.cn; Tel.: +86-158-8455-1418

**Abstract:** In order to study the interlayer shear behavior of bridge deck pavement, a numerical simulation was conducted to analyze the influence of varying interfacial conditions on shear stress at the bottom of pavement layers under the moving loading effect. Moreover, the shear strength of the different adhesive and waterproof adhesive materials was evaluated by conducting laboratory tests. The results showed that improving the bonding condition at the upper interlayer led to the reduction of the shear stress at the bottom of the pavement layers. With the increase of the friction coefficient of the upper interface to the full bonding state, the resulting shear stress at the bottom of the upper layer declined to the lowest value, which was about 35% of that of the full slip state. When the lower interlayer was in the full slip or partial bonding state, the resulting shear stress at the bottom of the lower layer decreased linearly with the increase of the friction coefficient of the upper interlayer. Moreover, once the contact state of the upper interlayer reached the full bonding state, the resulting shear stress at the bottom of the lower layer reached the minimum, which is about 88% of that of the full slip state. To improve the integrity and shear resistance of the bridge deck pavement structure, interlayer bonding should be strengthened. In this regard, the resin emulsified asphalt was determined as an appropriate adhesive material to be applied at the upper and lower interlayers. In addition, interlayer shear bond strength, regardless of the type of adhesive materials, was decreased with increasing temperature. Finally, statistical analysis results indicated that all factors of structure type, type of adhesive material, and temperature statistically have a significant effect on interlayer bond strength. The findings of this study could provide a theoretical basis and experimental support for improving the interlayer design and construction in the concrete bridge deck pavement structure.

**Keywords:** bridge deck pavement; interlayer bond strength; adhesive material; emulsified asphalt; shear stress

Citation: Chen, W.; Hui, B.; Rahman, A. Interlayer Shear Characteristics of Bridge Deck Pavement through Experimental and Numerical Analysis. *Materials* 2022, 15, 7001. https://doi.org/10.3390/ma15197001

Academic Editors: Xueyan Liu, Linbing Wang, Zhanping You, Yuqing Zhang and Changhong Zhou

Received: 30 August 2022
Accepted: 5 October 2022
Published: 9 October 2022

**Publisher's Note:** MDPI stays neutral with regard to jurisdictional claims in published maps and institutional affiliations.

**Copyright:** © 2022 by the authors. Licensee MDPI, Basel, Switzerland. This article is an open access article distributed under the terms and conditions of the Creative Commons Attribution (CC BY) license (https://creativecommons.org/licenses/by/4.0/).

## 1. Introduction

Due to direct contact with traffic loading and environmental circumstances, the characteristics of the bridge deck pavement play an important role in the performance of the bridge structure. In recent years, many studies have been conducted to improve the performance and functionality of bridge deck pavement. The type and properties of materials used in the bridge deck pavement are one of the key factors in the overall performance of the bridge deck pavement. Epoxy asphalt mixture is one of the most widely used pavement materials in the construction of many large-span steel bridges because of its excellent anti-deformation ability, fatigue resistance, water stability, and corrosion resistance. Many studies were conducted to improve the performance of this type of asphalt mixture in the bridge deck pavement. A novel study proposed an approach of using polyurethane (PU) and epoxy resin (EP) to prepare an epoxy/polyurethane (EPU) modified asphalt binder and mixture. The results showed that the low-temperature cracking resistance of EPU-modified asphalt mixture was significantly improved compared to epoxy asphalt mixture, exhibiting a good application prospect in flexible bridge deck pavement engineering [1]. Several methods such as optimizing the skeleton structure or developing a cold-mixed ultra-thin antiskid surface layer (UTASS) were also successfully devised to improve the skid resistance of the epoxy asphalt-based concrete (EAC) to be used as a surface layer for the bridge deck pavement [2,3].

To improve the low-temperature and fatigue performances of the bridge deck pavement and decrease its cost, Zhang et al. [4] developed an unsaturated polyester resin (UPR) modified asphalt mixture. Their experimental results showed that the laboratory performance of UPR modified asphalt mixture outperformed SBS modified asphalt mixture and epoxy-modified asphalt mixture. In addition, the cost of UPR-modified asphalt was only 62% of that of epoxy-modified asphalt. Other studies demonstrated that the incorporation of UPR-emulsion could solve the problems associated with durability reduction and service-life shortening of bridge deck pavement concrete [5]. The application of sustainable pavement materials to improve the longevity of orthotropic bridge deck pavement structures has been promising. For instance, characteristics of a newly developed thermosetting polyurethane modified asphalt binder (TPUA) revealed its excellent high-temperature deformation resistance and elastic recovery properties [6]. The problem of early damage to concrete bridge deck pavement in cold regions has become one of the major problems for bridges. To address this issue, Cheng and Shi [7] used nanotechnology material by the addition of Nano-$SiO_2$ into the concrete to enhance its durability. The test results indicated that the incorporation of nano-SiO2 greatly improved the four durability indexes, strength, frost resistance, resistance to $Cl^-$ ion permeability, and abrasion resistance of the modified concrete. In another attempt, a polyurethane concrete (PUC) consisting of a one-phase tough polyurethane binder as the matrix and dolerite aggregates as filler was successfully developed to be used in bridge deck pavement. The findings showed that PUC composites have strong low-temperature toughness, high dynamic stability, and above all excellent durability [8].

The evaluation of the mechanical performance of pavement bridge deck has been a matter of concern among researchers due to complex service conditions, such as composite structure design, massive vehicle loadings, environmental conditions, large vibration, and large deflection deformation. Wang et al. [9] conducted a numerical analysis using the finite element method (FEM) to study both the static and dynamic stress in bridge deck pavement. Their results showed that the suggested thickness for the upper layer pavement is 3.5 cm to 4.5 cm and the lower layer is 5 cm to 7 cm. Moreover, the suggested modulus for the pavement upper layer is 1600 to1900 MPa and 900 to 1000 MPa for the lower pavement layer. In recent years, the application of ground penetration radar (GPR) as a non-destructive testing (NDT) method has become a viable technique for proper monitoring and early diagnosis of distresses such as moisture damage in bridge deck pavement [10].

The quality of the bonding between pavement layers directly affects the integrity and performance of pavement structure. Previous studies [11–16] applied the FEM to analyze the mechanical responses of the pavement structure under different interlayer contact and loading conditions. The structure of a bridge deck pavement, as a typical layered structure, from top to bottom is generally composed of a surface layer, bonding layer, lower layer, waterproof bonding layer, and cement concrete bridge deck. It was found that, owing to the differences in material properties and defects during the construction of bridge deck pavement, the problem of insufficient bonding between layers is common, which will reduce the shear capacity of layers [17–23]. The bridge deck pavement is different from the road pavement in terms of stress state, environmental, and service conditions. Therefore, it is more prone to various distresses, and the performance of the pavement directly affects the durability of the bridge, safety, and comfort of driving [24,25]. The current research on the bonding performance of bridge deck pavement is mostly focused on the upper interlayer or the lower interlayer exclusively [19,26–29]. The factors affecting interlayer bonding performance considered in previous studies are mainly divided into internal factors (including type and dosage of bonding material [24,26,30], interface roughness [31]), and external factors (including ambient temperature, construction conditions, traffic load) [17,32,33]. Indoor tests and finite element analysis are commonly used as effective methods to study interlayer bonding performance [34]. Shear strength and tensile strength are often used as evaluation indicators [23]. The above-mentioned studies provide valuable experience in improving the overall quality of the interlayer bonding between pavement layers. However, these studies have mainly focused on the performance of bonding or waterproof bonding layers, and interlayer bonding behavior in the presence of two adjacent layers is not investigated thoroughly. In view of this research gap, this study aims at analyzing the influence of the adhesive layer and waterproof bonding layer on mechanical responses of the bridge deck pavement under dynamic loading through numerical simulation. Moreover, in order to improve the shear resistance between the layers of the pavement structure, the shear strength of different adhesive materials was studied through laboratory experiments, so as to provide experimental support for the rational material selection of interlayer treatment.

## 2. Numerical Simulation

The finite element method (FEM) is a popular numerical method for solving partial differential equations arising in engineering and mathematical modeling in two or three space variables. The FEM partitions a large system into smaller, simpler parts that are called finite elements to solve a problem. This is achieved by a particular space discretization in the space dimensions, which is implemented by the construction of a mesh of the object [35]. In this study, the FEM was utilized to study the influence of the adhesive layer and waterproof bonding layer of the concrete bridge deck pavement on the mechanical responses of the upper and lower layers of the pavement bridge deck.

*2.1. FEM Model Development*

2.1.1. Model Structure

The bridge deck pavement structure not only withstands the combined effect of the environmental circumstances (such as temperature, humidity, etc.) and the traffic loads, but also hinders the infiltration of harmful substances such as moisture and oil, which play a key role in the entire bridge structure. Owing to the particularity of the bridge deck pavement structure under the combined effect of external and internal factors, its stress and deformation are more complex than that of highway pavement or airport pavement [9]. For this reason, double-layered modified asphalt pavement structure is often utilized in bridge deck pavement structures. In this study, the upper layer of the pavement was selected as an open gradation friction course (OGFC) with a nominal maximum aggregate size (NMAS) of 13 mm, OGFC-13, and the lower layer of the pavement was asphalt concrete (AC) mixture with the NMAS of 13 mm, AC-13. The type of asphalt binder used in the mixture was a high viscoelastic modified asphalt. The structural system of the bridge deck pavement is shown in Figure 1.

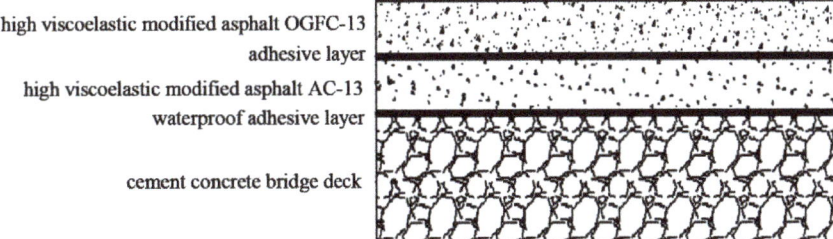

**Figure 1.** Bridge deck pavement structure.

The model has the dimensions of 6 m × 6 m on the horizontal plane, and the dimension of the vertical plane is equal to the thickness of the structural system. The distribution of the loading area is positioned in the middle of the surface of the model structure, as shown in Figure 2.

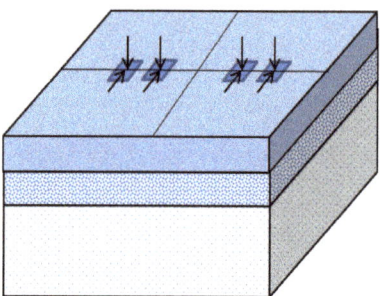

**Figure 2.** Schematic diagram and load arrangement of the bridge deck pavement.

Considering the symmetry of the load and structure, only 1/4 of the model was used for calculation and analysis. The boundary condition was fixed at the bottom of the bridge deck. In addition, no transverse horizontal displacement in the transverse direction of the bridge and no longitudinal horizontal displacement in the longitudinal direction of the bridge were assumed, as shown in Figure 3.

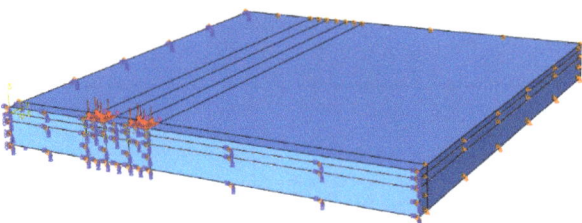

**Figure 3.** Boundary conditions of 1/4 three-dimensional FE model.

Element type and size are imperative for the accuracy of a finite element (FE) analysis. Twenty-node brick element with reduced integration (C3D8R) was adopted for meshing. The finest mesh was used around the loading area with rectangular surface contact, and the meshing size gradually widened as it recedes from the loading area to decrease the computation time (Figure 4). The analysis temperature was set to 25 °C, and other structural and material parameters are displayed in Table 1. It should be noted that all materials were characterized as elastic materials.

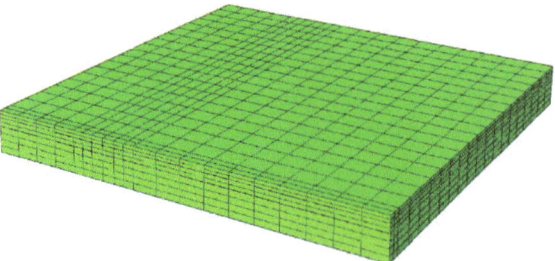

**Figure 4.** The meshing of 1/4 three-dimensional FE model.

Table 1. Properties of bridge deck pavement layers.

| Material | Thickness (cm) | Elastic Modulus (MPa) | Density (g·cm$^{-3}$) | Poisson Ratio |
|---|---|---|---|---|
| Upper layer | 4 | 2000 | 2.20 | 0.3 |
| Lower layer | 5 | 3000 | 2.20 | 0.3 |
| Concrete | 15 | 31,000 | 2.40 | 0.15 |

2.1.2. Loading Application

In this study, a moving 100 kN single axle load with dual tire assembly, which is the standard axle load for the structural design of asphalt pavement in China, was applied to the loading area. Moreover, the effects of horizontal and vertical loadings were considered concurrently. The horizontal load is obtained by multiplying the vertical load and the horizontal load coefficient. In this study, the coefficient of horizontal force was set as 0.5. In the mechanistic pavement design method, it is assumed that the approximate shape of the contact area for each tire consists of a rectangle and two semicircles with length L and width 0.6 L [36]. The length can be calculated using Equation (1):

$$L = \sqrt{\frac{A_c}{0.5227}} \quad (1)$$

where $A_c$ is the contact area calculated by dividing the load on each tire by the tire pressure. In the FE procedure, the contact area is simplified as an equivalent rectangular area with length 0.8712 L and width 0.6 L (Figure 5).

Figure 5. Schematic diagram of tire imprint.

In this study, the axle load is 100 kN and the tire pressure is 0.7 MPa. Therefore, it is calculated that the tire width is 15.68 cm, and the tire length is 22.78 cm. The schematic diagram of load distribution after a simplified calculation is shown in Figure 6.

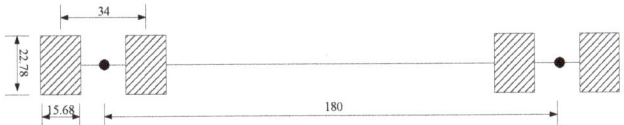

Figure 6. Schematic diagram of load arrangement (unit: cm).

Nonuniform vertical compression stresses have a single direction along the longitudinal contact length, and a half-sinusoidal pressure is assumed for its distribution, as shown in Equation (2).

$$P(t) = p \cdot \sin^2\left(\frac{\pi t}{T}\right) \quad (2)$$

The duration of load depends on the vehicle speed and the tire contact radius as follows:

$$T = \frac{12R}{v} \quad (3)$$

where $p$ is the maximum contact pressure which is 0.70 MPa; $T$ is the duration of load, s; $v$ is the vehicle speed, m·s$^{-1}$; and $R$ is the tire contact radius, m, which is 0.1065 m for standard axel load. The driving speed is set as 80 km/h, and the calculated load duration is 0.058 s.

### 2.1.3. Interfacial Conditions

For the bonding conditions of the adhesive layer (upper interface) and waterproof adhesive layer (lower interface) in bridge deck pavement structure, five different bonding conditions were considered as a function of the friction coefficient ($\mu$). The coefficient is used to characterize different bonding states, which are full slip ($\mu = 0$), full bonding ($\mu = \infty$), and partial bonding ($\mu = 0.25, 0.5,$ and $0.75$). On this basis, $\mu_1$ represents the bonding state of the upper interface and $\mu_2$ represents the bonding state of the lower interface. In this study, when the influence of the upper interlayer bonding conditions on the shear stress at the bottom of the pavement structure is concerned, it was assumed that the lower interlayer is in partial bonding condition (i.e., $\mu_2 = 0.75$) and vice versa.

## 3. Experimental Program

### 3.1. Materials

To improve the shear resistance between layers of bridge deck pavement and facilitate reasonable material selection, in this study, varying typical bonding materials were selected to determine their shear bond strength through the direct shear test. As shown in Figure 7, three different types of adhesive materials, resin emulsified asphalt, modified emulsified asphalt, and ordinary emulsified asphalt were selected as bonding materials between the layers of the bridge deck pavement structure. OGFC-10 mixture was used as the upper layer of the pavement and AC-13 mixture was employed as the lower layer of the pavement, overlaying the concrete bridge deck.

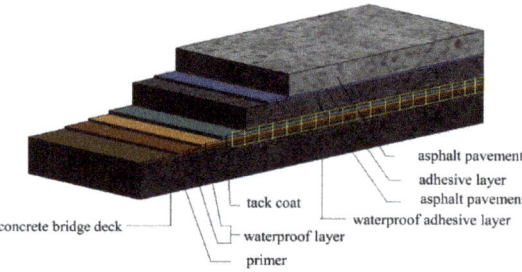

**Figure 7.** Schematic diagram of bridge deck pavement structure.

The aggregate gradation of the mixtures is presented in Table 2. In addition, basalt as coarse aggregate, machine-made sand as fine aggregate, alkaline mineral powder as filler, and a high viscoelastic modified asphalt as asphalt binder were utilized for preparing the asphalt mixtures. The bridge deck was made up of the C40 ordinary cement concrete, and the mix proportion was selected as cement: water: sand: gravel = 347.5:139:543:1396. The test temperature was set as room temperature at 25 °C.

**Table 2.** Design parameters of OGFC-10 and AC-13.

| Sieve Size (mm) | Passing Percentage (%) | | | | | | | | | Mineral Powder | Fiber (%) * | Asphalt Binder (%) * |
|---|---|---|---|---|---|---|---|---|---|---|---|---|
| | 13.2 | 9.5 | 4.75 | 2.36 | 1.18 | 0.6 | 0.3 | 0.15 | 0.075 | | | |
| OGFC-10 | 0 | 3 | 5 | 53 | 7 | 7 | 7 | 4 | 4 | 10 | 0.3 | 6.2 |
| AC-13 | 5 | 33 | 35 | 7 | 1 | 3 | 3 | 1 | 2 | 10 | 0.3 | 5.8 |

* Note: The amount of fiber and asphalt is the percentage of the total amount of mixture.

### 3.2. Specimen Preparation

For manufacturing of the double-layered asphalt specimens, first, the lower layer AC-13 mixture was prepared by 32 times single-sided compaction. The compacted specimen was then allowed to cool to room temperature. After cooling, an appropriate amount of bonding material was applied to the surface and then the coated specimen was set aside to cure. Then, the upper layer OGFC-10 loose mixture was formed by 32 times single-sided compaction.

Similarly, for the fabrication of double-layered composite specimens, first, the lower half concrete layer with a thickness of 32 mm was obtained as a core specimen and its surface was sandblasted. After cleaning the surface, an appropriate amount of bonding materials was applied to surface. Subsequently, the coated surface was set aside at room temperature for 2 h to allow the curing procedure completed. Following this, the AC-13 mixture was poured into the second half of the mold and prepared by 32 times single-sided compaction. To distinguish two interlayers from each other, the interlayer between the upper and lower asphalt layers is called the upper interlayer, and the interlayer between the lower asphalt layer and the concrete layer is called the lower interlayer. Images of two types of double-layered specimens are presented in Figure 8.

(a)        (b)

**Figure 8.** Double-layered specimen: (**a**) double-layered asphalt specimen (system I); (**b**) double-layered composite specimen (system II).

*3.3. Test Device*

A supplementary fixture was developed and attached to a Marshall testing machine to conduct the interlayer bonding evaluation of the double-layered specimens, as shown in Figure 9. The direct shear strength of the interface was reported as the test result.

**Figure 9.** The direct shear test set up.

## 4. Results and Discussion

*4.1. Simulation Results*

In this section, the effect of the bonding condition at the upper and lower interfaces on the pavement responses at the bottom of the upper and lower layers will be analyzed and discussed.

4.1.1. Shear Stress in the Upper Layer

(a) Effect of upper interlayer bonding condition

Figure 10 shows the change trends of the shear stress at the bottom of the upper layer against the upper interlayer bonding condition when the contact state of the lower interlayer changes from full slip to full bonding.

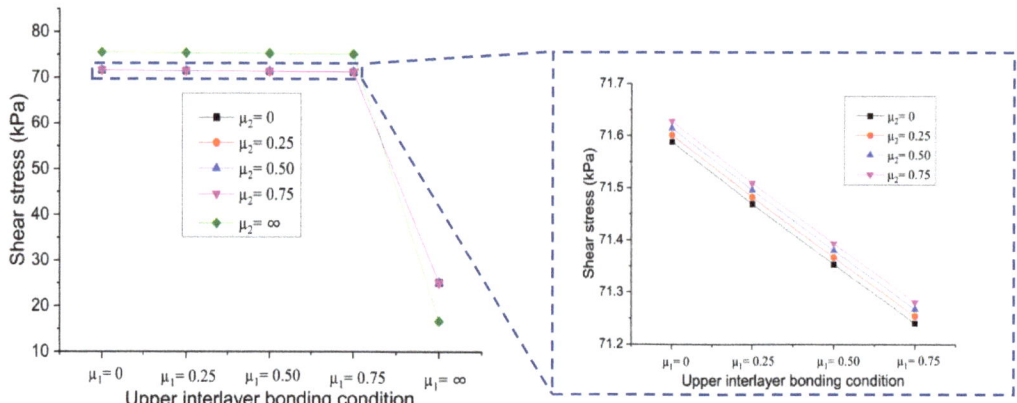

**Figure 10.** Influence of upper interlayer bonding condition on shear stress at the bottom of upper layer.

It can be seen that the shear stress at the bottom of the upper layer reduced with the increase of the friction coefficient at the upper interface, regardless of the bonding conditions at the lower interface. The results reveal that when the bonding condition of the lower interface is unchanged, the better bonding condition at upper interface causes the lesser shear damage to the interlayer generated because of loading. Therefore, there is a remote possibility of upper interface failure. In addition, when the bonding conditions at upper and lower interfaces are in a full bonding state, the resulting shear stress at the bottom of the upper pavement layer is minimum. When the lower interface is in the full slip and partial bonding states, the curves show a linear downward trend, and the rate of decline is almost the same. With the increase of the friction coefficient of the upper interface to the full bonding state, the resulting shear stress declined to the lowest point, which is about 35% of that of the full slip state. On the other hand, when the lower interface is in the full bonding state, the curve also shows a linear downward trend, but the downward rate is larger. In conclusion, different bonding conditions of the upper and lower interface produce different effects on the shear stress at the bottom of the upper layer.

(b) Effect of lower interlayer bonding condition

Figure 11 illustrates the changing trend of the shear stress at the bottom of the upper layer against the lower interlayer bonding condition when the contact state of the upper interlayer changes from full slip to full bonding.

It is evident that when the lower interlayer is in a full slip and partial bonding state, the resulting shear stress at the bottom of the upper layer gradually increased with increasing the friction coefficient of the lower interlayer. Moreover, when the lower interlayer is in full bonding condition, the shear stress reduced with the increase of the friction coefficient of the lower interlayer. On the other hand, when the upper interlayer is in the state of the full slip or partial bonding, the increase of the friction coefficient of the lower interlayer is not necessarily conducive to the shear resistance of the upper interlayer. Only when the friction coefficient of the upper interlayer increased to a certain extent did the increase of the friction coefficient of the lower interlayer lead to the resistance of the upper interlayer to shear stress.

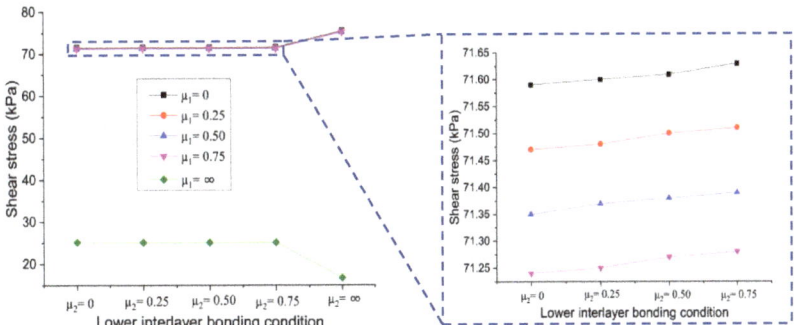

**Figure 11.** Influence of lower interlayer bonding condition on shear stress at the bottom of upper layer.

It is evident that when the lower interlayer is in a full slip and partial bonding state, the resulting shear stress at the bottom of the upper layer gradually increased with increasing the friction coefficient of the lower interlayer. Moreover, when the lower interlayer is in full bonding condition, the shear stress reduced with the increase of the friction coefficient of the lower interlayer. On the other hand, when the upper interlayer is in the state of the full slip or partial bonding, the increase of the friction coefficient of the lower interlayer is not necessarily conducive to the shear resistance of the upper interlayer. Only when the friction coefficient of the upper interlayer increased to a certain extent did the increase of the friction coefficient of the lower interlayer lead to the resistance of the upper interlayer to shear stress.

4.1.2. Shear Stress in the Lower Layer

(a) Effect of upper interlayer bonding condition

Figure 12 displays the changing trend of the shear stress at the bottom of the lower layer versus the upper interlayer bonding condition when the contact state of the lower interlayer is changes from full slip to full bonding.

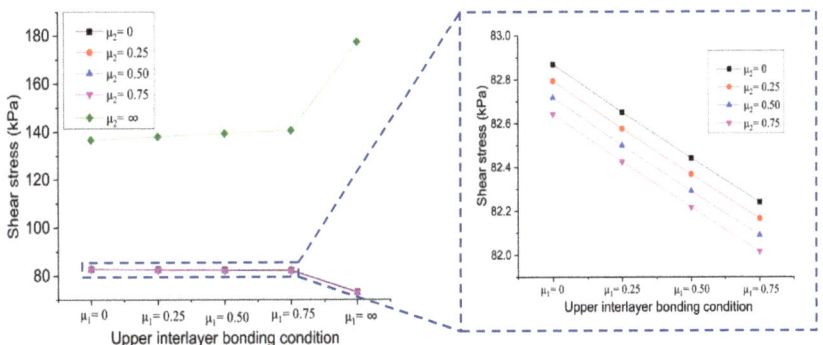

**Figure 12.** Influence of upper interlayer bonding condition on shear stress at the bottom of lower layer.

It can be seen that when the lower interlayer is in the full slip and partial bonding state, the shear stress at the bottom of the lower layer decreased linearly with the increase of the friction coefficient of the upper interlayer, and the decline rate of different curves was almost the same. When the contact state of the upper interlayer reached the full bonding state, the resulting shear stress reached the minimum, which is about 88% of that of the full slip state. When the lower interlayer is in full bonding condition, the shear stress gradually increased with the increase of the friction coefficient of the upper interlayer and reached the maximum value when the upper interlayer is in a full bonding state. In this case, the shear stress is about 1.3 times that of the full slip state. It can be inferred that varying bonding conditions at the upper interlayer will have different effects on the resulting shear stress at the bottom of the lower layer. In all cases, increasing the friction coefficient of the upper interlayer led to a decrease in shear stress at the bottom of the lower layer. However, when the bonding condition at the lower interlayer is in a full bonding state, increasing the friction coefficient at the upper interface will increase the shear stress at the bottom of the layer, which is not beneficial to the interlayer performance.

(b) Effect of lower interlayer bonding condition

Figure 13 shows the changing trend of the shear stress at the bottom of the lower layer versus the lower interlayer bonding condition when the contact state of the upper interlayer is changes from full slip to full bonding.

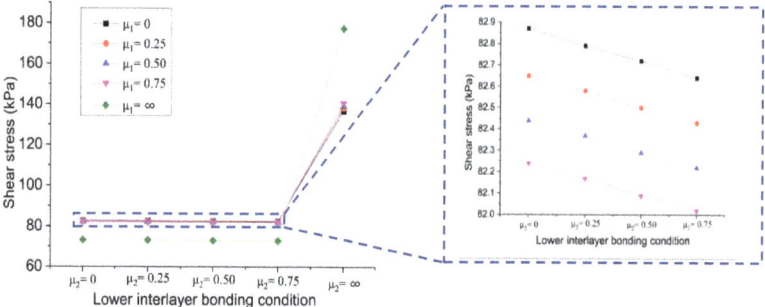

**Figure 13.** Influence of lower interlayer bonding condition on shear stress at the bottom of lower layer.

It is evident that the shear stress at the bottom of the lower layer decreased first and then increased with the increase of the friction coefficient of the lower interlayer, regardless of the contact state of the upper interlayer. Especially, when the bonding condition of the lower interlayer is full bonding, the resulting shear stress at the bottom of the lower layer was significantly increased, which is unfavorable to the overall performance of the bridge deck pavement structure.

To sum up, it can be inferred that no matter what contact state the lower interlayer is in, the shear stress at the bottom of the upper pavement layer gradually decreased with the increase of the friction coefficient of the upper interlayer, and the lower interlayer in different contact states had an impact on the change rate of the curve, but the impact was very small. In addition, only when the friction coefficient of the upper interlayer was large was the increase of the friction coefficient of the lower interlayer conducive to reducing the shear stress at the bottom of the upper pavement layer. Finally, different bonding conditions of the upper interlayer will have significantly different effects on the shear stress and at the bottom of the lower pavement layer. When the lower interlayer bonding condition is in full slip or partially bonding condition, then increasing the friction coefficient of the upper interlayer is conducive to reducing the shear stress at the bottom of the lower pavement layer. It was reported that better interface bonding condition could enhance the resistance to bottom-up fatigue cracking significantly for thin (40 mm-thick) deck pavements [32].

### 4.2. Experimental Results

Table 3 displays the interlayer bond strength results for each system and bonding material type under three temperature conditions. For each test condition, three replicates were prepared, and the average value was reported as test results. The effects of adhesive material and structure type on interlayer bonding performance were discussed in the following section.

Table 3. Interface bond strength, (MPa).

| System Type | Adhesive Material | Temperature (°C) | | |
|---|---|---|---|---|
| | | 0 | 25 | 70 |
| Type I | Resin emulsified asphalt | 1.52 | 0.73 | 0.03 |
| | Modified emulsified asphalt | 1.12 | 0.94 | 0.05 |
| | Ordinary emulsified asphalt | 1.38 | 0.75 | 0.03 |
| Type II | Resin emulsified asphalt | 1.54 | 0.56 | 0.03 |
| | Modified emulsified asphalt | 0.8 | 0.32 | 0.03 |
| | Ordinary emulsified asphalt | 1.25 | 0.43 | 0.02 |

4.2.1. Effect of Bonding Material

Figure 14 illustrates the variation curves of interlayer shear strength against temperature for three different adhesive materials in structure I and Structure II respectively.

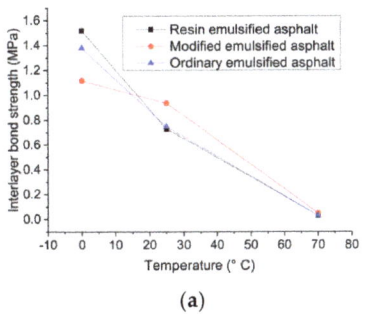
(a)

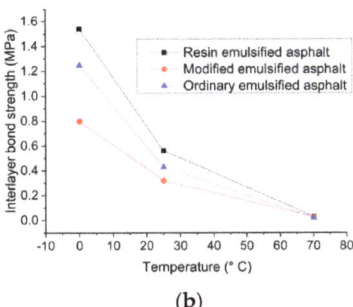
(b)

Figure 14. Interlayer shear strength of three adhesive materials in two structures: (a) Structure I; (b) Structure II.

It can be seen that the interlayer shear bond strength formed by the three adhesive layer materials decreased significantly with the increase of temperature in both bridge deck pavement structures, but there are significant differences in the changing trend of the interlayer bond strength with temperature. These results are in agreement with the findings of previous studies [37]. This indicates that the temperature sensitivity of the interlayer bonding materials exists in all kinds of bridge deck structures, and there are significant differences in the sensitivity of different interlayer treatments. Currently, most of the distresses caused by the interlayer problems are related to the excessive temperature sensitivity of the interlayer bonding materials. The excessive sensitivity leads to the sharp decrease of the direct shear strength of the interlayer, which cannot provide sufficient resistance to shear stress, and finally leads to the occurrence of corresponding distresses. For example, the direct shear strength of the interlayer formed by resin emulsified asphalt and ordinary emulsified asphalt materials in Figure 14a is 1.52 MPa and 1.38 MPa respectively when the temperature is 0 °C. It is evident that there are limited shear distresses at the interface at this stage, indicating that the interlayer has sufficient resistance to shear damage. However, when the temperature increased to 25 °C, the corresponding interlayer bonding strength dropped to 0.73 MPa and 0.75 MPa, respectively. In this condition, the interlayer bonding cannot fully resist resulting in shear damage. When the temperature continues to rise to 70 °C, the interlayer shear strength decreased to 0.03 MPa. At this stage, the bridge deck pavement structure experiences a high possibility of interlayer shear failure. When the severity of shear bonding failure is large, the ability of the interlayer to resist shear damage is seriously insufficient at this stage. To sum up, when an adhesive material is selected for interlayer treatment, interlayer bonding performance and the influence of temperature dependency of interlayer treatment should be considered concurrently.

It can also be seen from Figure 14 that, independent of structure type, when the temperature is normal or low, the interlayer shear strength formed by different adhesive materials is usually different at the same temperature. However, when the temperature increased to high elevation, all the interlayer shear strengths gradually tend to be consistent and reached close to 0 MPa. The reason is that all adhesive materials are temperature dependent. When the temperature is low, the adhesive material itself is stiff and has certain rigidity and strength. In this condition, the interlayer bond strength between the upper and lower layers of the bridge deck pavement has a certain resistance to damage. When the temperature gradually rises to the high temperature, the adhesive material is gradually softened into a plastic state or even a flowing state. In this case, the adhesive material does not play a bonding role, and the friction between the upper and lower layers of the interlayer is greatly weakened. Therefore, the interlayer bonding almost loses the ability to resist shear damage at high temperatures, which also explains why the bridge deck pavement structure is prone to delamination, rutting, and other distresses in high-temperature areas in summer. When selecting a reasonable adhesive material, not only the adhesive ability of the material but also the stability of the interfacial bonding performance is equally important. Zhou et al. [38] also demonstrated that the resin-type waterproof adhesive layer materials have good impermeability, bonding performance, temperature stability, and construction damage resistance in bridge deck pavement.

4.2.2. Effect of Structure Type

Figure 15 shows the changing trend of the interlayer shear strength formed by three adhesive materials against the temperature in structures I and II.

It can be seen that the interlayer shear strength of the two bridge deck pavement structures, regardless of the type of adhesive material, decreased significantly with the increase in temperature, but the changing trend of the same material in different structures was different. For example, in Figure 15b, although the two curves exhibited an obvious downward trend with the increase in temperature, structure I was less sensitive at low temperature, and the intensity of the decline rate was moderate. On the other hand, the sensitivity of interlayer bond strength is greater at high temperatures, and the decline rate is sharp. However, the interlayer bond strength of structure II is more prone to low temperature, and the bonding strength decreases significantly. When the temperature is higher, the temperature sensitivity is smaller, and the interlayer bonding strength decreases moderately. The results demonstrate that the degree of temperature dependency of the interlayer bond strength formed by the same adhesive material in different structures is not comparable, leading to the obvious variation trend of the interlayer bond shear properties. Consequently, when a bonding material exhibits a good bond performance in a certain pavement structure, it is not necessarily suitable for other pavement structures. In conclusion, when the bonding performance of certain adhesive material in different pavement structures is compared, the evaluation of performance under various working conditions should be considered rather than comparing the performance, for example, at a certain temperature.

It can also be found from Figure 1 that no matter what kind of adhesive material is utilized, the direct shear performance of the interlayer in two structures is not identical at the same temperature, and the degree of difference is not consistent at different temperatures. Moreover, the results show that the same adhesive material will exhibit different bonding abilities in different structures, resulting in significant differences in the interlayer shear properties. Therefore, for the same adhesive material, the type of pavement structure is also one of the important factors affecting the shear resistance of the interlayer bonding interface in bridge deck pavement. For instance, in Figure 15b, when the temperature is 0 °C, the interlayer shear strength of structure I is 1.12 Mpa while that of structure II is 0.8 Mpa. In this case, the interlayer shear strength of structure II is only 71% of structure I at 0 °C. When the temperature is 25 °C, the shear strength of structure I is 0.94 Mpa and that of structure II is 0.32 Mpa, which is only 34% of structure I. This indicates that the modified emulsified asphalt is suitable for structure I but not for structure II. Therefore, a good interlayer bonding performance is not the superposition of excellent pavement structure and good bonding material, but the optimal combination of a pavement structure and a bonding material. In practical engineering, it is necessary to select the appropriate bonding material according to the specific pavement structure, not necessarily the one with the strongest bonding strength.

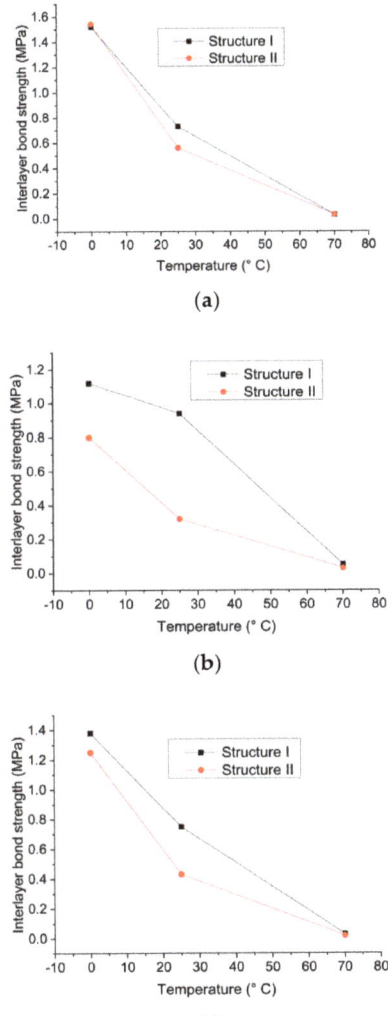

Figure 15. Interlayer shear strength of the same adhesive material in each structure: (a) Resin emulsified asphalt; (b) Modified emulsified asphalt; (c) Ordinary emulsified asphalt.

In conclusion, there are differences in the interlayer shear performance whether different interlayer treatments are applied under the same pavement structure, or the same interlayer treatment is applied under different pavement structures. Hence, it is imperative to ensure that the interlayer bonding has sufficient resistance against shear failure.

### 4.2.3. Analysis of Variance

Based on the above results it can be found that structure type, type of adhesive material, and temperature affect interlayer bond strength. However, the significance level of each factor is unknown. For this reason, the statistical analysis of variance (ANOVA) was used to determine the significance of each factor for interlayer bond strength. The analysis was conducted according to data obtained from the test program in this study. ANOVA sets up a collection of statistical procedures to quantitatively specify whether certain parameters and their combinations affect particular responses, which is interlayer bond strength in this study.

The $p$-value is the minimum level of significance at which the factor is considered significant in influencing the response. In this study, the $p$-value of 0.05, i.e., a confidence level of 95% was considered. Table 4 presents the ANOVA results for interlayer bond strength.

**Table 4.** ANOVA analysis for interlayer bond strength.

| Factor | DF [a] | SS [b] | MS [c] | F-Value [d] | p-Value [e] |
|---|---|---|---|---|---|
| System type | 1 | 0.399 | 0.399 | 93.042 | ≤0.001 |
| Adhesive material | 2 | 0.353 | 0.176 | 41.123 | ≤0.001 |
| Temperature | 2 | 13.860 | 6.930 | 1616.517 | ≤0.001 |
| Error | 36 | 0.154 | 0.004 | | |
| Total | 53 | 15.932 | | | |

[a] Degrees of freedom. [b] Sum of squares. [c] Mean square, which is the SS divided by DF. [d] Ratio of mean squares. It is used to determine the $p$-value. [e] Factor is significant ($p$-value < 0.05).

As shown in Table 4, $p$-values of all factors are all less than the significant level of 0.05. In other words, all studied factors have a significant effect on the interlayer bond strength. The F-value of each factor signifies that the significant effect of each factor on interlayer bond strength in the order of importance ranks as temperature, type of system, and type of adhesive material type.

## 5. Conclusions

The following conclusions can be drawn from the numerical and experimental results of this study:

(1) Regardless of the bonding condition of the lower interlayer, the shear stress at the bottom of the upper layer gradually decreased with the increase of the friction coefficient of the upper layer, and the bonding condition of the lower interlayer had an insignificant effect on the produced shear stress. Only when the friction coefficient of the upper interlayer was large was the increase of the friction coefficient of the lower interlayer conducive to reducing the shear stress at the bottom of the upper layer.

(2) Different bonding conditions of the upper interlayer produced different effects on the shear stress at the bottom of the lower layer. Only when the lower interlayer was in a full slip or partial bonding state was increasing the friction coefficient of the upper interlayer conducive to reducing the shear stress at the bottom of the lower layer.

(3) Regardless of the bonding condition of the upper interlayer, with the increase of the friction coefficient of the lower interlayer, the shear stress at the bottom of the lower layer reduced first and then increased.

(4) When a bonding material is selected for interlayer treatment, interlayer bonding performance and the influence of temperature dependency of interlayer treatment should be considered concurrently. The bridge deck pavement structure is prone to delamination, rutting, and other distresses in high-temperature areas in summer. Therefore, when a bonding material is selected for interlayer treatment, not only the adhesive ability of the material but also the stability of the interfacial bonding performance is equally important.

(5) When the bonding performance of certain adhesive material in different pavement structures is compared, the evaluation of performance under various working conditions should be considered rather than comparing the performance for example at a certain temperature. In practical engineering, it is imperative to select the appropriate bonding material according to the specific pavement structure, not necessarily the one with the strongest bonding strength.

(6) Results of statistical analysis revealed that all factors, including structure type, type of adhesive material, and temperature statistically have a significant effect on interlayer bonding strength.

**Author Contributions:** Conceptualization: W.C. and A.R.; Methodology: W.C. and B.H.; Software: W.C.; Laboratory experiment: W.C. and A.R.; Formal analysis: W.C. and A.R.; investigation: W.C. and B.H.; Writing—original draft preparation: W.C.; writing—review and editing: B.H. and A.R. All authors have read and agreed to the published version of the manuscript.

**Funding:** The research was funded by Fundamental Research Funds for the Central Universities, Southwest Jiaotong University, (No. 2682022CX002) and CHD (No. 300102212210); Sichuan Science and Technology Program (No. 2021JDTD0023); Sichuan Science and Technology Program (No. 2021YJ0065); and Inner Mongolia Transportation Research Project (No. NJ-2021-17).

**Institutional Review Board Statement:** Not applicable.

**Informed Consent Statement:** Not applicable.

**Data Availability Statement:** Not applicable.

**Acknowledgments:** The authors appreciate the Key Laboratory for Highway Engineering of Sichuan Province for providing the equipment and premises.

**Conflicts of Interest:** The authors declare no conflict of interest.

# References

1. Zhang, Z.; Sun, J.; Huang, Z.; Wang, F.; Jia, M.; Lv, W.; Ye, J. A laboratory study of epoxy/polyurethane modified asphalt binders and mixtures suitable for flexible bridge deck pavement. *Constr. Build. Mater.* **2020**, *274*, 122084. [CrossRef]
2. Qian, Z.-D.; Liu, Y.; Liu, C.-B.; Zheng, D. Design and skid resistance evaluation of skeleton-dense epoxy asphalt mixture for steel bridge deck pavement. *Constr. Build. Mater.* **2016**, *114*, 851–863. [CrossRef]
3. Liu, Y.; Qian, Z.; Shi, X.; Zhang, Y.; Ren, H. Developing cold-mixed epoxy resin-based ultra-thin antiskid surface layer for steel bridge deck pavement. *Constr. Build. Mater.* **2021**, *291*, 123366. [CrossRef]
4. Zhang, H.; Zhang, G.; Han, F.; Zhang, Z.; Lv, W. A lab study to develop a bridge deck pavement using bisphenol A unsaturated polyester resin modified asphalt mixture. *Constr. Build. Mater.* **2018**, *159*, 83–98. [CrossRef]
5. Zhang, Z.; Zhang, H.; Liu, T.; Lv, W. Study on the micro-mechanism and structure of unsaturated polyester resin modified concrete for bridge deck pavement. *Constr. Build. Mater.* **2021**, *289*, 123174. [CrossRef]
6. Yang, F.; Cong, L.; Li, Z.; Yuan, J.; Guo, G.; Tan, L. Study on preparation and performance of a thermosetting polyurethane modified asphalt binder for bridge deck pavements. *Constr. Build. Mater.* **2022**, *326*, 126784. [CrossRef]
7. Cheng, Y.; Shi, Z. Experimental Study on Nano-SiO2 Improving Concrete Durability of Bridge Deck Pavement in Cold Regions. *Adv. Civ. Eng.* **2019**, *2019*, 5284913. [CrossRef] [PubMed]
8. Jiang, Z.; Tang, C.; Yang, J.; You, Y.; Lv, Z. A lab study to develop polyurethane concrete for bridge deck pavement. *Int. J. Pavement Eng.* **2022**, *23*, 1404–1412. [CrossRef]
9. Wang, L.; Hou, Y.; Zhang, L.; Liu, G. A combined static-and-dynamics mechanics analysis on the bridge deck pavement. *J. Clean. Prod.* **2017**, *166*, 209–220. [CrossRef]
10. Zhang, J.; Zhang, C.; Lu, Y.; Zheng, T.; Dong, Z.; Tian, Y.; Jia, Y. In-situ recognition of moisture damage in bridge deck asphalt pavement with time-frequency features of GPR signal. *Constr. Build. Mater.* **2020**, *244*, 118295. [CrossRef]
11. Chun, S.; Kim, K.; Greene, J.; Choubane, B. Evaluation of interlayer bonding condition on structural response characteristics of asphalt pavement using finite element analysis and full-scale field tests. *Constr. Build. Mater.* **2015**, *96*, 307–318. [CrossRef]
12. Wu, S.; Chen, H.; Zhang, J.; Zhang, Z. Effects of interlayer bonding conditions between semi-rigid base layer and asphalt layer on mechanical responses of asphalt pavement structure. *Int. J. Pavement Res. Technol.* **2017**, *10*, 274–281. [CrossRef]
13. Wang, X.; Ma, X. Responses of Semi-Rigid Base Asphalt Pavement with Interlayer Contact Bonding Model. *Adv. Civ. Eng.* **2020**, *2020*, 8841139. [CrossRef]
14. Li, S.; Tang, L.; Yao, K. Comparison of Two Typical Professional Programs for Mechanical Analysis of Interlayer Bonding of Asphalt Pavement Structure. *Adv. Mater. Sci. Eng.* **2020**, *2020*, 5850627. [CrossRef] [PubMed]

15. Wang, X.; Feng, J.; Wang, H.; Hong, S.; Zheng, S. Stress Regression Analysis of Asphalt Concrete Deck Pavement Based on Orthogonal Experimental Design and Interlayer Contact. *IOP Conf. Ser. Mater. Sci. Eng.* **2018**, *322*, 042026. [CrossRef]
16. He, C.; Hu, J.; Zhu, Y.; Lu, H. Mechanical Analysis for Asphalt Paving on Steel Bridge Using Different Interlayer Contact Condition. In *ICCTP 2011: Towards Sustainable Transportation System*; ASCE: Reston, VA, USA, 2011.
17. Liu, Y.; Qian, Z.; Yin, Y.; Ren, H. Investigation on Interlayer Behaviors of a Double-Layered Heterogeneous Asphalt Pavement Structure for Steel Bridge Deck. *J. Mater. Civ. Eng.* **2022**, *34*, 04022062. [CrossRef]
18. Jin, W.; Zhao, Y.; Wang, W.; He, F. Performance Evaluation and Optimization of Waterproof Adhesive Layer for Concrete Bridge Deck in Seasonal Frozen Region Using AHP. *Adv. Mater. Sci. Eng.* **2021**, *2021*, 5555535. [CrossRef]
19. Fu, J.; Shen, A.; Yuan, Z. Properties of Different Waterproof Bonding Layer Systems for Cement Concrete Bridge Deck Pavement. *Coatings* **2022**, *12*, 308. [CrossRef]
20. Zhang, M.; Hao, P.; Men, G.; Liu, N.; Yuan, G. Research on the compatibility of waterproof layer materials and asphalt mixture for steel bridge deck. *Constr. Build. Mater.* **2020**, *269*, 121346. [CrossRef]
21. Xu, Y.; Lv, X.; Ma, C.; Liang, F.; Qi, J.; Chou, Z.; Xu, S. Shear Fatigue Performance of Epoxy Resin Waterproof Adhesive Layer on Steel Bridge Deck Pavement. *Front. Mater.* **2021**, *7*, 618073. [CrossRef]
22. Cao, M.M.; Huang, W.Q.; Lu, Y.; Tan, Q.Q. Test and Evaluation Method of lnterlaminar Shear Performance of Composite Pavement. *J. Highway Transp. Res. Dev.* **2018**, *12*, 33–43. (In English) [CrossRef]
23. Liu, Y.; Wu, J.; Chen, J. Mechanical properties of a waterproofing adhesive layer used on concrete bridges under heavy traffic and temperature loading. *Int. J. Adhes. Adhes.* **2014**, *48*, 102–109. [CrossRef]
24. Ai, C.; Huang, H.; Rahman, A.; An, S. Establishment of a new approach to optimized selection of steel bridge deck waterproof bonding materials composite system. *Constr. Build. Mater.* **2020**, *264*, 120269. [CrossRef]
25. Liu, H.; Wu, S.; Han, J. Research on shear strength and fatigue characteristics of bridge deck pavement bonding layers. In Proceedings of the 2010 International Conference on Mechanic Automation and Control Engineering, Wuhan, China, 26–28 June 2010; pp. 970–972.
26. Yin, Y. *Research on Interlayer Adhesion Performance of "EA+SMA" Steel Deck Pavement Structure*; Southeastern University: Lakeland, FL, USA, 2020.
27. He, Q.; Zhang, H.; Li, J.; Duan, H. Performance evaluation of polyurethane/epoxy resin modified asphalt as adhesive layer material for steel-UHPC composite bridge deck pavements. *Constr. Build. Mater.* **2021**, *291*, 123364. [CrossRef]
28. Guo, L.; Zeng, G. Study on mechanical properties of typical steel bridge deck pavement waterproof bonding system. *J. Phys. Conf. Ser.* **2021**, *1802*, 22018. [CrossRef]
29. Qian, G.; Li, S.; Yu, H.; Gong, X. Interlaminar Bonding Properties on Cement Concrete Deck and Phosphorous Slag Asphalt Pavement. *Materials* **2019**, *12*, 1427. [CrossRef]
30. Qiu, Y.; An, S.; Rahman, A.; Ai, C. Evaluation and optimization of bridge deck waterproof bonding system using multi-objective grey target decision method. *Road Mater. Pavement Des.* **2019**, *21*, 1844–1858. [CrossRef]
31. Chen, Z.; Xu, W.; Zhao, J.; An, L.; Wang, F.; Du, Z.; Chen, Q. Experimental Study of the Factors Influencing the Performance of the Bonding Interface between Epoxy Asphalt Concrete Pavement and a Steel Bridge Deck. *Buildings* **2022**, *12*, 477. [CrossRef]
32. Kim, T.W.; Baek, J.; Lee, H.J.; Lee, S.Y. Effect of pavement design parameters on the behaviors of orthotropic steel bridge deck pavements under traffic loading. *Int. J. Pavement Eng.* **2014**, *15*, 471–482. [CrossRef]
33. Wang, X.; Zhang, C.; Sun, R. Response analysis of orthotropic steel deck pavement based on interlayer contact bonding condition. *Sci. Rep.* **2021**, *11*, 23692. [CrossRef] [PubMed]
34. Xu, Q.; Zhou, Q.; Medina, C.; Chang, G.K.; Rozycki, D.K. Experimental and numerical analysis of a waterproofing adhesive layer used on concrete-bridge decks. *Int. J. Adhes. Adhes.* **2009**, *29*, 525–534. [CrossRef]
35. Logan, D.L.; Chaudhry, K.K.; Singh, P. *A First Course in the Finite Element Method*, 4th ed.; Cengage Learning: Stamford, CT, USA, 2011; p.798.
36. Yang, H.H. *Pavement Analysis and Design*, 2nd ed.; Pearson Education International: Upper Saddle River, NJ, USA, 2012.
37. Liu, Y.; Qian, Z.; Gong, M.; Huang, Q.; Ren, H. Interlayer residual stress analysis of steel bridge deck pavement during gussasphalt pavement paving. *Constr. Build. Mater.* **2022**, *324*, 126624. [CrossRef]
38. Zhou, L.; Zhang, D.; Li, X.; Gao, Z.; Chen, Q.; Wang, C. Overview: Application of Resin Waterproof Adhesive Materials in Bridge Deck Pavement in China. *Adv. Civ. Eng.* **2022**, *2022*, 2320374. [CrossRef]

Article

# Experimental Investigation of the Performance of a Hybrid Self-Healing System in Porous Asphalt under Fatigue Loadings

Shi Xu [1,2,*], Xueyan Liu [2], Amir Tabaković [2,3,4] and Erik Schlangen [2]

1. Hubei Key Laboratory of Roadway Bridge and Structure Engineering, Wuhan University of Technology, Wuhan 430070, China
2. Civil Engineering and Geosciences, Delft University of Technology, 2628CN Delft, The Netherlands; x.liu@tudelft.nl (X.L.); amir.tabakovic@TUDublin.ie (A.T.); Erik.Schlangen@tudelft.nl (E.S.)
3. Centre for Research in Engineering Surface Technology (CREST), Technological University Dublin, D08CKP1 Dublin, Ireland
4. School of Civil Engineering, University College Dublin, D04K3H4 Dublin, Ireland
* Correspondence: S.Xu-1@tudelft.nl

**Abstract:** Self-healing asphalt, which is designed to achieve autonomic damage repair in asphalt pavement, offers a great life-extension prospect and therefore not only reduces pavement maintenance costs but also saves energy and reduces $CO_2$ emissions. The combined asphalt self-healing system, incorporating both encapsulated rejuvenator and induction heating, can heal cracks with melted binder and aged binder rejuvenation, and the synergistic effect of the two technologies shows significant advantages in healing efficiency over the single self-healing method. This study explores the fatigue life extension prospect of the combined healing system in porous asphalt. To this aim, porous asphalt (PA) test specimens with various healing systems were prepared, including: (i) the capsule healing system, (ii) the induction healing system, (iii) the combined healing system and (iv) a reference system (without extrinsic healing). The fatigue properties of the PA samples were characterized by an indirect tensile fatigue test and a four-point bending fatigue test. Additionally, a 24-h rest period was designed to activate the built-in self-healing system(s) in the PA. Finally, a damaging and healing programme was employed to evaluate the fatigue damage healing efficiency of these systems. The results indicate that all these self-healing systems can extend the fatigue life of porous asphalt, while in the combined healing system, the gradual healing effect of the released rejuvenator from the capsules may contribute to a better induction healing effect in the damaging and healing cycles.

**Keywords:** self-healing asphalt; fatigue life; induction heating; calcium alginate capsules; combined healing system

Citation: Xu, S.; Liu, X.; Tabaković, A.; Schlangen, E. Experimental Investigation of the Performance of a Hybrid Self-Healing System in Porous Asphalt under Fatigue Loadings. *Materials* **2021**, *14*, 3415. https://doi.org/10.3390/ma14123415

Academic Editor: Angelo Marcello Tarantino

Received: 16 May 2021
Accepted: 16 June 2021
Published: 20 June 2021

**Publisher's Note:** MDPI stays neutral with regard to jurisdictional claims in published maps and institutional affiliations.

**Copyright:** © 2021 by the authors. Licensee MDPI, Basel, Switzerland. This article is an open access article distributed under the terms and conditions of the Creative Commons Attribution (CC BY) license (https://creativecommons.org/licenses/by/4.0/).

## 1. Introduction

In the Netherlands, the concept of zeer open asfaltbeton (ZOAB), which is known as porous asphalt (PA) in the rest of the world, was first applied in 1972 [1]. With a void content above 20%, PA shows advantages in noise reduction, comfortable driving and reduction of splash and spray during rainfall, which has resulted in it being implemented quickly in asphalt pavement design in both the Netherlands and worldwide [2–4].

Microcracking is one of the most common early distresses in PA which deteriorates the stone-to-stone contact and can develop into macroscopic damages (e.g., ravelling), and the healing of microcracks is considered as the key factor in delaying or preventing ravelling in PA, therefore extending the lifespan of PA [5]. Consistent with the basic principle of general self-healing materials, the crack healing in asphalt pavement relies on the subsequent generation of a 'mobile phase' which gradually results in crack closure during the rest period. Figure 1 illustrates an asphalt crack healing event: when a crack occurs (Figure 1a), the subsequent generation of a 'mobile phase' (Figure 1b), triggered either by the intrinsic healing capacity of bitumen or by external stimuli, can heal the crack with the flow of bitumen or mastic (Figure 1c). After crack closure, the previously mobile material is immobilised again, resulting in the regain of mechanical bonding (Figure 1d) [6].

**Figure 1.** The basic principle of crack healing in asphalt pavement: (**a**) a crack generated in asphalt mastic; (**b**) the 'mobile phase' induced at the crack face; (**c**) closure of the crack by the 'mobile phase' and (**d**) immobilisation after healing (Hager et al., 2010).

At the early stage of a PA pavement, microcrack healing can be achieved with the intrinsic healing capacity of the bitumen. However, this bitumen-intrinsic healing capacity diminishes with bitumen ageing, which not only reduces the temperature susceptibility of bitumen but also leads to a lower bitumen ductility [7,8]. At that moment, extrinsic healing methods can be employed to induce the 'mobile phase' to achieve microcrack healing. The thermally induced healing method heats the asphalt mixture with microwave or induction energy, therefore allowing the bitumen to flow to heal the crack [9–15]. The embedded rejuvenator encapsulation method offers in situ rejuvenation at the cracking site; as such, the rejuvenated bitumen will regain the intrinsic healing capacity and gradually heal the crack driven by capillary flow [16–22]. Both methods have been demonstrated to not only improve the crack recovery in asphalt pavement but also increase its fatigue life [23].

The authors of an earlier study [24,25] have investigated the calcium alginate capsules healing system, which is illustrated in Figure 2. The calcium alginate capsules encapsulating the rejuvenator were prepared and optimized, and their crack healing effect was demonstrated in bituminous materials [24,26]. It was also found that the healing efficiency of the induction heating technique could be largely reduced by asphalt ageing and gradient healing [27–29]. Additionally, the induction healing system was introduced to work together with the calcium alginate capsules healing system, and this combined asphalt healing system could not only combine the advantages from both systems but also create synergistic effects, hence offering a better healing prospect [27]. Results from previous findings successfully evaluate the efficiency of each self-healing system in the healing of one major propagating crack; however, the performance of these self-healing systems under cyclic fatigue loadings, especially for the calcium alginate capsules healing system and the combined healing system, is still unknown.

The self-healing performance of an asphalt mixture under fatigue loadings can be investigated using the four-point bending fatigue test (4PB) and the indirect tensile fatigue test (ITF):

- Based on 4PB, Liu et al. [30] evaluated induction healing effect on the fatigue damage in PA. Liu et al. discovered that asphalt beams incorporated with steel wool fibres not only exhibited a higher fatigue resistance in the first 4PB cycle but also gained higher stiffness and showed significantly longer fatigue life when induction heating was introduced during the rest period. Liu et al. indicated that the induction healing rate is highly applied microstrain dependent and the optimum induction heating temperature is 85 °C. Based on these findings, Liu et al. believed that the durability of PA pavement can be improved by induction heating.
- Tabaković et al. [31] employed a 4PB and healing programme to investigate the fatigue damage recovery prospect of alginate fibres in a full asphalt mix. Tabaković et al. found that, after a 20-h healing period at 20 °C, the asphalt beams incorporated with alginate fibres showed a higher stiffness recovery than the controlled beams, indicating that the alginate fibres are a promising approach to improve the self-healing capacity

of asphalt pavement. Tabaković et al. further indicated that 4PB is the most suitable test for evaluating the performance of self-healing asphalt.

- The 4PB was also used by Sun et al. [32] to study the fatigue behaviour of self-healing asphalt with melamine urea formaldehyde (MUF) microcapsules. Both the modulus recovery ratio and fatigue life extension ratio were used to evaluate the healing efficiency of MUF microcapsules, and the results indicate that the addition of MUF microcapsules can improve both ratios of the asphalt mixture, thus achieving a better healing performance.
- Menozzi et al. [33] used an ITF and healing programme to examine the induction healing efficiency on an asphalt mixture. The damage and healing in the asphalt mixture were characterized with computed tomography tests. Menozzi et al. reported that the lifetime of Marshall test samples subjected to fatigue damage can be extended with the induction healing.

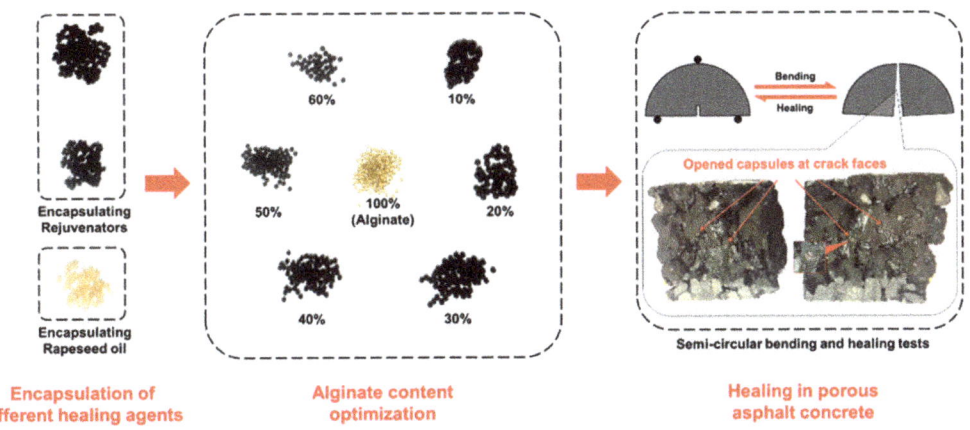

**Figure 2.** Development of the calcium alginate capsules healing system.

The findings above show that both 4PB and ITF can be used to characterize the self-healing behaviour of asphalt mixture under fatigue loadings if proper rest periods are introduced based on the built-in extrinsic healing methods.

The main objective of this study focused on the fatigue damage healing prospect of PA mixtures incorporated with various healing systems, namely the capsule healing system, the induction healing system, the combined healing system and a reference mix (without extrinsic healing). The fatigue behaviour of PA with various healing systems was studied using ITF and 4PB. A 4PB damaging and healing programme was carried out to evaluate the fatigue damage healing efficiency of various healing systems, and the fatigue healing index was obtained from the development of the damage rate, which considers changes in both stiffness and fatigue life. Moreover, the influence of asphalt ageing on the fatigue damage healing via the induction heating method was investigated by including a test group of PA mixture (incorporated with the induction healing system) without an extra ageing process. The research methodology is illustrated in Figure 3.

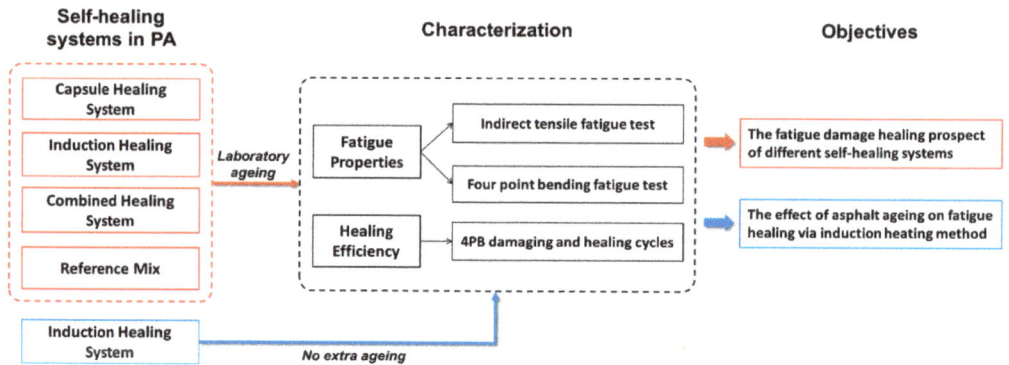

**Figure 3.** The research methodology of this study.

## 2. Materials and Methods

In this section, the built-in healing systems and the preparation method of the PA cylinder and beam specimens are presented. Then, the ITF and 4PB test setups used in this study are introduced, followed by the healing procedure and the damaging and healing programme, which was finally used to evaluate the fatigue damage healing efficiency of different healing systems.

### 2.1. Porous Asphalt Sample Preparation

The calcium alginate capsules and the steel fibres were used to build the self-healing systems in PA. The calcium alginate capsules used in this study were prepared in Microlab, TUDelft, Delft, the Netherlands and the microscopic images in Figure 4 show that the capsules had a diameter of 1.95 mm and a honeycomb-like structure [24,26]. Aiming to improve the conductivity of PA to achieve induction healing, steel fibres were used in the induction healing system and combined healing system. Steel fibres with a density of 7.6 g/cm$^3$, an average length of 1.4 mm, a diameter of 40 μm and a resistivity of $7 \times 10^{-7}$ Ω·cm were provided by Heijmans Infra BV, Rosmalen, The Netherlands.

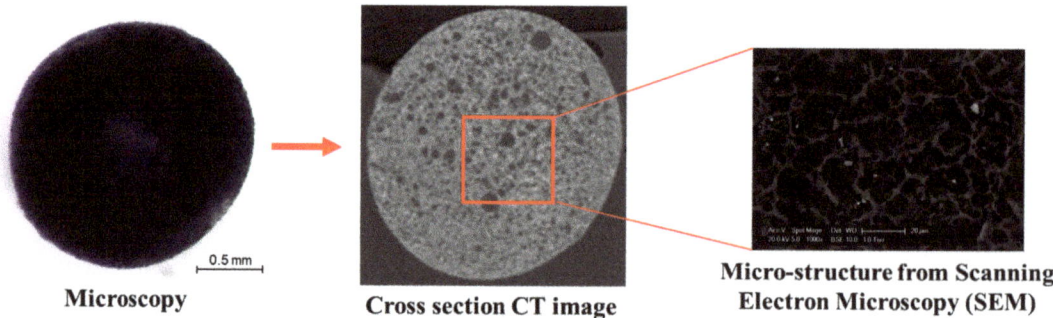

**Figure 4.** Images of a single calcium alginate capsule. Reproduced with permission from refs. [24,26]. Copyright 2018 Elsevier Ltd.

To obtain test specimens for ITF and 4PB, the PA mixture was mixed with a rotating drum mixer and then compacted into slabs with a roller compactor. The materials and mix composition of the PA mixture incorporated with various healing systems can be referred to in a previous study [27]. After compaction, a laboratory ageing process was used to simulate the condition when healing was needed (after years of serving) [27]. Hence, based on the ageing levels and the built-in healing systems, five different PA mixture groups were derived and the detailed group information is presented in Table 1.

**Table 1.** PA mixture group information.

| PA Mixture Group Name | Laboratory Ageing | Built-In Healing Systems |
|---|---|---|
| Induction healing (fresh mixture) | No | Induction |
| Capsule healing (aged mixture) | Yes | Capsules |
| Induction healing (aged mixture) | Yes | Induction |
| Combined healing (aged mixture) | Yes | Capsules and induction |
| No healing (aged mixture) | Yes | None |

Two types of PA slab were fabricated for the study of various healing systems in which Slab_type_1 has the dimensions of 500 × 500 × 50 mm and Slab_type_2 has the dimensions of 600 × 400 × 80 mm. The Slab_type_1 was used for the drilling of cylinder specimens for ITF which had a diameter of 100 mm and a height of 50 mm. The Slab_type_2 was used to produce beam specimens for the 4PB which had dimensions of 400 × 50 × 50 mm.

Figure 5 shows schematic diagrams of the detailed test sample drilling/cutting process. As shown in Figure 5, a minimum of nine PA cylinders were drilled from Slab_type_1 (Figure 5a), and four PA beams were cut from Slab_type_2 (Figure 5b). During the drilling/cutting process, a 5-cm edge around the slabs (the light grey area) is ignored to avoid edge effect from the compaction process.

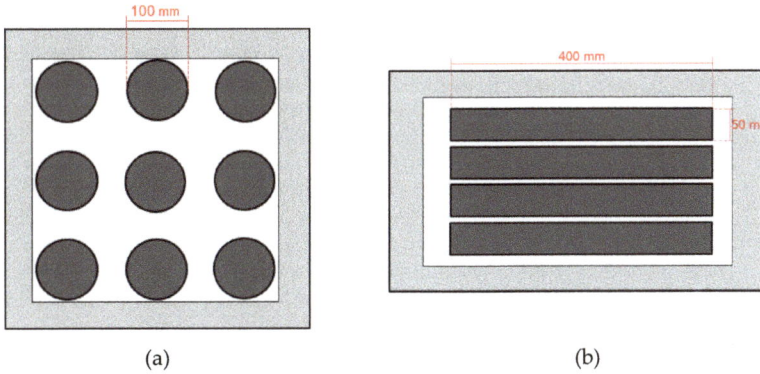

**Figure 5.** Cylinder and beam samples' detailed drilling/cutting schematic: (**a**) cylinder samples drilled from Slab_type_1 and (**b**) beam samples cut from Slab_type_2.

*2.2. Indirect Tensile Fatigue Test*

ITF tests aimed to evaluate the fatigue life of a PA sample by recording the total number of continuous loading cycles that the sample can bear, and the loading mode is selected as the stress control. Figure 6a shows the loading configuration schematic for ITF. Following the European standard EN 12697-24, the ITF was carried out by applying a continuous haversine fatigue loading with a peak value of 400 N and the loading frequency of 8 Hz. The ITF was performed in a temperature chamber of 5 °C to avoid permanent deformation upon loadings. The tests were terminated at the point of the full failure and the number of fatigue loadings that led to the sample failure was recorded. Figure 6b shows the data acquired from the ITF, where the red dashed line in the graph shows the maximum number of ITF, which indicates the fatigue life of the test specimen.

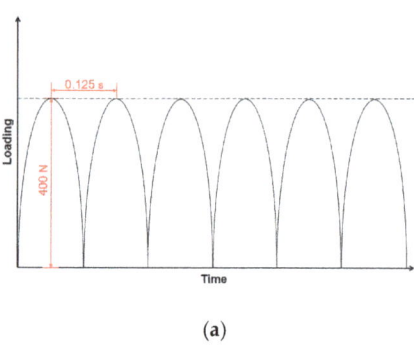 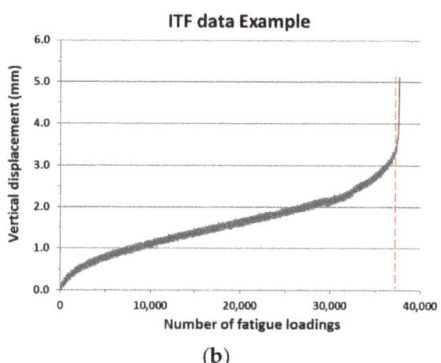

(a) (b)

**Figure 6.** The ITF test: (**a**) the loading configuration schematic and (**b**) ITF test data example, where the red dashed line indicates the fatigue life.

### 2.3. Four-Point Bending Fatigue Test

During the service life of asphalt pavement, the asphalt layer will be subjected to a high number of bending loads, which leads to fatigue damage, and 4PB is regarded as the most representative laboratory method which can be used to determine the fatigue performance of an asphalt mixture under controlled bending loads [34,35]. In this paper, 4PB aims to investigate the fatigue behaviour of PA beams incorporating various extrinsic healing systems under strain-controlled fatigue loadings. Followed by a rest period, the fatigue damage healing capacity of the PA beams was evaluated. Figure 7 shows the 4PB fatigue test samples, test setup and loading configuration. Figure 7a shows the testing beams kept on a plain wooden board and stored in the storage room at 5 °C. Figure 7b shows the 4PB testing setup where the middle of the beam is subjected to a continuous sine shaped loading by the inner two clamps with strain control. Figure 7c shows the schematic of the 4PB test schematic, and Figure 7d shows the loading configuration in which the loading frequency is 8 Hz. The 4PB fatigue tests were performed in a temperature chamber of 20 °C, and the maximum strain was set as 400 $\mu\varepsilon$. The number of load cycles at the time when stiffness modulus decreases to 50% of its initial level is regarded as the fatigue life of a 4PB fatigue test (EN 12697-24).

 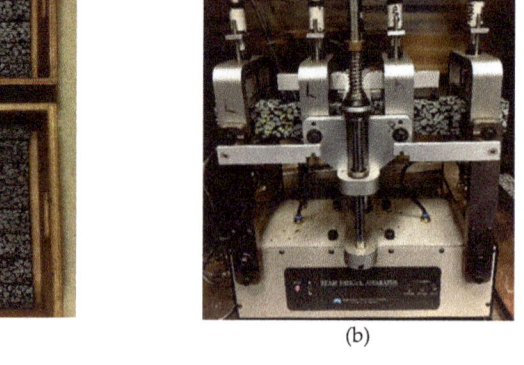

(a) (b)

**Figure 7.** *Cont.*

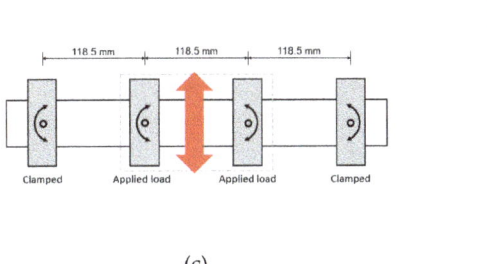
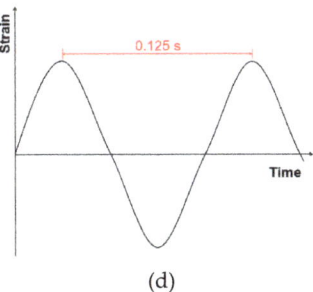

(c)                         (d)

**Figure 7.** 4PB fatigue test: (**a**) beam samples, (**b**) 4PB testing schematic, (**c**) testing setup and (**d**) loading configuration.

During a 4PB fatigue test, the changes in flexural stiffness of the beam specimen are recorded with the increase of fatigue loadings. Figure 8 shows the data acquired from the 4PB fatigue test, which shows the measured flexural stiffness with time. The red dashed line in the graph shows the average decreasing rate of the flexural stiffness whose slope indicates the damage rate ($D$) of the beam specimen. The damage rate ($D$) in a 4PB fatigue test indicates the average decreasing rate of the flexural stiffness of a beam specimen, which means that a higher damage rate results in a faster damaging process and therefore displays a lower fatigue resistance [34]. The damage rate of a 4PB test considers both the change in stiffness and the time it takes to make this change, which provides a more comprehensive method to illustrate 4PB fatigue behaviour. The following equations can be used for the calculation of the damage rate ($D$):

$$S_t = \frac{S_0}{2} \tag{1}$$

$$t = \frac{N_f}{f} \tag{2}$$

$$D = \frac{S_0 - S_t}{t} = \frac{4S_0}{N_f} \tag{3}$$

where:
$S_0$ is the initial flexural stiffness (MPa);
$S_t$ is 50% of the initial flexural stiffness at time $t$ (MPa);
$t$ is the time to reach 50% of the initial flexural stiffness (s);
$N_f$ is the number of loadings to reach 50% of the initial flexural stiffness;
$f$ is the fatigue loading frequency which is 8 Hz;
$D$ is the damage rate (MPa/s).

The total number of fatigue loadings ($N_f$), the flexural stiffness (S) and the damage rate ($D$) are the key parameters that are used to illustrate the fatigue life, stability and fatigue resistance of the beam specimens, respectively, in the 4PB fatigue test.

*2.4. Healing Procedure*

The healing effect of the extrinsic healing technology for self-healing asphalt is largely affected by the rest period as well as the provided environmental conditions, such as time, temperature, humidity, etc. The importance of a rest period in asphalt healing has been proven by the asphalt service life extension in both laboratory testing and field application [36,37]. As such, a proper healing procedure for each healing system needs to be designed. To this aim, a 24-h rest period is designed, which is illustrated in Figure 9. The rest period begins with the 20-h healing period where the majority of healing actions take place by activating the built-in healing system so that the damage healing process is largely accelerated.

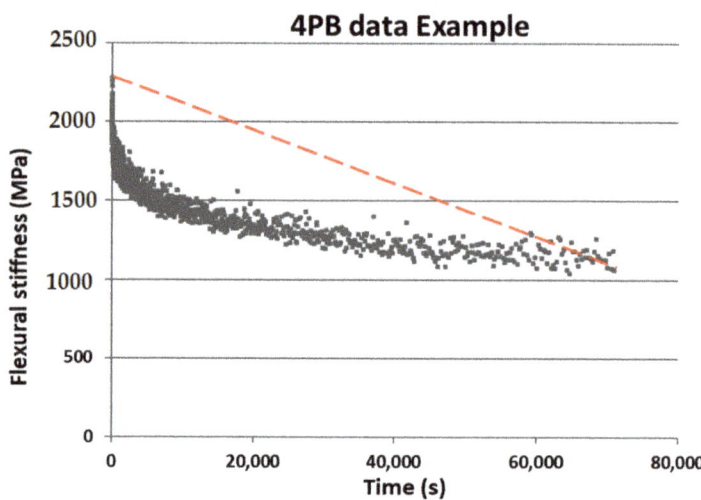

**Figure 8.** 4PB test data example, where the slope of the red dashed line shows the damage rate.

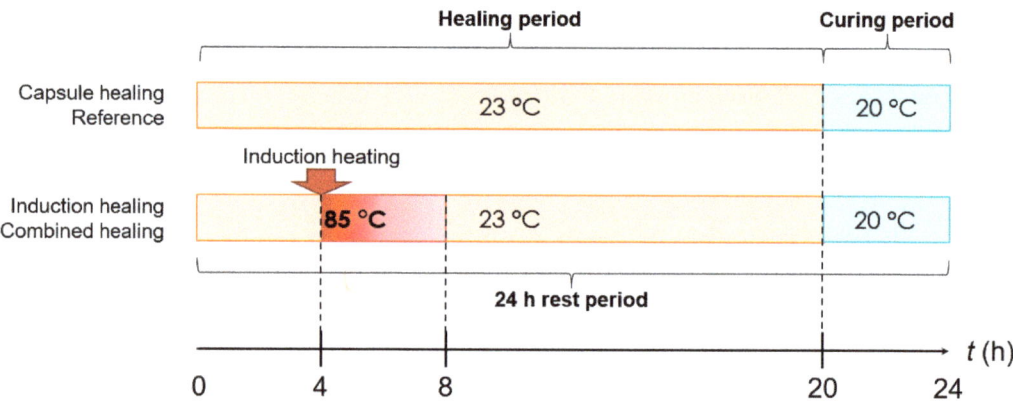

**Figure 9.** 24-h rest period for the healing of various healing systems.

Figure 9 shows that, for the healing of damaged samples with the capsule healing system and without a healing system, the whole healing period is conditioned in a temperature chamber at 23 °C on a plain surface. However, for the healing of damaged samples with the induction healing system and the combined healing system, the first 4 h is conditioned at 23 °C to allow the damaged sample to reach the ambient temperature. After that, the induction heating is applied to increase the sample's surface temperature to 85 °C, followed by 16 h conditioned in the temperature chamber at 23 °C to cool down and be further healed. After the healing period, the sample is cured in a temperature chamber at 20 °C for 4 h to meet the test temperature for the next 4PB round.

To avoid permanent deformation during the healing process, constant confinement is created for all samples throughout the 24-h rest period. The 4PB fatigue test is sensitive to the deformations of the tested specimen, and as such, the beam specimens need to be carefully confined to avoid permanent deformation or even loss of particles during induction heating, transportation and the rest period. The adjustable wooden boxes were made to provide the confinement for beam specimens throughout their 24-h rest period. The confining process for beam specimens is presented in Figure 10. Figure 10a shows that the PA beam specimen is placed in the corner of a wooden frame fixed on three sides. Afterwards, a wooden bar is added to cover the front side of the beam, and a piece of wood with a suitable size is placed at the right side of the beam (Figure 10b). Finally, the wooden box is wrapped with tape to ensure that the adjustable two pieces of wood are closely secured against the beam sample (Figure 10c). Figure 10d shows the image of a beam specimen confined in the wooden box.

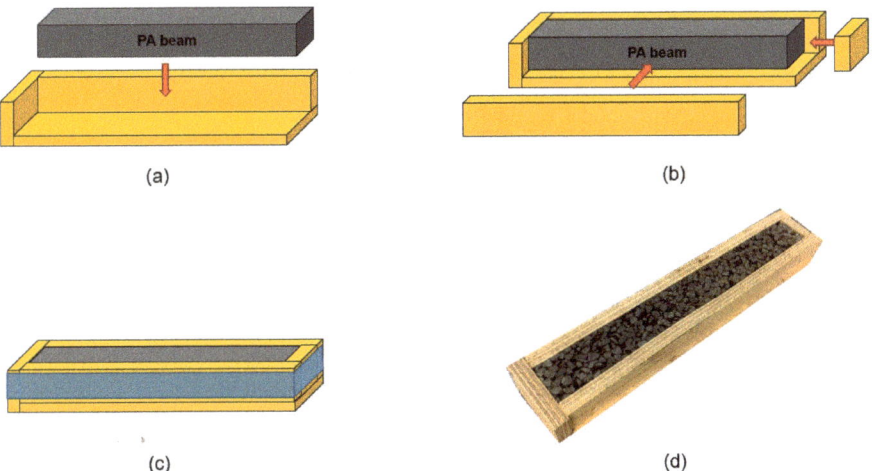

**Figure 10.** The adjustable wooden box for the confining of beam specimen: (**a**), (**b**) and (**c**) show the schematic confining process for a beam specimen, while (**d**) shows the image of a beam specimen confined in the wooden box.

### 2.5. Damaging and Healing Programme

To evaluate the fatigue damage healing efficiency of PA samples from each mixture group, a damaging and healing programme was designed based on 4PB, which is illustrated in Figure 11. First, the initial property of the testing sample was measured by 4PB. Then, the 24-h rest period was provided based on the built-in healing system, therefore completing a damaging and healing cycle. The damaging and healing cycle(s) continued until the testing sample fully failed (fractured into two parts).

## Damaging and healing programme

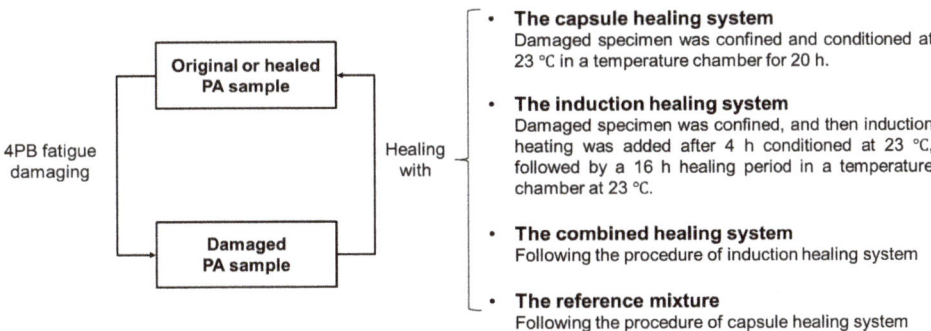

**Figure 11.** The damaging and healing programme for PA samples with various healing systems.

The fatigue healing index is used to illustrate the fatigue healing effect of a PA mixture. The damage rate (D) acquired from the 4PB fatigue tests is used to characterize the durability of a beam specimen under fatigue loadings, and a higher damage rate refers to a lower performance in the PA durability. The fatigue healing index (FHI) can be calculated with the following equation [34]:

$$FHI = \frac{C_x}{C_1} \times 100\% \qquad (4)$$

where:

$FHI$ is the fatigue healing index (%);
$D_1$ is the initial damage rate (MPa/s);
$D_x$ is the damage rate measured from the $x$ test cycle (MPa/s).

## 3. Results and Discussion

### 3.1. The Fracture Faces of PA Mixture Containing Capsules

In the previous study, the calcium alginate capsules were found in two pieces at the fracture faces of the PA mixture sample after the semicircular bending test, demonstrating that these capsules can be opened by the crack propagation thus releasing the rejuvenator. Similar phenomena were observed on the fracture faces of the PA mixture containing capsules after the fatigue test. Figure 12 shows the opened capsules on the fracture faces of the PA samples after fatigue loadings. Although the fracture faces were crushed upon fatigue loadings, broken capsules can be observed on the fracture faces of a cylinder specimen after ITF (Figure 12a) and a beam specimen after 4PB damaging and healing tests (Figure 12b). This finding indicates that the calcium alginate capsules encapsulating the rejuvenator embedded in the PA mixture can be opened upon fatigue loadings, potentially from vehicles, which means that this capsule healing system is qualified as an in situ rejuvenator delivery mechanism to achieve damage self-repair in the long-term service life of PA pavement.

(a)        (b)

**Figure 12.** The opened capsules on fracture faces after fatigue loadings: (**a**) fracture faces of a cylinder specimen after ITF and (**b**) fracture face of a beam specimen after 4PB damaging and healing tests.

### 3.2. Indirect Tensile Fatigue Results

In the ITF test, the ITF fatigue life of a PA sample is illustrated by the number of loadings that leads to failure. It is also noted that induction heating or a rest period was not applied throughout the continuous fatigue loadings. Figure 13 presents the indirect tensile fatigue test results for all PA mixture groups. The PA samples without laboratory ageing show the least fatigue life, which means that the laboratory ageing process improves the ITF fatigue life, and similar findings were reported by other researchers [38,39]. This might be because the ageing increases the stiffness of PA samples, which improves the samples' resistance to deformations under fatigue loadings.

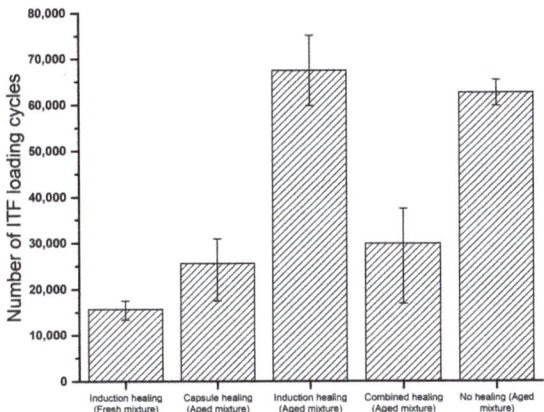

**Figure 13.** Indirect tensile fatigue results of cylinder specimens from all PA mixture groups.

The ITF test results also indicate that the incorporation of the induction healing system can extend the ITF fatigue life of PA samples, which can be seen from the results of the aged mixture groups between induction healing and no healing and between combined healing and capsule healing. This might be due to the reinforcing effect from steel fibres, which was also found in the indirect tensile stiffness of these PA samples from the previous study [27]. A similar ITF fatigue life extension effect of the steel fibres was also reported by Liu [40].

However, the PA samples with embedded capsules showed much less ITF fatigue life, which indicates that the presence of calcium alginate capsules will reduce the ITF fatigue life of the cylinder specimens under continuous fatigue loadings. This might be due to the released rejuvenator, either from the opened capsules by microcracking or from being squeezed out by the fatigue loadings, which could develop in two ways:

1. The released rejuvenator worked. The encapsulated rejuvenator released from capsules and softened the aged binder to reduce the stiffness of the PA sample and, finally, resulted in a reduction in ITF fatigue life that behaved like the PA samples with a fresh mixture;
2. The released rejuvenator did not work. The rejuvenator released upon continuous fatigue loadings but was not able to diffuse into the aged binder at a low temperature (5 °C). In this case, the rejuvenator would be located at the damage site in a liquid phase which might cause slippage, and this could be amplified under indirect tensile fatigue loadings.

*3.3. PB Damaging and Healing Test Results*

3.3.1. Effect of Asphalt Ageing on Induction Healing

Asphalt ageing makes the binder stiffer, and this will result in a reduction of the asphalt healing effect with the induction heating method (Xu, Shi et al., 2018). In this study, the influence of asphalt ageing on the fatigue damage healing with the induction heating method was investigated, and the results are presented in Figure 14. Figure 14a shows that the aged PA mixture has a longer fatigue life than the fresh PA mixture in the damaging and healing test cycles, which is similar to the ITF test results, while the difference is not that significant.

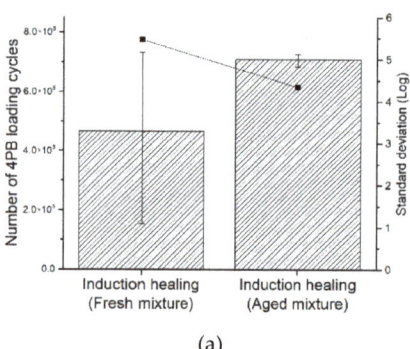
(a)

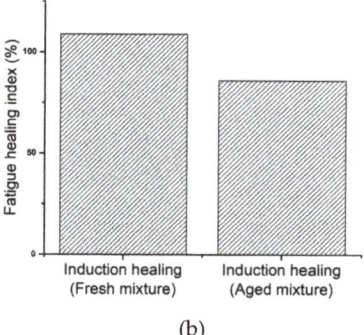
(b)

**Figure 14.** Four-point bending fatigue test results for the induction healing system in PA mixture with different ageing levels: (**a**) total number of 4PB fatigue loading cycles and (**b**) the fatigue healing index.

Figure 14b shows that the induction healing system has a higher fatigue healing index on the fresh mixture, which is 108.9%. This means the stiffness decreasing rate of the beam specimens with fresh mixture became even slower after the induction healing process, while the fatigue healing index for the aged mixture is 86.2%. As a result, the induction heating approach has a promising fatigue damage healing effect especially on a fresh mixture, and this healing effect is better than the PA mixture treated with the laboratory ageing process. Although the longer fatigue life of the aged beams is confusing, the fatigue healing index calculated from the damage rate successfully illustrates the decreasing of induction healing efficiency due to asphalt ageing.

3.3.2. The Healing Effect of Four Asphalt Healing Systems

Figure 15 presents the 4PB fatigue test results for various healing systems incorporated in the aged PA mixture, which illustrates the fatigue behaviours of each healing system in the damaging and healing cycles.

Figure 15a shows the 4PB fatigue test results for the capsule healing system. Damages that took place in the first 4PB fatigue test were recovered in the rest period, which allowed the beam specimen to regain a part of the lost flexural stiffness. However, due to the softening effect from the released rejuvenator, samples with the capsule healing system showed a continuous reduction in both 4PB fatigue life and flexural stiffness with the increase of testing cycles. Furthermore, the beam specimens with the capsule healing system showed a large variety in the 4PB fatigue life results, which indicates that the calcium alginate capsules have an unstable impact on the total 4PB fatigue life tested from the 4PB fatigue test cycles.

Figure 15b shows the 4PB fatigue test results for the induction healing system. The induction healing system showed more stable healing than the capsule healing system, which lies in the recovery of both flexural stiffness and 4PB fatigue life. The results acquired from the third fatigue test cycles could still have an average maximum flexural stiffness of 1845 MPa, which is more than 85% of the average initial stiffness, which is 2080 MPa. Furthermore, the three beam specimens with induction healing systems showed a very similar 4PB fatigue life, as opposed to a large variety for the specimens with the capsule healing system.

Figure 15c shows the 4PB fatigue test results for the combined healing system in which the fatigue behaviours of both the capsule healing system and induction healing system are found. The combined healing system showed an effective recovery on the flexural stiffness from the rest period and a significant 4PB fatigue life extension effect. However, the combined healing system leads to different 4PB fatigue life extensions of the three beam specimens, which is similar to the capsule healing system.

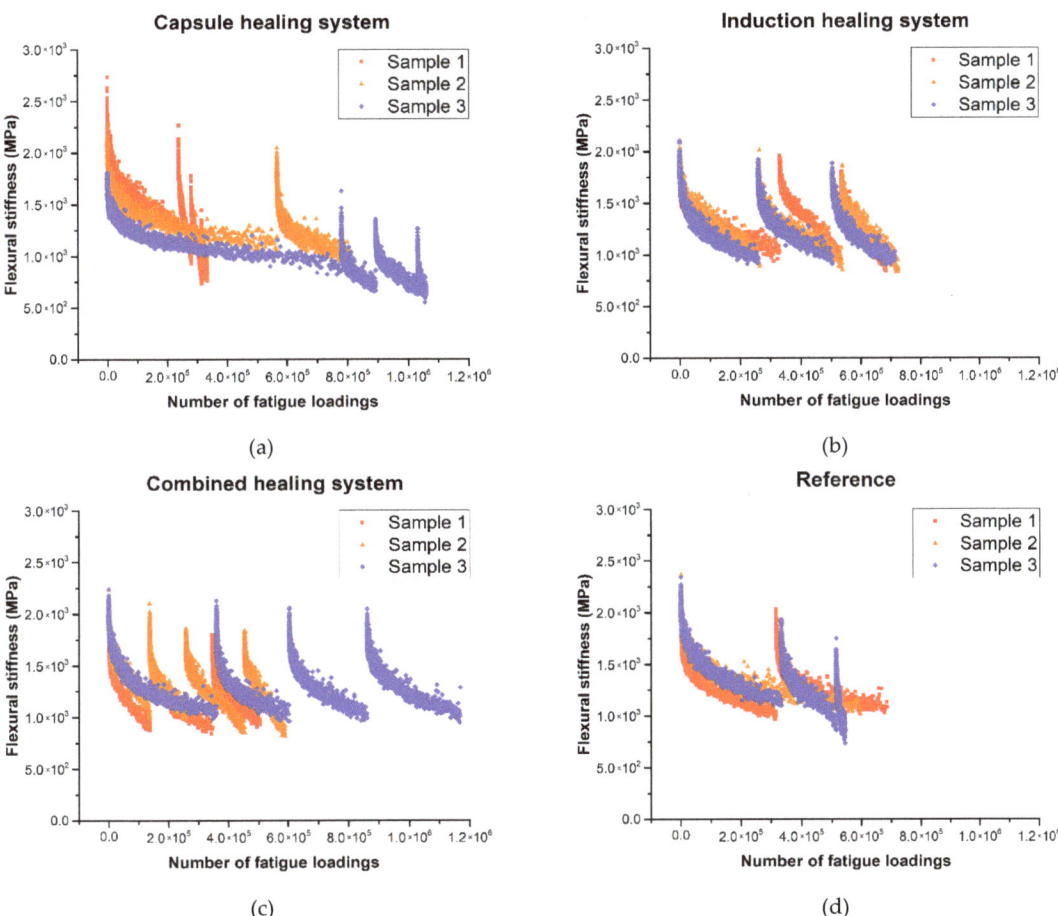

Figure 15. Four-point bending fatigue test results for the aged mixture with various healing systems: (**a**) the capsule healing system, (**b**) the induction healing system, (**c**) the combined healing system and (**d**) the reference mix.

For the reference beams without healing systems, the fatigue damage healing actions still took place during the 24-h rest period, which showed a notable extension of 4PB fatigue life as well as a recovery of stiffness for Sample 2 and Sample 3 (Figure 15d). However, the reference beams have a lower number of average possible healing cycles, and the healing effect is limited compare to beams with a built-in healing system.

Figure 16 shows the summary of the total number of 4PB fatigue loading cycles which leads to the failure of the beams. In Figure 16, among the aged PA samples with various healing systems, beams from the reference group showed the lowest number of loading cycles, which means that the incorporation of these healing systems may result in an increased 4PB fatigue life.

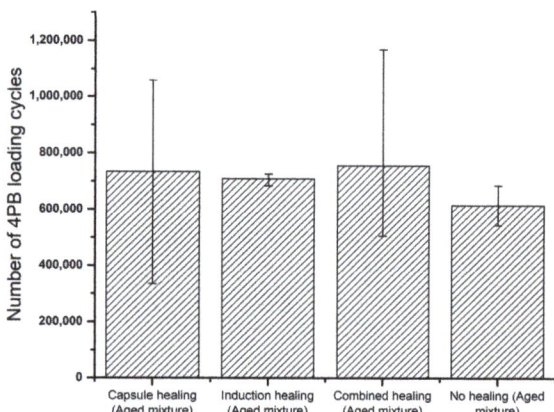

**Figure 16.** Number of 4PB fatigue loadings leading to failure of all beam specimens.

It is also indicated that the PA samples with the combined healing system have the highest fatigue life; however, the data dispersion is large, which is also found in the results for the capsule healing system. It might be the localised rejuvenation effect from the embedded capsules that softened the aged material at the damage site, therefore slowing down the development of microcracks under the repeated fatigue loadings where the rejuvenation took place. As a result, the fatigue damage healing with the capsule healing system is determined by the distribution of the opened capsules, whereby leading to the variety in 4PB fatigue life.

Figure 17 summarises the developments of flexural stiffness with the 4PB fatigue loadings for all the aged beams. Despite some scattered results from the capsule healing system and the reference group, the general trend for the stiffness of beam specimens developed throughout the damaging and healing programme reflects the stiffness stability of a healing system. The flexural stiffness of the combined healing system decreased slowly, followed by the induction healing system and the capsule healing system, while the reference group showed much faster decrease of flexural stiffness under fatigue loadings. These findings can be better illustrated with the slopes of trendlines in Figure 17, which decrease from the combined healing system (red) to the induction healing system (purple), the capsule healing system (blue) and the reference group (grey). Hence, the combined healing system shows an advantage in stiffness recovery under fatigue loadings over the other healing systems.

For all aged PA mixtures, the damage rate acquired from each 4PB damaging and healing cycle is presented in Figure 18. The capsule healing with the aged mixture group showed an increasing trend in damage rate during the 4PB fatigue test cycles, which means that the beams incorporated only with the capsule healing system more easily lost the regained stiffness in the following fatigue test cycles. Compared to the reference (no healing) group, the capsule healing system showed two advantages in fatigue damage recovery: much lower damage rate after two rest periods and one extra potential healing cycle. This might be because the released rejuvenator showed a more significant damage healing effect after two rest periods, and then stimulated the healing of microcracks in beam specimens to be able to conduct the fourth 4PB fatigue test before failure.

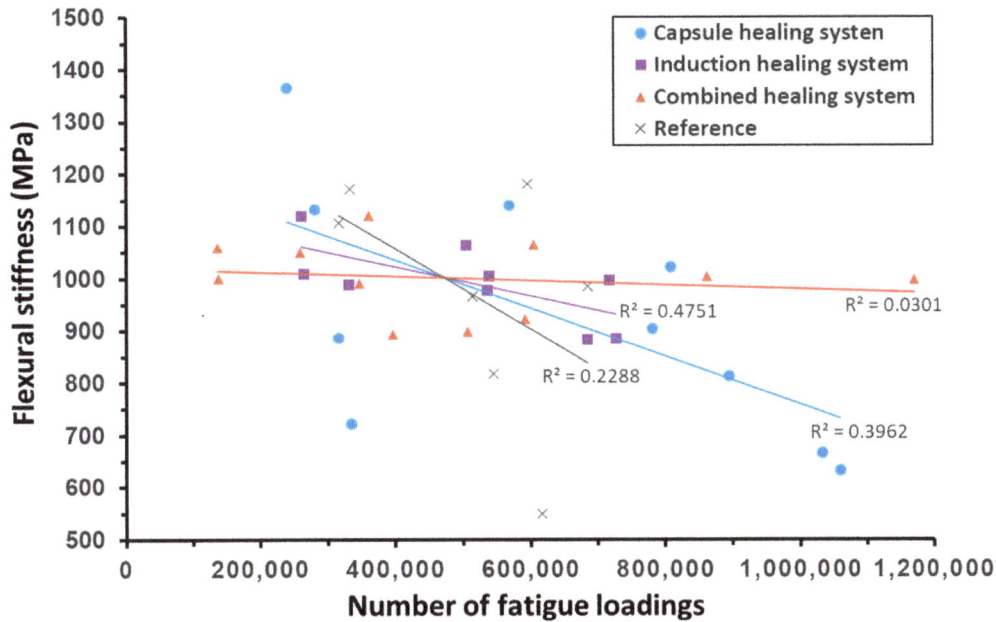

**Figure 17.** Development of flexural stiffness for the various healing systems.

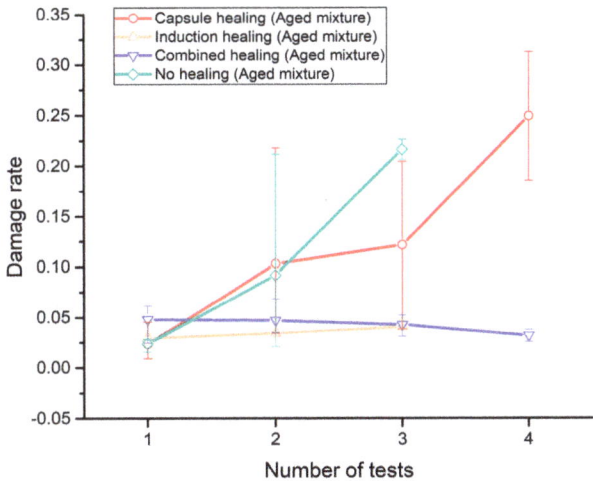

**Figure 18.** The damage rate acquired from each 4PB damaging and healing cycle.

The damage rate for the induction healing (aged mixture) group slightly increased with the increase of fatigue test cycles (Figure 18); however, it is much lower than the capsule healing group and the reference (no healing) group in the second and the third 4PB fatigue test cycles, which indicates the advantage of the induction healing system in fatigue damage healing.

The combined healing system showed a more stable and durable healing effect than the induction healing system, which not only showed a decreased trend in damage rate but also survived four 4PB fatigue test cycles. As a result, the combined healing system demonstrated the best performance in fatigue damage healing in comparison to the single extrinsic healing systems (capsule healing and induction healing) and the intrinsic asphalt healing.

### 3.4. Fatigue Healing Index

The fatigue healing index (FHI) acquired from the damaging and healing programme for all four healing systems is shown in Figure 19. Compared to the no healing group, the capsule healing group showed a lower FHI in the first healing, but a higher FHI in the second healing, and could achieve effective healing for three cycles. This indicates that the capsule healing system can improve the healing capacity of an aged PA mixture and achieve a more durable fatigue behaviour in the 4PB test series. The induction healing (aged mixture) group has a much higher fatigue healing index than the capsule haling group in the first two healing events, which indicates that induction heating has a much better fatigue damage healing effect than capsule healing. However, the induction healing system could not provide effective healing for the third time. When the capsule healing system and induction healing system are combined, the fatigue damage healing effect is significantly improved (Figure 19). In contrast to the capsule healing system and the induction healing system, the combined healing system shows a much higher FHI during all testing cycles, and these values even increase after every healing event. A possible explanation is the gradual healing effect from the calcium alginate capsules whose compartmented rejuvenator is gradually released upon the fatigue loadings, so that the induction healing effect is enhanced due to aged binder rejuvenation time after time. Finally, the terrific fatigue damage healing effect from the combined healing system is demonstrated.

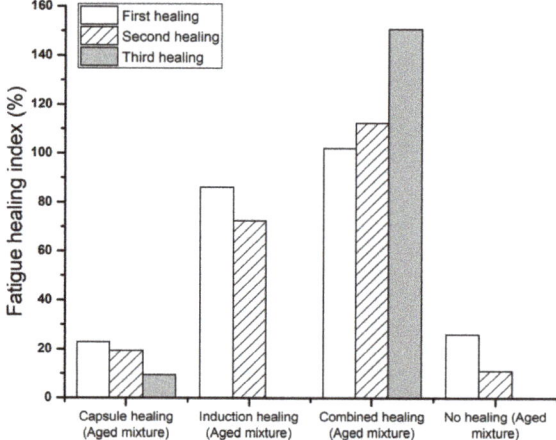

**Figure 19.** The fatigue healing index for various healing systems.

### 4. Conclusions

This paper presents a study on the fatigue damage healing prospects of porous asphalt incorporated with four different self-healing systems. The following conclusions can be drawn:
- The ITF test results indicate that the calcium alginate capsules show a negative effect under the stress-controlled fatigue life, which might be due to the fact that the released rejuvenator from the capsules softens the aged material and results in higher deformations under stress-controlled fatigue loadings. It could also be possible that the released rejuvenator can hardly diffuse and rejuvenate at 5 °C under continuous loadings.
- The rest period plays an important role in determining the healing effect of all asphalt self-healing systems. For the capsule healing system, the curing temperature should

not be too low. For the induction healing system, stable confinement is recommended to avoid permanent deformation under a high temperature (85 °C).
- Induction heating provides stable stiffness recovery and a reliable 4PB fatigue life extension effect in the long-term damaging and healing cycles. This finding agrees with the conclusions from existing research [40].
- The beam specimens with the capsule healing system showed the longest fatigue life in the damaging and healing cycles, which might because the rejuvenator released from the capsules improves the healing capacity of the aged materials, thus resulting in a more durable fatigue behaviour in the test series. The scattered results might be caused by the random capsule distribution which determines whether the healing takes place on the damage site or not.
- The combined healing system showed the best performance in fatigue life, stability and fatigue resistance, which points to the healing effect from the calcium alginate capsules whose compartmented rejuvenator is gradually released upon the fatigue loadings, so that the induction healing effect is enhanced due to aged binder rejuvenation time after time. As such, the terrific fatigue damage healing effect from the combined healing system is demonstrated.
- Additionally, by comparing the results between the induction healing system on the aged mixture and the fresh mixture, it turns out that ageing actually contributes to longer fatigue life but results in a higher damage rate and lower healing index.

Among all four asphalt healing systems, the combined healing system is demonstrated to be the most promising method to extend the service life of porous asphalt pavement based on the results from the four-point bending fatigue tests. For future research, it is strongly recommended to further optimize the capsule healing system and the combined healing system by reducing, for example, the calcium alginate capsules' diameter to improve distribution in the asphalt mixture. This will generate progress towards field applications of the combined self-healing system in asphalt pavement.

**Author Contributions:** S.X.: Investigation, Methodology, Data curation, Writing—original draft. X.L.: Supervision, Writing—review and editing. A.T.: Supervision, Writing—review and editing. E.S.: Conceptualization, Writing—review and editing. All authors have read and agreed to the published version of the manuscript.

**Funding:** This work was funded by the National Key R&D Program of China (2017YFE0111600) and a scholarship from the China Scholarship Council (No. 201506950066).

**Institutional Review Board Statement:** Not applicable.

**Informed Consent Statement:** Not applicable.

**Data Availability Statement:** The raw/processed data required to reproduce these findings cannot be shared at this time as the data also form part of an ongoing study.

**Acknowledgments:** Support from Heijmans is greatly appreciated. The authors also wish to thank the technicians from the section of pavement engineering of TUDelft for their help with the project.

**Conflicts of Interest:** The authors declare no conflict of interest.

# References

1. Van Der Zwan, J.T.; Goeman, T.; Gruis, H.; Swart, J.; Oldenburger, R. Porous asphalt wearing courses in the Netherlands: State of the art review. *Transp. Res. Rec.* **1990**, *1265*, 95–110.
2. Bolzan, P.; Nicholls, J.; Huber, G. Searching for superior performing porous asphalt wearing courses. In Proceedings of the 80th Transportation Research Board Annual Meeting, Washington, DC, USA, 7–11 January 2001.
3. McDaniel, R.; Thornton, W. Field evaluation of a porous friction course for noise control. In Proceedings of the Annual Meeting of the Transportation Research Board, Washington, DC, USA, 9–13 January 2005.
4. Swart, J. Experiences with porous asphalt in the Netherlands. In Proceedings of the European Conference on Porous Asphalt, Madrid, Spain, 12–14 March 1997.
5. Qiu, J. Self Healing of Asphalt Mixtures: Towards a Better Understanding of the Mechanism. PhD Thesis, Delft University of Technology, Delft, The Netherlands, 2012.

6. Hager, M.D.; Greil, P.; Leyens, C.; van der Zwaag, S.; Schubert, U.S. Self-healing materials. *Adv. Mater.* **2010**, *22*, 5424–5430. [CrossRef] [PubMed]
7. Behnood, A. Application of rejuvenators to improve the rheological and mechanical properties of asphalt binders and mixtures: A review. *J. Clean. Prod.* **2019**, *231*, 171–182. [CrossRef]
8. Hou, X.; Xiao, F.; Wang, J.; Amirkhanian, S. Identification of asphalt aging characterization by spectrophotometry technique. *Fuel* **2018**, *226*, 230–239. [CrossRef]
9. Apostolidis, P.; Liu, X.; Scarpas, A.; Kasbergen, C.; van de Ven, M. Advanced evaluation of asphalt mortar for induction healing purposes. *Constr. Build. Mater.* **2016**, *126*, 9–25. [CrossRef]
10. Norambuena-Contreras, J.; Garcia, A. Self-healing of asphalt mixture by microwave and induction heating. *Mater. Des.* **2016**, *106*, 404–414. [CrossRef]
11. Wang, H.; Yang, J.; Liao, H.; Chen, X. Electrical and mechanical properties of asphalt concrete containing conductive fibers and fillers. *Constr. Build. Mater.* **2016**, *122*, 184–190. [CrossRef]
12. Xu, S.; Liu, X.; Tabakovic, A.; Schlangen, E. The Prospect of Microwave Heating: Towards a Faster and Deeper Crack Healing in Asphalt Pavement. *Processes* **2021**, *9*, 507. [CrossRef]
13. Liu, K.; Fu, C.; Xu, P.; Li, S.; Huang, M. An eco-friendliness inductive asphalt mixture comprising waste steel shavings and waste ferrites. *J. Clean. Prod.* **2020**, *283*, 124639. [CrossRef]
14. Wang, Z.; Dai, Q.; Porter, D.; You, Z. Investigation of microwave healing performance of electrically conductive carbon fiber modified asphalt mixture beams. *Constr. Build. Mater.* **2016**, *126*, 1012–1019. [CrossRef]
15. Xu, H.; Wu, S.; Li, H.; Zhao, Y.; Lv, Y. Study on Recycling of Steel Slags Used as Coarse and Fine Aggregates in Induction Healing Asphalt Concretes. *Materials* **2020**, *13*, 889. [CrossRef]
16. Shu, B.; Wu, S.; Dong, L.; Wang, Q.; Liu, Q. Microfluidic synthesis of ca-alginate microcapsules for self-healing of bituminous binder. *Materials* **2018**, *11*, 630. [CrossRef]
17. Shirzad, S.; Hassan, M.M.; Aguirre, M.A.; Mohammad, L.N.; Daly, W.H. Evaluation of sunflower oil as a rejuvenator and its microencapsulation as a healing agent. *J. Mater. Civ. Eng.* **2016**, *28*, 04016116. [CrossRef]
18. Su, J.-F.; Qiu, J.; Schlangen, E.; Wang, Y.-Y. Investigation the possibility of a new approach of using microcapsules containing waste cooking oil: In situ rejuvenation for aged bitumen. *Constr. Build. Mater.* **2015**, *74*, 83–92. [CrossRef]
19. Wan, P.; Liu, Q.; Wu, S.; Zhao, Z.; Chen, S.; Zou, Y.; Rao, W.; Yu, X. A novel microwave induced oil release pattern of calcium alginate/ nano-Fe $_3$O$_4$ composite capsules for asphalt self-healing. *J. Clean. Prod.* **2021**, *297*, 126721. [CrossRef]
20. Su, J.-F.; Qiu, J.; Schlangen, E. Stability investigation of self-healing microcapsules containing rejuvenator for bitumen. *Polym. Degrad. Stab.* **2013**, *98*, 1205–1215. [CrossRef]
21. Al-Mansoori, T.; Norambuena-Contreras, J.; Micaelo, R.; Garcia, A. Self-healing of asphalt mastic by the action of polymeric capsules containing rejuvenators. *Constr. Build. Mater.* **2018**, *161*, 330–339. [CrossRef]
22. Shu, B.; Zhou, M.; Yang, T.; Li, Y.; Ma, Y.; Liu, K.; Bao, S.; Barbieri, D.; Wu, S. The Properties of Different Healing Agents Considering the Micro-Self-Healing Process of Asphalt with Encapsulations. *Materials* **2021**, *14*, 16. [CrossRef] [PubMed]
23. Ayar, P.; Moreno-Navarro, F.; Rubio-Gámez, M.C. The healing capability of asphalt pavements: A state of the art review. *J. Clean. Prod.* **2016**, *113*, 28–40. [CrossRef]
24. Xu, S.; Tabaković, A.; Liu, X.; Schlangen, E. Calcium alginate capsules encapsulating rejuvenator as healing system for asphalt mastic. *Constr. Build. Mater.* **2018**, *169*, 379–387. [CrossRef]
25. Xu, S.; Liu, X.; Tabakovi, A.; Lin, P.; Zhang, Y.; Nahar, S.; Lommerts, B.J.; Schlangen, E. The role of rejuvenators in embedded damage healing for asphalt pavement. *Mater. Des.* **2021**, *202*, 109564. [CrossRef]
26. Xu, S.; Tabaković, A.; Liu, X.; Palin, D.; Schlangen, E. Optimization of the Calcium Alginate Capsules for Self-Healing Asphalt. *Appl. Sci.* **2019**, *9*, 468. [CrossRef]
27. Xu, S.; Liu, X.; Tabaković, A.; Schlangen, E. A novel self-healing system: Towards a sustainable porous asphalt. *J. Clean. Prod.* **2020**, *259*, 120815. [CrossRef]
28. Li, H.; Yu, J.; Wu, S.; Liu, Q.; Li, B.; Li, Y.; Wu, Y. Study on the gradient heating and healing behaviors of asphalt concrete induced by induction heating. *Constr. Build. Mater.* **2019**, *208*, 638–645. [CrossRef]
29. Dinh, B.H.; Park, D.-W.; Le, T.H.M. Effect of rejuvenators on the crack healing performance of recycled asphalt pavement by induction heating. *Constr. Build. Mater.* **2018**, *164*, 246–254. [CrossRef]
30. Liu, Q.; Schlangen, E.; van de Ven, M.; van Bochove, G.; van Montfort, J. Evaluation of the induction healing effect of porous asphalt concrete through four point bending fatigue test. *Constr. Build. Mater.* **2012**, *29*, 403–409. [CrossRef]
31. Tabaković, A.; Schuyffel, L.; Karač, A.; Schlangen, E. An evaluation of the efficiency of compartmented alginate fibres encapsulating a rejuvenator as an asphalt pavement healing system. *Appl. Sci.* **2017**, *7*, 647. [CrossRef]
32. Sun, D.; Li, B.; Ye, F.; Zhu, X.; Lu, T.; Tian, Y. Fatigue behavior of microcapsule-induced self-healing asphalt concrete. *J. Clean. Prod.* **2018**, *188*, 466–476. [CrossRef]
33. Menozzi, A.; Garcia, A.; Partl, M.N.; Tebaldi, G.; Schuetz, P. Induction healing of fatigue damage in asphalt test samples. *Constr. Build. Mater.* **2015**, *74*, 162–168. [CrossRef]
34. Huang, M.; Huang, W. Laboratory investigation on fatigue performance of modified asphalt concretes considering healing. *Constr. Build. Mater.* **2016**, *113*, 68–76. [CrossRef]

35. Pramesti, F.; Molenaar, A.; Van de Ven, M. The prediction of fatigue life based on four point bending test. *Procedia Eng.* **2013**, *54*, 851–862. [CrossRef]
36. Francken, L. Fatigue performance of a bituminous road mix under realistic test conditions. *Transp. Res. Rec.* **1979**, *712*, 30–37.
37. Sun, G.; Sun, D.; Guarin, A.; Ma, J.; Chen, F.; Ghafooriroozbahany, E. Low temperature self-healing character of asphalt mixtures under different fatigue damage degrees. *Constr. Build. Mater.* **2019**, *223*, 870–882. [CrossRef]
38. Kavussi, A.; Qazizadeh, M.J. Fatigue characterization of asphalt mixes containing electric arc furnace (EAF) steel slag subjected to long term aging. *Constr. Build. Mater.* **2014**, *72*, 158–166. [CrossRef]
39. Vallerga, B.; Finn, F.; Hicks, R. Effect of asphalt aging on the fatigue properties of asphalt concrete. In Proceedings of the Annual Meeting of the Transportation Research Board, Washington, DC, USA, January 1967.
40. Liu, Q. Induction Healing of Porous Asphalt Concrete. Ph.D. Thesis, Delft University of Technology, Delft, The Netherlands, 2012.

*Article*

# Key Performance Analysis of Emulsified Asphalt Cold Recycling Mixtures of the Middle Layer of Pavement Structure

Jun Li, Mingliang Li * and Hao Wu

Research Center of Road, Research Institute of Highway Ministry of Transport, Beijing 100088, China
* Correspondence: ml.li@rioh.cn; Tel.: +86-010-62079235

**Abstract:** In the maintenance engineering of asphalt pavement, it is often encountered that both the surface and middle layers are damaged and need to be maintained. The cold in-place recycling technology can be used to simultaneously treat multi-layer diseases and reduce the waste of pavement materials. The cold in-place recycling mixture is rarely used for high layer of pavement structure in high-grade highway. In the supporting practical engineering, the emulsified asphalt cold in-place recycling mixtures were paved as the middle layer of pavement structure by the laying of an overlay. In order to comprehensively evaluate the material performances, coring samples were drilled after cold recycling pavement opening to traffic, and different performance tests were carried out based on the coring samples. The newly paved SMA mixtures were set as the control group. The high temperature stability of cold recycling mixture was analyzed by dynamic creep test and MMLS3 accelerated loading test. Then, the cracking resistance of cold recycling mixture was studied by semi-circular bending test. Finally, the effect of curing time on splitting strength of cold recycling mixture was measured, and the moisture susceptibility was analyzed by dry–wet splitting test and freeze–thaw splitting test. The test results showed that the high temperature stability of cold recycling mixture was worse than SMA mixture. For the cold recycling mixture, the deformation value at the early stage and deformation rate at the stable stage were larger than SMA mixture in the accelerated loading process, and shear failure at high temperature occurred earlier. The cracking resistance of cold recycling mixture was worse than SMA mixture because of the aging effect of the old asphalt and adverse influence of the added cement binder. The effect of curing time on splitting strength of cold recycling mixture was significant, and two stable periods of early strength were, respectively, reached after curing 3 days and 7 days. The indexes of moisture susceptibility, including dry–wet splitting strength ratio and freeze–thaw splitting strength ratio, were obviously lower than that of SMA mixture, and the test values not up to the standard requirement existed. For the emulsified asphalt cold in-place recycling mixture, the improvement of material performances should be focused on, especially the moisture susceptibility. In the research, the emulsified asphalt cold in-place recycling mixtures were acceptably used as the middle layer of maintenance pavement structure. The reliable discussions were summarized based on coring samples collected from real-life road sections. The case can provide guidance and reference for similar engineering applications.

**Keywords:** road engineering; cold in-place recycling; emulsified asphalt; high temperature stability; cracking resistance; moisture susceptibility

**Citation:** Li, J.; Li, M.; Wu, H. Key Performance Analysis of Emulsified Asphalt Cold Recycling Mixtures of the Middle Layer of Pavement Structure. *Materials* **2023**, *16*, 1613. https://doi.org/10.3390/ma16041613

Academic Editor: Francesco Canestrari

Received: 11 December 2022
Revised: 5 February 2023
Accepted: 13 February 2023
Published: 15 February 2023

**Copyright:** © 2023 by the authors. Licensee MDPI, Basel, Switzerland. This article is an open access article distributed under the terms and conditions of the Creative Commons Attribution (CC BY) license (https:// creativecommons.org/licenses/by/ 4.0/).

## 1. Introduction

Asphalt pavement durability is affected by many factors, including vehicle loading and climate, during the process of transportation, which makes its pavement performance decay continuously until it needs to be repaired because of pavement diseases. The performance of the upper layer or the middle layer of some pavement sections is seriously attenuated. As a result, a large number of reclaimed asphalt pavement (RAP) materials will be produced to cause waste because of the use of conventional milling and resurfacing methods for maintenance, which does not conform to the maintenance concept of green and environment-friendly protection. For pavements that need multi-layer maintenance, it is economical and effective to repair pavement diseases and improve the overall structural performance by using the cold in-place recycling technology to carry out multi-layer synchronous recycling and subsequent construction of a new overlay, however, in which the cold recycling layer is used as the middle layer of the maintenance pavement structure [1,2]. In comparison with the conventional low-layer cold recycling, the high-layer cold recycling mixture takes more vehicle loading and poses higher requirements for comprehensive pavement performance of cold recycling mixtures [3,4].

Many factors influence the pavement performance of cold recycling mixtures. Therefore, reasonable design of material composition characteristics, such as raw material, volume characteristics, and key control parameters, can give full play to the performance of cold recycling mixtures [5–9]. RAP material is the biggest contributor of cold recycling mixture as its properties, including field moisture content, asphalt binder condition, content, source, aggregate gradation, etc., are closely related to the performances the of cold recycling mixture [10–13]. In the field of soil or recycled demolition wastes stabilization for subbase and base, asphalt emulsion is widely used, and the adhesive force of the mixture is enhanced after the demulsification [14,15]. Asphalt emulsion is one of the commonly used binder materials for cold recycling mixtures. The polymer type, ionic charge, and demulsification rate of emulsified asphalt all have significant influence on the performance of cold recycling mixture [16]. Different types of asphalt emulsion have corresponding applicability, so it is necessary to determine the optimal emulsion type according to the application scenarios. The cementitious stabilization agent is added into cold recycling mixtures for the objective of increasing the bearing strength and the compressive strength [17]. Cement is the most commonly used among all of the cementitious stabilization agents, but the moderate amount of the agent should be determined based on a proper mix design to achieve the best in-service performance [18]. The influences of curing conditions on the consolidation behavior of cold recycling mixtures were analyzed, in the laboratory and on site, respectively [19,20]. With the idea that the curing temperature has an important influence on the consolidation behavior of the cold recycling mixture, the measure was proposed to accelerate its consolidation by heating [21]. The improvement of pavement performance also stands out as one of the key directions in the research of cold recycling mixtures. The pavement loading test or triaxial compression test was used to test the rutting resistance of cold recycling mixtures with different amounts of cementing materials, leading to the conclusion that the rutting resistance is the best when the amount of cementing materials ranges between 2% and 2.5% [22,23]. The cold recycling mixtures used for pavement should also have good durability to avoid pavement diseases that will shorten the service life of the pavement. The cracking resistance and fatigue resistance of the asphalt mixture can be effectively characterized by fracture energy and flexibility index calculated by the semi-circular bending test [24–28]. The indirect tensile fatigue test was used to study the fatigue resistance of cold recycling mixtures [29]. The moisture susceptibility of cold recycling mixtures under immersion or freeze–thaw condition is also very important, which can be improved by optimizing the gradation and adding the appropriate amount of emulsified asphalt, cement, and fibers [30–34].

The cold in-place recycling mixture is mainly used for low-grade highway surface layers or high-grade highway base layers, and relevant research has been studied by scholars at home and abroad. The cold in-place recycling mixture is rarely applied to high-grade highway high-layer surface layers because of the limitations of the material performances and construction quality of on site. In addition, the research subjects of relevant studies are mostly cold recycling mixtures prepared in the laboratory, with few scholars conducting systematic research on the service performance of cold recycling pavements in service in practical engineering. The durability is directly affected by the rutting resistance, cracking resistance, and moisture susceptibility of the cold recycling mixtures in the duration of their service. In this paper, the application effects of cold recycling mixtures in the middle layer of pavement structure are comprehensively evaluated. The research subjects consisted of two sections of emulsified asphalt cold in-place recycling pavement. By drilling core samples on the pavement, various pavement performances of the cold recycling mixture were analyzed through different test methods.

## 2. Test Design

*2.1. Cold in-Place Recycling Maintenance Scheme*

Two sections of emulsified asphalt cold in-place recycling pavement were selected in expressway practical engineering. The construction was completed in July 2019 with the pavement core drilling conducted in December 2019, followed by related tests. The maintenance scheme, that milling of 1 cm on the upper layer of the original pavement and cold recycling of the upper and middle layers synchronously followed by the laying of an overlay, was designed in the traffic lane, based on which the cold recycling layer was used as the middle layer of the maintenance pavement structure. For the cold recycling mixture, besides adding 3.5% of SBR emulsified asphalt (by mass of RAP materials, the same below), 2.0% of cement and 2.79% of water (ensuring the optimum moisture content) were added into the cold recycling mixture. These additives can improve the workability and enhance the comprehensive performances of the cold recycling mixture. The emergency lane was directly paved with an overlay due to its original good performance. The maintenance schemes are shown in Figure 1. In the figures, SMA stands for stone matrix asphalt, and AC stands for asphalt concrete. The numbers of 13, 16, 20, and 25 represent nominal maximum aggregate size of 13.2 mm, 16 mm, 19 mm, and 26.5 mm, commonly used in China.

*2.2. Test Methods*

2.2.1. High-Temperature Stability Test

(1) Dynamic creep test

A dynamic creep test was conducted based on asphalt mixture performance tester in accordance with the standard of AASHTO T 378-17 [35]. The test temperature was set at 60 °C initially and then adjusted to 55 °C due to the rapid destruction of the core samples. The load was applied at 0.7 MPa, with a half-sine wave as the loading waveform and a loading cycle of 1 s (consisting of a half-sine pressure load for 0.1 s and an interval for 0.9 s).

The dynamic creep test curve was composed of three phases: migration, stabilization, and damage. The model formulas for each stage were given in Equations (1)–(3). In the migration phase, the accumulation of permanent deformation was rapid, but the accumulation rate decreased slowly. In the stabilization phase, the accumulation rate of permanent deformation remained generally constant, while the deformation accumulation was slow. As for the damage phase, the deformation accumulation began to grow fast with a sharply rising growth rate. The number of repeated load actions for the third phase was determined as the flow number (FN), representing the inflection point at which the permanent deformation of the asphalt mixture entered a rapid-developing phase. In this paper, the FN was adopted as a dynamic creep test indicator for evaluating the ultimate high-temperature stability of asphalt mixtures.

$$\varepsilon_p = a \times N^b \tag{1}$$

$$\varepsilon_p = \varepsilon_{ps} + c \times (N - N_{ps}) \tag{2}$$

$$\varepsilon_p = \varepsilon_{st} + d \times (e^{f(N-N_{st})} - 1) \tag{3}$$

where $\varepsilon_p$ donates the accumulative permanent strain, $\varepsilon_{ps}$ represents the permanent strain at the beginning of the second phase, $\varepsilon_{st}$ stands for the permanent strain at the beginning of the third phase, $N$ means the number of load actions, $N_{ps}$ corresponds to the number of load actions at the beginning of the second phase. $N_{st}$ is the number of load actions at the beginning of the third phase, and the letters $a$, $b$, $c$, $d$, $e$, and $f$ are material constants related to the test conditions.

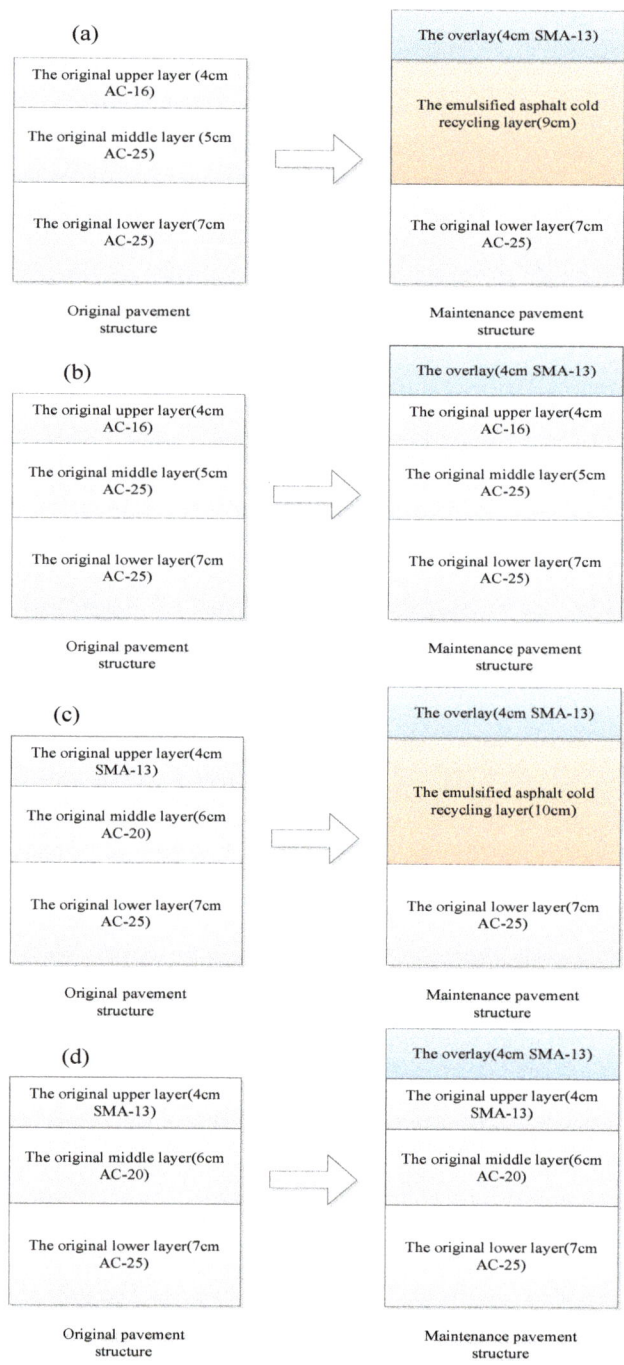

**Figure 1.** Different maintenance schemes of asphalt pavement. (**a**) Maintenance scheme of traffic lane (Section 1); (**b**) maintenance scheme of emergency lane (Section 1); (**c**) maintenance scheme of traffic lane (Section 2); (**d**) maintenance scheme of emergency lane (Section 2).

(2) MMLS3 accelerated loading test

The accelerated loading test was carried out based on 1/3 model mobile load simulator (MMLS3). Simulated loading was performed with a maximum load of 2.7 KN, equivalent to a 0.7 MPa load. The test was conducted at a maximum loading speed of 9 km/h (7200 times/h) and in a water bath heating environment of 60 °C. More details of the test setup are available in the relevant literature [36]. The rutting depth of the samples at different numbers of loading was recorded to reflect the high temperature and anti-deformation of the asphalt mixture. The indicators of the MMLS3 accelerated loading test included the deformation after 100,000 times of loading ($RD_{10}$) and the deformation ratio between 100,000 and 200,000 times of loading ($DS_{20-10}$), with the former characterizing the high-temperature stability of the asphalt mixture at the early stage of the loading while the latter characterizing the long-term high-temperature stability of the asphalt mixture. The calculation for $DS_{20-10}$ is shown in Equation (4).

$$DS_{20-10} = \frac{RD_{20} - RD_{10}}{20 - 10} \times 100 \tag{4}$$

where $DS_{20-10}$ is the deformation rate between 100,000 and 200,000 times of loading ($10^{-3}$ μm/time), $RD_{10}$ donates the deformation after 100,000 times of loading (mm), and $RD_{20}$ means the deformation after 200,000 times of loading (mm).

2.2.2. Anti-Cracking Performance Test

A semi-circular bending (SCB) test was conducted at the temperature of 15 °C and a loading rate of 50 mm/min in accordance with the standard of AASHTO TP 124-16, with the typical test loading curve shown in Figure 2 (from TP 124-16, AASHTO Provisional Standards, published by the American Association of State Highway and Transportation Officials, Washington, DC, USA, used with permission) [37]. In Figure 2, the letter $u_1$ represents the intersection of the post-peak slope with the displacement-axis. A straight line is drawn connecting the inflection point and displacement axis with a slope $m$. The letter $u_{final}$ means displacement at the 0.1 kN cut-off load. The intersection of the arrow with the displacement-axis stands for the displacement at peak load. The test indicators consisted of fracture energy and flexibility index (FI). The damage form of the specimen in the SCB test is similar to the cracking process of asphalt pavement. The SCB test can be used to predict the crack propagation law of asphalt pavement. The fracture energy represents energy required to create a unit surface area of a crack. Therefore, the calculated fracture energy indicates an asphalt mixture's overall capacity to resist cracking-related damage. Generally, a mixture with higher fracture energy can resist greater stresses with higher damage resistance. A greater value of both indicators implied better anti-cracking performance of the asphalt mixture. Fracture energy $G_f$ reflected the total energy absorbed from the material from the state of intactness to fracture, which was calculated by the ratio of the fracture power to the toughness zone area in Equation (5).

$$G_f = \frac{W_f}{Area_{lig}} \tag{5}$$

where $G_f$ represents the fracture energy (J/m$^2$). $Area_{lig}$ means the toughness zone area (m$^2$), which was calculated as shown in Equation (6):

$$Area_{lig} = (r - a) \times t \tag{6}$$

where the letter $r$ is the radius of the sample (m). The letter $a$ means the crack length (m). The letter $t$ stands for sample thickness (m).

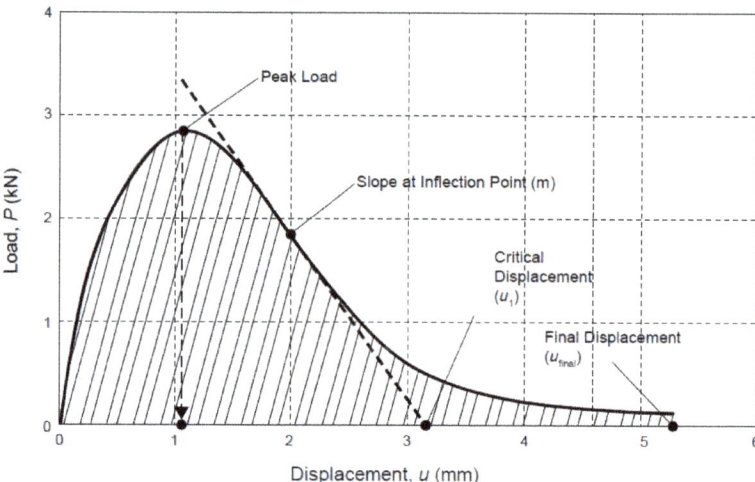

**Figure 2.** Typical test loading curve of SCB test [37].

$W_f$ means the work of fracture (J), which can be calculated using the integral equation below.

$$W_f = \int P du \tag{7}$$

where the letter $P$ means the applied load (N). The letter $u$ is the average displacement of the load (m). The letter $du$ represents differentiation of the displacement $u$.

The flexibility index was calculated as follows:

$$\text{FI} = \frac{G_f}{|m|} \times A \tag{8}$$

where FI indicates the flexibility index (dimensionless). $G_f$ means the fracture energy (J/m$^2$). $|m|$ is the absolute value of the inflection point slope of the load displacement curve after the peak (kN/mm). The letter $A$ denotes the unit conversion coefficient, which is 0.01.

2.2.3. Moisture Susceptibility Test

The moisture susceptibility of emulsified asphalt cold recycling mixtures was evaluated via the wet–dry splitting test and freeze–thaw splitting test. The former adopted the wet–dry splitting strength ratio as the evaluation indicator, which is the percentage of the splitting strength in water immersion for 24 h to that in normal conditions (Equation (9)). The splitting test under normal conditions was carried out based on the T0716 in accordance with Chinese specification of JTG E20-2011 [38]. In the 24 h water immersion splitting test, the samples were completely immersed in a constant temperature water bath at 25 °C for 22 h in advance, after which the splitting strength was tested according to the requirements of the splitting test under normal conditions. The wet–dry splitting strength ratio of emulsified asphalt cold recycling mixtures should not be less than 80% for heavy-load and above transportation purposes.

$$TSR_1 = \frac{\overline{R}_{\text{wet}}}{\overline{R}_{\text{dry}}} \times 100 \tag{9}$$

where $TSR_1$ means the dry–wet splitting strength ratio (%). $\overline{R}_{\text{wet}}$ represents the average value of splitting tensile strength of effective samples after immersion curing (MPa). $\overline{R}_{\text{dry}}$ denotes the average value of splitting tensile strength of effective samples after normal curing (MPa).

As the evaluation indicator for the freeze–thaw splitting test, the freeze–thaw splitting strength ratio was calculated in Equation (10) and performed based on the T0729. The freeze–thaw splitting strength ratio of emulsified asphalt cold recycling mixtures should not be less than 75% for heavy-load and above transportation purposes.

$$TSR_2 = \frac{\overline{R}_{T2}}{\overline{R}_{T1}} \times 100 \tag{10}$$

where $TSR_2$ is the freeze–thaw splitting strength ratio (%). $\overline{R}_{T2}$ indicates the average value of splitting tensile strength of effective samples after freeze–thaw cycles (MPa). $\overline{R}_{T1}$ stands for the average value of splitting tensile strength of effective samples without freeze–thaw cycles (MPa).

### 2.3. Coring Sample Schemes

The coring sample schemes for different tests are shown in Table 1. The core samples for the SCB test were cut into two semi-circles. The appearance of coring samples is shown in Figure 3.

**Table 1.** Coring sample schemes of different tests.

| Test Type | Core Sample Size | Section 1 | | Section 2 | |
|---|---|---|---|---|---|
| | | Traffic Lane | Emergence Lane | Traffic Lane | Emergence Lane |
| Dynamic creep test | Diameter/cm | 10 | 10 | 10 | 10 |
| | Height/cm | 13 | 13 | 14 | 14 |
| | Component | Overlay + recycling layer | Overlay + original upper layer + original middle layer | Overlay + recycling layer | Overlay + original upper layer + original middle layer |
| MMLS3 loading test | Diameter/cm | 15 | 15 | 15 | 15 |
| | Height/cm | 10 | 10 | 10 | 10 |
| | Component | Overlay + part of recycling layer | Overlay + original upper layer + part of original middle layer | Overlay + part of recycling layer | Overlay + original upper layer + part of original middle layer |
| SCB test | Diameter/cm | 15 | 15 | 15 | 15 |
| | Height/cm | 4 | 5 | 4 | 4 | 4 | 5 | 4 | 4 |
| | Component | Overlay | Part of recycling layer | Overlay | Original upper layer | Overlay | Part of recycling layer | Overlay | Original upper layer |
| Splitting test | Diameter/cm | 10 | 10 | 10 | 10 |
| | Height/cm | 4 | 5 | 4 | 4 | 4 | 5 | 4 | 4 |
| | Component | Overlay | Part of recycling layer | Overlay | Original upper layer | Overlay | Part of recycling layer | Overlay | Original upper layer |

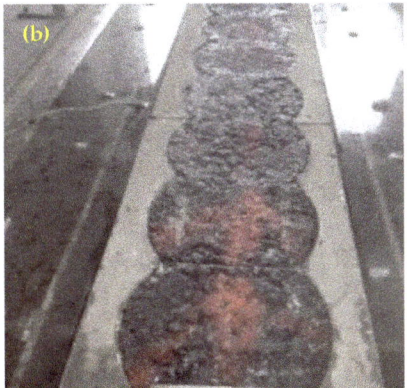

**Figure 3.** *Cont.*

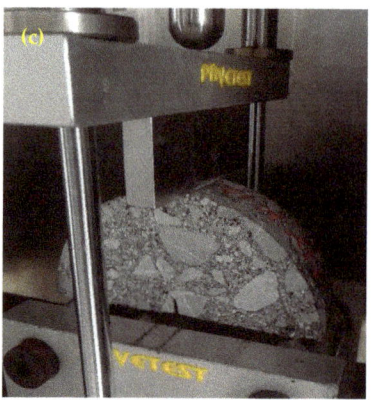

**Figure 3.** Appearance of core samples during different tests. (**a**) Dynamic creep test; (**b**) accelerated loading test; (**c**) semi-circular bending test; (**d**) splitting test.

## 3. Test Results and Analysis

### 3.1. Analysis of High-Temperature Stability

(1) Dynamic creep test

Table 2 presents the dynamic creep test results of core samples of different types of mixtures. In Table 2, for example 1#, the "#" stands for a symbol of specimen number, distinguishing that "1" is not a numerical value for test analysis (unless stated, the same below).

**Table 2.** Dynamic creep test results of different core samples.

| Section Type | Position | Mixture | Test Temperature/°C | Loading Cycles (Number) | FN (Number) |
|---|---|---|---|---|---|
| Section 1 | Traffic lane | Recycling layer | 60 | 42 | 8 |
|  | Emergence lane | Overlay of SMA-13 | 55 | 601 | 314 |
| Section 2 | Traffic lane | Recycling layer -1# | 55 | 57 | 10 |
|  | Traffic lane | Recycling layer -2# | 55 | 64 | 11 |
|  | Emergence lane | Overlay of SMA-13 | 55 | 900 | 564 |

According to Table 2, the FN of overlay of SMA-13 stood at 439 times on average. Under a test temperature of 60 °C, the FN of cold recycling mixtures only reached 8, and the core samples of mixtures were damaged quickly. The FN increased to only 10~11, even when the temperature was changed to 55 °C. Therefore, the high-temperature stability of cold recycling mixtures was much lower than that of SMA overlay mixtures.

(2) MMLS3 loading test

Figure 4 provides the MMLS3 accelerated loading test results of core samples of different types of mixtures, and Figure 5 presents the appearance of the core samples of cold recycling mixtures after loading. The deformation $RD_{10}$ (after 100,000 times of loading) and the deformation rate $DS_{20-10}$ (between 100,000 times and 200,000 times of loading) were calculated as shown in Table 3.

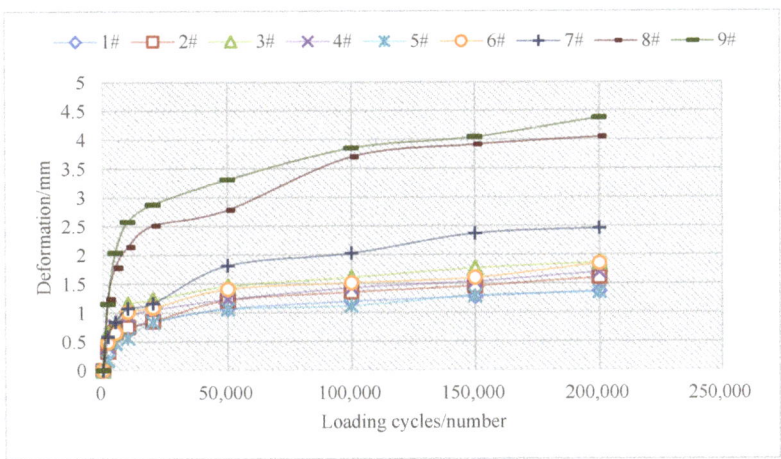

**Figure 4.** MMLS3 loading test results.

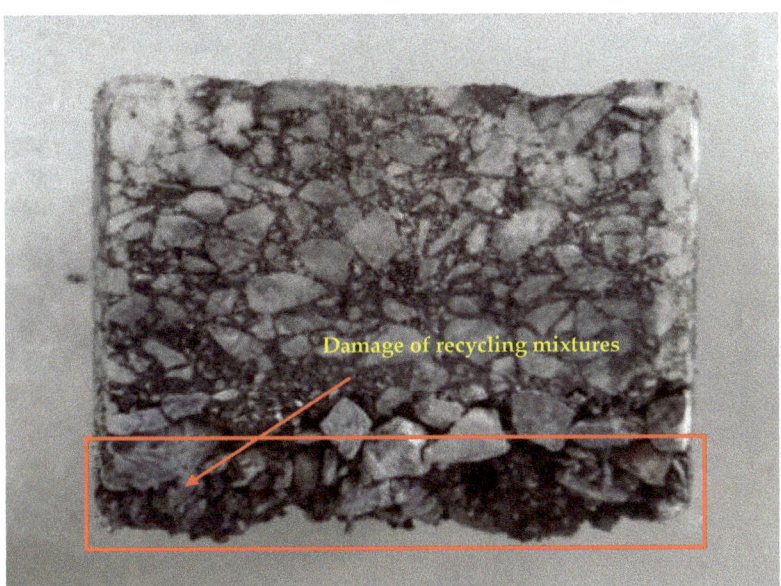

**Figure 5.** Appearance of cold recycling mixture after loading.

**Table 3.** Calculation result of $RD_{10}$, $DS_{20-10}$.

| Evaluation Index | Mixtures of SMA-13 Overlay | | | | | | | Cold Recycling Mixtures | |
|---|---|---|---|---|---|---|---|---|---|
| | Section 1 | | | | Section 2 | | | Section 1 | Section 2 |
| | 1# | 2# | 3# | 4# | 5# | 6# | 7# | 8# | 9# |
| $RD_{10}$ (mm) | 1.20 | 1.35 | 1.61 | 1.43 | 1.12 | 1.51 | 2.03 | 3.70 | 3.85 |
| $DS_{20-10}$ /($10^{-3}$ μm/cycle) | 1.7 | 2.5 | 2.6 | 2.7 | 2.4 | 3.4 | 4.3 | 3.4 | 5.3 |

In Figure 4, although the deformation of both cold recycling mixtures and SMA-13 overlay mixtures was below 5 mm after 200,000 times of loading, this was not much large. The cold recycling mixtures started to drop particles under high temperature and hydrodynamic pressure, though complete loosening did not take place under the restraint of the mold. When the mold was removed at the end of the test, however, serious loosening occurred in the cold recycling mixtures (Figure 5). In addition, the high-temperature stability of SMA-13 overlay mixtures was better than that of cold recycling mixtures according to indicators $RD_{10}$ and $DS_{20-10}$, with the $RD_{10}$ average value (1.5 mm) of SMA-13 overlay mixtures smaller than that (3.8 mm) of cold recycling mixtures and the $DS_{20-10}$ average value ($2.8 \times 10^{-3}$ μm/cycle) of the former also smaller than that ($4.4 \times 10^{-3}$ μm/cycle) of the latter.

In summary, emulsified asphalt cold recycling mixtures were inferior to SMA-13 overlay mixtures in terms of high-temperature stability, which was attributed to the use of styrene-butadiene-styrene (SBS) modified asphalt and skeleton-dense gradation of the latter. Emulsified asphalt cold recycling mixtures were regenerated with the original upper layer (modified asphalt) and the middle layer (ordinary asphalt) and mixed with some modified emulsified asphalt and cement. Regardless of their asphalt performance or gradation composition, they were worse than the newly paved SMA mixtures. Furthermore, due to the slow moisture evaporation and cement hydration of emulsified asphalt, the strength of cold recycling mixtures formed in a longer time. Therefore, the poor strength was also a cause for their worse high-temperature performance compared to the SMA mixtures.

*3.2. Analysis of Anti-Cracking Performance*

Figure 6 illustrates the SCB test results of core samples of different types of mixtures.

As shown in Figure 6, the best anti-cracking performance of the newly paved SMA mixtures was achieved with an average fracture energy of 2849 J/m$^2$ and an average FI of 19.9. In contrast, the average fracture energy and FI of the cold recycling mixtures were 1696 J/m$^2$ and 8.3, respectively, markedly lower than the newly paved SMA mixtures in fracture energy. Moreover, the average fracture energy and FI of the original AC layer mixtures was 1754 J/m$^2$ and 5.6, respectively, while those of original SMA layer mixtures reached 1388 J/m$^2$ and 5.0, respectively. Therefore, cold recycling mixtures were similar to original layer mixtures in anti-cracking performance.

As for the newly paved SMA mixtures, new SBS modified asphalt was adopted as the cementing material, featuring a high asphalt content and large filler-asphalt ratio, both of which can improve the anti-cracking performance. The serious aging condition of asphalt in emulsified asphalt cold recycling mixtures, together with the added cement, led to the decline in their anti-cracking performance.

*3.3. Analysis of Moisture Susceptibility*

3.3.1. Effects of Curing Duration on Splitting Strength

Firstly, a set of core samples was drilled and taken every two days within eight days after the cold recycling layer of emulsified asphalt was formed, followed by the test of splitting strength, to analyze the effects of curing duration on the splitting strength of cold recycling mixtures. The splitting strength test results of the core samples of the cold recycling layer at 15 °C after different curing durations are given in Figure 7.

According to Figure 7, the splitting strength of cold recycling mixtures grew gradually as the duration of curing prolonged, suggesting that a proper curing duration was necessary for ensuring sufficient mechanical strength of cold recycling mixtures. The splitting strength development of cold recycling mixtures exhibited two phases depending on the formation law of splitting strength. The cold recycling mixtures welcomed the first stabilization phase on 3 d and the second one on 7 d of curing. A benchmark was set with the splitting strength of 0.60 MPa on 7 d of curing, thus the splitting strength on 1 d was 0.21 MPa, only 35% of the benchmark. The splitting strength stood at 0.39 MPa on 3 d, reaching 65% of the benchmark and achieving the early-stage strength to some degree.

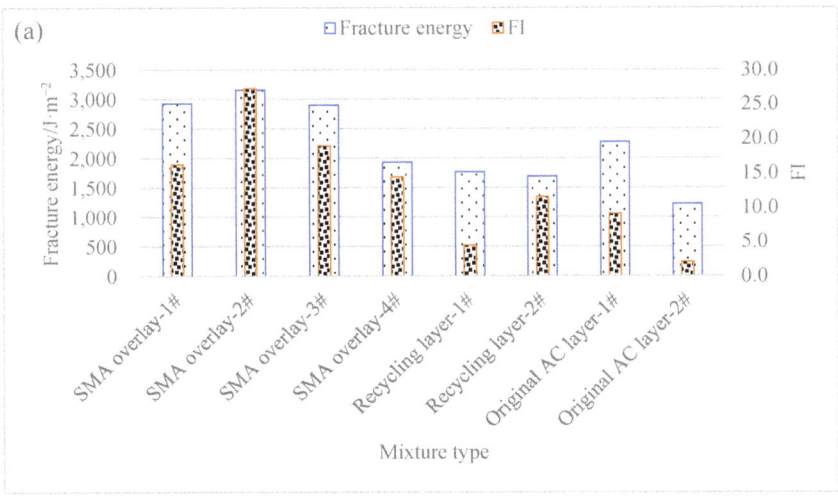

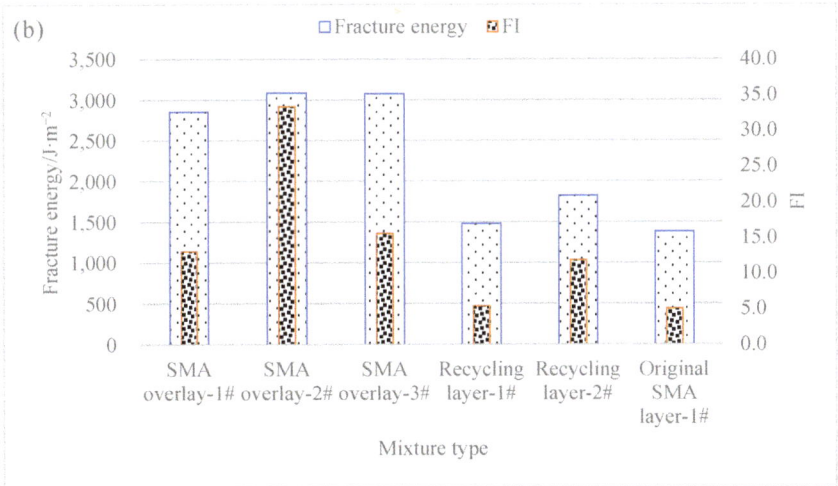

**Figure 6.** Semi-circular bending test results. (**a**) Fracture energy and FI of Section 1; (**b**) fracture energy and flexibility index of Section 2.

3.3.2. Moisture Susceptibility of Core Samples

Figures 8 and 9 show the dry–wet splitting strength ratio and the freeze–thaw splitting strength of different types of mixture of core samples.

As shown in Figure 8, the dry–wet splitting strength ratios of core samples of different types of mixtures varied greatly. In Pavement Section 1, the dry–wet splitting strength ratio was 74.1% for the cold recycling layer, 115% for the SMA overlay on average, and 88.7% for the original AC layer. In Pavement Section 2, the dry–wet splitting strength ratio was 80.1% for the cold recycling layer, 98.3% for the SMA overlay on average, and 75% for the original SMA layer. Therefore, the dry–wet splitting strength ratio of the emulsified asphalt cold recycling layer was significantly smaller than that of the SMA overlay but was close to that of the original pavement.

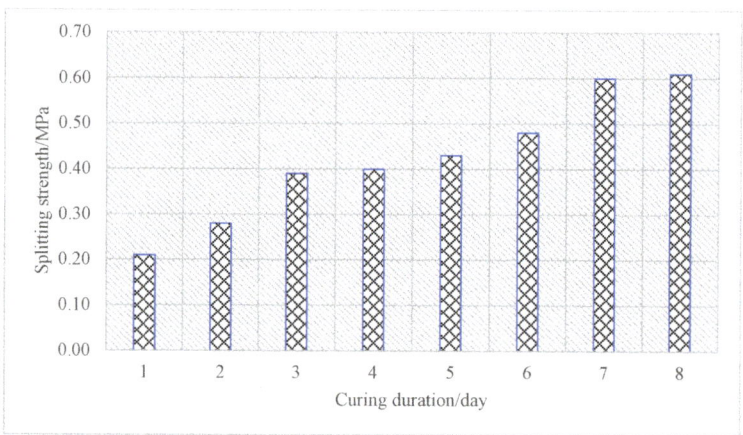

**Figure 7.** Splitting strength of cold recycling mixture under different curing duration.

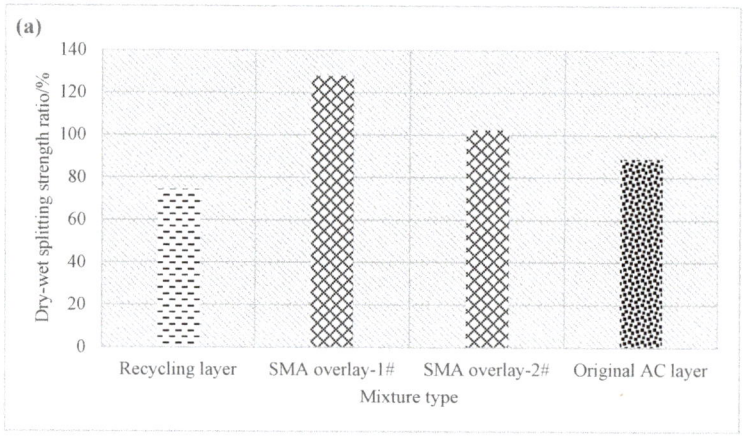

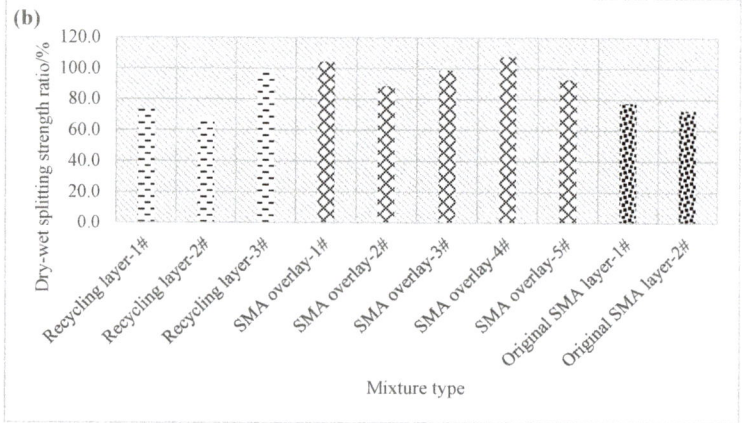

**Figure 8.** Dry–wet splitting test results. (**a**) Dry–wet splitting strength ratio of Section 1; (**b**) dry–wet splitting strength ratio of Section 2.

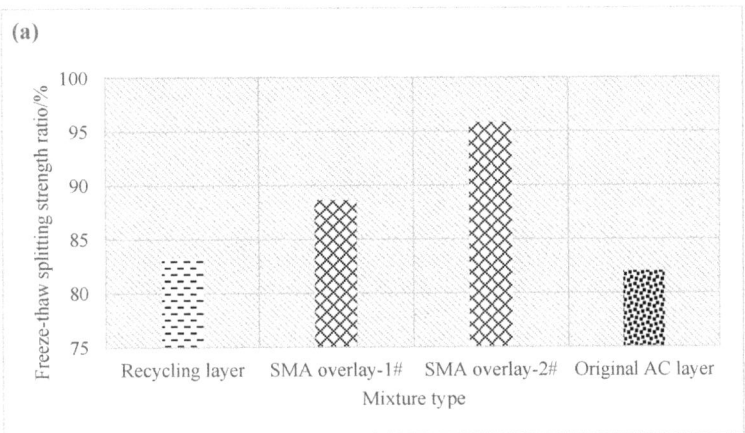

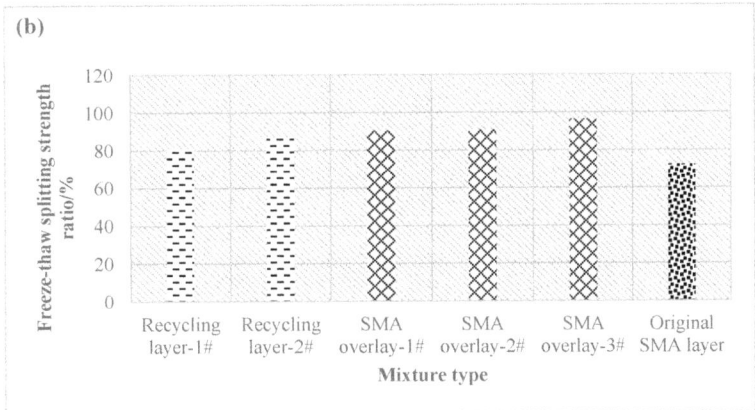

**Figure 9.** Freeze–thaw splitting test results. (**a**) Freeze–thaw splitting strength ratio of Section 1; (**b**) freeze–thaw splitting strength ratio of Section 2.

According to Figure 9, there was a wide difference among the freeze–thaw splitting strength ratios of the core samples of different types of mixtures as well. In Pavement Section 1, the freeze–thaw splitting strength ratio was 83.2% for the cold recycling layer, 92.2% for the SMA pavement layer on average, and 82% for the original AC layer. In Pavement Section 2, the freeze–thaw splitting strength ratio was 83.9% for the cold recycling layer, 92.4% for the SMA pavement layer on average, and 72.2% for the original SMA layer. As a result, the freeze–thaw splitting strength ratio of the emulsified asphalt cold recycling layer was also dramatically smaller than that of the SMA overlay and merely similar to that of the original pavement.

Based on the above analysis, the emulsified asphalt cold recycling mixture was much inferior to the newly paved SMA overlay but close to the original pavement in terms of moisture susceptibility, indicating that the cold in-place recycling technology cannot improve the moisture susceptibility of the original pavement. This was mainly attributed to the serious aging condition of the original asphalt in the cold recycling mixture, poor gradation and significant variability of the mixture, and unfavorable factors such as difficulties in the control of construction quality for large thickness recycling of the original pavement. Such adverse factors would lead to limited or failed effects of cold recycling technology on the moisture susceptibility of the original pavement. In the newly paved SMA overlay, the fresh SBS modified asphalt was used as the cementing material. In addition, the traits of a high asphalt content, a high filler–asphalt ratio, and the added fibers were all beneficial to the enhancement of moisture susceptibility.

According to the Specification of JTG/T 5521-2019 in China, in case of the use of emulsified asphalt cold recycling mixtures for heavy-load and above transportation purposes, the dry–wet splitting strength ratio shall be greater than 80% and the freeze–thaw splitting strength ratio shall exceed 75%. As shown by the test results of core samples drilled on site from the emulsified asphalt cold recycling pavement, the dry–wet splitting strength ratio of Pavement Section 1 did not meet the requirements, while that of Pavement Section 2 was merely qualified. However, the freeze–thaw splitting strength ratio of both sections satisfied the requirements. The wet–dry splitting strength ratio and the freeze–thaw splitting strength ratio, both of which are key control indicators for the quality of cold recycling mixtures, significantly affect the moisture susceptibility and anti-loosening performance of cold recycling pavements after they are put into service. The road project had been put into service for 6 months when the cores were taken, so the strength of the cold recycling mixtures had been further improved compared with the time when the pavement initially came into use. Nevertheless, the samples still failed the standards, indicating that the improvement of moisture susceptibility is a key issue for the use of cold recycling mixtures in high-layer recycling of asphalt pavement.

## 4. Conclusions

Two sections of emulsified asphalt cold in-place recycling pavement were selected in expressway practical engineering. By drilling core samples on the pavement, various pavement performances of the cold recycling mixture were analyzed through different test methods. The conclusions drawn are summarized as follows.

(1) The cold recycling layer was used as the middle layer of the maintenance pavement structure in high grade highway. Regardless of rutting resistance, cracking resistance, or moisture susceptibility, the comprehensive performances of cold recycling mixture were inferior to the newly paved SMA mixture.

(2) After opening to traffic 5 months, the cold recycling mixture of core samples loosened during the process of loading test, and the dry–wet splitting strength ratio failed the standards. The slow strength formation resulted in insufficient durability of the cold recycling mixture under high temperature and water immersion.

(3) The cement was used as a stabilizer in the cold recycling mixture. Adding a proper amount of stabilizer can improve the comprehensive performances of the cold recycling mixture. However, due to the performance attenuation and gradation deterioration of RAP materials, it is difficult for existing stabilizers to improve the performances to the level of fresh mixture. New types of stabilizers need to be developed to produce high performance cold recycling mixtures. Moreover, strength formation has a significant effect on the comprehensive performances of the cold recycling mixture. In the research, slow-setting emulsified asphalt and ordinary Portland cement were used. The effect of rapid-setting emulsified asphalt on the performances of cold recycling mixtures should be focused on, along with the early-strength type of cement. The size and voids of coring samples have significant influence on the test results. The volume parameters of coring samples should be considered to ensure more reliable test results.

**Author Contributions:** J.L.: data curation, visualization, investigation, and writing—original draft; M.L.: methodology, validation, formal analysis, and writing—reviewing and editing; H.W.: data curation and investigation. All authors have read and agreed to the published version of the manuscript.

**Funding:** This research was funded by the key project of building China's strength in transportation pilot tasks from research institute of highway ministry of transport of China (No. 2022-C301).

**Institutional Review Board Statement:** Not applicable.

**Informed Consent Statement:** Not applicable.

**Data Availability Statement:** Not applicable.

**Conflicts of Interest:** The authors declare no conflict of interest.

## References

1. Xiao, F.; Yao, S.; Wang, J.; Li, X.; Amirkhanian, S. A literature review on cold recycling technology of asphalt pavement. *Constr. Build. Mater.* **2018**, *180*, 579–604. [CrossRef]
2. Charmot, S.; Teh, S.Y.; Haris, R.E.A.; Ayob, M.A.; Ramzi, M.R.; Kamal, D.D.M.; Atan, A. Field performance of bitumen emulsion Cold Central Plant Recycling (CCPR) mixture with same day and delayed overlay compared with traditional rehabilitation procedures. *Case Stud. Constr. Mater.* **2022**, *17*, 01365. [CrossRef]
3. Marinković, M.; Zavadskas, E.K.; Matić, B.; Jovanović, S.; Das, D.K.; Sremac, S. Application of wasted and recycled materials for production of stabilized layers of road structures. *Buildings* **2022**, *12*, 552. [CrossRef]
4. Vaitkus, A.; Gražulytė, J.; Baltrušaitis, A.; Židanavičiūtė, J.; Čygas, D. Long-term performance of pavement structures with cold in-place recycled base course. *Balt. J. Road Bridge Eng.* **2021**, *16*, 48–65. [CrossRef]
5. Liu, Z.; Sun, L.; Zhai, J.; Huang, W. A review of design methods for cold in-place recycling asphalt mixtures: Design processes, key parameters, and evaluation. *J. Cleaner Prod.* **2022**, *370*, 133530. [CrossRef]
6. Saidi, A.; Ali, A.; Lein, W.; Mehta, Y. A balanced mix design method for selecting the optimum binder content of cold in-place recycling asphalt mixtures. *Transp. Res. Rec.* **2019**, *2673*, 526–539. [CrossRef]
7. Ayala, F.C.; Sebaaly, P.E.; Hand, A.J.; Hajj, E.Y.; Baumgardner, G. Performance characteristics of cold in-place recycling mixtures. *J. Mater. Civ. Eng.* **2021**, *33*, 04021264. [CrossRef]
8. Schroeder, R.L. The use of recycled materials in highway construction. *Public Roads* **1994**, *58*, 32–41.
9. Kuchiishi, A.K.; Vasconcelos, K.; Bariani Bernucci, L.L. Effect of mixture composition on the mechanical behaviour of cold recycled asphalt mixtures. *Int. J. Pavement Eng.* **2021**, *22*, 984–994. [CrossRef]
10. Euch Khay, S.E.; Euch Ben Said, S.E.; Loulizi, A.; Neji, J. Laboratory investigation of cement-treated reclaimed asphalt pavement material. *J. Mater. Civ. Eng.* **2015**, *27*, 04014192. [CrossRef]
11. Ma, B.; Wang, H.; Wei, D. Performance of RAP in the system of cold inplace recycling of asphalt pavement. *J. Wuhan Univ. Technol. Mater. Sci. Ed.* **2011**, *26*, 1211–1214. [CrossRef]
12. Kim, Y.; Lee, H.D. Performance evaluation of Cold In-Place Recycling mixtures using emulsified asphalt based on dynamic modulus, flow number, flow time, and raveling loss. *KSCE J. Civ. Eng.* **2012**, *16*, 586–593. [CrossRef]
13. Ameri, M.; Behnood, A. Laboratory studies to investigate the properties of CIR mixes containing steel slag as a substitute for virgin aggregates. *Constr. Build. Mater.* **2012**, *26*, 475–480. [CrossRef]
14. Andavan, S.; Kumar, B.M. Case study on soil stabilization by using bitumen emulsions—A review. *Mater. Today Proc.* **2020**, *22*, 1200–1202. [CrossRef]
15. Yaghoubi, E.; Ghorbani, B.; Saberian, M.; van Staden, R.; Guerrieri, M.; Fragomeni, S. Permanent deformation response of demolition wastes stabilised with bitumen emulsion as pavement base/subbase. *Transp. Geotech.* **2023**, *39*, 100934. [CrossRef]
16. Gao, L.; Ni, F.; Braham, A.; Luo, H. Mixed-mode cracking behavior of cold recycled mixes with emulsion using arcan configuration. *Constr. Build. Mater.* **2014**, *55*, 415–422. [CrossRef]
17. Wang, H.F.; Ma, B.G.; Yin, X.B. Mechanical property effect of $Na_2SO_4$ on cement-reclaimed asphalt pavement mixture. *Adv. Mater. Res.* **2010**, *150–151*, 1209–1213. [CrossRef]
18. Gao, L.; Ni, F.; Charmot, S.; Li, Q. High-temperature performance of multilayer pavement with cold in-place recycling mixtures. *Road Mater. Pavement Des.* **2014**, *15*, 804–819. [CrossRef]
19. Ogbo, C.; Dave, E.; Sias, J. Laboratory investigation of factors affecting the evolution of curing in cold in-place recycled materials. *Transp. Res. Rec.* **2022**, *2676*, 28–40. [CrossRef]
20. Graziani, A.; Grilli, A.; Mignini, C.; Balzi, A. Assessing the field curing behavior of cold recycled asphalt mixtures. *Adv. Mater. Sci. Eng.* **2022**, *2022*, 4157090. [CrossRef]
21. Pérez, I.; Gómez-Meijide, B.; Pasandín, A.R.; García, A.; Airey, G. Enhancement of curing properties of cold in-place recycling asphalt mixtures by induction heating. *Int. J. Pavement Eng.* **2021**, *22*, 355–368. [CrossRef]
22. Saidi, A.; Ali, A.; Mehta, Y.; Decarlo, C.J.; Elshaer, M. Field assessment of cold in-place recycled asphalt mixtures using accelerated pavement testing. *J. Transp. Eng. Part B Pavements* **2022**, *148*, 04022035. [CrossRef]
23. Orosa, P.; Perez, I.; Pasandin, A.R. Evaluation of the shear and permanent deformation properties of cold in-place recycled mixtures with bitumen emulsion using triaxial tests. *Constr. Build. Mater.* **2022**, *328*, 127054. [CrossRef]
24. Teshale, E.Z.; Rettner, D.; Hartleib, A.; Kriesel, D. Application of laboratory asphalt cracking tests to cold in-place recycled mixtures. *Road Mater. Pavement Des.* **2017**, *18*, 79–97. [CrossRef]
25. Kaseer, F.; Yin, F.; Arámbula-Mercado, E.; Martin, A.E.; Daniel, J.S.; Salari, S. Development of an index to evaluate the cracking potential of asphalt mixtures using the semi-circular bending test. *Constr. Build. Mater.* **2018**, *167*, 286–298. [CrossRef]
26. Safazadeh, F.; Romero, P.; Mohammad Asib, A.S.; VanFrank, K. Methods to evaluate intermediate temperature properties of asphalt mixtures by the semi-circular bending (SCB) test. *Road Mater. Pavement Des.* **2022**, *23*, 1694–1706. [CrossRef]
27. Saha, G.; Biligiri, K.P. Novel procedural pragmatics of dynamic Semi-Circular Bending test for fatigue evaluation of asphalt mixtures. *Road Mater. Pavement Des.* **2019**, *20*, 454–461. [CrossRef]
28. Sabouri, M.; Wegman, D.E. Performance evaluation of cold in-place recycling materials through a simple semi-circular bending test. *Road Mater. Pavement Des.* **2022**, *23*, 1–15. [CrossRef]

29. Dolzycki, B.; Szydlowski, C.; Jaczewski, M. The influence of combination of binding agents on fatigue properties of deep cold in-place recycled mixtures in Indirect Tensile Fatigue Test (ITFT). *Constr. Build. Mater.* **2020**, *239*, 117825. [CrossRef]
30. Zhao, H.; Ren, J.; Chen, Z.; Luan, H.; Yi, J. Freeze and thaw field investigation of foamed asphalt cold recycling mixture in cold region. *Case Stud. Constr. Mater.* **2021**, *15*, 00710. [CrossRef]
31. Lyu, Z.; Shen, A.; Qin, X.; Yang, X.; Li, Y. Grey target optimization and the mechanism of cold recycled asphalt mixture with comprehensive performance. *Constr. Build. Mater.* **2019**, *198*, 269–277. [CrossRef]
32. Cheng, P.; Yi, J.; Chen, Z.; Luan, H.; Feng, D. Influence factors of strength and performance of foamed asphalt cold recycled mixture. *Road Mater. Pavement Des.* **2022**, *23*, 461–476. [CrossRef]
33. Wang, D.; Guo, T.; Chang, H.; Yao, X.; Chen, Y.; Wang, T. Research on the performance of regenerant modified cold recycled mixture with asphalt emulsions. *Sustainability* **2021**, *13*, 7284. [CrossRef]
34. Du, S. Effect of different fibres on the performance properties of cold recycled mixture with asphalt emulsion. *Int. J. Pavement Eng.* **2021**, *23*, 3444–3453. [CrossRef]
35. *AASHTO T 378-17*; Standard Method of Test for Determining the Dynamic MODULUS and Flow number for Asphalt Mixtures Using the Asphalt Mixture Performance Tester (AMPT). American Association of State Highway and Transportation Officials: Washington, DC, USA, 2017.
36. Bhattacharjee, S.; Gould, J.; Mallick, R.B.; Hugo, F. An evaluation of use of accelerated loading equipment for determination of fatigue response of asphalt pavement in laboratory. *Int. J. Pavement Eng.* **2004**, *5*, 61–79. [CrossRef]
37. *AASHTO TP 124-16*; Standard Method of Test For Determining the Fracture Potential of Asphalt Mixtures Using Semicircular Bend Geometry (Scb) At Intermediate Temperature. American Association of State Highway and Transportation Officials: Washington, DC, USA, 2016.
38. *JTG E20-2011*; Standard Test Methods of Bitumen and Bituminous Mixtures for Highway Engineering. People's Communications Press: Beijing, China, 2011.

**Disclaimer/Publisher's Note:** The statements, opinions and data contained in all publications are solely those of the individual author(s) and contributor(s) and not of MDPI and/or the editor(s). MDPI and/or the editor(s) disclaim responsibility for any injury to people or property resulting from any ideas, methods, instructions or products referred to in the content.

Article

# Dosage Effect of Wet-Process Tuff Silt Powder as an Alternative Material of Sand on the Performance of Reactive Powder Concrete

Yanxia Cai [1,2,3], Zhi Lin [4], Jingrui Zhang [5,*], Kaiji Lu [1,2,3], Linbing Wang [6], Yue Zhao [7] and Qianlong Huang [1,2,3,*]

1. Beijing Zhonglu Gaoke Highway Technology Co., Ltd., Beijing 100088, China; 13311530926@163.com (Y.C.); kj.lu@rioh.cn (K.L.)
2. Research and Development Center of Transport Industry of New Materials, Technologies Application for Highway Construction and Maintenance, Beijing 100088, China
3. Research Institute of Highway Ministry of Transport, Beijing 100088, China
4. Zhejiang Comm Mining Co., Ltd., Zhoushan 316000, China; nblinzhi3291@163.com
5. School of Civil Engineering, Hebei University of Engineering, Handan 056038, China
6. Department Civil & Environmental Engineering, Virginia Tech, Blacksburg, VA 24061, USA; wangl@vt.edu
7. Beijing General Research Institute of Mining and Metallurgy, Beijing 100160, China; zhaoyue@bgrimm.com
* Correspondence: z18335368596@163.com (J.Z.); hql806438934@163.com (Q.H.)

**Abstract:** A large amount of stone powder is produced during the production of machine-made sand. This research aims to study the effect of wet-process tuff silt powder (WTSP) dosages (as an alternative sand material to utilize waste stone powder and reduce environmental hazards) on reactive powder concrete's (RPC) mechanical performance. The physical and chemical properties of WTSP were analyzed as per relevant standards. This study prepared RPC samples with various WTSP content (0%, 6%, 12%, and 18%) to replace quartz sand at the same water–binder ratio (0.14) and allowed the samples to cure for 3 days, 7 days and 28 days prior to unconfined compression testing and flexural testing. Scanning electron microscopy (SEM) and Mercury Intrusion Porosimetry (MIP) testing were also carried out to observe the evolution of macroscopic properties in response to replacing part of quartz sand with the same amount of WTSP. The results show that the developed flexural and unconfined compressive strength (UCS) decreases slowly with a greater dosage of WTSP. However, when the WTSP content is 12% or less, the RPC made with WTSP satisfies the industrial application threshold regarding mechanical properties. For RPC samples containing more than 12% WTSP, the UCS and flexural strength showed a dramatic drop. Thus 12% of WTSP content was deemed the maximum and the corresponding UCS of 104.6 MPa and flexural strength of 12 MPa for 28 days of curing were the optimums. The microscopic characteristics indicate that the addition of WTSP can effectively fill the large pores in the RPC micro-structure, hence reducing the porosity of RPC. Furthermore, the WTSP can react with the cementitious material to form calcium aluminate during the hydration process, further strengthening the interface. The alkaline calcium carbonate in WTSP could improve the interfacial adhesion and make the structure stronger.

**Keywords:** wet-process tuff silt powder; RPC; mechanical properties; microstructure

## 1. Introduction

Sand and gravel are considered one of the most consumed non-renewable natural resources, which are vital to many industries, especially in building and road construction [1,2]. Since 2018, up to 50 billion tonnes of sand and gravel have been mined each year globally, leading to riverbank collapse, sinking deltas, coastal erosion, and biodiversity loss and causing environmental damage worldwide [3]. In recent years, natural sand and gravel have been gradually depleting with large-scale exploitation [4]. With the urbanization of many developing countries (especially China), the sand and gravel production has risen to a high level, and the price of sand and gravel has increased dramatically over the last decade, more than ten times in some areas, due to the shortage of sand and gravel [5]. In addition, with the awareness of sustainable development, to protect the natural ecological environment, various regions have introduced policies to restrict or ban natural sand mining [6–9]. As an alternative resource to natural sand, machine-made sand, obtained from crushing larger aggregate pieces into sand-sized aggregate particles, has been promoted and widely used in recent decades [10–12].

In 2020, China's total machine-made sand and gravel production was 19 billion tons [4,13–15]. According to processes, there are three main machine-made aggregate production methods: wet production, dry production, and dry–wet combination production. In Zhejiang, China, one artificial sand company used the wet method for machine-made sand production as the wet production method has the characteristics of a clean aggregate surface and low environmental pollution [16]. Waste material, called stone powder, is obtained from crushing larger aggregate pieces during sand production. In general, stone powder production accounts for about 10% of the total machine-made sand generation.

The accumulated waste material (stone powder) leads to serious environmental issues and requires a large number of land resources to deposit waste stone powder [17]. The utilization and treatment of stone powder have become a massive challenge for the sustainable development of China's building industry.

Reactive powder concrete (RPC) is a new type of cement-based composite material developed by French scholar Richard et al., in 1993 [18,19]. Its basic preparation principle is based on the closest packing theory, removing coarse aggregate, taking fine quartz sand as aggregate, and adding a high-efficiency water reducer and an appropriate amount of steel fiber. The main raw materials also include fine mineral admixtures such as silica fume and fly ash, Thus, internal defects are reduced to improve the compressive strength, toughness and durability of concrete [20,21]. The cover plate made of RPC has the characteristics of high strength, high durability, lightweight, easy modeling, high service life and remarkable long-term benefits but its cost is high. At present, RPC products such as RPC sidewalk cover plates and RPC cable trench cover plates are not widely used.

Studies have shown that stone powder can be used as an auxiliary cementing material or fine aggregate to replace part of the cement or fine aggregate to enhance the performance of concrete. Chadli Mounira et al. [22] added 30% of granite waste stone powder to reactive powder concrete (RPC) to replace quartz sand and improve RPC's mechanical properties. Fatih Hattatoglu et al. [23] used ahlat stone powder (ASP) as ignimbrite powder instead of cement in the production of RPC. The binder percentages obtaining the highest compressive and flexural strength were determined by changing the amounts of cement, so as to reduce the cost of the RPC. Yu et al. [24] replaced machine-made sand with granite porphyry powder to improve concrete's mechanical properties and durability. Jiang and Wang [25] illustrated that using natural river sand as aggregate and stone powder together as micro-aggregate to prepare RPC is feasible. The addition of stone powder can reduce the pore size and porosity of RPC. Fan [26] explored the feasibility of using marble powder and granite powder as cement admixtures and found that adding stone powder and granite powder would shorten the setting time of cement. When the amount of marble powder does not exceed 10%, it is beneficial to improve the fluidity of cement mortar. The contribution rate of marble powder to the strength of cement mortar is higher than that of granite powder, and the tuff rock powder is not used as a stone powder for concrete. Yang et al. [27] replaced fly ash with stone powder and studied the workability and compressive strength of concrete, which provided a certain reference for the rational use of stone powder in concrete. Yuan [28] conducted experiments to evaluate the stone powder content effects on concrete compressive strength and flexural strength, slump, chloride ion diffusion coefficient, and freeze–thaw resistance. The results show that when the content of stone powder is 5~10%, adding stone powder can effectively improve compressive strength, flexural strength, slump, freeze–thaw resistance, and other mechanical properties of concrete.

Currently, the research on stone powder mainly focuses on applying granite stone powder, limestone powder, marble powder and other wastes as mineral admixtures or replacing parts of cement and sand in concrete [29]. However, there are few studies on the wet-processed tuff silt powder (WTSP) and the application of WTSP in high-performance concrete. The addition of WTSP partially replacing quartz sand in RPC can realize the resource utilization of waste.

In this paper, the effects of WTSP as an alternative material of sand on the mechanical properties and microstructures of RPC were studied. The WTSP is used to replace part of quartz sand in RPC, which reduces the amount of quartz sand, hence reducing the RPC's preparation cost and promoting the resource utilization of waste stone powder. It is also of great significance for alleviating environmental pollution and provides a reference basis for applying stone powder in RPC.

## 2. Material

### 2.1. Cementitious Material

Ordinary Portland Cement (OPC 42.5) from a local manufacturer was used as the cementitious material. The physical properties and chemical composition of the cement, determined as per relevant standards are shown in Tables A1 and A2 in the Appendix A. This OPC has a density of 3.12 g/cm$^3$ and a specific surface area of 355 m$^2$/kg. The measured 28-day standard Flexural and Compressive strengths of the cement paste were 9.3 MPa and 53.2 MPa, respectively. The chemical composition of OPC mainly consists of carbon oxide (CaO) and silicon dioxide ($SiO_2$), with mass fractions of 63.57% and 20.58%, respectively.

Silica fume, microbeads powder, and mineral filler from the same manufacturer were used as additional cementitious material to prepare reactive powder concrete. Their physical and chemical properties are illustrated in Tables A3–A5 in the Appendix A. The silica fume is SF93 type silica fume, where the $SiO_2$ content is 94.61%, and the specific surface area is 18,648 m$^2$/kg. The microbeads powder adopts an uffa2.0-type microbeads powder; mineral powder adopts an S95-type mineral powder with a density of 2.9 g/cm$^3$.

### 2.2. Aggregate

Quartz sand from a local manufacturer was used as the original aggregate material, which will be partially replaced by WTSP. The chemical composition of the quartz sand is summarised in Table A6 in the Appendix A, which indicates that the quartz sand consists of 98.21% of silicon dioxide ($SiO_2$). The particle size distribution shows that more than 90% of quartz sand is retained between 0.85 and 0.425 mm sieves, and the quartz sand's fineness modulus (FM) is 3.11.

A manufactured WTSP from Zhejiang Comm Mining Co., Ltd., Zhoushan, China, wet-collected during mining production, was used as an alternative aggregate material to replace quartz sand. Its properties are shown in Table A7 in the Appendix A. The particle size distribution of the WTSP was measured by Malvern Mastersizer 2000 (Malvern Instruments Ltd., Malvern, UK) and displayed in Figure 1.

An X-ray fluorescence test (XRF) and X-ray diffraction test (XRD) were used for the chemical and mineralogical analysis of the WTSP. As shown in Table A8 in the Appendix A, the mineralogical composition of the stone power consists of feldspar (45.9%), quartz (20.6%), kaolinite (14.5%), montmorillonite (7.7%), calcite (5.9%), illite (4%), chlorite (1.1%) and pyrite (0.2%). As shown in Table A9 in the Appendix A, the WTSP's chemical composition mainly consists of silicon dioxide ($SiO_2$), Aluminium oxide ($Al_2O_3$), Potassium oxide ($K_2O$), and ferric oxide($Fe_2O_3$), with mass fractions of 61.51%, 16.40%, 4.91% and 4.28%, respectively; The corrosion analysis results in Table A10 in the Appendix A indicate the stone power has a pH value of 8.95 and contains $HCO_3^-$, $Ca^{2+}$, $Cl^-$, $SO_4^{2-}$.

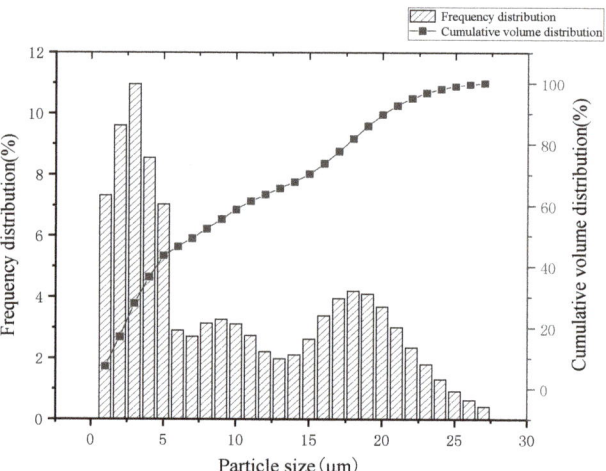

**Figure 1.** The frequency division and cumulative volume distribution of WTSP particle size.

From Figure 1 and Tables A7–A10 in the Appendix A, it can be seen that the average particle size of WTSP is 9.38 μm, and the small particle size helps to achieve compact filling of concrete and increase strength. The main crystalline minerals are quartz, feldspar, and kaolinite, which account for 81% of the total phase composition, and the main chemical components are $SiO_2$ and $Al_2O_3$, with their content up to 77.92%; through the analysis of corrosive substances, the pH of the WTSP is weakly alkaline. The content of corrosive substances is low. The compressive strength ratio of 7 days and 28 days shows that the WTSP has a certain activity, indicating that it is feasible to use WTSP to prepare reactive powder concrete.

*2.3. Additives, Fiber, and the Mixing Water*

The copper-plated flat steel fiber with a length of 13 mm, a diameter of 0.18 mm, and tensile strength of 3105 MPa was used to improve the strength of RPC.

A salt high-performance water reducing agent (Polycarboxylate superplasticizer), produced by Sika (China) Co., Ltd., Suzhou, China, was used to maintain the flowability of mixed slurry and the water-reducing efficiency of the agent is 33%.

Tap water was used for the slurry preparations.

## 3. Methodology

*3.1. Mix Designs*

To study the dosage influence of WTSP as an alternative sand material on reactive powder concrete performance, a total of 5 mix designs, as outlined in Table 1, were examined. Various stone power contents (0%, 6%, 12%, and 18%) were used to replace quartz sand at the same water–binder ratio (0.14). Sample A-0 prepared with 0% WTSP replacement was used as a reference specimen, which provides the standard unconfined compressive strength and flexural strength of RPC.

The dosage of the water reducing agent was used to adjust the fluidity of various mixed slurries; the fluidity of all RPC slurries should be greater than 180 mm. The amount of water-reducing agent needed shows an increasing trend with higher WTSP dosage. This is because the average particle size of WTSP and quartz sand of the same quality is much smaller than that of quartz sand. The surface area and the water demand ratio are large to achieve the same fluidity effect, which leads to an increase in the amount of water-reducing agent.

**Table 1.** Mixing ratio of wet WTSP instead of quartz sand (kg/m$^3$).

|  | Cement | Silica Fume | Mineral Filler | Microbead Powder | Quartz Sand | Steel Fiber | WTSP | 2% Desulfurization Gypsum | Defoamer | Water Reducing Agent | Water |
|---|---|---|---|---|---|---|---|---|---|---|---|
| A-0 | 650 | 180 | 120 | 200 | 1250 | 40 | 0 | 0 | 1 | 32 | 158 |
| A-1 | 650 | 180 | 120 | 200 | 1175 | 40 | 75 | 0 | 1 | 40 | 158 |
| A-2 | 650 | 180 | 120 | 200 | 1100 | 40 | 150 | 0 | 1 | 58 | 158 |
| A-3 | 650 | 180 | 120 | 200 | 1025 | 40 | 225 | 0 | 1 | 77 | 158 |
| A-4 | 650 | 180 | 120 | 200 | 1025 | 40 | 225 | 4.5 | 1 | 77 | 158 |

As illustrated in Table 1, an increasing WTSP content leads to a higher water-reducing agent requirement. The dosage of the water reducing agent is adjusted with the fluidity as the reference, and the fluidity of RPC is between 180–200 mm.

*3.2. Mixing, Forming and Curing*

The designed amount of aggregates (quartz sand, WTSP), steel fibers, and cementitious materials (cement, silica fume, mineral powder, fly ash) were blended in dry form in accordance with the selected mix designs outlined in Table 1. First, the dry materials were mixed for 4 min. The required amount of water and admixtures (water-reducing agent, defoaming agent, desulfurization gypsum) was added to the mixer and mixed for approximately 8 min to obtain slurries of uniform consistency. The prepared mixed slurry was poured into the test molds (two types of molds were used in this study, cubic molds with a dimension of 100 mm × 100 mm × 100 mm for UCS, prism molds with a dimension of 40 mm × 40 mm × 160 mm for flexural strength) and then vibrated on the vibrating table for 2 min to remove any entrapped air.

After 24 h of curing at atmosphere conditions, the specimens were demolded and moved into the accelerated curing box for curing, with an ambient temperature increased from 25 °C to 75 °C, at a heating rate of 10 °C/hour, and the temperature is kept at 75 °C for 24 h. After that, the specimens were naturally cooled in the curing box for 24 h and cured in atmosphere condition for designed dates (3 days, 7 days and 28 days).

*3.3. Test Method*

3.3.1. Unconfined Compressive Strength

Unconfined compressive strength tests were conducted for 100 mm × 100 mm × 100 mm cubic samples (see Figure 2) at designed curing periods using a Micro-electro-hydraulic servo pressure testing machine (HYE-2000, Hebei Sanyu Weiye Testing Machine Co., Ltd., Beijing, China) with a maximum capacity of 2000 kN (see Figure 3) at 1.2 MPa/s loading rate, in accordance with the Standards for Test Methods of Physical and Mechanical Properties of Concrete (GB/T 50081–2019). The number of cube samples at each predetermined curing period is 3.

**Figure 2.** Cubic samples for unconfined compressive strength tests.

**Figure 3.** Electro-hydraulic servo pressure testing machine.

3.3.2. Flexural Strength

A 40 mm × 40 mm × 160 mm prism sample (see Figure 4) was used for the flexural strength test (FS). The flexural strength of samples at the predetermined curing periods (3 days, 7 days, and 28 days) was determined via a fully automatic concrete flexural, compressive loading machine (YAW-300D, Zhejiang Schlikor equipment manufacturing Co., Ltd., Shaoxing, China), according to the Standards for Test Methods of Physical and Mechanical Properties of Concrete (GB/T 50081-2019), with a maximum capacity of 2000 kN (see Figure 5), at a load loading rate of 0.08 MPa/s. The number of beam samples at each predetermined curing period is 3.

**Figure 4.** Prism samples for flexural strength tests.

**Figure 5.** Fully automatic concrete flexural compressive all-in-one.

### 3.3.3. Microstructure

Scanning Electron Microscopy (SEM) is an observational means between transmission electron microscopy and optical microscopy. It uses a narrow-focused high-energy electron beam to scan the sample, acquire various physical information through the interaction between the beam and the material, and collect, amplify, and re-image the information to characterize the microscopic morphology of the material. The microscopic morphology of the samples cured for 3 days and 28 days were observed with a Japanese electron S4800 scanning electron microscope (see Figure 6).

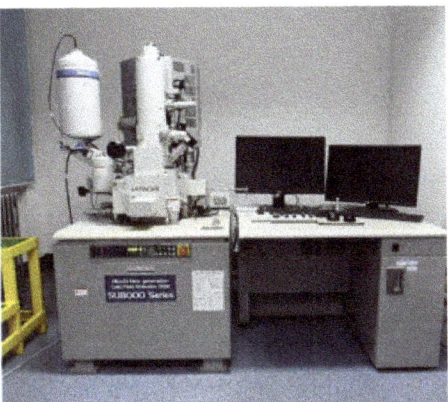

**Figure 6.** Field emission electron microscope.

Mercury intrusion porosimetry (MIP) was conducted to understand the pore volume of the corresponding pore size by measuring the amount of mercury entering the pores under different external pressures. The pore structure of the samples cured for 28 days was analyzed by a high-performance automatic mercury porosimeter AutoPore V9620 (Software Version 2.03.00, Micromeritics Instrument Corporation, Atlanta, GA, USA) produced by McMerritik, USA. The differential curves of porosity and pore size distribution were automatically analyzed and recorded.

## 4. Analysis of Test Results

### 4.1. Unconfined Compressive Strength of RPC

The UCS result of RPC samples prepared with various WTSP contents to replace quartz sand is shown in Figure 7. Overall, the compressive strength of RPC gradually increases with the increase in the curing period. With the increase in WTSP content, the compressive strength of RPC showed a downward trend. By adding 2% desulfurized gypsum to the A-4 sample, the compressive strength of 3 days and 7 days increased significantly, and so did the compressive strength of 28 days. The increase in strength is not apparent. This is because the $SO_3$ content in RPC is increased after adding desulfurized gypsum. Its stimulating effect and crystallization effect on sulfate activity will further improve the pozzolanic, morphological, and filling effect of other admixtures, thereby increasing the RPC mechanical properties [30]. A certain amount of desulfurized gypsum can increase the RPC's early strength and slightly improve the RPC's long-term strength [31–33]. The 28-day compressive strength of sample A-1 is 110.3 MPa, 11.9% lower than that of the reference sample A-0. The 28-day compressive strength of sample A-2 is 104.6 MPa, 14.1% lower than that of reference sample A-0. The 28-day compressive strength of sample A-3 is 91.6 MPa, 26.8% lower than that of reference sample A-0.

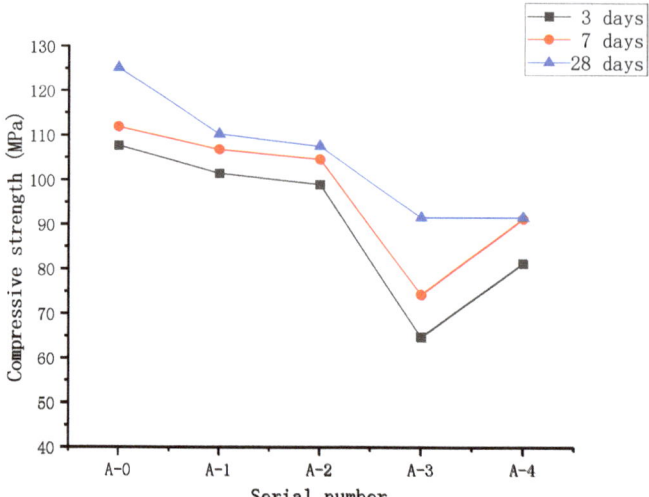

**Figure 7.** The unconfined compressive strength of RPC with WTSP replacing part of quartz sand.

The downward trend of RPC strength with the increase in WTSP content is mainly due to the low water content in RPC slurries. The used water-reducing agent maintains the fluidity without adding water for pumping purposes. However, the WTSP has a large specific surface area. It absorbs water and leads to insufficient water for the hydration process of cementitious material, the dispersion of WTSP within the cement matrix is poor [34,35], decreasing the UCS of RPC [36,37]. In addition, there are certain impurities in the WTSP, and the mud content is significant. With the increase in the WTSP content, the increasing amount of impurities and mud in WTSP might decelerate the hydration process in RPC [38].

On the other hand, the WTSP has a lower silicon–alumina oxide content than the quartz sand, and the strength of WTSP is lower than that of quartz sand, leading to a total lower strength. By replacing quartz sand with WTSP, the aggregate gradation of the prepared mixture was changed, hence influencing the development of RPC compressive strength.

### 4.2. Flexural Strength

Figure 8 illustrates the flexural strength of RPC samples prepared with various WTSP contents to replace quartz sand. In general, for samples at the curing period of 3 days, 7 days, and 28 days, the flexural strength decreased with the increasing amount of wet-processed WTSP. The cement hydration reaction is delayed due to the increase in WTSP replacement and the increasing amount of water-reducing agent. Furthermore, the addition of 2% desulfurized gypsum, sample A-4, shows a retarding effect on the binder hydration, and the setting time is 1.6 times that of the control group, which further affects the flexural strength of RPC [39]. When the curing age is 28 days, the flexural strength of sample A-1 is significantly lower than the flexural strength of the reference group by 14.6%, and the flexural strength is 12.3 MPa.

With the increase in WTSP content, the flexural strength further decreases and was found rather marginal. Compared with the flexural strength of the reference group, the flexural strength of sample A–2 and sample A–3 decreased by 16.7% and 18.1%, and the flexural strength was 12.0 MPa and 11.8 MPa, respectively.

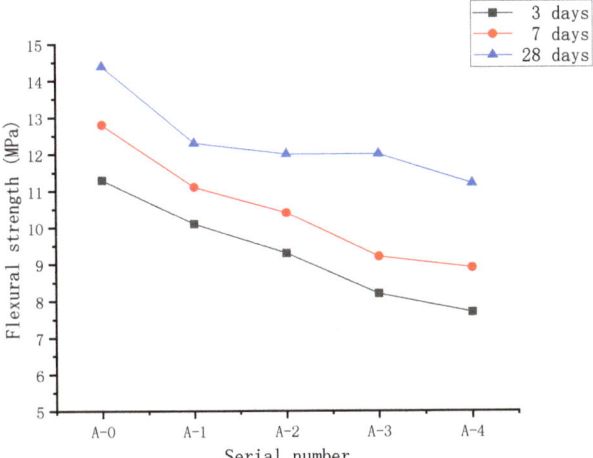

**Figure 8.** The flexural strength of RPC with WTSP replacing part of quartz sand.

Replacing part of the quartz sand with WTSP will adversely affect the flexural strength of RPC. The downward trend of flexural strength is due to the particular amount of mud in the WTSP. With the increase in the amount of WTSP, the mud content also increases, which affects the hydration of cement. On the other hand, the Zeta potential of WTSP is greater than that of cement, and the adsorption rate of water reducer is higher than that of cement, which leads to the increase in the amount of water reducer [6], the increase in the amount of water-reducing agent affects the RPC flexural strength as well. It can be found from Figure 8 that the replacement of part of quartz sand by WTSP has an effect on the flexural strength of RPC. This is due to the active $SiO_2$ and $Al_2O_3$ components in WTSP. It could react with the cement hydration product Ca (OH)$_2$ to generate more Calcium Silicate Hydrate (C-S-H) gel, this is conducive to the development of the flexural strength of RPC. In addition, the WTSP contains a large amount of rough and fine particles, which changes the gradation of aggregates and improves the cohesion between particles and the section compactness. The improved cohesion between particles and good cross-sectional compactness have certain positive effects on enhancing the flexural strength of concrete [40,41].

*4.3. Analysis of Micro Characteristics*

In order to further study the influence of WTSP dosage on the mechanical properties of RPC, the specimens with various mix designs at the age of 3 days and 28 days were analyzed by scanning electron microscopy (SEM). Figures 9 and 10 illustrate the SEM micrographs for the samples A-0 to A-4 cured for 3 days and 28 days, respectively.

Figure 9a shows that the SEM micrograph for reference sample A-0 has obvious cracks and pores with a loose structure, and exposed quartz sand could be observed. Figure 9b shows that a small amount of spindle-shaped Monosulfur calcium sulfoaluminate hydrate (AFm) crystals could be observed in A-1 samples after 3 days of hydration. However, a small amount of delicate pores and some un-hydrated Ca (OH)$_2$ crystals are also distributed on the micrograph. When the WTSP content is 12%, as shown in Figure 9c, the SEM micrographs of sample A-2 contain hexagonal plate-shaped calcium trisulfide hydrated calcium sulfoaluminate (AFt) crystals. When the WTSP content is 18%, in Figure 9d, a small amount of Ca (OH)$_2$ crystals are found in the micrograph of sample A-3 and a dense microstructure could be found on the surface of sample A-3. After adding 2% desulfurized gypsum, in Figure 9e, the hydration produced rod-shaped AFt crystals could be observed. Overall, in Figure 9, for all RPC samples cured for 3 days, the SEM micrographs all showed un-hydrated substances, which indicates the uncompleted cement hydration process at an early age and influences the overall UCS and flexural strength. In Figure 10a, the structure of the interface zone of the A-0 sample is loose and lamellar Ca(OH)$_2$ crystals are locally layered. The structural morphology of the interface area between the A-1 sample in Figure 10b and the A-4 sample in Figure 10e is obviously loose, and the exposed quartz sand particles can also be seen.

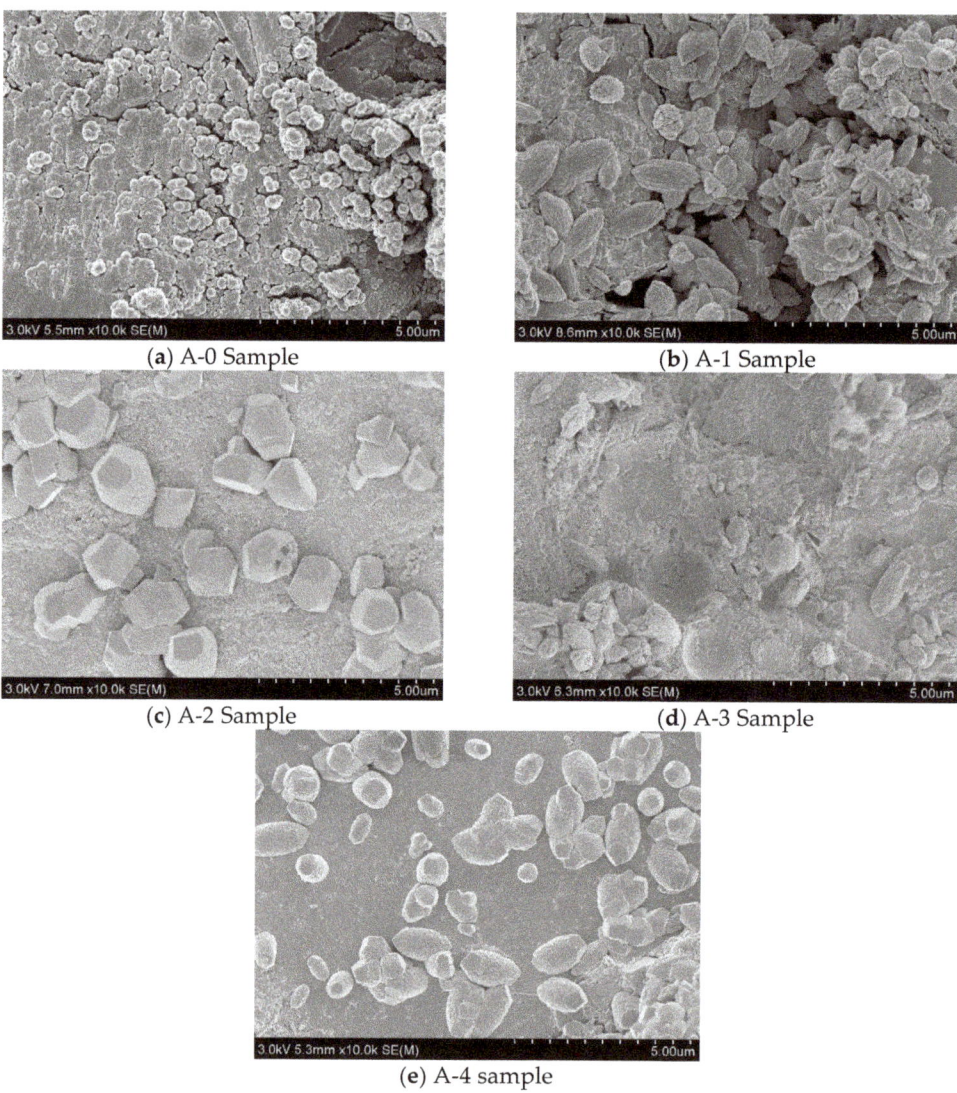

(a) A-0 Sample  (b) A-1 Sample
(c) A-2 Sample  (d) A-3 Sample
(e) A-4 sample

**Figure 9.** Three-day SEM photo of RPC sample.

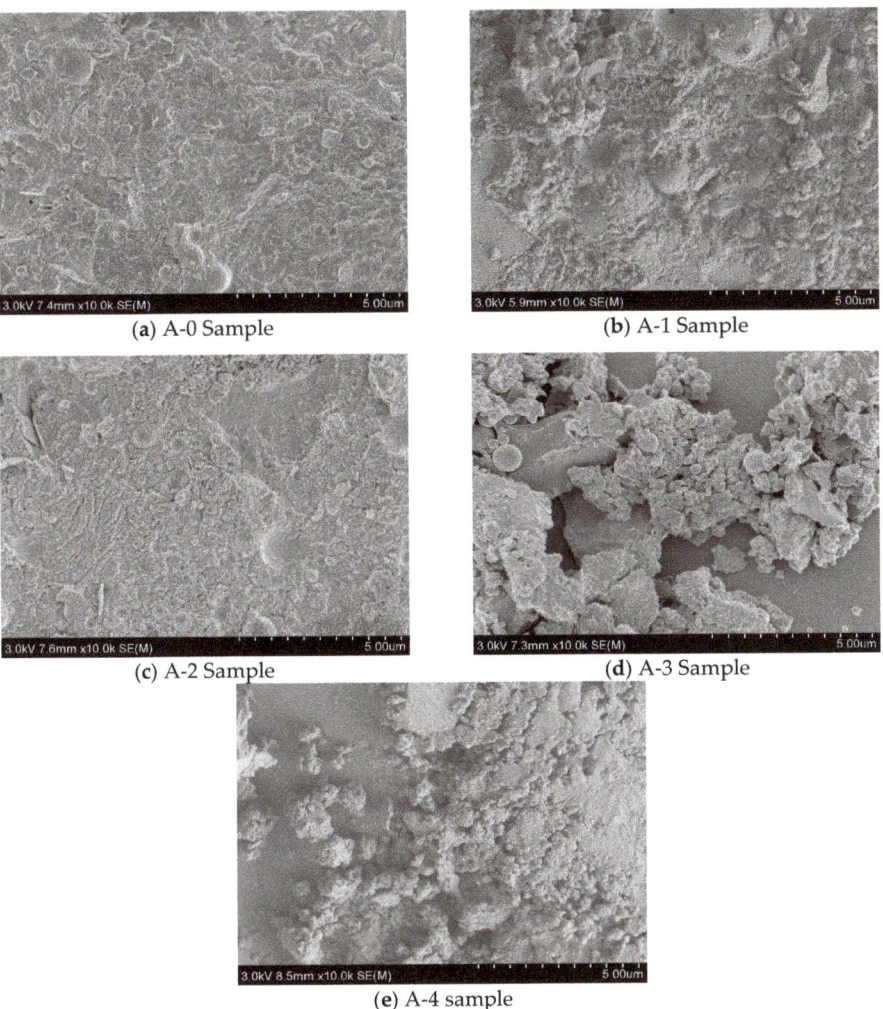

**Figure 10.** Twenty-eighth day SEM photo of RPC sample.

In Figure 10c, there are a few granular and lamellar Ca (OH)$_2$ crystals in the interface area of the A-2 sample, and the interface area is dense. Figure 10d shows rod-shaped C-S-H and a small amount of granular Ca (OH)$_2$ crystal interface area in the A-3 sample interface area. The results show that WTSP has a nucleation effect, due to the small size and high surface energy of WTSP, this enhances the nucleation effect and the deposition of hydration products on its surface so as to form a denser microstructure [42–44]. The WTSP has a pozzolanic effect, it can chemically react with tricalcium aluminate (C$_3$A) and tetracalcium ferric aluminate (C$_4$AF) in the cement during cement hydration to form hydrated calcium aluminate. The formation of alkaline calcium carbonate in wet-processed WTSP makes the hydration products tricalcium silicate (C$_3$S) and CaCO$_3$ have a denser micro-interface structure, while Ca (OH)$_2$ forms crystals on the surface of CaCO$_3$, which makes the Ca (OH)$_2$ grains refined, thereby improving the interfacial adhesion and making the RPC more compact, thus improve RPC flexural strength [45–47]. However, when the content of WTSP exceeds 12%, the micro-aggregate ratio of RPC deviates from the optimal value. The large mass fraction of free WTSP particles and impurities affect the bonding between the quartz sand and the cementing material, resulting in a decrease in strength.

*4.4. Analysis of Pore Structure Characteristics*

The pore structure characteristic of RPC samples cured for 28 days was also tested using the Mercury Intrusion Porosimetry (MIP). Figure 11 illustrates the cumulative porosity of sample A0–A4 cured for 28 days. As shown in Figure 11, the porosity of the sample A0–A3 is 4.6%, 3.7%, 3.4%, and 4.4%, when the WTSP content is 0%, 6%, 12%, and 18%, respectively. The addition of WTSP increases the number of fine particles in the RPC sample, filling the pores in the sample and forming a dense microstructure, reducing the porosity of the RPC sample. Figure 12 illustrates the pore size distribution of RPC samples A0–A4 cured for 28 days. When the WTSP content is 0%, 6%, 12%, and 18%, the average pore diameter of RPC is 41.43 nm, 27.04 nm, 24.30 nm, and 28.73 nm, respectively. From Figure 11 to Figure 12, it can be found that when the WTSP content is 12%, the larger diameter pores (>20 nm) in the RPC are significantly reduced, the concrete density is improved, and the porosity is reduced. However, when the WTSP content is more than 12%, the larger pores in RPC samples have already been filled where the nano-pores remain uninfluenced as the WTSP particles are more significant than the nano-pores. The increasing amount of WTSP will not further improve the micro-structure of the RPC sample [48,49]. Furthermore, the impurity content in the WTSP will cause a decrease in the compressive strength of the RPC, and the excess amount of WTSP may lead to the interfacial microcracks, resulting in a decrease in overall RPC UCS [41,50]. Hence, when the content of WTSP exceeds 12%, the addition of WTSP will not play a positive role in the mechanical properties development of RPC.

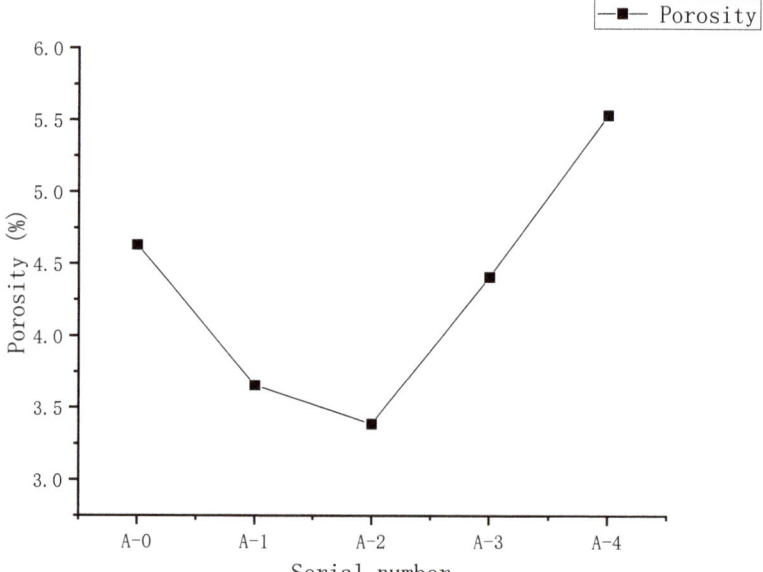

**Figure 11.** Twenty-eight-day porosity of RPC samples with different numbers.

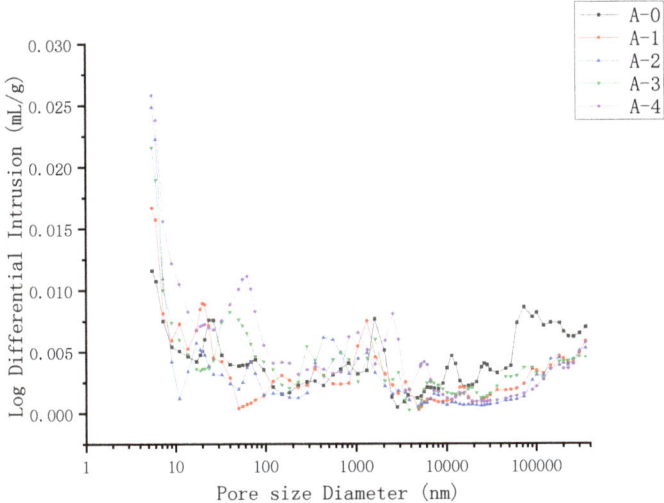

**Figure 12.** Twenty-eight-day pore size distribution of RPC samples with different numbers.

## 5. Conclusions

This research presents an experimental study of the effect of WTSP dosages on reactive powder concrete's performance. Unconfined compressive strength and flexural strength were measured for RPC samples of all five mixes with three curing periods (3 days, 7 days and 28 days). Scanning electron microscopy and mercury intrusion porosimetry (MIP) testing was also carried out for selected RPC samples to identify the influence of WTSP content on the microstructural properties of RPC. Based on the results obtained, the following conclusions can be drawn from this study:

(1) The main crystalline minerals of WTSP are quartz, feldspar, and kaolinite, accounting for 81% of the total phase composition. The main chemical components are $SiO_2$ and $Al_2O_3$, the content is as high as 77.92%, and the average particle size is 9.38 μm. The WTSP does not contain any corrosive substances. Through the 7-day and 28-day UCS tests, it was found that the WTSP has a certain activity for the hydration of cementitious material. The fine particles in WTSP could fill the large pores in RPC, leading to a dense microstructure. Hence, it is feasible to use solid waste WTSP to prepare reactive powder concrete.

(2) Under a constant water–binder ratio, when replacing the quartz sand with the same quality of the WTSP, the UCS of RPC gradually increases with the increase in curing age. With the increase in WTSP content, the UCS of RPC showed a downward trend and was found rather marginal. By adding 2% desulfurized gypsum to the A-4 sample, the compressive strength on day 3 and day 7 are significantly improved. However, the 28-day UCS remains unchanged. The RPC flexural strength shows a downward trend as the WTSP dosage increases. The flexural strength does not increase significantly by adding 2% desulfurized gypsum to the A-4 sample. RPC prepared with WTSP content within 12% can meet mechanical properties requirements [51]. At this content, the compressive strength of the RPC cube is 104.6 MPa, and the flexural strength is 12.0 MPa.

(3) The WTSP can react with tricalcium aluminate ($C_3A$) and tetracalcium ferric aluminate ($C_4AF$) to form hydrated calcium aluminate during the hydration process. The alkaline calcium carbonate in WTSP makes the structure of the micro-interface between hydration product tricalcium silicate ($C_3S$) and $CaCO_3$ denser. Moreover, $Ca(OH)_2$ generates crystals on the surface of $CaCO_3$, which makes the $Ca(OH)_2$ grains refined, thereby improving the interfacial adhesion and making the RPC structure denser.

(4) The incorporated WTSP helps fill the larger diameter pores in RPC, hence improving the RPC's pore structure and reducing the RPC's porosity. When the content of WTSP is 12%, the total porosity of the sample is 0.0122 mL/g, which is reduced by 27.5% compared with reference sample A-0.

In this study, a WTSP–RPC preparation technology is proposed. Although it causes the loss of RPC mechanical properties, it still meets the requirements of RPC mechanical properties. It is a good choice to apply it to the RPC cover plate. In future work, through the laboratory test, the simplification of the WTSP–RPC curing process and the durability of WTSP–RPC will be studied.

**Author Contributions:** Methodology, Q.H. and J.Z.; investigation, Y.C., Z.L. and Q.H.; formal analysis, Y.C., Q.H., Z.L. and J.Z.; project administration, Y.C. and Q.H.; writing—original draft preparation, Y.C., Q.H. and J.Z.; writing—review and editing, K.L., L.W. and Y.Z.; resource support, Y.C., K.L., Z.L. and L.W. All authors have read and agreed to the published version of the manuscript.

**Funding:** This research received no external funding.

**Institutional Review Board Statement:** Not applicable.

**Informed Consent Statement:** Not applicable.

**Data Availability Statement:** Data are contained within the article.

**Conflicts of Interest:** The authors declare no conflict of interest.

## Appendix A

Table A1. Physical properties of Ordinary Portland Cement.

| Specific Surface Area | Density | Standard Consistency | Soundness | Setting Time (min) | | Flexural Strength (MPa) | | | Compressive Strength (MPa) | | |
| --- | --- | --- | --- | --- | --- | --- | --- | --- | --- | --- | --- |
| | | | | First Set | Final Set | 3 days | 7 days | 28 days | 3 days | 7 days | 28 days |
| m²/kg | g/cm³ | % | Boiling Method | | | | | | | | |
| 355 | 3.12 | 25.4 | qualified | 99 | 158 | 6.4 | 7.3 | 9.3 | 27.0 | 37.8 | 53.2 |

Table A2. Chemical composition of Ordinary Portland Cement.

| Element | $SiO_2$ | $Al_2O_3$ | $Fe_2O_3$ | CaO | MgO | $SO_3$ | $Na_2Oeq$ | f-CaO | Loss on Ignition | $Cl^-$ |
| --- | --- | --- | --- | --- | --- | --- | --- | --- | --- | --- |
| Content (%) | 20.58 | 4.97 | 3.76 | 63.57 | 2.29 | 2.00 | 0.53 | 0.75 | 1.40 | 0.026 |

Table A3. Physical and chemical properties of silica fume.

| Detection Indicator | Total Alkali Content (%) | $SiO_2$ Content (%) | Chlorine Content (%) | Moisture Content (%) | Ignition Loss (%) | Water Demand Ratio (%) | Specific Surface Area m²/kg | 7 days Activity Index | Radioactivity |
| --- | --- | --- | --- | --- | --- | --- | --- | --- | --- |
| Test results | 0.88 | 94.61 | 0.016 | 1.4 | 1.26 | 112 | 18,648 | 120 | none |

Table A4. Physical and chemical properties of microbeads.

| Detection Indicator | 45 μm Sieve Residue (%) | Water Demand Ratio (%) | Ignition Loss (%) | $SO_3$ Content (%) | f-CaO (%) | Moisture Content (%) |
| --- | --- | --- | --- | --- | --- | --- |
| Test results | 12.5 | 98 | 2.5 | 0.1 | 0 | 0.1 |

Table A5. Physical and chemical properties of mineral powder.

| Detection Indicator | Density (g/cm$^3$) | Specific Surface Area (m$^2$/kg) | Ignition Loss (%) | 7 days Activity Index | 28 days Activity Index | Flow Ratio (%) | Moisture Content (%) |
|---|---|---|---|---|---|---|---|
| Test results | 2.9 | 455 | 2.3 | 78 | 96 | 100 | 0.3 |

Table A6. Chemical composition analysis of quartz sand.

| Element | $SiO_2$ | $Al_2O_3$ | $Fe_2O_3$ | $K_2O$ | $Na_2O$ |
|---|---|---|---|---|---|
| Content (%) | 98.21 | 0.83 | 0.03 | 0.02 | 0.02 |

Table A7. Properties of WTSP.

| Performance | Density (g/cm$^3$) | Methylene Blue Value | Surface Area/Volume (cm$^2$/cm$^3$) | The Average Particle Size (μm) | 7 days Activity Index (%) | 28 days Activity Index (%) | Water Demand Ratio (%) | pH |
|---|---|---|---|---|---|---|---|---|
| Test results | 2.561 | 9 | 14,225 | 9.38 | 55.3 | 59.3 | 117.2 | 8.95 |

Table A8. Phase analysis of WTSP.

| Phase | Quartz | Feldspar | Calcite | Kaolinite | Chlorite | Lllite | Montmorillonite | Pyrite |
|---|---|---|---|---|---|---|---|---|
| Content (%) | 20.6 | 45.9 | 5.9 | 14.5 | 1.1 | 4 | 7.7 | 0.2 |

Table A9. Chemical composition of WTSP.

| Element | CaO | MgO | $SiO_2$ | $Al_2O_3$ | $Fe_2O_3$ | FeO | $TiO_2$ | $Na_2O$ | MnO | $K_2O$ | $P_2O_5$ | $H_2O^+$ | $SO_3$ |
|---|---|---|---|---|---|---|---|---|---|---|---|---|---|
| Content (%) | 3.20 | 0.44 | 61.52 | 16.40 | 4.28 | 1.95 | 0.49 | 2.28 | 0.40 | 4.91 | 0.06 | 2.51 | <0.1 |

Table A10. Corrosion analysis of WTSP.

| Test Items | pH | $Ca^{2+}$ | $Mg^{2+}$ | $Cl^-$ | $CO_3^{2-}$ | $HCO_3^-$ | $SO_4^{2-}$ |
|---|---|---|---|---|---|---|---|
| Content (mg/kg) | 8.95 | 0.081 | 0.012 | 0.054 | 0.000 | 0.308 | 0.078 |

## References

1. Su, X. Research on C55 box girder concrete based on quartz sandstone aggregate. *China Foreign Highw.* **2018**, *38*, 315–318.
2. Chen, X.; Zhen, F. Research on the application of recycled aggregate concrete in road base course. *China Foreign Highw.* **2020**, *40*, 249–252.
3. Yang, Z.; Wang, S.; Wei, Z.; Han, L.; Mo, M. Research on the application of stone powder in concrete. *China Cem.* **2021**, *7*, 101–102.
4. The future trend of sand and gravel industry and some issues that enterprises should pay attention to. *Jiangxi Build. Mater.* **2021**, *7*, 309–311.
5. Hu, Y.; Zhang, P.; Zhao, J. Current problems and countermeasures facing my country's sand and gravel industry. *China Build. Mater.* **2021**, *1*, 128–130.
6. Sun, R.; Wang, Z.; Huang, F.; Yi, Z.; Yuan, Z.; Xie, Y.; Li, H. Research on the properties of stone powder-cement composite cementitious materials with different lithology. *Mater. Guide* **2021**, *35*, 211–215.
7. Wang, X.; Chen, H.; Zhan, Q. Experimental study on the influence of stone powder content on the performance of cement mortar. *China Build. Mater. Sci. Technol.* **2021**, *30*, 52–54.
8. Yang, C.; Li, C.; Zhao, Y. Research on the effect of different polymer fillers in concrete on its properties. *Plast. Technol.* **2021**, *49*, 39–42.
9. Wei, Y.; Huang, X. Research on mechanical and shrinkage properties of machine-made sand concrete. *J. Jiaxing Univ.* **2021**, *33*, 94–97.
10. Wang, J.; Zhou, H.; Ge, C.; Chen, Y. Influence of stone powder on the working performance and mechanical properties of high-strength machine-made sand concrete. *J. Drain. Irrig. Mach. Eng.* **2021**, *39*, 804–810.
11. Li, D.; Chi, H.; Hu, Y.; He, J. The influence of ground machine-made sand tailings on concrete performance. *Highway* **2021**, *66*, 342–346.

12. Liu, J.; Liu, L. Experimental study on the strength and elastic modulus of machine-made sand high-strength concrete. *Build. Struct.* **2020**, *50*, 57, 96–99.
13. Liu, J. Current status and suggestions for the development and utilisation of sand and gravel aggregates in Gansu Province. *China Min. Ind.* **2020**, *29*, 8–10.
14. Xi, J. Research on the status quo of my country's sand and gravel industry industrial chain. *Economist* **2017**, *2*, 28–30.
15. Huang, Z. The application status of construction sand and gravel aggregates and analysis of the utilisation of ore waste residues. *Green Build. Mater.* **2021**, *5*, 7–8.
16. Li, W.; Tang, W.; Zhao, D. Analysis of the characteristics of the combined dry and wet production process of machine-made aggregates. *Constr. Sci. Technol.* **2020**, *4*, 63–65.
17. Hao, T.; Tian, W.; Cao, L.; Leng, L.; Ye, W.; Tang, C. Influence of a large amount of waste granite powder on the mechanical properties of mortar. *Concrete* **2021**, *103–106*, 111.
18. Richard, P.; Cheyrezy, M. Composition of reactive powder concrete. *Cem. Concr. Res.* **1995**, *25*, 1501–1511. [CrossRef]
19. Richard, P.; Cheyrezy, M. Reactive Powder Concrete with High Ductility and 200~800 MPa Compressive Strength. *Spec. Publ.* **1994**, *144*, 507–518.
20. Li, X.; Yang, C.; Zhou, Q. Study on strength and fluidity of reactive powder concrete based on orthogonal test. *Silic. Bull.* **2019**, *38*, 1201–1210.
21. Liu, J.; Wang, D. Effect of curing on properties of mineral fine powder reactive powder concrete. *J. Wuhan Univ. Technol.* **2009**, *31*, 100–103.
22. Chadli, M.; Tebbal, N.; Mellas, M. Impact of elevated temperatures on the behavior and microstructure of reactive powder concrete. *Constr. Build. Mater.* **2021**, *300*, 124031. [CrossRef]
23. Hattatoglu, F.; Bakis, A. Usability of ignimbrite powder in reactive powder concrete road pavement. *Road Mater. Pavement Des.* **2017**, *18*, 1448–1459. [CrossRef]
24. Yu, H.; Liu, T.; Wang, H.; Xie, C.; Li, S. Research on the influence of granite porphyry powder content on the properties and microstructure of concrete. *J. Jilin Univ.* **2021**, *11*, 1–13.
25. Jiang, J.; Wan, H. Experimental study on the preparation of RPC using river sand and stone powder. *Concrete* **2015**, *8*, 151–154, 158.
26. Fan, P. Research on the influence of waste stone powder on the performance of cement mortar. *Fujian Constr. Sci. Technol.* **2020**, *5*, 42–45.
27. Yang, K.; Huang, X.; Wang, Z.; Ding, H. The application of machine-made sand to collect stone powder in concrete. *Sichuan Build. Mater.* **2021**, *47*, 16–18.
28. Yuan, X. Analysis of the influence of machine-made sand powder content on concrete performance. *Eng. Mach. Maint.* **2021**, *3*, 182–184.
29. Cheng, Q. Study on the mechanical properties and microstructure of RPC mixed with granite powder. *Concr. Cem. Prod.* **2016**, *8*, 8–13, 19.
30. Qian, D.; Sun, L. The effect of desulfurised gypsum on the mechanical properties of composite cementitious concrete. *Non-Met. Miner.* **2013**, *36*, 63–65.
31. Qian, D. Research on the influence of desulfurised gypsum on the workability of high performance concrete. *Concrete* **2013**, *8*, 76–78, 82.
32. Tan, P.; Wang, H. The influence of desulfurised gypsum content on concrete performance. *Compr. Util. Miner. Resour.* **2010**, *1*, 46–48.
33. Shi, H.; Cai, Y. The effect of desulfurised gypsum on the properties of slag cement. *Cem. Technol.* **2006**, *1*, 26–30.
34. Ramezani, M.; Kim, Y.H.; Sun, Z. Mechanical properties of carbon-nanotube-reinforced cementitious materials: Database and statistical analysis. *Mag. Concr. Res.* **2020**, *72*, 1047–1071. [CrossRef]
35. Ramezani, M.; Kim, Y.H.; Sun, Z. Probabilistic model for flexural strength of carbon nanotube reinforced cement-based materials. *Compos. Struct.* **2020**, *253*, 112748. [CrossRef]
36. Dong, X.; Li, W.; Wang, K.; Yang, K. Analysis of the influence of concrete admixtures on concrete performance. *Green Build. Mater.* **2021**, *11*, 11–12.
37. Qin, L. Analysis of the influence of polycarboxylic acid water reducer on the compressive strength of concrete. *Chem. Manag.* **2021**, *32*, 157–158.
38. Pan, W.H. Study on the influence of mud content in sand on the mixing performance, mechanical properties and durability of C30 self compacting concrete. *Guangdong Build. Mater.* **2020**, *36*, 12–14.
39. Qian, D.; Zhang, R. Research on the influence of desulfurised gypsum on the performance of high performance concrete. *Concr. Cem. Prod.* **2013**, *4*, 10–13.
40. Chen, F.; Zhang, Y.; Liang, J.; Tang, Y.; Rong, H. Research on the anti-skid durability of pavement machine-made sand cement concrete. *Concrete* **2021**, *10*, 44–47.
41. Wang, H.; Guo, J. The influence of stone powder content on the performance of high performance concrete. *Concrete* **2021**, *8*, 75–78.
42. Ramezani, M.; Kim, Y.H.; Sun, Z. Modeling the mechanical properties of cementitious materials containing CNTs. *Cem. Concr. Compos.* **2019**, *104*, 103347. [CrossRef]

43. Ramezani, M.; Kim, Y.H.; Sun, Z. Elastic modulus formulation of cementitious materials incorporating carbon nanotubes: Probabilistic approach. *Constr. Build. Mater.* **2021**, *274*, 122092. [CrossRef]
44. Ramezani, M.; Dehghani, A.; Sherif, M.M. Carbon nanotube reinforced cementitious composites: A comprehensive review. *Constr. Build. Mater.* **2022**, *315*, 125100. [CrossRef]
45. Xue, G.; Wang, C.; Zhang, J.; Lu, X.; Liu, Z.; Zhang, T. Influence of different particle size of quartz powder on the strength and microstructure analysis of reactive powder concrete. *J. Shantou Univ.* **2021**, *36*, 50–58.
46. Chen, L.; Zhou, C.; Jiang, C. The effect of metakaolin on the mechanical properties and microstructure of reactive powder concrete. *Bull. Chin. Ceram. Soc.* **2021**, *40*, 1162–1169.
47. Sun, K.; Liu, Y.; Liu, Z.; Zhang, K. The influence of forming pressure on the compressive strength and microstructure of 200 MPa reactive powder concrete. *Concrete* **2014**, *1*, 28–30, 34.
48. Xing, J.; Yu, Q.; Quan, S.; Li, Y.; Liu, C.; Deng, Z.; Zhang, Y. Research on the influence of waste rock powder on the properties of 3days printing cement mortar. *Concr. Cem. Prod.* **2021**, *12*, 1–5.
49. Yang, H.; Chen, C.; Fan, Z. The effect of granite machine-made sand and gravel powder on the performance of C80 high-strength and high-performance concrete and its mechanism. *Water Transp. Eng.* **2021**, *11*, 13–20, 49.
50. Chen, X. Research on the mechanism of the influence of stone powder on the fluidity of cement mortar. *Fujian Build. Mater.* **2021**, *8*, 20–23.
51. General Administration of Quality Supervision, Inspection and Quarantine of the People's Republic of China. *China National Standardization Administration Reactive Powder Concrete*; China Standards Press: Beijing, China, 2015.

*Article*

# Fast-Acquiring High-Quality Prony Series Parameters of Asphalt Concrete through Viscoelastic Continuous Spectral Models

Yan Zhang [1] and Yiren Sun [2,*]

[1] City Institute, Dalian University of Technology, Dalian 116600, China; yanzhang_dut@163.com
[2] School of Transportation and Logistics, Dalian University of Technology, Dalian 116024, China
* Correspondence: sunyiren@dlut.edu.cn

**Abstract:** Prony series representations have been extensively applied to characterizing the time-domain linear viscoelastic (LVE) material functions for asphalt concrete. However, existing methods that can generate high-quality Prony series parameters (i.e., discrete spectra) mostly involve complicated programming algorithms, which poses a challenge for quick access of Prony series parameters. Also, very limited research has been devoted to establishing methods for simultaneously determining both retardation and relaxation spectra. To resolve these issues, this study presented a practical approach to fast acquiring high-quality Prony series parameters for both relaxation modulus and creep compliance of asphalt concrete by using the complex modulus test data. The approach adopts the analytical representations of the continuous relaxation and retardation spectra from the Havriliak-Negami (HN) and 2S2P1D complex modulus models to directly determine the discrete spectra, and the elastic constants, $E_e$ and $D_g$, for both LVE modulus and compliance functions are further calculated by fitting the corresponding generalized Maxwell model representations to smoothed data from the storage modulus representations of the HN and 2S2P1D complex modulus models. In this way, all the procedures in the proposed method can be easily implemented in Microsoft Excel. The results showed that the HN and 2S2P1D models yielded slightly different continuous spectral patterns at shorter relaxation times and longer retardation times. However, at the region covered by the test data, the continuous spectra of the two complex modulus models were very close to each other. Thus, the two models can generate comparable Prony series parameters within the time or frequency range covered by the test data. Considering that the quality of the resulting Prony series parameters are closely related to the master curve models used for presmoothing, the HN and 2S2P1D models were compared with the conventional Sigmoidal model. Additionally, the Black diagram was recommended for examining the quality of the complex modulus test data before constructing the master curves.

**Keywords:** asphalt concrete; Prony series; Havriliak-Negami (HN) model; 2S2P1D model; continuous relaxation and retardation spectra

**Citation:** Zhang, Y.; Sun, Y. Fast-Acquiring High-Quality Prony Series Parameters of Asphalt Concrete through Viscoelastic Continuous Spectral Models. *Materials* 2022, *15*, 716. https://doi.org/10.3390/ma15030716

**Academic Editor:** Francesco Canestrari

Received: 6 December 2021
Accepted: 17 January 2022
Published: 18 January 2022

**Publisher's Note:** MDPI stays neutral with regard to jurisdictional claims in published maps and institutional affiliations.

**Copyright:** © 2022 by the authors. Licensee MDPI, Basel, Switzerland. This article is an open access article distributed under the terms and conditions of the Creative Commons Attribution (CC BY) license (https://creativecommons.org/licenses/by/4.0/).

## 1. Introduction

Asphalt concrete, which has been paved on most roadways in the world, is a typical particulate composite with a viscoelastic matrix. In engineering applications, it is commonly regarded as a linear viscoelastic (LVE) material [1–3]. As such, many mechanical tests based on the LVE theory, like the static relaxation and creep tests and the dynamic complex modulus test, can be used for characterizing its LVE behavior. Theoretically, the properties from these tests such as the relaxation modulus, creep compliance and complex modulus are equivalent [4–6]; however, for a practical purpose, the uniaxial compressive complex modulus test has been widely accepted as a standard LVE material characterization test. After the complex modulus test data is obtained, it is usually required to extract the LVE information from the test data through mathematical models.

In the LVE theory, the generalized Maxwell model and generalized Kelvin model appear to be the most commonly used models for describing both time- and frequency-domain material functions, and they have been implemented into many commercial numerical simulation programs, e.g., ABAQUS, ANSYS and COMSOL Multiphysics [7–11]. This can be primarily attributed to their high computational efficiency and wide applicability [3,12,13]. The two models are composed of linear springs and dashpots linked in different configurations, and mathematically yield the so-called Prony series expressions for the relaxation modulus and creep compliance in time domain [3,12,14]. The Prony series expressions are not only very convenient to be converted analytically into the frequency-domain complex modulus and compliance, but also can considerably facilitate the computation of the convolution integrals for the LVE constitutive equations due to the presence of decaying exponential terms. Therefore, accurate and efficient identification of the Prony series parameters (i.e., discrete relaxation and retardation spectra) is crucial to the subsequent performance analysis and prediction of asphalt pavement or mixtures.

To date, researchers have proposed various methods for determining the Prony series parameters. Several representative approaches that apply directly to raw data in the time or frequency domain have been widely used for LVE materials, e.g., the collocation method by Schapery [15], the multidata method by Cost and Becker [16], and the windowing method by Tschoegl and Emri [17]. Nonetheless, these classic schemes would encounter difficulties when utilized for asphalt concrete. Two major issues, namely negative spectrum strengths and local spectrum oscillations, occur frequently due to the narrowband nature of the Prony series terms and significant scatters in test data. To address these problems, presmoothing techniques have been introduced by using broadband functions, like the Sigmoidal model [2], power-law series [3], Huet-Sayegh model [18], Havriliak-Negami (HN) model [14] and 2S2P1D model [19,20]. The use of these broadband functions not only improves the quality of the test data, but facilitates the data shift in accordance with the time-temperature superposition principle (TTSP) during the construction of master curves. In view of the equivalence of the LVE material functions, some interconversion algorithms [13,21,22] were also presented to calculate the Prony series parameters of the relaxation functions from the retardation functions, or vice versa.

On the other hand, the continuous spectrum-based methods attract increasing attention from the asphalt paving research community in that they are able to eliminate negative spectrum strengths and excessive parameters. Levenberg [23] developed a continuous relaxation spectrum model for asphalt concrete by using a lognormal distribution function. However, this model is symmetrical on the logarithmic timescale and thus may not be appropriate for all mixtures. Zhao et al. [2] established a confining pressure dependent continuous relaxation spectrum by considering the relationship between the relaxation spectrum and storage modulus. Luo et al. [24] and Lv et al. [25] respectively deduced continuous relaxation spectra from a modified power law-based relaxation modulus model and a generalized Sigmoidal model-based storage modulus model. Nevertheless, these works were all concentrated on the relaxation spectrum, and thus may be inconvenient for those who need fast solutions for both retardation and relaxation functions. Aiming at this issue, Sun et al. [26] presented a numerical approach to determining a continuous spectrum from the other. Bhattacharjee et al. [27] and Zhang et al. [28] calculated the two continuous spectra from storage modulus and storage compliance separately based on the Sigmoidal function and the generalized Sigmoidal function; however, due to the inconsistency of the model parameters of the storage modulus and storage compliance, the LVE relationship cannot be strictly satisfied.

Although there have been so many methods developed for determining the Prony series parameters as mentioned above, most of them involve complicated programming algorithms, which poses a challenge for quick access of Prony series parameters. Furthermore, very limited research has been devoted to establishing approaches for determining both retardation and relaxation spectra at the same time. To deal with these problems, this study gave a practical approach by adopting analytical representations of the continuous relaxation and retardation spectra from two complex-valued models, and all the procedures in the method can be easily implemented in Microsoft Excel.

## 2. Materials and Complex Modulus Test

Two dense-graded asphalt mixtures, denoted as Mix-13.2 and Mix-9.5 herein, were prepared for the complex modulus testing. Mix-13.2 had a nominal maximum aggregate size (NMAS) of 13.2 mm. The coarse aggregates and asphalt binder used were limestone and PG 58-22 unmodified asphalt, respectively. Mix-9.5 had a NMAS of 9.5 mm. The coarse aggregates and asphalt binder used were granite and PG 64-22 neat asphalt, respectively. The asphalt contents of Mix-13.2 and Mix-9.5 were 3.9% and 5.7%, respectively. Figure 1 presents the aggregate gradations of the two asphalt mixtures.

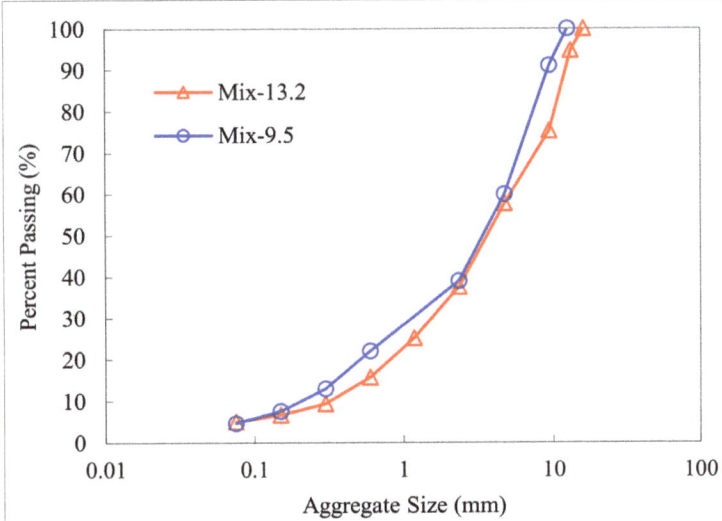

**Figure 1.** Aggregate gradations of Mix-13.2 and Mix-9.5.

The complex modulus tests were performed on all specimens of the two mixtures in accordance with the standard testing method AASHTO T342 [29]. The two mixtures were first compacted using the Superpave Gyratory Compactor and then trimmed into the final cylindrical specimens (150 mm in height and 100 mm in diameter) containing an air void content of $4 \pm 1\%$. For each mixture, three replicate specimens were fabricated.

The complex modulus testing was conducted on a universal testing machine (UTM). A stress-controlled compressive mode was employed for all the complex modulus tests. For Mix-13.2, five testing temperatures ($-10$, 5, 20, 35 and 50 °C) and seven loading frequencies (0.1, 0.5, 1, 5, 10, 20 and 25 Hz) were adopted, and for Mix-9.5, five testing temperatures ($-16$, 4, 24, 40 and 50 °C) and six loading frequencies (0.1, 0.5, 1, 5, 10 and 25 Hz) were adopted. During testing, the strain was kept between 50~150 με and the accumulated strain was controlled below 1500 με to ensure LVE measurements. By means of the obtained stress and strain data, two quantities, i.e., the dynamic modulus $|E^*|$ and phase angle $\varphi$, can be calculated as follows:

$$|E^*| = \frac{\sigma_0}{\varepsilon_0} \qquad (1)$$

$$\varphi = \frac{\Delta t}{t_\mathrm{p}} \times 360° \qquad (2)$$

where $\sigma_0$ and $\varepsilon_0$ are the amplitudes of the axial stress and strain; $\Delta t$ is the time lag of the strain curve behind the stress curve; $t_\mathrm{p}$ is the loading period. The dynamic modulus $|E^*|$, which is the absolute value of the complex modulus $E^*$, characterizes the resistance to deformation of a viscoelastic material, whereas the phase angle $\varphi$ characterizes the extent to which the viscoelastic material behaves like a viscous liquid ($\varphi = 90°$) or an elastic solid ($\varphi = 0°$).

## 3. Methodology

### 3.1. Viscoelastic Master Curve Models

In this study, two models, i.e., the HN model and the 2S2P1D model, were employed to build the complex modulus master curves of the asphalt mixtures. The HN model has five parameters and is represented by [30]:

$$E^*(\omega) = E_g + \frac{E_e - E_g}{\left[1 + (i\omega\tau_0)^\alpha\right]^\beta} \quad (3)$$

where $E_g$ is the glassy modulus; $E_e$ is the equilibrium modulus; $\omega = 2\pi f$ is the angular frequency; $f$ is the frequency; $i = \sqrt{-1}$; $\alpha$, $\beta$ and $\tau_0$ are model parameters and they respectively control the width, asymmetry and horizontal position of the relaxation spectrum.

The real part $E'$ and imaginary part $E''$ of the complex modulus $E^*$ have the following relationship with the dynamic modulus $|E^*|$ and phase angle $\varphi$:

$$E^* = E' + iE'' = |E^*|\cos\varphi + i|E^*|\sin\varphi \quad (4)$$

where $E'$ is the storage modulus; $E''$ is the loss modulus. Thus, the dynamic modulus $|E^*|$ and phase angle $\varphi$ can be calculated using the storage modulus $E'$ and the loss modulus $E''$, as follows

$$|E^*| = \sqrt{E'^2 + E''^2} \text{ and } \varphi = \arctan\frac{E''}{E'} \quad (5)$$

From the HN model, the representations of the storage modulus and loss modulus can be analytically separated out according to De Moivre's formula, as the following [30]:

$$E' = E_g + \frac{(E_e - E_g)\cos(\beta\psi)}{\left[1 + 2\omega^\alpha\tau_0^\alpha\cos(\alpha\pi/2) + \omega^{2\alpha}\tau_0^{2\alpha}\right]^{\beta/2}} \quad (6)$$

$$E'' = \frac{(E_g - E_e)\sin(\beta\psi)}{\left[1 + 2\omega^\alpha\tau_0^\alpha\cos(\alpha\pi/2) + \omega^{2\alpha}\tau_0^{2\alpha}\right]^{\beta/2}} \quad (7)$$

$$\psi = \arctan\frac{\omega^\alpha\tau_0^\alpha\sin(\alpha\pi/2)}{1 + \omega^\alpha\tau_0^\alpha\cos(\alpha\pi/2)} \quad (8)$$

Obviously, with the analytical expressions of the storage and loss moduli, those for the dynamic modulus and phase angle are also available according to Equation (5).

The 2S2P1D model, composed of two spring elements, two parabolic elements and a dashpot element, possesses seven parameters and has the following mathematical form [19]:

$$E^*(\omega) = E_e + \frac{E_g - E_e}{1 + \alpha(i\omega\tau_0)^{-k} + (i\omega\tau_0)^{-h} + (i\omega\beta\tau_0)^{-1}} \quad (9)$$

where $\alpha$, $k$ and $h$ ($0 < k < h < 1$) are the parameters of the two parabolic elements; $\tau_0$ is a temperature-dependent parameter; $\beta$ is the parameter associated with the Newtonian viscosity of the dashpot element, $\eta = (E_g - E_e)\beta\tau_0$.

According to De Moivre's formula, the storage and loss moduli representations of the 2S2P1D model can also be derived, as follows:

$$E' = E_e + \frac{(E_g - E_e)(1 + A)}{(1 + A)^2 + B^2} \quad (10)$$

$$E'' = \frac{(E_e - E_g)B}{(1+A)^2 + B^2} \tag{11}$$

$$A = \alpha(\omega\tau_0)^{-k}\cos(k\pi/2) + (\omega\tau_0)^{-h}\cos(h\pi/2) \tag{12}$$

$$B = -\alpha(\omega\tau_0)^{-k}\sin(k\pi/2) - (\omega\tau_0)^{-h}\sin(h\pi/2) - (\omega\beta\tau_0)^{-1} \tag{13}$$

Since the storage and loss moduli of the HN and 2S2P1D models can all be derived analytically from the corresponding complex-valued models, they accurately meet the Kronig–Kramers relation that correlates the real and imaginary parts of the response to a harmonic load to each other theoretically [31].

Besides, in the viscoelastic theory, when the representation of the complex modulus $E^*$ is known, the complex compliance $D^*$ can be analytically obtained by taking the inverse of the complex modulus, as follows:

$$D^* = D' - iD'' = 1/E^* \tag{14}$$

$$D' = \frac{E'}{E'^2 + E''^2} \tag{15}$$

$$D'' = \frac{E''}{E'^2 + E''^2} \tag{16}$$

where $D'$ is the storage compliance; $D''$ is the loss compliance.

### 3.2. Continuous Relaxation and Retardation Spectra

For a LVE material, the modulus functions in the time and frequency domains can be uniformly expressed using the continuous relaxation spectrum $H(\rho)$ through integral forms [31]:

$$E'(\omega) = E_e + \int_{-\infty}^{\infty} H(\rho)\frac{\omega^2\rho^2}{1+\omega^2\rho^2}d\ln\rho \tag{17}$$

$$E''(\omega) = \int_{-\infty}^{\infty} H(\rho)\frac{\omega\rho}{1+\omega^2\rho^2}d\ln\rho \tag{18}$$

$$E(t) = E_e + \int_{-\infty}^{\infty} H(\rho)e^{-t/\rho}d\ln\rho \tag{19}$$

where $\rho$ is the relaxation time; $E(t)$ is the relaxation modulus; $t$ is the loading time.

Similarly, the compliance functions of a LVE material in the time and frequency domains can be uniformly expressed using the continuous retardation spectrum $L(\tau)$ through integral forms [31]:

$$D'(\omega) = D_g + \int_{-\infty}^{\infty} L(\tau)\frac{1}{1+\omega^2\tau^2}d\ln\tau \tag{20}$$

$$D''(\omega) = \int_{-\infty}^{\infty} L(\tau)\frac{\omega\tau}{1+\omega^2\tau^2}d\ln\tau \tag{21}$$

$$D(t) = D_g + \int_{-\infty}^{\infty} L(\tau)\left(1 - e^{-t/\tau}\right)d\ln\tau \tag{22}$$

where $\tau$ is the retardation time; $D(t)$ is the creep compliance.

The continuous relaxation spectrum $H(\rho)$ and continuous retardation spectrum $L(\tau)$ essentially contain identical time- and frequency-dependent material information; thus, they are equivalent of characterizing the LVE behavior of a material. As can be seen from Equations (17)–(22), once the continuous relaxation and retardation spectra are determined, both modulus and compliance functions in the time and frequency domain can be attained.

In accordance with the LVE theory, the continuous spectra have the following relationships with the complex modulus $E^*$ [31]:

$$H(\rho) = \pm \pi^{-1} \mathrm{Im} E^*(i\omega)\big|_{i\omega \to \rho^{-1} e^{\pm i\pi}} = \pm \pi^{-1} \mathrm{Im} E^*\left(\rho^{-1} e^{\pm i\pi}\right) \tag{23}$$

$$L(\tau) = \mp \pi^{-1} \mathrm{Im} D^*(i\omega)\big|_{i\omega \to \tau^{-1} e^{\pm i\pi}} = \mp \pi^{-1} \mathrm{Im} D^*\left(\tau^{-1} e^{\pm i\pi}\right) = \mp \pi^{-1} \mathrm{Im}\left[E^*\left(\tau^{-1} e^{\pm i\pi}\right)\right]^{-1} \tag{24}$$

where Im represents the operation of retaining the imaginary part of a complex-valued function.

For the HN model, Havriliak and Negami [30] presented the analytical expression of $H(\rho)$ through Equation (23), as follows:

$$H(\rho) = \frac{(E_g - E_e)(\rho/\tau_0)^{\alpha\beta} \sin(\beta\phi)}{\pi \left[1 + (\rho/\tau_0)^{2\alpha} + 2(\rho/\tau_0)^{\alpha} \cos(\alpha\pi)\right]^{\beta/2}} \tag{25}$$

$$\phi = \arctan \frac{\sin(\alpha\pi)}{(\rho/\tau_0)^{\alpha} + \cos(\alpha\pi)} \tag{26}$$

Analogously, Sun et al. [26] derived the close-form solution for $L(\tau)$ of the HN model through Equation (24), as follows:

$$L(\tau) = \frac{\Omega(E_g - E_e) \sin(\beta\phi)}{\pi \left\{ \left[(E_e - E_g)\cos(\beta\phi) + E_g\Omega\right]^2 + \left[(E_e - E_g)\sin(\beta\phi)\right]^2 \right\}} \tag{27}$$

$$\Omega = \left[1 + (\tau_0/\tau)^{2\alpha} + 2(\tau_0/\tau)^{\alpha} \cos(\alpha\pi)\right]^{\beta/2} \tag{28}$$

For the 2S2P1D model, Alavi et al. [32] obtained the analytical representation of $H(\rho)$ through Equation (23), as follows:

$$H(\rho) = \frac{(E_g - E_e)X}{\pi(X^2 + Y^2)} \tag{29}$$

$$X = 1 + \alpha \tau_0^{-k} \rho^k \cos(k\pi) + \tau_0^{-h} \rho^h \cos(h\pi) - \beta^{-1} \tau_0^{-1} \rho \tag{30}$$

$$Y = \alpha \tau_0^{-k} \rho^k \sin(k\pi) + \tau_0^{-h} \rho^h \sin(h\pi) \tag{31}$$

Sun et al. [33] successfully deduced the analytical expression for $L(\tau)$ of the 2S2P1D, as follows:

$$L(\tau) = \frac{Y(E_g - E_e)(X^2 + Y^2)}{\pi \left\{ \left[E_e(X^2 + Y^2) + X(E_g - E_e)\right]^2 + \left[Y(E_g - E_e)\right]^2 \right\}} \tag{32}$$

It is noted that $\rho$ in $X$ and $Y$ of Equation (32) should be replaced by $\tau$.

Evidently, for the HN and 2S2P1D models, the corresponding relaxation modulus $E(t)$ and creep compliance $D(t)$ in the time domain can be readily calculated with the continuous relaxation and retardation spectra according to Equations (19) and (22). Further, in terms of the Boltzmann superposition integrals, the constitutive relationships for the LVE material can be determined [31].

### 3.3. Construction of Master Curves

Asphalt concrete is a typical thermorheologically simple material in the LVE region; therefore, the master curves for various LVE material functions in both frequency and time domains can be constructed in accordance with the time–temperature superposition principle (TTSP). During this process, viscoelastic test data measured at different temperatures is shifted horizontally along the frequency or time axis on the logarithmic scale, thus generating a smooth master curve at a given reference temperature $T_r$. By means of the constructed master curve, the LVE behavior over a wider range of loading time or frequency than that offered by the test instrument can be predicted. The reduced angular frequency $\omega_r$ and reduced time $t_r$ for the shifted test data are represented by:

$$\omega_r = \omega \times \alpha_T \tag{33}$$

$$t_r = \frac{t}{\alpha_T} \tag{34}$$

where $\alpha_T$ is the time-temperature shift factor. The time–temperature shift factors can be represented using a function of temperature, e.g., the Williams–Landel–Ferry (WLF) or the Arrhenius equation, or in a non-functional form. To avoid the effect of the functional expression of $\alpha_T$, the non-functional method was adopted for constructing the master curve of the complex modulus in the present study.

The parameters of the complex modulus model and time-temperature shift factors were determined simultaneously through a nonlinear optimization process. To fully extract the LVE information, both dynamic modulus and phase angle test data were taken into account, and the target error function to minimize was as the following:

$$F = \frac{1}{N}\sqrt{\sum_{i=1}^{N}\left(\frac{|E^*_{m,i}|-|E^*_{c,i}|}{|E^*_{m,i}|}\right)^2} + \frac{1}{N}\sqrt{\sum_{i=1}^{N}\left(\frac{\varphi_{m,i}-\varphi_{c,i}}{\varphi_{m,i}}\right)^2} \tag{35}$$

where $N$ is the number of the dynamic modulus or phase angle data points; $|E^*_{m,i}|$ and $\varphi_{m,i}$ are the measured values for the dynamic modulus and phase angle, respectively; $|E^*_{c,i}|$ and $\varphi_{c,i}$ are the calculated values for the dynamic modulus and phase angle from the master curve model used, respectively. The optimization operation can be easily completed using the Solver in Microsoft Excel. Before this, initial values for both master curve model parameters and shift factors should be given. The reference temperatures for Mix-13.2 and Mix-9.5 were set to 20 and 24 °C, respectively.

### 3.4. Determination of Prony Series Parameters

As stated above, once the parameters of the complex-valued models, like HN and 2S2P1D models, are known, the corresponding $H(\rho)$ and $L(\tau)$ can be automatically determined due to the existence of their analytical expressions with the same parameters as the original complex modulus models. Although all the modulus and compliance functions can further be straightforward calculated with $H(\rho)$ and $L(\tau)$ through Equations (17)–(22), the integral forms based on the continuous spectra are actually inconvenient to implement in numerical simulation techniques, e.g., the finite element method. Instead, the Prony series expressions on the basis of discrete spectra have been extensively utilized due to their advantage at computation efficiency.

The relaxation modulus expression derived from the generalized Maxwell model and the creep compliance expression derived from the generalized Kelvin model are two typical Prony series representations. For the generalized Maxwell model, the modulus functions can be formulated by [31]:

$$E(t) = E_e + \sum_{j=1}^{n} E_j e^{-t/\rho_j} \tag{36}$$

$$E'(\omega) = E_e + \sum_{j=1}^{n} E_j \frac{\omega^2 \rho_j^2}{1+\omega^2 \rho_j^2} \tag{37}$$

$$E''(\omega) = \sum_{j=1}^{n} E_j \frac{\omega \rho_j}{1+\omega^2 \rho_j^2} \tag{38}$$

where $E_j$ is the modulus of the spring or the relaxation strength; $\rho_j = \eta_j/E_j$ is the discrete relaxation time; $\eta_j$ is the viscosity of the dashpot; the set of Prony series parameters $[\rho_j, E_j]$ is called the discrete relaxation spectrum.

For the generalized Kelvin model, the compliance functions can be represented by [31]:

$$D(t) = D_g + \sum_{j=1}^{n} D_j \left(1 - e^{-t/\tau_j}\right) \tag{39}$$

$$D'(\omega) = D_g + \sum_{j=1}^{n} D_j \frac{1}{1+\omega^2 \tau_j^2} \tag{40}$$

$$D''(\omega) = \sum_{j=1}^{n} D_j \frac{\omega \tau_j}{1+\omega^2 \tau_j^2} \tag{41}$$

where $D_j$ is the compliance of the spring or the retardation strength; $\tau_j = \lambda_j D_j$ is the discrete retardation time; $\lambda_j$ is the viscosity of the dashpot; the set of Prony series parameters $[\tau_j, D_j]$ is called the discrete retardation spectrum.

In fact, when the discrete relaxation and retardation spectra become infinitely dense, they evolve into the so-called continuous relaxation and retardation spectra. As such, Equations (36)–(41) can be interpreted as discretizations of Equations (17)–(22). For the storage modulus $E'(\omega)$, $H(\rho)d\ln\rho$ in Equation (17) represents the contribution of the model to the modulus function in the interval of $\ln\rho$ and $\ln\rho+d\ln\rho$, which leads to the following derivation:

$$\begin{aligned} E'(\omega) &= E_e + \int_{-\infty}^{\infty} H(\rho) \frac{\omega^2 \rho^2}{1+\omega^2 \rho^2} d\ln\rho \\ &\approx E_e + \sum_{j=1}^{n} \left[H(\rho_j) \times \Delta \ln \rho_j\right] \frac{\omega^2 \rho_j^2}{1+\omega^2 \rho_j^2} \\ &= E_e + \sum_{j=1}^{n} E_j \frac{\omega^2 \rho_j^2}{1+\omega^2 \rho_j^2} \end{aligned} \tag{42}$$

Likewise, for the storage compliance $D'(\omega)$, the integral form based on the continuous retardation spectrum and the series expression based on the discrete retardation spectrum have the following relationship:

$$\begin{aligned} D'(\omega) &= D_g + \int_{-\infty}^{\infty} L(\tau) \frac{1}{1+\omega^2 \tau^2} d\ln\tau \\ &\approx D_g + \sum_{j=1}^{n} \left[L(\tau_j) \times \Delta \ln \tau_j\right] \frac{1}{1+\omega^2 \tau_j^2} \\ &= D_g + \sum_{j=1}^{n} D_j \frac{1}{1+\omega^2 \tau_j^2} \end{aligned} \tag{43}$$

During the determination of the Prony series parameters, the discrete time constants, $\rho_j$ and $\tau_j$, are commonly preselected. Specifically, they are set to values with equal intervals on the logarithmic scale according to Equation (44):

$$\rho_i = \tau_i = b \times 10^{d+i/M} \tag{44}$$

where $b$ and $d$ are specified according to the logarithmic time range covered by the shifted test data, and generally $b = 1$; $M$ is the number of the discrete times assumed in each decade on the logarithmic scale.

It can be observed that with the discrete time constants ($\rho_j$ and $\tau_j$) known, the Prony series coefficients, namely the relaxation and retardation strengths ($E_j$ and $D_j$), can be quickly and easily calculated using the following equations:

$$E_i = H(\rho_i) \times \Delta \ln \rho_i \tag{45}$$

$$D_i = L(\tau_i) \times \Delta \ln \tau_i \tag{46}$$

$$\Delta \ln \rho_i = \Delta \ln \tau_i = \frac{1}{M} \ln 10 \tag{47}$$

Finally, the remaining two elastic constants $E_e$ and $D_g$ can be determined by fitting Equations (37) and (40) to the corresponding real part expression of the original complex modulus model, like Equation (6) or (10), over the range covered by the shifted test data through the Excel Solver. In such a manner, all the Prony series parameters can be fast acquired. Actually, $E_e$ and $D_g$ have an analytical relationship, as follows:

$$D_g = \frac{1}{E_g} = \frac{1}{E_e + \sum_{j=1}^{n} E_j} \tag{48}$$

Therefore, once the elastic constant $E_e$ along with the discrete relaxation strengths $E_j$ is available, $D_g$ can be obtained accordingly.

## 4. Results and Discussion
### 4.1. Examination of Test Data Quality of Asphalt Concrete

Before constructing the master curves, it is crucial to examine the quality of the complex modulus measurements. In the present study, the Black diagram [34] was employed to conduct this manipulation, in which the dynamic modulus $|E^*|$ is plotted against the phase angle $\varphi$ in a single plane. Since for a thermorheologically simple material, all the components of the complex modulus are the functions of the reduced angular frequency, any two of them can form a unique curve in a complex plane. In the Black diagram, the angular frequency axis can be treated as an additional axis perpendicular to the complex plane in accordance with the right-hand rule. Thus, the testing temperatures would have no effect on the analysis of the overlapping behavior of the test data during the construction of master curves in the Black diagram. A smoother Black curve generally represents a higher quality of the test data. In such a manner, the Black diagram allows an effective and efficient detection of inconsistency with thermorheological simplicity.

Figure 2 shows the resulting Black diagrams for the two asphalt mixtures. It can be observed that in both diagrams, the complex modulus test data obtained at different temperatures basically formed unique curves, indicating the compliance with thermorheological simplicity under the test conditions as well as the applicability of the TTSP. In addition, the test results at lower temperatures exhibited better overlapping behavior, whereas those at higher temperatures showed slightly higher dispersion. This is mainly because nonlinear behaviors (e.g., the viscoplastic deformation) of asphalt concrete occur more easily at higher temperatures, which impact the measurement of LVE responses of the material to a certain degree.

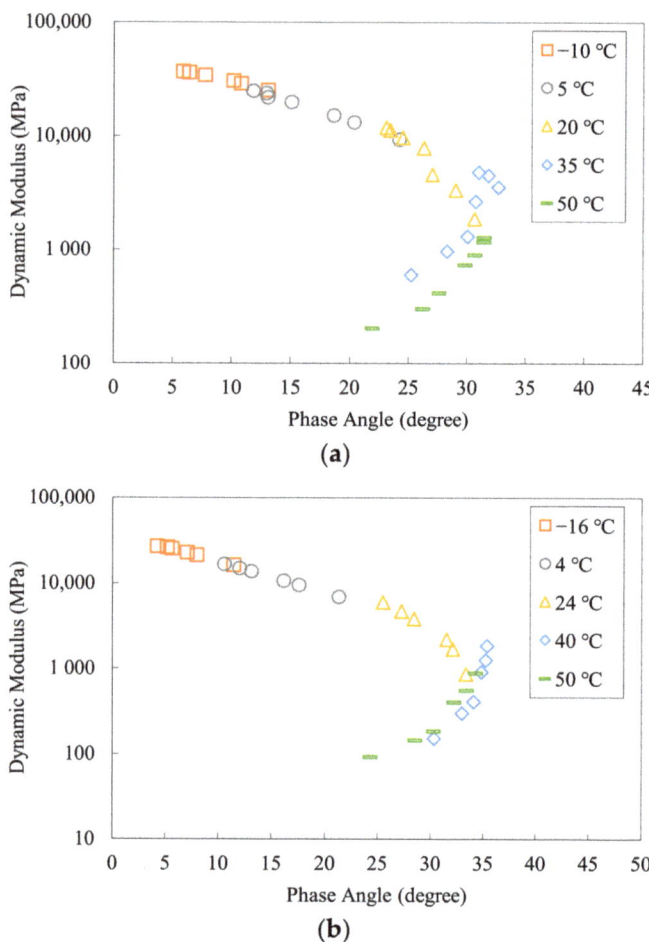

**Figure 2.** Black diagram of the complex modulus test data set: (**a**) Mix-13.2; (**b**) Mix-9.5.

*4.2. Analysis of Results from the Developed Method*

Figures 3 and 4 present the master curves of the dynamic modulus and phase angle respectively developed from the HN and 2S2P1D models. As observed, both models fitted to the test data of the two mixtures very well. Tables 1 and 2 list the resulting model parameters and fitting errors. For Mix-13.2, the two complex modulus models contributed to very close fitting errors, whereas for Mix-9.5, the 2S2P1D model yielded slightly lower fitting error than that from the HN model. This may be because the 2S2P1D model has more parameters and thus higher flexibility. Besides, the time-temperature shift factors calculated using the HN and 2S2P1D methods were found very close as well, as shown in Figure 5.

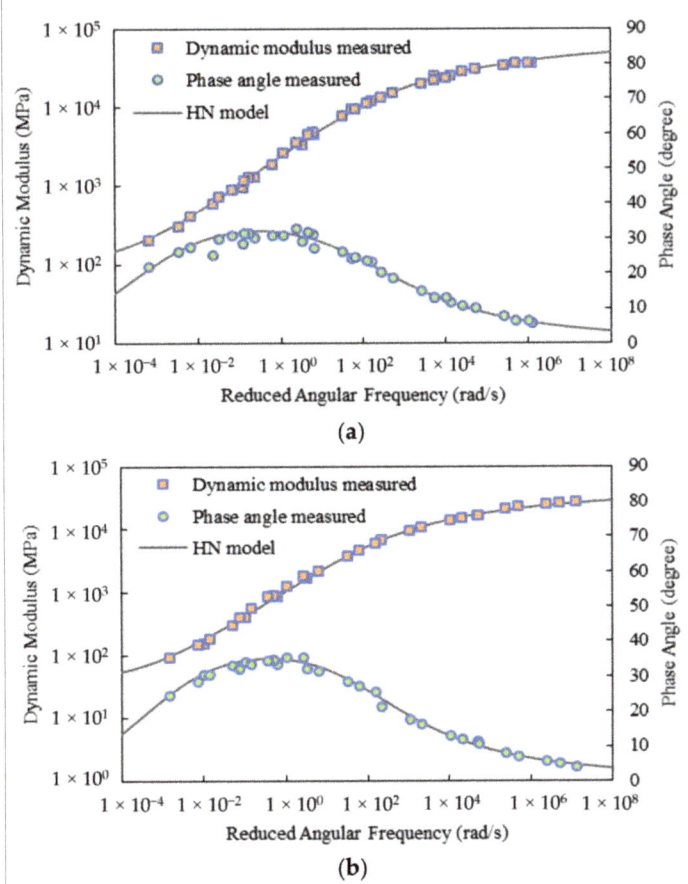

**Figure 3.** Master curves of dynamic modulus and phase angle from the HN model: (**a**) Mix-13.2; (**b**) Mix-9.5.

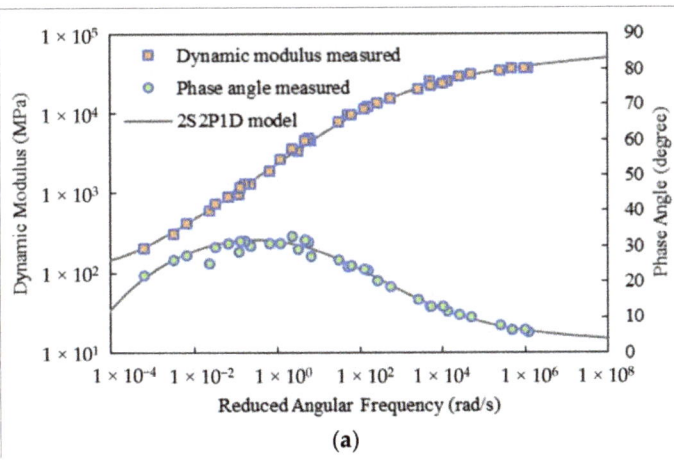

**Figure 4.** *Cont.*

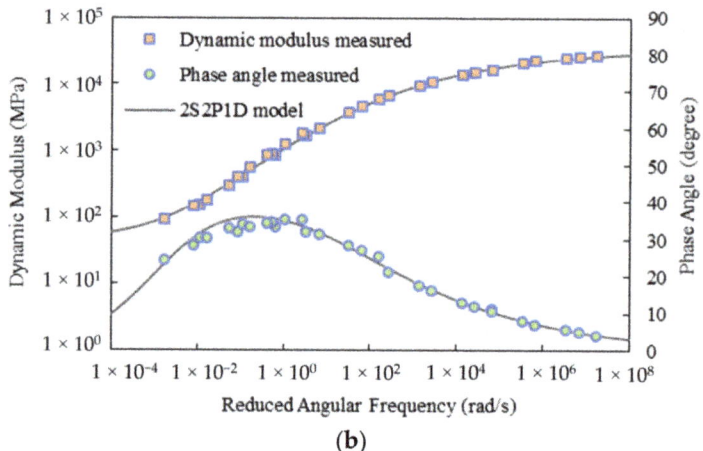

(b)

**Figure 4.** Master curves of dynamic modulus and phase angle from the 2S2P1D model: (**a**) Mix-13.2; (**b**) Mix-9.5.

**Table 1.** HN model parameters and fitting errors.

| Mix Type | $T_r/°C$ | $E_g$/MPa | $E_e$/MPa | α | β | $τ_0$/s | F/% |
|---|---|---|---|---|---|---|---|
| Mix-13.2 | 20 | 73,132 | 92.0 | 0.398 | 0.193 | 0.013 | 1.943 |
| Mix-9.5 | 24 | 43,707 | 37.5 | 0.431 | 0.175 | 0.011 | 2.461 |

**Table 2.** 2S2PD model parameters and fitting errors.

| Mix Type | $T_r/°C$ | $E_g$/MPa | $E_e$/MPa | α | k | h | β | $τ_0$/s | F/% |
|---|---|---|---|---|---|---|---|---|---|
| Mix-13.2 | 20 | 80,329 | 126.2 | 1.805 | 0.104 | 0.412 | 38,400 | $2.485 \times 10^{-4}$ | 1.961 |
| Mix-9.5 | 24 | 34,132 | 49.1 | 2.329 | 0.205 | 0.493 | 41,666 | $1.558 \times 10^{-3}$ | 2.204 |

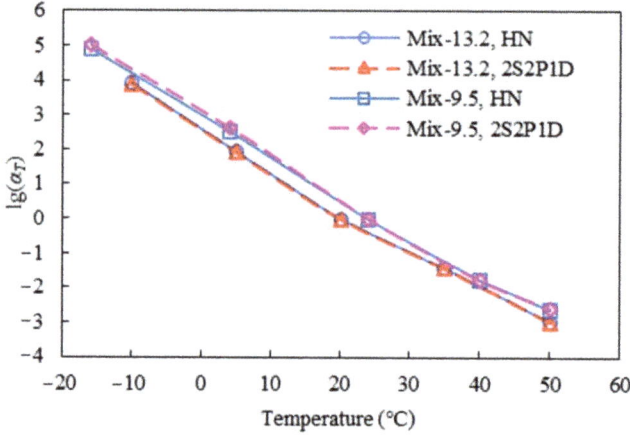

**Figure 5.** Time-temperature shift factors obtained using the HN and 2S2P1D methods.

As mentioned previously, with the obtained complex modulus model parameters, both continuous relaxation and retardation spectra can be analytically developed [see Equations (25)–(32)]. Figure 6 shows the continuous spectra of the two asphalt mixtures. It can be observed that for both mixtures, the HN and 2S2P1D models exhibited slightly different continuous spectral patterns, particularly at shorter relaxation times and longer retardation times. However, at the time range of $10^{-8}$ to $10^4$ s, which is approximately corresponding to the angular frequency range of $10^{-4}$ to $10^8$ rad/s, that is, the region mostly covered by the test data (Figures 3 and 4), the continuous spectra for the HN and 2S2P1D models were very close to each other.

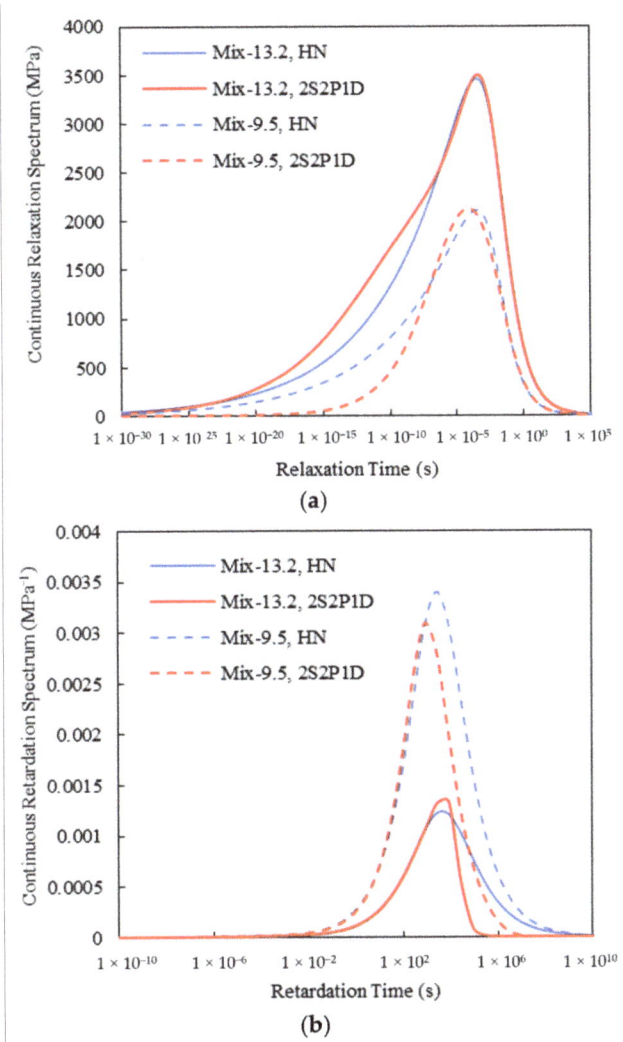

**Figure 6.** Continuous spectra of the two asphalt mixtures: (a) Relaxation; (b) Retardation.

Based on the continuous spectra developed, the corresponding discrete spectra can be fast determined using Equations (44)–(47). Although the relaxation and retardation times of the discrete spectra can be preset at any time regions of interest with any widths, it is a common practice that they are selected at regions covered by test data [3,12,35]. In this way, the numbers of the Prony series parameters can be reduced reasonably without losing significant computation accuracy. Consequently, the range of the discrete spectra was selected at $10^{-8}$ to $10^4$ s in this study.

Figures 7 and 8 display the calculated discrete relaxation and retardation spectra for the two mixtures. Three densities of the discrete spectrum lines, namely, $M$ = 1, 2 and 3, were considered. As can be seen, the resulting discrete spectra from both HN and 2S2P1D models were very smooth without any local oscillations. Also, since the discrete spectrum strengths were all calculated from the corresponding positive continuous spectra, no negative strength values were produced.

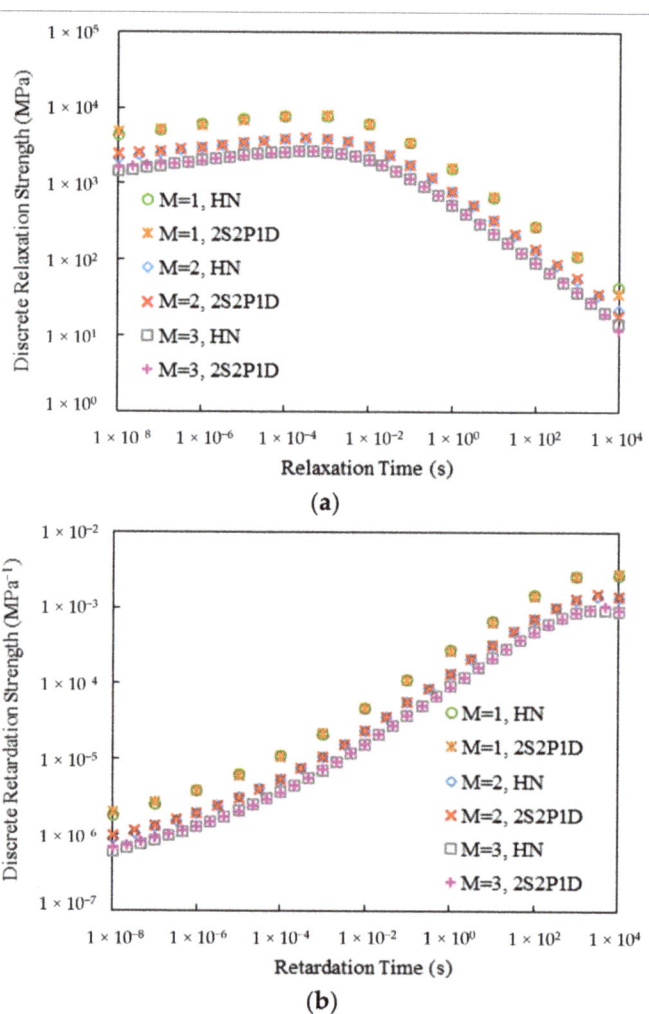

**Figure 7.** Discrete spectra of Mix-13.2: (**a**) Relaxation; (**b**) Retardation

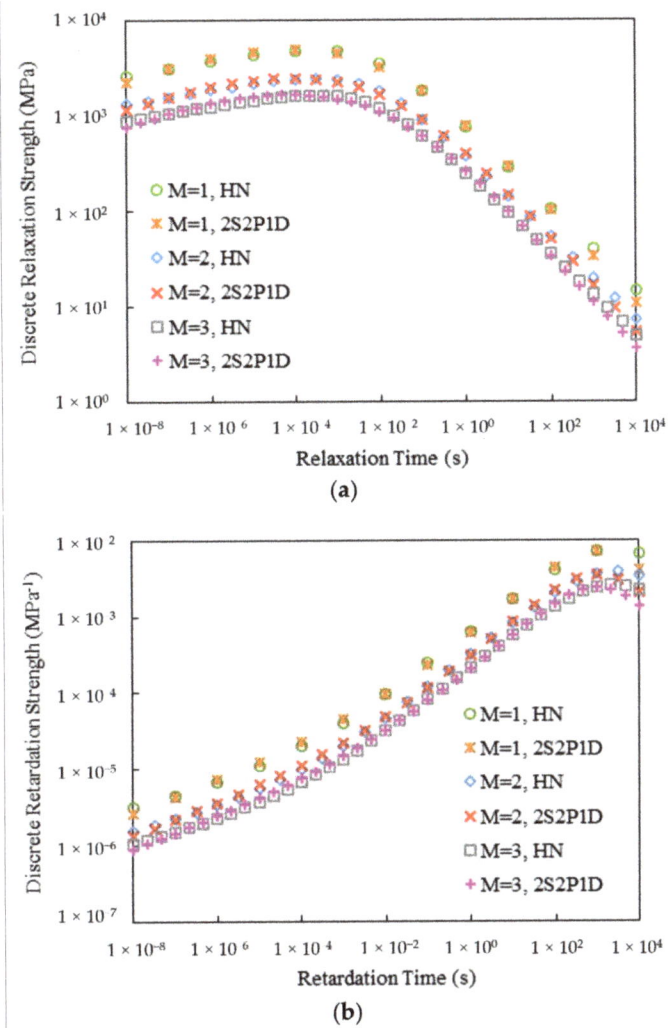

**Figure 8.** Discrete spectra of Mix-9.5: (**a**) Relaxation; (**b**) Retardation.

To establish the Prony series representations for the relaxation modulus $E(t)$ and creep compliance $D(t)$ in Equations (36) and (39), the elastic constants $E_e$ and $D_g$ need to be further determined. To this end, Equation (37) for the storage modulus $E'$ was fitted to Equations (6) and (10) separately for the real parts of the HN and 2S2P1D models. Before fitting, smoothed data points were generated from Equations (6) and (10), equally spaced on the logarithmic scale within the region covered by the test data. With $E_e$ determined, $D_g$ can be fast obtained by Equation (48).

Figure 9 gives the developed master curves of the storage modulus for Mix-9.5 using the discrete relaxation spectrum, i.e., using the generalized Maxwell model, with $M = 1$ from the HN and 2S2P1D models, respectively. Obviously, both methods yielded satisfactory results over the region where the spectrum lines were selected. Similar observations were made for Mix-13.2. It should be mentioned that, traditionally, one spectrum line per decade ($M = 1$) is extensively accepted for generating the Prony series representation. The higher density of the spectrum lines would generate higher accuracy for fitting but would produce more Prony series parameters. Thus, in the following sections, only the results for $M = 1$ are presented.

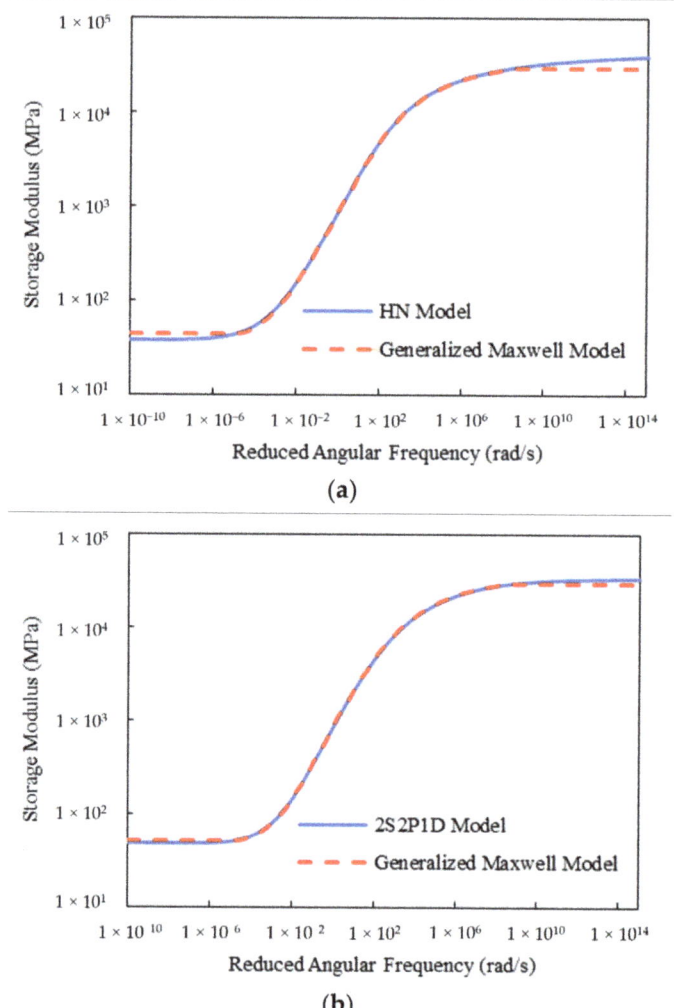

**Figure 9.** Master curves of storage modulus for Mix-9.5 using the discrete relaxation spectrum with $M = 1$ from: (**a**) HN model; (**b**) 2S2P1D model.

Figure 10 gives the master curves of the relaxation modulus and creep compliance for Mix-9.5 in the Prony series forms from the HN and 2S2P1D models. To verify the quality of the calculated Prony series parameters, the corresponding curves developed through the continuous spectra are also presented. Figure 11 gives the relative errors between the master curves from the Prony series forms and continuous spectra for Mix-9.5. It should be mentioned that since the spectrum lines were selected only at the time range covered by the test data, only the relative errors at $10^{-8}$ to $10^{4}$ s were calculated. To achieve the infinite integrals in Equations (19) and (22), an integral interval of $10^{-40}$ to $10^{+40}$ s was employed to approximately represent the infinite one through the trapezoidal rule, in which 100 increments per decade were equidistantly selected on the logarithmic time scale. It can be seen that the curves from the Prony series parameters were in good agreement with those from the continuous spectra for both HN and 2S2P1D models over the region where the spectrum lines were selected, thus demonstrating the effectiveness of the proposed method in this study. Equally desirable results were also found for Mix-13.2.

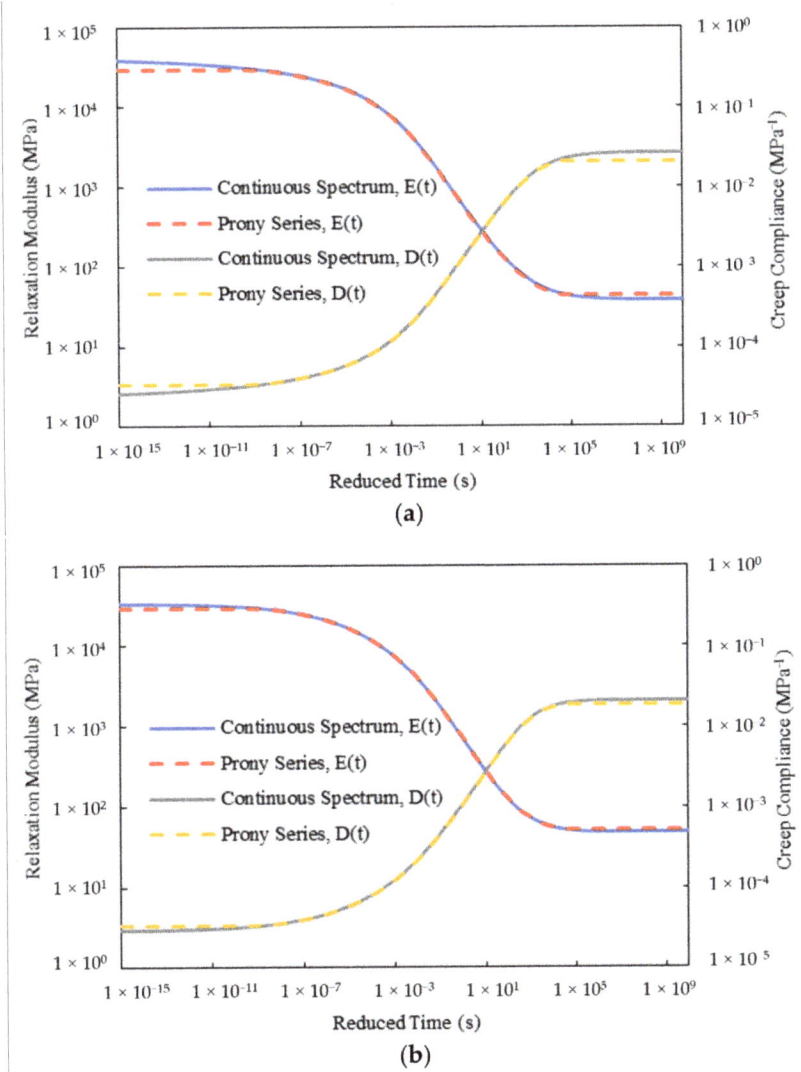

**Figure 10.** Master curves of relaxation modulus and creep compliance for Mix-9.5 in the Prony series forms from: (**a**) HN model; (**b**) 2S2P1D model.

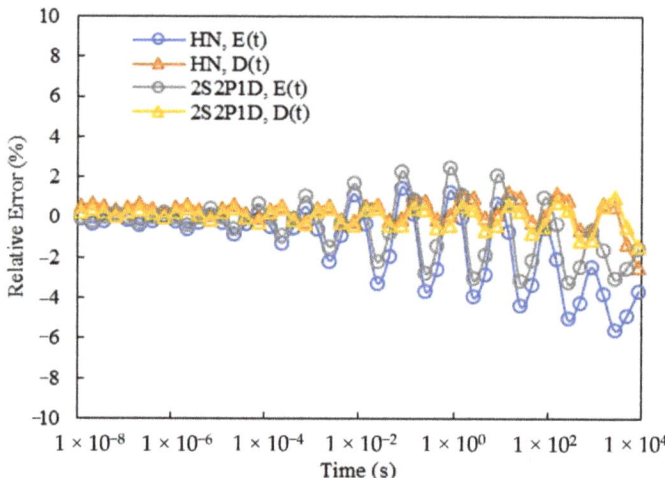

**Figure 11.** Relative errors between the master curves from Prony series forms and continuous spectra for Mix-9.5.

### 4.3. Comparison to the Conventional Sigmoidal Model Method

Considering that the quality of the resulting Prony series parameters are dependent on the master curve models used for presmoothing, the results obtained were compared with those from the Sigmoidal model, which has been adopted by MEPDG [36]. The Sigmoidal model with four parameters can be expressed by:

$$\lg(|E^*|) = a_1 + \frac{a_2}{1 + e^{a_3 + a_4 \lg \omega}} \tag{49}$$

where $a_1$ is the on the minimum logarithmic value of the dynamic modulus; $a_2$ is the difference of the maximum and minimum logarithmic values of the dynamic modulus; $a_3$ and $a_4$ are model parameters governing the curve shape.

Unlike the HN and 2S2P1D models, the Sigmoidal function is a real-valued model for the dynamic modulus, and thus does not have an accurate analytical model for the corresponding phase angle. To deal with this issue, Rowe [37] developed a representation for the phase angle using an approximate Kronig–Kramers relation [38], as follows:

$$\varphi \approx 90 \times \frac{\mathrm{d}\lg(|E^*|)}{\mathrm{d}\lg \omega} = -90 a_2 a_4 \frac{e^{a_3 + a_4 \lg \omega}}{\left(1 + e^{a_3 + a_4 \lg \omega}\right)^2} \tag{50}$$

Table 3 shows the calculated Sigmoidal model parameters and fitting errors for the two mixtures. It can be observed that both HN and 2S2P1D models generated lower fitting errors than the Sigmoidal model, indicating their higher applicability to the complex modulus test data. To gain an in-depth insight into their advantages, the master curves of the dynamic modulus and phase angle for the three models were plotted in Figures 12 and 13. It can be found that for the dynamic modulus, the curves from both HN and 2S2P1D models are non-centrosymmetric, that is, they offer asymmetric inflection points, whereas the Sigmoidal model is centrosymmetric on the log-log scale. As a result, the HN and 2S2P1D models exhibit higher flexibility than the Sigmoidal model in modeling the dynamic modulus of asphalt concrete. Additionally, the phase angle master curves of the HN and 2S2P1D models are non-axisymmetric, while that for the Sigmoidal is axisymmetric. Evidently, non-axisymmetric curves are more suitable for simulating the phase angle test data of asphalt concrete.

**Table 3.** Sigmoidal model parameters and fitting errors.

| Mix Type | $T_r/°C$ | $a_1$ | $a_2$ | $a_3$ | $a_4$ | $F/\%$ |
|---|---|---|---|---|---|---|
| Mix-13.2 | 20 | 1.647 | 3.089 | −0.246 | −0.460 | 2.089 |
| Mix-9.5 | 24 | 1.152 | 3.370 | −0.233 | −0.459 | 2.471 |

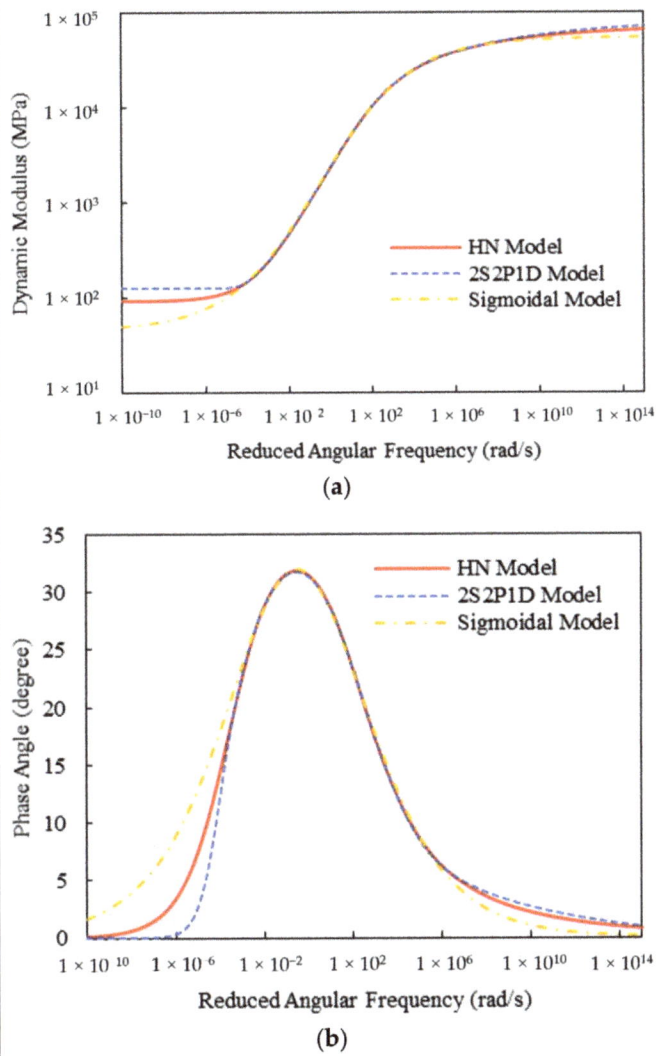

**Figure 12.** Comparison of master curves of dynamic modulus and phase angle for Mix-13.2: (**a**) dynamic modulus; (**b**) phase angle.

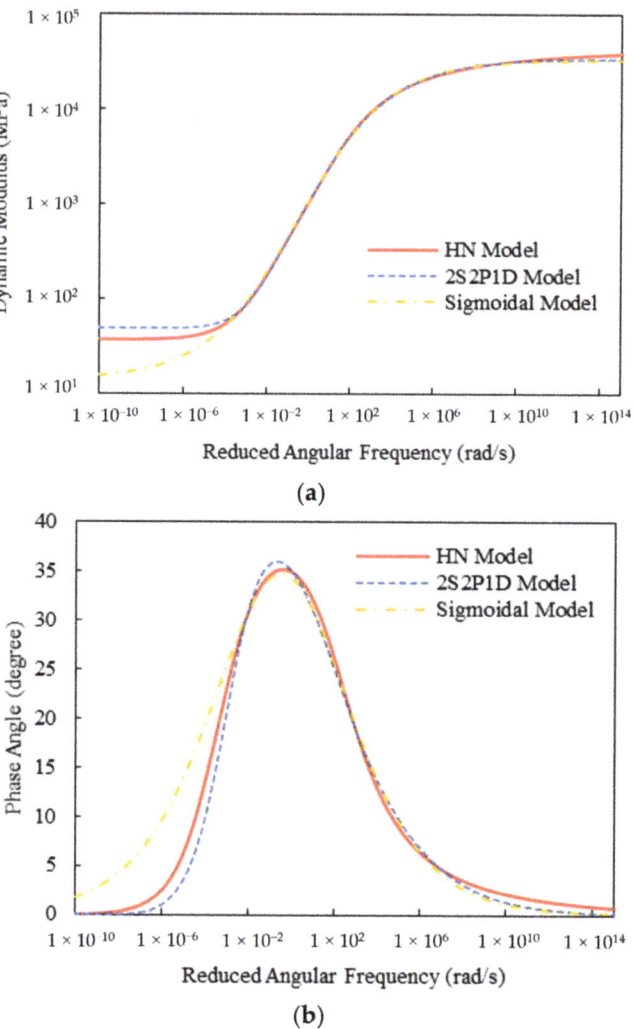

**Figure 13.** Comparison of master curves of dynamic modulus and phase angle for Mix-9.5: (a) dynamic modulus; (b) phase angle.

As a representation for the dynamic modulus, the Sigmoidal model does not have corresponding close-form solutions for $H(\rho)$ or $L(\tau)$. Thus, the Prony series parameters for the relaxation modulus and creep compliance cannot be analytically yielded. To obtain the Prony series parameters, only the numerical approach can be used, in which the storage modulus representation from the generalized Maxwell model in Equation (37) is directly fitted to smoothed data produced from Equations (49) and (50). Similarly, the Prony series for the creep compliance also needs to be numerically computed. In this regard, the complex-valued models adopted in this study, like the HN and 2S2P1D models, have the prominent advantage over real-valued models.

Figure 14 displays the Black diagrams plotted using the Sigmoidal model for the two asphalt mixtures. For a comparison purpose, the generalized Maxwell model developed using the storage modulus data generated from the Sigmoidal model is also shown. To guarantee a good consistency of the generalized Maxwell model to the smoothed storage modulus data, the recursive fitting method developed by Sun et al. [14] was utilized. It can be clearly seen from Figure 14 that the curves from the two models diverge around the peaks of the phase angle, which indicates a noncompliance of the Sigmoidal model method with the LVE theory. This is ascribed to the use of approximate Kronig–Kramers relation. In this connection, the HN and 2S2P1D models employed in the presented approach can accurately satisfy the Kronig–Kramers relation due to the presence of the analytical representations of both the real and imaginary parts of the complex modulus.

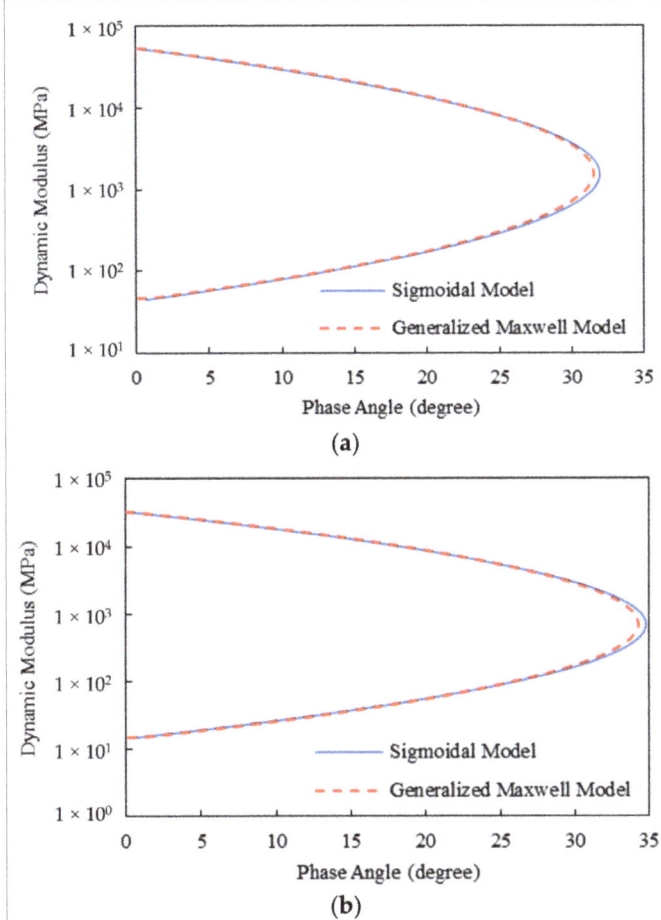

**Figure 14.** Black diagrams from the Sigmoidal model for the two asphalt mixtures: (**a**) Mix-13.2; (**b**) Mix-9.5.

## 5. Summary and Conclusions

This study presented a practical approach to fast acquiring high-quality Prony series parameters for both relaxation modulus and creep compliance of asphalt concrete based on the complex modulus test data. The approach can directly determine Prony series parameters through the analytical representations of the continuous relaxation and retardation spectra from the HN and 2S2P1D complex modulus models. With the model parameters determined in constructing dynamic modulus and phase angle master curves, the Prony series parameters can be immediately obtained with required accuracy. The elastic constants, $E_e$ and $D_g$, for both LVE modulus and compliance functions can further be readily calculated through the smoothed data from the storage modulus representations of the HN and 2S2P1D complex modulus models. To offer an in-depth interpretation for the approach, the performance of the HN, 2S2P1D and conventional Sigmoidal models in fitting the complex modulus master curves were compared. Based on the results and analysis from this study, main conclusions can be drawn as follows:

(1) The HN and 2S2P1D models yielded slightly different continuous spectral patterns at shorter relaxation times and longer retardation times. However, at the region covered by the test data, the continuous spectra of the two complex modulus models were very close to each other. Thus, the two models can generate comparable Prony series parameters within the time or frequency range covered by test data.

(2) By means of the positive analytical expressions of the continuous spectra, local spectrum oscillations and undesirable negative spectrum strengths were successfully eliminated, thus generating high-quality Prony series parameters.

(3) The HN and 2S2P1D models provide non-centrosymmetric curve patterns for the dynamic modulus master curves on the log-log scale and non-axisymmetric curve patterns for the phase angle master curves on the logarithmic angular frequency scale. Therefore, they performed better than the traditional Sigmoidal model in fitting to the complex modulus test data.

(4) The Black diagram is recommended for examining the quality of the complex modulus test data before constructing the master curves, because it can effectively avoid the effect of testing temperatures.

(5) The analytical expressions of the storage and loss moduli for both HN and 2S2P1D models accurately meet the Kronig–Kramers relation, and therefore the master curves constructed are consistent with the LVE theory.

(6) All the procedures in the proposed method can be easily achieved even only by Microsoft Excel, successfully avoiding sophisticated expertise for programming in implementation process. Thus, the proposed method furnishes a practical way to fast acquiring high-quality Prony series parameters.

Further studies are required to develop predictive models of the complex modulus master curve of asphalt concrete based on the HN and 2S2P1D models by using statistical relationships between the model parameters and the constituent and volumetric properties of asphalt concrete, and the work is ongoing.

**Author Contributions:** Conceptualization, Y.Z. and Y.S.; methodology, Y.Z. and Y.S.; writing—original draft preparation, Y.Z.; writing—review and editing, Y.S. All authors have read and agreed to the published version of the manuscript.

**Funding:** This research was funded by Doctoral Start-up Foundation of Liaoning Province (2019-BS-048) and Fundamental Research Funds for the Central Universities [DUT19RC(4)023].

**Institutional Review Board Statement:** Not applicable.

**Informed Consent Statement:** Not applicable.

**Data Availability Statement:** Data sharing is not applicable to this article.

**Acknowledgments:** This study was sponsored by Doctoral Start-up Foundation of Liaoning Province (2019-BS-048) and Fundamental Research Funds for the Central Universities [DUT19RC(4)023]. The supports are gratefully acknowledged.

**Conflicts of Interest:** The authors declare no conflict of interest.

## References

1. Hernandez-Fernandez, N.; Ossa-Lopez, A.; Harvey, J.T. Viscoelastic characterisation of asphalt concrete under different loading conditions. *Int. J. Pavement Eng.* **2021**, 1–14. [CrossRef]
2. Zhao, Y.; Liu, H.; Liu, W. Characterization of linear viscoelastic properties of asphalt concrete subjected to confining pressure. *Mech. Time-Depend. Mater.* **2012**, *17*, 449–463. [CrossRef]

3. Park, S.W.; Kim, Y.R. Fitting Prony-series viscoelastic models with power-law presmoothing. *J. Mater. Civ. Eng.* **2001**, *13*, 26–32. [CrossRef]
4. Nguyen, H.T.T.; Do, T.-T.; Tran, V.-T.; Phan, T.-N.; Pham, T.-A.; Nguyen, M.L. Determination of creep compliance of asphalt mixtures at intermediate and high temperature using creep-recovery test. *Road Mater. Pavement Des.* **2021**, *22* (Suppl. 1), S514–S535. [CrossRef]
5. Nguyen, H.T.T.; Nguyen, D.-L.; Tran, V.-T.; Nguyen, M.-L. Finite element implementation of Huet-Sayegh and 2S2P1D models for analysis of asphalt pavement structures in time domain. *Road Mater. Pavement Des.* **2020**, *23*, 22–46. [CrossRef]
6. Li, L.; Wu, C.; Cheng, Y.; Ai, Y.; Li, H.; Tan, X. Comparative analysis of viscoelastic properties of open graded friction course under dynamic and static loads. *Polymers* **2021**, *13*, 1250. [CrossRef]
7. Hristov, J.Y. Linear viscoelastic responses: The Prony decomposition naturally leads into the Caputo-Fabrizio fractional operator. *Front. Phys.* **2018**, *6*, 135. [CrossRef]
8. Xu, Q.; Engquist, B.; Solaimanian, M.; Yan, K. A new nonlinear viscoelastic model and mathematical solution of solids for improving prediction accuracy. *Sci. Rep.* **2020**, *10*, 2202. [CrossRef] [PubMed]
9. Gu, L.; Zhang, W.; Ma, T.; Qiu, X.; Xu, J. Numerical simulation of viscoelastic behavior of asphalt mixture using fractional constitutive model. *J. Eng. Mech.* **2021**, *147*, 04021027. [CrossRef]
10. London, J.G.; Shen, R.; Waisman, H. Temperature-dependent viscoelastic model for asphalt–concrete implemented within a novel nonlocal damage framework. *J. Eng. Mech.* **2020**, *146*, 04019119. [CrossRef]
11. Yu, D.; Yu, X.; Gu, Y. Establishment of linkages between empirical and mechanical models for asphalt mixtures through relaxation spectra determination. *Constr. Build. Mater.* **2020**, *242*, 118095. [CrossRef]
12. Zhao, Y.; Ni, Y.; Zeng, W. A consistent approach for characterising asphalt concrete based on generalised Maxwell or Kelvin model. *Road Mater. Pavement Des.* **2014**, *15*, 674–690. [CrossRef]
13. Park, S.W.; Schapery, R.A. Methods of interconversion between linear viscoelastic material functions. Part I—A numerical method based on Prony series. *Int. J. Solids Struct.* **1999**, *36*, 1653–1675. [CrossRef]
14. Sun, Y.; Huang, B.; Chen, J. A unified procedure for rapidly determining asphalt concrete discrete relaxation and retardation spectra. *Constr. Build. Mater.* **2015**, *93*, 35–48. [CrossRef]
15. Schapery, R.A. *A Simple Collocation Method for Fitting Viscoelastic Models to Experimental Data*. GALCIT SM 61-32A; California Institute of Technology: Pasadena, CA, USA, 1961.
16. Cost, T.L.; Becker, E.B. A multidata method of approximate Laplace transform inversion. *Int. J. Numer. Meth. Eng.* **1970**, *2*, 207–219. [CrossRef]
17. Tschoegl, N.W.; Emri, I. Generating line spectra from experimental responses. Part II: Storage and loss functions. *Rheol. Acta* **1993**, *32*, 322–327. [CrossRef]
18. Xu, Q.; Solaimanian, M. Modelling linear viscoelastic properties of asphalt concrete by the Huet–Sayegh model. *Int. J. Pavement Eng.* **2009**, *10*, 401–422. [CrossRef]
19. Olard, F.; di Benedetto, H. General "2S2P1D" model and relation between the linear viscoelastic behaviours of bituminous binders and mixes. *Road Mater. Pavement Des.* **2003**, *4*, 185–224.
20. Tiouajni, S.; di Benedetto, H.; Sauzéat, C.; Pouget, S. Approximation of Linear Viscoelastic Model in the 3 Dimensional Case with Mechanical Analogues of Finite Size. *Road Mater. Pavement Des.* **2011**, *12*, 897–930.
21. Tschoegl, N.W.; Emri, I. Generating line spectra from experimental responses. III. Interconversion between relaxation and retardation behavior. *Int. J. Polym. Mater.* **1992**, *18*, 117–127. [CrossRef]
22. Schapery, R.A.; Park, S.W. Methods of interconversion between linear viscoelastic material functions. Part II—An approximate analytical method. *Int. J. Solids Struct.* **1999**, *36*, 1677–1699. [CrossRef]
23. Levenberg, E. Smoothing asphalt concrete complex modulus test data. *J. Mater. Civ. Eng.* **2011**, *23*, 606–611. [CrossRef]
24. Luo, R.; Lv, H.; Liu, H. Development of Prony series models based on continuous relaxation spectrums for relaxation moduli determined using creep tests. *Constr. Build. Mater.* **2018**, *168*, 758–770. [CrossRef]
25. Lv, H.; Liu, H.; Tan, Y.; Sun, Z. Improved methodology for identifying Prony series coefficients based on continuous relaxation spectrum method. *Mater. Struct.* **2019**, *52*, 86. [CrossRef]
26. Sun, Y.; Chen, J.; Huang, B. Characterization of asphalt concrete linear viscoelastic behavior utilizing Havriliak–Negami complex modulus model. *Constr. Build. Mater.* **2015**, *99*, 226–234. [CrossRef]
27. Bhattacharjee, S.; Swamy, A.K.; Daniel, J.S. Continuous relaxation and retardation spectrum method for viscoelastic characterization of asphalt concrete. *Mech. Time-Depend. Mater.* **2011**, *16*, 287–305. [CrossRef]
28. Zhang, F.; Wang, L.; Li, C.; Xing, Y. The discrete and continuous retardation and relaxation spectrum method for viscoelastic characterization of warm mix crumb rubber-modified asphalt mixtures. *Materials* **2020**, *13*, 3723. [CrossRef] [PubMed]
29. AASHTO. *Standard Method of Test for Determining Dynamic Modulus of Hot Mix Asphalt (HMA)*, AASHTO T 342-11; AASHTO: Washington, DC, USA, 2011.
30. Havriliak, S.; Negami, S. A complex plane representation of dielectric and mechanical relaxation processes in some polymers. *Polymer* **1967**, *8*, 161–210. [CrossRef]
31. Tschoegl, N.W. *The Phenomenological Theory of Linear Viscoelastic Behavior: An Introduction*; Springer: New York, NY, USA, 1989.
32. Alavi, M.Z.; Hajj, E.Y.; Morian, N.E. Approach for quantifying the effect of binder oxidative aging on the viscoelastic properties of asphalt mixtures. *Transp. Res. Rec.* **2013**, *2373*, 109–120. [CrossRef]

33. Sun, Y.; Huang, B.; Chen, J.; Jia, X.; Ding, Y. Characterizing rheological behavior of asphalt binder over a complete range of pavement service loading frequency and temperature. *Constr. Build. Mater.* **2016**, *123*, 661–672. [CrossRef]
34. Airey, G.D. Use of black diagrams to identify inconsistencies in rheological data. *Road Mater. Pavement Des.* **2011**, *3*, 403–424. [CrossRef]
35. Mun, S.; Geem, Z.W. Determination of viscoelastic and damage properties of hot mix asphalt concrete using a harmony search algorithm. *Mech. Mater.* **2009**, *41*, 339–353. [CrossRef]
36. ARA (Applied Research Associates), ERES Consultants Division. *Guide for Mechanistic-Empirical Design of New and Rehabilitated Pavement Structures, NCHRP 1-37A Final Report*; ERES Consultants Division, Transportation Research Board: Washington, DC, USA, 2004.
37. Rowe, G. Phase angle determination and interrelationships within bituminous materials. In Proceedings of the 7th International RILEM Symposium on Advanced Testing and Characterization of Bituminous Materials, Rhodes, Greece, 27–29 May 2009; pp. 43–52.
38. Booij, H.C.; Thoone, G.P.J.M. Generalization of Kramers-Kronig transforms and some approximations of relations between viscoelastic quantities. *Rheol. Acta* **1982**, *21*, 15–24. [CrossRef]

Article

# Viscoelastic Damage Characteristics of Asphalt Mixtures Using Fractional Rheology

Qipeng Zhang [1,2], Xingyu Gu [1,*], Zilu Yu [1], Jia Liang [1] and Qiao Dong [1]

1. School of Transportation, Southeast University, Nanjing 211189, China; zhangqipeng@seu.edu.cn (Q.Z.); 220215104@seu.edu.cn (Z.Y.); liangjiahs@seu.edu.cn (J.L.); qiaodong@seu.edu.cn (Q.D.)
2. National Demonstration Center for Experimental Road and Traffic Engineering Education (Southeast University), Nanjing 211189, China
* Correspondence: guxingyu1976@seu.edu.cn; Tel.: +86-13915936786

**Citation:** Zhang, Q.; Gu, X.; Yu, Z.; Liang, J.; Dong, Q. Viscoelastic Damage Characteristics of Asphalt Mixtures Using Fractional Rheology. *Materials* **2021**, *14*, 5892. https://doi.org/10.3390/ma14195892

Academic Editor: Francesco Canestrari

Received: 11 September 2021
Accepted: 4 October 2021
Published: 8 October 2021

**Publisher's Note:** MDPI stays neutral with regard to jurisdictional claims in published maps and institutional affiliations.

**Copyright:** © 2021 by the authors. Licensee MDPI, Basel, Switzerland. This article is an open access article distributed under the terms and conditions of the Creative Commons Attribution (CC BY) license (https://creativecommons.org/licenses/by/4.0/).

**Abstract:** The mechanical behavior of asphalt mixtures at high stress levels are characterized by nonlinear viscoelasticity and damage evolution. A nonlinear damage constitutive model considering the existence of creep hardening and creep damage mechanisms in the entire creep process is proposed in this study by adopting the fractional rheology theory to characterize the three-stage creep process of mixtures. A series of uniaxial compressive creep tests under various stresses were conducted at different temperatures to verify the model. The results indicated that the model predictions were in good agreement with the creep tests. The relationship between the model parameters and applied stresses was established, and the stress range in which the mixture exhibited only creep consolidation was obtained. The damage to the asphalt mixture was initiated in the steady stage; however, it developed in the tertiary stage. A two-parameter Weibull distribution function was used to describe the evolution between the damage values and damage strains at different stress levels and temperatures. The correlation coefficients were greater than 0.99 at different temperatures, indicating that a unified damage evolution model could be established. Thus, the parameters of the unified model were related to material properties and temperature, independent of the stress levels applied to the mixtures.

**Keywords:** pavement materials; asphalt mixture; compressive creep; damage evolution; fractional rheology theory; viscoelasticity

## 1. Introduction

Asphalt mixtures are considered as heterogeneous pavement materials typically composed of asphalt, aggregate, admixtures, and air voids. Furthermore, asphalt mixtures are the primary materials used worldwide for constructing pavement layers because of their remarkable advantages, such as short construction periods, long service lives, recyclability, ease of maintenance, and comfortable driving. Although aggregates, which are a type of rigid material, account for approximately 90% of all components within the asphalt mixtures, such mixtures exhibit complex rheological properties. Asphalt mixtures exhibit both viscous (fluid) and elastic (solid) behavior [1–3], and their mechanical properties are highly dependent on temperature, loading stress level, loading time, and loading speed [4,5]. The theoretical and experimental studies on the mechanical behaviors of asphalt mixtures provide a crucial basis for designing pavement layers [6,7]. Creep behavior is amongst the mechanical characteristics of asphalt mixtures [8], and it is strongly related to the rutting formation of the pavement structure [3,9]. Therefore, it is important to establish a constitutive model that can accurately predict the creep behavior of asphalt mixtures.

The improvement of constitutive models is gaining increasing interest in the field of rheology in order to better describe the mechanical behaviors of complex rheological materials. Furthermore, several rheological models of asphalt mixtures were proposed in the past [10–19]. These models can be divided into integral constitutive and differential constitutive models [15,20]. The integral viscoelastic constitutive model, which considers the loading history and memory of materials, is developed based on methods such as the Boltzmann superposition principle or the Leaderman modified superposition principle. Compared with multiple integral models, single integral models have fewer material parameters and simpler forms, which makes them easier to use in practical engineering. Schapery's model, which is widely applied to characterize the viscoelastic properties of asphalt mixtures, is a single integral constitutive model [21]. Im et al. [22] described the viscoelastic and viscoplastic behavior of mixtures by combining Schapery's viscoelastic model and Perzyna viscoplasticity with a generalized Drucker–Prager yield surface.

Classical integer differential constitutive models, such as the Maxwell model, Kelvin model, Burgers model, and generalized models, comprise springs and dashpots in series or in parallel. The mathematical expressions of the differential constitutive models can correspond to the mechanical models that have advantages of intuitionism and simplicity. The classical integer derivative constitutive models are widely used to analyze the creep behavior of asphalt mixtures. The Burgers model provides a better description of mechanical behavior than the Maxwell and Kelvin–Voigt models [23]. However, Cheng et al. [24] reported that, compared to the Burgers model, the generalized Kelvin (GK) model and the generalized Maxwell (GM) model have a better performance in describing the creep performance of asphalt mixtures. Although the classical constitutive models can describe the viscoelastic properties of asphalt mixtures, they yield a large error in describing the initial stage of the creep performance [13]. With the development of viscoelasticity theory, the fractional derivative operator that can describe the historical memory dependence and spatially wide relevance of a material meets the research requirements for examining the complex viscoelastic, non-Newtonian fluid and the porous media mechanics [25]. Furthermore, the fractional calculus theory achieves great success in describing the rheological behavior of asphalt materials [13–15,26,27]. Lagos-Varas et al. [13] established a viscoelastic model using fractional-order derivatives, and unlike the Burgers model, the fractional order model can predict the elastic jump observed at the beginning of the creep modulus. Celauro et al. [28] demonstrated that the fractional-order Burgers model can accurately characterize the creep/recovery response of asphalt mixtures. In addition, the fractional derivative model has the advantage of exhibiting a simpler expression, utilizing fewer parameters, and providing more accurate results. Furthermore analysis revealed that a fractional derivative model comprising two elements of fractional derivatives in series with a Maxwell element can be used to model the rheological behavior of asphalt binders. This includes the dynamic viscoelastic behavior, static creep, and relaxation characteristics [27].

The entire creep process of asphalt mixtures is characterized into three stages: primary (decelerated stage), steady (stationary stage), and tertiary (accelerated stage). Furthermore, different deformation mechanisms occur throughout the creep process of asphalt mixtures at various stresses. At a lower stress level, the asphalt mixture is gradually compacted and the aggregates are gradually stacked to form a stable skeleton structure, which results in a "consolidation effect" [19]. This deformation mechanism is known as the creep hardening mechanism [18], and the asphalt mortar resists deformation together with the skeleton structure during the procedure. However, at higher stress levels, the stress in the asphalt mixture exceeds the frictional resistance between the aggregates and the bonding between the aggregate and asphalt mortar, which results in the fragmentation of the asphalt mixture and the destruction of the skeleton. This consequently accelerates the viscoelastic flow rate and creep failure of the asphalt mixture in the tertiary stage. Here, the deformation mechanism is called the damage softening mechanism [18]. Linear models (e.g., the ordinary Burgers model, generalized Kelvin model, and fractional-order Burgers model) cannot describe the tertiary stage of creep.

One potential approach is to add a plastic element into the classic differential models to construct viscoelastoplastic models to accurately describe the three-stage creep behavior of asphalt mixtures at high stresses. The Nishihara model, which adds a plastic element to the Burgers model, was developed to describe the tertiary stage of creep and was used in the mechanical response analysis of bituminous materials [19]. A nonlinear viscoelastic-plastic creep model for asphalt mixtures was developed using variable-order fractional calculus, where time-varying viscoplastic elements were used to describe the tertiary stage of creep performance [15]. In addition, damage occurs in the asphalt mixture during the creep process. Therefore, another feasible approach is coupling a continuum damage evolution law with a linear viscoelastic model. Sun et al. [29] developed a damage constitutive model on the basis of the generalized Burgers model and the Perzyna viscoplastic flow theory, combined with a modified Rabotnov damage theory, which can better reflect the three stages of deformation in creep testing and the hardening and softening effects in the constant strain rate compression testing. Zhang et al. [18] suggested that the characteristics of the three-stage permanent deformation were attributed to a competition between the damage softening and strain hardening effect, and they proposed a viscoelastoplastic damage mechanics model wherein the damage and hardening variables were introduced to modify the Burger's model for describing the deformation of asphalt mixtures.

Most existing creep damage models are formed by the integral order difference constitutive model coupled directly with the damage factor. These models have the disadvantage of many material parameters and overly complex expressions, which are not conducive to their generalization in practical engineering. Furthermore, these models directly couple the strain caused by damage (called the damage strain) to the viscoelastic strain of the asphalt mixture, which is not conducive to studying the strain development mechanism and damage characteristics during the creep process. Therefore, this study aims to propose a creep damage model based on the fractional derivative operator that considers both the creep hardening mechanism and damage softening mechanism to characterize and analyze the creep process of asphalt mixtures at various stress levels and temperatures. Simultaneously, a unified damage evolution model is constructed for different stress levels, and the development of the strain and damage characteristics of the asphalt mixture throughout the creep process is analyzed. Finally, a method for the statistical quantification of the damage evolution of the asphalt mixture is proposed.

The paper is structured in the following way: a nonlinear viscoelastic creep damage model adopting the fractional rheology theory is developed in Part 2. The compression creep tests are conducted on AC-13 asphalt mixtures in Part 3. In Part 4, the experimental results are described and the proposed model is experimentally validated. Meanwhile, the model parameters and the creep damage evolution of the asphalt mixtures are analyzed in Part 4. Finally, several conclusions are drawn in Part 5.

## 2. Theoretical Background
### 2.1. Modelling of Creep Hardening Mechanism

In the classical integer-order differential viscoelastic model, the elastic mechanical behavior of the material is modeled with a Hook spring, and the viscous behavior is modeled with a Newton dashpot. The viscoelastic behavior of the asphalt mixtures is achieved by various combinations of Hook springs and Newton dashpots in series or in parallel. Gu et al. [26] adopted the Abel spring-pot element to describe the viscoelastic behavior based on the fractional order calculus theory for better describing the complex viscoelastic mechanical behavior of the materials. The fractional order viscoelastic element (Abel spring-pot element) is shown in Figure 1 and the stress–strain relation of the Abel spring-pot element is expressed as in [26]:

$$\sigma(t) = \xi D^r \varepsilon(t) (0 \leq r \leq 1), \tag{1}$$

where $\sigma$, $t$, $\varepsilon$, $\xi$, and $D^r$ denote the nominal stress, function variable, strain, material coefficient, and differential operator of the Riemann–Liouville fractional calculus [30], respectively, which is defined as:

$$D^r f(t) = \frac{d^r f(t)}{dt^r} = \frac{1}{\Gamma(1-r)} \frac{d}{dt} \int_0^t \frac{f(\tau)}{(t-\tau)^r} d\tau (0 \leq r \leq 1), \tag{2}$$

where $r$ denotes the fractional order of differentiation, $\Gamma(r)$ denotes the gamma function, and $\Gamma(r) = \int_0^\infty t^{r-1} e^{-t} dt$. According to Equations (1) and (2), the spring-pot reduces to the Hook spring when $r = 0$, whereas it reduces to the Newton dashpot when $r = 1$. Thus, the spring-pot can be regarded as a type of fractional differential viscoelastic element.

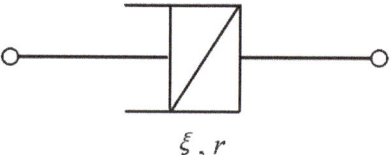

**Figure 1.** Schematic of fractional viscoelastic element (Abel spring-pot element).

The strain can be obtained from Equation (1) as:

$$\varepsilon(t) = \frac{1}{\xi} D^{-r} \sigma(t), \tag{3}$$

where $D^{-r}$ denotes the integral operator of the Riemann–Liouville fractional calculus [30], which is defined as:

$$D^{-r} f(t) = \frac{d^{-r} f(t)}{dt^{-r}} = \int_0^t \frac{(t-\tau)^{r-1}}{\Gamma(r)} f(\tau) d\tau \, (0 \le r \le 1), \tag{4}$$

In the case of the creep, under a constant stress level $\sigma(t) = \sigma_0 H(t)$, the creep strain can be obtained from Equations (3) and (4) as:

$$\varepsilon(t) = \frac{\sigma_0 t^r}{\xi \Gamma(r+1)} = \frac{\sigma_0}{K} t^r \, (0 \le r \le 1), \tag{5}$$

where $K$ is related to the coefficient $\xi$ and the fractional order $r$ of the material, and $K = \xi \Gamma(r+1)$. $H(t)$ denotes the Heaviside step function, that is, $H(t) = \begin{cases} 1 & t > 0 \\ 0 & t \le 0 \end{cases}$. According to Equation (5), $K$ is related to the deformation of the material and referred to here as the deformation factor.

Find the derivative of Equation (5) with respect to $t$:

$$\dot{\varepsilon} = \frac{d\varepsilon}{dt} = \frac{\sigma_0 r}{K} t^{r-1} (0 \le r \le 1). \tag{6}$$

Equation (6) indicates that the strain rate of the material, which decreases as time increases, and is consistent with the decrease in the strain rate at the decelerating stage of creep. When time $t$ is sufficiently large, the strain rate converges to zero, which is consistent with the consolidation effect at the lower stress levels [19]. Figure 2 shows the Abel model used to describe the consolidation effect and the decelerated stage of creep. (The test data in the figure were randomly selected from the creep tests in Section 3 to illustrate the applicability of the model.) Figure 2 shows that the Abel model can characterize both the hardening effect and the primary stage of creep performance very effectively. Thus, the Abel spring-pot model can be applied to characterize the creep hardening mechanism of asphalt mixtures.

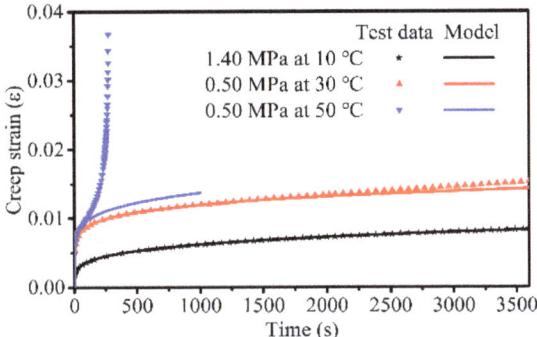

**Figure 2.** Consolidation effect and decelerating stage of the asphalt mixture: Abel model vs. test.

The load is kept constant ($\sigma(t) = \sigma_0 H(t)$) for an assigned time $\bar{t}$, where the viscoelastic material enters the creep phase. When $t > \bar{t}$, the load is removed and the material enters the recovery phase, and the stress is zero. The stress in the creep recovery phase can be expressed as:

$$\sigma(t) = \sigma_0 H(t) - \sigma_0 H(t - \bar{t}). \tag{7}$$

where $\bar{t}$ denotes the loading time.

The creep-recovery strain can be obtained from Equations (5) and (7) as:

$$\varepsilon(t) = \frac{\sigma_0}{K}\left\{t^r - H(t - \bar{t}) \cdot |t - \bar{t}|^r\right\} \ (0 \leq r \leq 1). \tag{8}$$

Equation (8) shows that for aviscoelastic materials, there are always irrecoverable strains in the material, no matter how much stress is applied. In other words, viscoelastic materials have permanent strains at any time after being stressed.

### 2.2. Modeling of Damage Softening Mechanism

There is a damage effect throughout the creep of the asphalt mixture, and the damage evolution results in a nonlinear mechanical behavior. Herein, the damage variable $D$ is used to describe the deterioration of the Abel spring-pot element, and damage evolution is related to the stress levels and its own distributed microdefects in the asphalt mixture. Based on the Kachanov damage theory [31], the damage evolution can be defined as:

$$\frac{dD}{dt} = C\sigma^q(1 - D)^{-v}, \tag{9}$$

where $C$, $u$, and $v$ represent the temperature-dependent material parameters, and $0 \leq D \leq 1$.

Taking the integral of Equation (9) with the initial condition $D = 0$ at $t = 0+$ and the critical condition $D = 1$ at $t = t_f$ yields the damage evolution as [19]:

$$D = 1 - \left[1 - (1 + v)\int_0^t C\sigma^u dt\right]^{1/(1+v)} \ \left(0 \leq t \leq t_f\right), \tag{10}$$

where $t_f$ denotes the damage-induced failure time.

For $\sigma(t) = \sigma_0 H(t)$, the damage evolution of creep can be written as:

$$D = 1 - \left(1 - t/t_f\right)^{1/(v+1)}, \tag{11}$$

where $t_f$ can be expressed as:

$$t_f = \frac{1}{C(1 + v)\sigma_0^u}, \tag{12}$$

According to the principle of the strain equivalence of the continuum damage mechanics [32], a constitutive relationship of the damaged material is obtained by replacing the nominal stress $\sigma$ with the effective stress $\overline{\sigma}$. The relationship between $\sigma$ and $\overline{\sigma}$ is expressed as:

$$\overline{\sigma} = \frac{\sigma}{1-D},\qquad(13)$$

Combining Equations (5), (11), and (13), the creep strain, considering the damage evolution based on the fractional derivative operator, can be expressed as:

$$\varepsilon(t) = \frac{\sigma_0}{K\cdot(1-D)}\cdot t^r = \frac{\sigma_0 t^r}{K}\cdot\left(1-t/t_f\right)^{-1/(v+1)} = \frac{\sigma_0}{K'}\cdot t^r\,(0\le r\le 1),\qquad(14)$$

where $K' = K\cdot(1-D)$. As can be seen from Equation (14), the damage causes the deformation coefficient to no longer be constant, and as the damage accumulates, the deformation coefficient deteriorates and the strain increases until the specimen is damaged or fails.

In this study, the creep strain, when considering the damage, is decomposed into a strain caused by the damage and the strain from the creep hardening mechanism in the undamaged state. The strain induced by the damage (referred to as the damage strain in this study) can be expressed as:

$$\varepsilon_d(t) = \frac{\sigma_0 t^r}{K(1-D)} - \frac{\sigma_0 t^r}{K} = \left(\frac{1}{1-D}-1\right)\cdot\frac{\sigma_0 t^r}{K}\,(0\le r\le 1),\qquad(15)$$

where $\varepsilon_d$ denotes the damage strain, and this can be applied to characterize the damage softening mechanism of the asphalt mixtures.

### 2.3. Modeling of Fractional Derivative Creep Damage Model

A simple creep damage model is proposed by combining a fractional derivative model that describes the creep hardening mechanism with a fractional derivative damage strain that describes the damage deterioration mechanism, as shown in Figure 3. Luo et al. [33] described the asymmetry of the dynamic viscoelastic properties of asphalt mixtures using a modified fractional Zener model, where the two spring-pot elements have different fractional orders $r$ and the same material coefficient $\xi$. In this study, it is suggested that not only should the two spring-pot elements describing the two mechanisms have different $r$, the coefficients $\xi$ should also not be the same to better study the creep performance of asphalt mixtures. The fractional derivative creep damage model can then be expressed as:

$$\varepsilon(t) = \frac{\sigma_0 t^r}{K_1} + \frac{\sigma_0 t^\alpha}{K_2}\left[\left(1-t/t_f\right)^{-1/(v+1)}-1\right],\qquad(16)$$

where $r$ and $\alpha$ denote the fractional orders of the two spring-pot elements, and $\xi_1$ and $\xi_2$ denote the coefficients of the two spring pots.

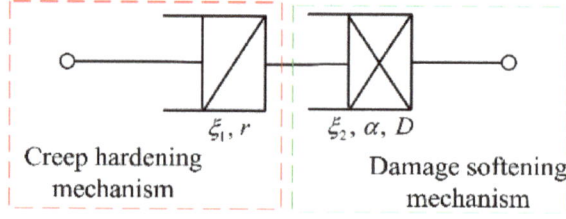

**Figure 3.** Schematic of fractional creep damage model.

## 3. Uniaxial Compression Static Creep Test

### 3.1. Material

An AH-70 virgin binder was employed to form the AC-13 asphalt mixture, which contained basalt aggregates. The gradation of the asphalt mixture was designed according to the Standard Test Methods of Bitumen and Bituminous Mixtures for Highway Engineering (JTG E20-2011), as shown in Figure 4. The performance tests of the asphalt binder and basalt aggregate were carried out according to JTG E20-2011 and JTG E42-2005 (Test Methods of Aggregate for Highway Engineering), respectively. The performances of the asphalt binder and basalt aggregates are presented in Tables 1 and 2. The optimum asphalt–aggregate mass ratio was determined to be 4.9% according to JTG E20-2011; the air voids in this study were set to 4.0 ± 0.2%. The Saint Venant principle clearly indicated that the larger the ratio of the height to the diameter of the specimen subjected to uniaxial stress, the greater the area beyond the end of the specimen that approached the uniaxial state. Therefore, this study adopted a gyratory compaction method to prepare a specimen with a height to diameter ratio of 1.5 for attenuating the effect of end friction on the stress state of the specimen. First, a certain amount of the mixture was compacted into a cylindrical specimen (diameter = 150 mm, height = 170 mm) using a gyratory compactor. The amount of asphalt mixture required to form the cylindrical specimen was calculated based on the maximum theoretical density, void ratio and specimen volume, and the amount should be such that the height of the specimen formed reached plus or minus 3 mm of the desired height. In this study, the amount of the mixture required to form a specimen was 7694 g. The parameters of the gyratory compactor were set according to the Superpave design method, where the internal rotation angle is 1.16° ± 0.02°, the vertical pressure was 600 ± 18 kPa, and the rotation rate was 30 r/min ± 0.5 r/min. Meanwhile, this study chose to set the required number of rotational compaction (100) as the end condition of rotational compaction. After compaction, the mold was demolded to obtain the specimen. Then, the formed specimen was core drilled, and the ends of the specimen were removed to obtain a cylindrical specimen (diameter = 100 mm, height = 150 mm).

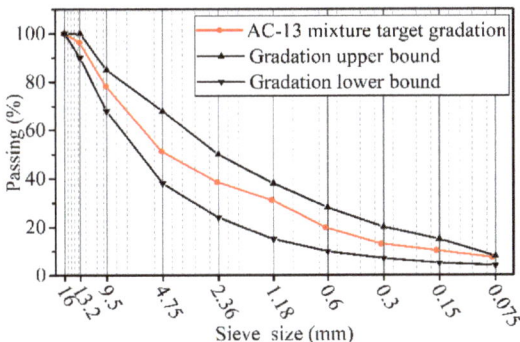

Figure 4. Gradations of the AC-13 asphalt mixture.

Table 1. Performances of asphalt binder.

| Type | Penetration (25 °C, 0.1 mm) | Ductility (15 °C, cm) | Softening Point (°C) | Brinell Viscosity (60 °C) | Elastic Recovery (%) |
|---|---|---|---|---|---|
| Virgin binder | 70.1 | 60 | 47 | 285 | 10 |
| Requirements | 60–80 | ≥40 | ≥45 | ≥160 | - |
| Specification | T0604 2011 | T0605 2011 | T0606 2011 | T0625 2011 | T0662 2000 |

Table 2. Performances of basalt aggregates.

| Type | Crushing Value (%) | Los Angeles Abrasion Value (%) | Needle and Flaky Particle Content (%) | Water Absorption Rate (%) | Apparent Density (g/cm$^3$) |
|---|---|---|---|---|---|
| Basalt aggregates | 13.5 | 14.2 | 9.8 | 0.4 | 2.93 |
| Requirements | ≤26 | ≤28 | ≤15 | ≤2.0 | >2.6 |
| Specification | T0316 2005 | T0317 2005 | T0312 2005 | T0307 2005 | T0304 2005 |

### 3.2. Methodology

Creep tests were conducted on AC-13 asphalt mixtures at various stress levels and temperatures to comprehensively analyze the viscoelastic damage behavior of asphalt mixes. Various researchers selected different test temperatures and test stress levels. Ghorbani et al. [7,9] assessed the temperature effects of mixtures at 5, 20, 35, and 50 °C. Cheng et al. [24] analyzed the viscoelastic properties of different mixtures from 10 to 50 °C at 10 °C increments. Luo et al. [15] studied the viscoelastic properties of asphalt mixtures at a room temperature of 25 °C only. This study was based on the work of Cheng et al. [24] who analyzed the creep properties of the AC-13 asphalt mixture from 10 to 50 °C at 20 °C increments. Considering the temperature, the loading capacity of the instrument, and the loading time, the stress levels selected at different temperatures were different for exhibiting both the consolidation effect and the tertiary stage of creep performance at different temperatures. For example, at 50 °C and 0.8 MPa, the asphalt mixture exhibited the three stages of creep, while, at 10 °C, the asphalt mixture only exhibited a consolidation effect at 0.8 MPa. To allow the asphalt mixture to exhibit the three stages of creep, it required a larger stress level. Therefore, in this study, in order to study the change of viscoelasticity under different stresses, and to consider the comparison of viscoelasticity under the same stress at different temperatures, the creep test conditions shown in Table 3 were selected.

Table 3. Creep experiment conditions of asphalt mixtures.

| Testing Temperature/°C | Stress Levels /MPa |
|---|---|
| 10 | 0.8, 1.4, 2.0, 2.5, 2.8 |
| 30 | 0.2, 0.5, 0.65, 0.8, 0.7 (validation) |
| 50 | 0.2, 0.35, 0.5, 0.8 |

The specimen was kept under the testing temperature for more than 4 h before the experiment to ensure a consistent temperature throughout the specimen and to eliminate the effects of temperature inhomogeneity. Before the creep test, the specimen was preloaded with 0.05 MPa and held for 60 s to eliminate the effect of mechanical errors. Double-greased membranes was used to reduce the effect of friction between the clamps and the specimen. All tests were carried out with the UTM-25 tester (IPC Australia), which could apply a maximum stress of 2.8 MPa to the specimens. All preset stress levels were maintained for 3600 s at 30 and 50 °C. However, at 10 °C, the stress duration was increased to 10,000 s to ensure that the specimen exhibited the tertiary stage of creep at 2.8 and 2.5 MPa.

## 4. Test Results and Analysis
### 4.1. Test Results

Three tests were performed at each stress level, and the results were averaged. The averaged creep strain vs. time curves for the applied stress levels at different temperatures are shown in Figure 5.

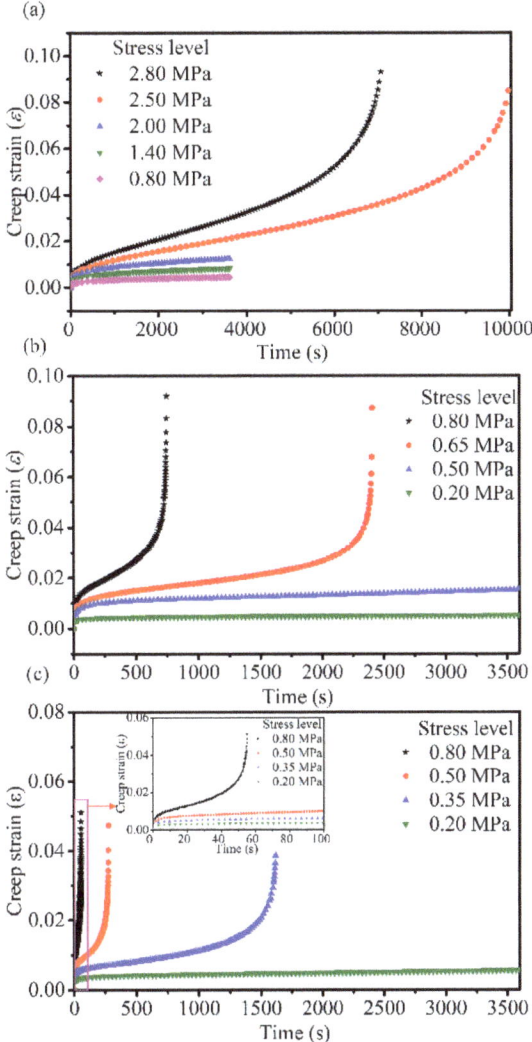

**Figure 5.** Creep strain curves of asphalt mixtures under different stress levels at (**a**) 10 °C, (**b**) 30 °C, and (**c**) 50 °C.

Figure 5 shows that the creep deformation of the asphalt mixtures is stress- and temperature-dependent. Here, the creep curve at 30 °C and 0.8 MPa is used as an example illustrate the three stages of creep and the time at which each stage begins. The curve of three stages of creep and the curve of creep rate are shown in Figure 6a,b, respectively. The creep rate curve was obtained by deriving the creep curve versus time. In Figure 6, $t_1$ and $t_2$ represent the start times of the steady stage and tertiary stage, respectively. Additionally, during this time, the creep rate remains almost constant and the value reaches a minimum. Therefore, the change in creep rate can be used to determine which creep stage the asphalt mixture is in. Combining Figures 5 and 6, at all temperatures, the creep process under larger stress levels exhibited all three stages of creep, whereas under smaller stress levels it only exhibited decelerated and stationary stages (i.e., consolidation effects). During the primary stage, the strain rate of the asphalt mixture gradually decreased with increasing in loading time; this is referred to as creep hardening. Creep hardening is different from strain hardening, which refers to the phenomenon wherein the material enters plastic deformation and the strength increases with an increase in deformation. The prerequisite for strain hardening is that material stresses should reach the yield strength and produce irreversible plastic deformation, i.e., the hardening is attributed to the strengthening of the yield strength. However, even under lower stress, the asphalt mixture exhibited varying degrees of decay creep; thus, creep hardening in the decelerating stage was not attributed to strain hardening, but to the compaction effect of the asphalt mixture under compressive stress. In this stage, the asphalt mixture was gradually compacted under the compressive stress, the voids were reduced and the strength gradually increased. In the accelerated stage, a rapid increase in deformation in addition to an increase in the strain rate with time was observed because of the presence and evolution of the damage, and this eventually led to a creep failure in the asphalt mixture. The strain rate at the accelerated creep stage was related to the stress and temperature: the higher the stress, the greater the strain rate; the higher the temperature, the greater the strain rate.

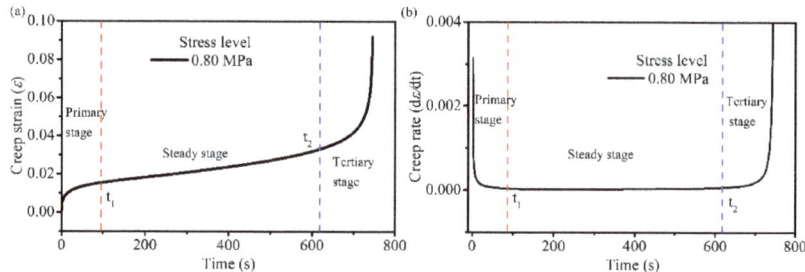

**Figure 6.** Schematic of the three stages (**a**) curve of three stages of creep, (**b**) curve of creep rate of asphalt mixture.

Thus, based on the testing results of the creep performance, this study considers that there are two mechanisms in the entire creep process based on the creep testing results: creep hardening and creep damage deterioration. When the creep hardening mechanism is dominant, the strain rate of the asphalt mixture gradually decreases with an increase in loading time, and the creep decays. When the damage deterioration mechanism is dominant, the strain rate increases with an increase in loading time, and the creep exhibits an accelerated state. When the two mechanisms occur in proximity, the strain rate remains constant and asphalt mixture is in a stable state. The point at which the steady stage enters the accelerated stage is called the flow time. The flow time depends on the level of stress applied; the higher the stress level, the earlier the damage deterioration mechanism occurs, and thus the flow time is shorter.

### 4.2. Model Verification and Parameter Determination

We verify whether the proposed creep damage model can accurately describe and reproduce all creep behavior presented in the tests. Equation (16) is determined by a curve fitting to the creep test data using the Levenberg–Marquardt optimization algorithm in 1stop. The parameters of the proposed model are listed in Table 4; the testing data of creep tests under various conditions and the simulation results of the proposed model are illustrated in Figure 7. The parameters of the correlation coefficients $R^2$ are greater than 0.998, which indicates that the fractional derivative creep damage model can accurately characterize the three-stage creep process of asphalt mixtures under various conditions.

Table 4 indicates that, at lower temperatures (10 °C), when the asphalt mixtures are subjected to smaller stress levels (0.8–2.0 MPa), the deformation factor $K_1$, which describes the creep hardening mechanism, remains approximately constant and its value is much greater than at the other temperatures. This indicates that the deformation factor $K_1$ is independent of the stress levels at lower stress levels (0.8–2.0 MPa) and the asphalt mixture has greater strength at lower temperatures (10 °C). This is most likely due to the fact that asphalt mortars have a greater strength and exhibit elastic behavior at lower temperatures. Additionally, in the smaller stress range, the asphalt mixture has a smaller strain, and the compaction within the asphalt mixture is not sufficient to change the skeletal structural strength of the mixture, so the deformation factor $K_1$ remains constant. As only consolidation effects are exhibited within the asphalt mix at these small stresses, these smaller stress ranges (0.8–2.0 MPa at 10 °C and <0.2 MPa at 30 °C) are referred to as creep consolidation stress (CCT) ranges. The creep consolidation stress range is temperature-dependent. At higher temperatures (30 °C) the asphalt mixtures exhibit a creep consolidation only at a smaller stress level (0.2 MPa); while, at lower temperatures (10 °C), the asphalt mixtures exhibit creep consolidation behavior over a wide stress range (0.8–2.0 MPa).

Table 4. The parameters of the proposed constitutive models.

| Testing Tmperature /°C | Stress /MPa | $r$ | $K_1$/MPa·s$^r$ | $\alpha$ | $K_2$/MPa·s$^\alpha$ | $t_f$ /s | $v$ | $R^2$ |
|---|---|---|---|---|---|---|---|---|
| 10 | 2.80 | 0.271 | 1453.211 | 0.029 | 18.589 | 7350.634 | 10.072 | 0.999 |
|  | 2.50 | 0.244 | 1316.923 | 0.023 | 19.8676 | 10,172.405 | 10.103 | 0.998 |
|  | 2.00 | 0.219 | 1113.468 | $3.849 \times 10^{-6}$ | $4.027 \times 10^{-4}$ | $8.577 \times 10^8$ | 9.987 | 0.999 |
|  | 1.40 | 0.216 | 1084.970 | 0 | $1.407 \times 10^{-7}$ | $3.792 \times 10^{13}$ | 9.779 | 0.999 |
|  | 0.80 | 0.212 | 1116.156 | 0 | $7.878 \times 10^{-12}$ | $7.661 \times 10^{18}$ | 10.085 | 0.999 |
| 30 | 0.80 | 0.225 | 141.968 | 0.396 | 310.243 | 746.492 | 5.174 | 0.999 |
|  | 0.65 | 0.165 | 125.024 | 0.350 | 352.432 | 2407.442 | 5.267 | 0.999 |
|  | 0.50 | 0.131 | 104.158 | 0.326 | 386.826 | 10,525.528 | 5.361 | 0.998 |
|  | 0.20 | 0.080 | 73.064 | 0 | 2.196 | $1.8014 \times 10^6$ | 5.248 | 0.999 |
| 50 | 0.80 | 0.254 | 144.683 | 0.517 | 455.567 | 55.552 | 2.558 | 0.999 |
|  | 0.50 | 0.147 | 106.803 | 0.441 | 542.917 | 278.371 | 2.599 | 0.999 |
|  | 0.35 | 0.115 | 96.704 | 0.419 | 587.824 | 1643.618 | 2.583 | 0.999 |
|  | 0.20 | 0.080 | 79.432 | 0.411 | 624.145 | 13,912.083 | 2.560 | 0.998 |

In the creep consolidation state, the fractional order $r$ increases as the stress increases and the parameters ($\alpha$, K2), describing the damage mechanism, are close to zero. It follows from Equation (16) that when the deformation factor $K_2$ tends to be zero, the damage must be sufficiently small to enable the proposed constitutive model to describe the creep properties. The strain induced by the damage is too small relative to the strain of the hardening mechanism. Therefore, the Abel model can be applied to characterize the entire creep process of an asphalt mixture under smaller stress levels. This once again shows that the Abel model can describe the consolidation effect of the asphalt mixes. Under other creep conditions, both the fractional order $r$ and deformation factor $K_1$ increase as the stress increases for the hardening mechanism. The fractional order increases because the flow rate of the asphalt mortar in the asphalt mixture increases as the stress increases, which causes the asphalt mixture to gradually exhibit a viscous behavior. The increase in the deformation factor indicates that the resistance to deformation increases with the increasing stress. As the stress increases in the hardening mechanism, the faster the asphalt mixture is compacted, the faster the increase in the skeletal strength between the aggregates, and the smaller the deformation. For the softening mechanism, the fractional order $\alpha$ increases with the increasing stress, whereas the deformation factor $K_2$ decreases with the increasing stress. The fractional order increases because the flow rate of the asphalt mortar increases as the stress increases. The decrease in the deformation factor is attributed to the presence of damage in the softening mechanism; the greater the damage to the inter-aggregate asphalt mortar as the stress increases, and the greater the deformation. For the same stress level, asphalt mixtures exhibit different viscoelastic properties at different temperatures. For example, $r$ increases and $K_1$ decreases as the temperature increases for the stress of 0.8 MPa. This is because, at higher temperatures, the asphalt mortar exhibits a stronger viscosity, while at lower temperatures the asphalt exhibits a stronger elasticity. When the model is applied to describe the three stages of creep, i.e., when the hardening and deterioration mechanisms are present, the fractional order of the damage softening mechanism is larger compared to the hardening mechanisms at higher temperatures (30 °C and 50 °C). This is due to the presence of damage, which makes the asphalt mixture exhibit stronger viscosity. The deformation factor of the damage softening mechanism is also larger compared to the hardening mechanism. This is because the asphalt mixture is compacted in the primary stage, which makes its resistance to deformation increase when it enters the softening mechanism. However, at 10 °C, the parameter $\alpha$ describing the damage mechanism is small, indicating that the damage of the asphalt mixture at this temperature tends to occur more closely to the elastic damage.

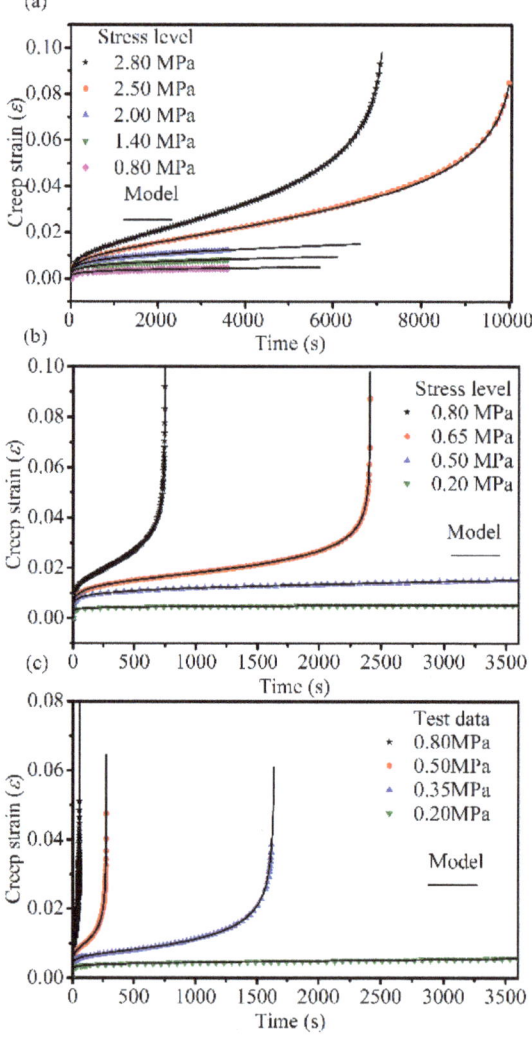

Figure 7. Comparison between the tests data of the creep tests and the fractional derivative creep damage model predicted creep curves at (a) 10 °C, (b) 30 °C, and (c) 50 °C.

## 4.3. Analysis of the Relationship between Model Parameters and Stress Levels

As can be seen from Table 4, several parameters in the fractional derivative creep damage model are related to both loading stress levels and temperatures. Figures 8–10 illustrate the distribution of the parameters of various stress levels at different temperatures. Table 4 indicates that the material parameter $v$ is essentially equal for different stresses at the same temperature; therefore, parameter $v$ at a given temperature is considered the average of the parameter $v$ at different stresses, as indicated in Figures 8f, 9f and 10f. When the asphalt mixtures were subjected to linear viscoelastic stress levels, the parameters $K_1$ remained approximately constant, and parameters $\alpha$ and $K_2$ were close to zero. Thus, the parameters $K_1$, $\alpha$, and $K_2$ were considered constants in the range of the linear viscoelastic stresses. Under other stress levels, fractional orders $r$ and $\alpha$ showed a good first-order exponential decay distribution regarding the stress levels, and the deformation factors, $K_1$ and $K_2$, showed a good linear distribution with the stress levels, as indicated in Figures 8–10. The linear viscoelastic stress range could also be obtained using $K_1$, as shown in Figures 9b and 10b. The intersection point of the two straight lines, which described the relationship between $K_1$ and the applied stress, was the point of demarcation between the linear viscoelastic range and the nonlinear viscoelastic range. When the stress level is less than the intersection point, the asphalt mixture is in the linear viscoelastic range and vice versa in the nonlinear viscoelastic range. The creep failure times at different stresses were fitted using Equation (12) to obtain the material parameters that describe the damage at different temperatures; the results are shown in Figures 8e, 9e and 10e.

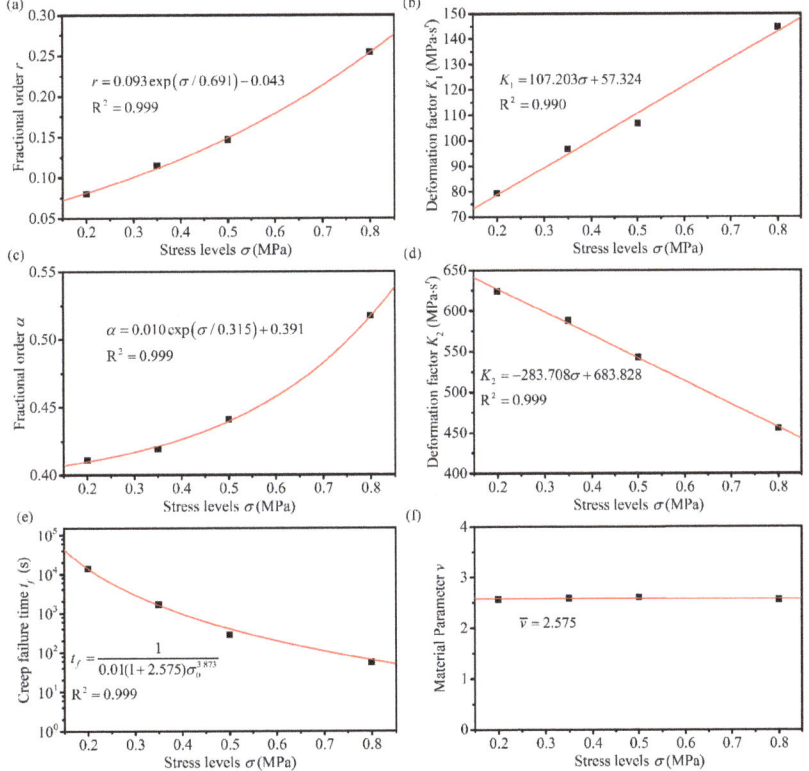

**Figure 8.** Variation of the proposed model parameters with stress levels at 50 °C (**a**) $r$, (**b**) $K_1$, (**c**) $\alpha$, (**d**) $K_2$, (**e**) $t_f$, and (**f**) $v$.

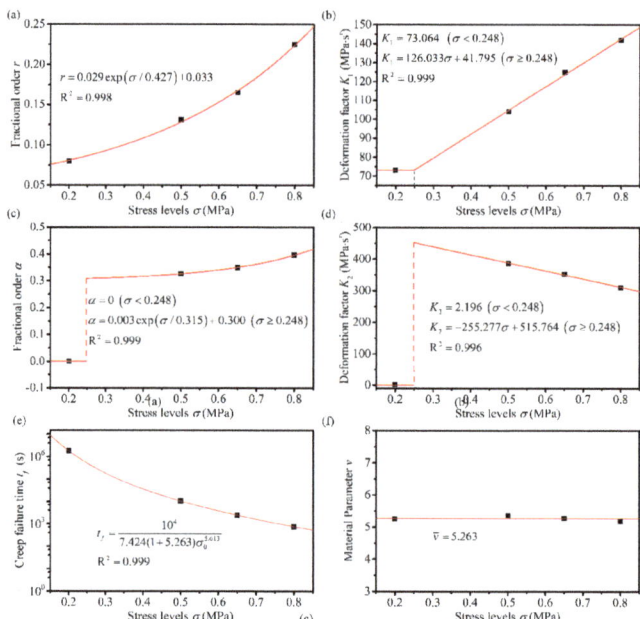

**Figure 9.** Variation of the proposed model parameters with stress levels at 30 °C (**a**) $r$, (**b**) $K_1$, (**c**) $\alpha$, (**d**) $K_2$, (**e**) $t_f$, and (**f**) $v$.

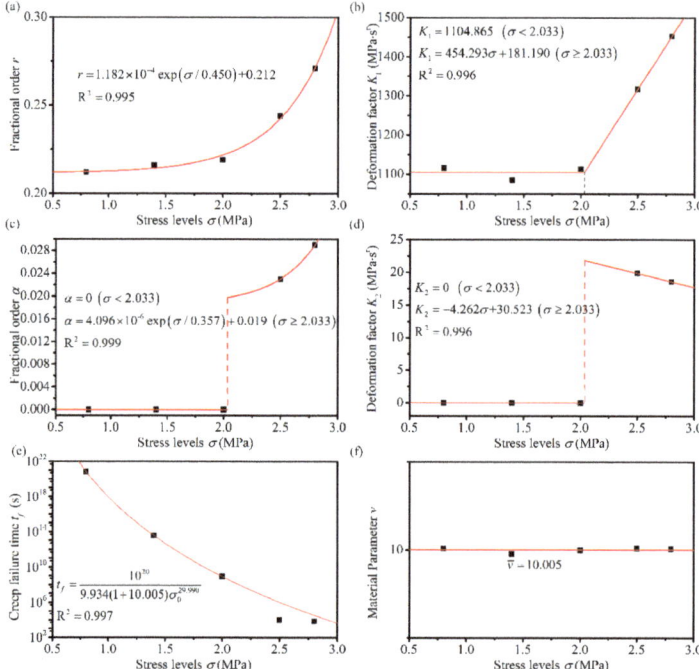

**Figure 10.** Variation of the proposed model parameters with stress levels at 10 °C (**a**) $r$, (**b**) $K_1$, (**c**) $\alpha$, (**d**) $K_2$, (**e**) $t_f$, and (**f**) $v$.

Based on the relationship between the model parameters and stresses given above, the creep curves could be predicted for any stress level at the three temperatures. Generally, the standard load was taken as 0.7 MPa when designing the pavement structure. Therefore, the creep behavior of the asphalt mixture was predicted and compared with the experimental data at 30 °C and 0.7 MPa, as shown in Figure 11. It was revealed that the predicted values were in good agreement with the experimental results.

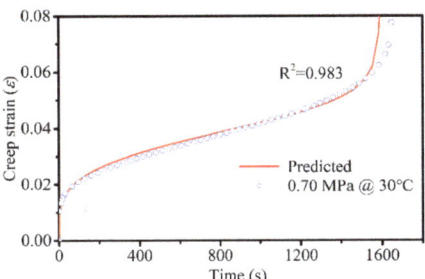

**Figure 11.** Comparison between the test data of the creep and predicted values at 30 °C and 0.7 MPa.

### 4.4. Analysis of Damage Evolution

It is believed that the fractional derivative creep damage model provides a good description of the three-stage creep process in asphalt mixtures. Once the parameters of the proposed constitutive model are identified, the proposed damage evolution model can be applied to calculate the damage value $D$. The damage variation curves with a loading time at various stresses and temperatures are shown in Figure 12.

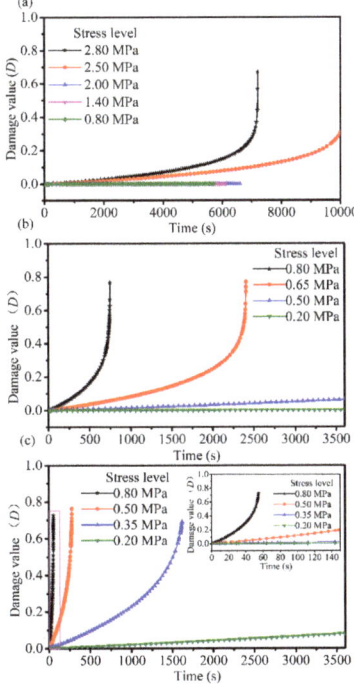

**Figure 12.** Variation curves of damage values with time under different compressive stress levels at (**a**) 10 °C, (**b**) 30 °C, and (**c**) 50 °C.

As indicated in Figure 12, at a given temperature, under higher stresses, the damage value D increases in an approximately linear fashion during the steady stage and rapidly during the tertiary stage. In contrast, at lower stresses, the damage value $D$ remains at virtually zero throughout the creep process. Regardless of the applied stresses, the asphalt mixture exhibited negligible damage during the primary stage [16,34]. This is because, in the initial stages of loading, the asphalt mixture is compacted, which makes the mixture structurally hardened, at which point no damage occurs to the asphalt mixture. On a macroscopic scale, asphalt mixtures show a decrease in the void fraction and an increase in the resistance to deformation; this indicates creep hardening. Over time, the asphalt mixture structure continues to stabilize, and the flow of the asphalt mortar tends to slow down. However, the asphalt mortar and coarse aggregate interface are gradually damaged and develop into microcracks. Macroscopically, the asphalt mixture exhibits a stable void ratio and constant strain rate. As the asphalt mortar flow deformation continues to accumulate, microcracks coalesce to form macro cracks, the mineral skeleton of the asphalt mixture begins to destabilize, and the asphalt mixture enters an accelerated creep phase. Macroscopically, it is characterized by cracks in the asphalt mixture and a rapid reduction in load-bearing capacity until the mixture breaks down in the creep phase; in general, damage to the asphalt mixture is initiated in the steady stage and developed in the tertiary stage, which is consistent with the works reported by Zeng et al. [16] and Al-rub et al. [34]. The experimental data and model damage analysis show the feasability of the damage evolution proposed in this study.

Equation (11) considers only the time dependence of the damage evolution, assuming that the damage only develops over time; it does not consider the relationship between the damage evolution and the relevant mechanical quantities such as stress or strain. Equations (11) and (16) show that each moment t corresponds to a damage value and a creep strain value. Therefore, through creep tests, the relationship curve between the damage value and strain can be established for different stress levels. At different stress levels, the damage values corresponding to the strain values at each moment of the decelerated stage are small and essentially close to zero. The strain caused by the damage (damage strain, $\varepsilon_d$) is obtained by subtracting the strain in the decelerated stage from the total creep strain in the tests. According to Equation (16), $\varepsilon_d$ can be expressed as:

$$\varepsilon_d(t) = \varepsilon(t) - \frac{\sigma_0 t^r}{K_1}. \tag{17}$$

Based on the time dependence of the damage evolution and damage values obtained from the creep experiments, the relationship curves between the damage value D and $\varepsilon_d$ for different stress levels at different temperatures is established in this study, as shown in Figure 13.

Figure 13 shows that the relationship curves between the creep $D$ and the $\varepsilon_d$ of the asphalt mixture at different stress levels overlap. Therefore, a unified damage evolution model between $D$ and $\varepsilon_d$ can be developed, where the damage parameters in the damage evolution relationship are only material- and temperature-dependent, independent of the magnitude of the stress levels. Katsuki et al. [4] used the Weibull distribution function to describe the variation in the damage with the strain, and verified the feasibility of the model using the experimental computed tomography images from Tashman at al. [35]. Therefore, in this study, the Weibull distribution function is used to describe the evolution of the damage with $\varepsilon_d$, and the model expression is shown in Equation (18). By adopting the proposed damage model (Equation (18)) to fit the experimental data under different levels of stress for different temperatures in Figure 13, the damage model curve shown in Figure 13 is obtained. The parameters are listed in Table 5. The parameters of the correlation coefficient $R^2$ are greater than 0.992, which shows that the Weibull distribution function can be used to describe the damage evolution and demonstrate the feasibility of establishing a unified damage evolution model for different stress levels. Statistically speaking, the microelement strength of randomly distributed microdefects within the material obeys the Weibull distribution under $\varepsilon_d$:

$$D = 1 - \exp\left[-\left(\frac{\varepsilon_d}{p}\right)^q\right], \qquad (18)$$

where $p$ and $q$ are temperature-dependent material parameters, respectively.

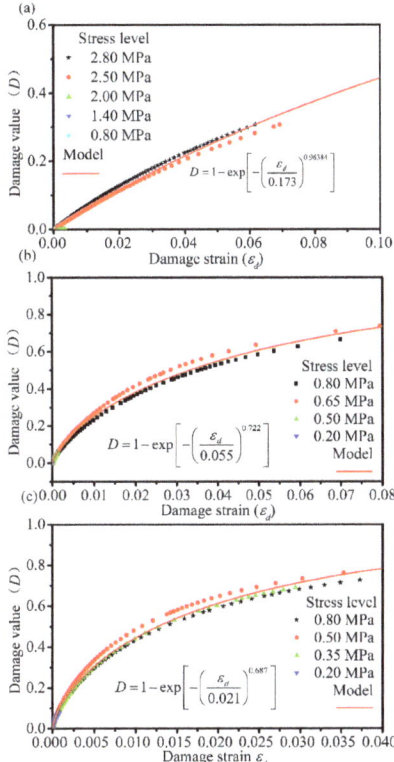

**Figure 13.** The relation curve between creep damage value and damage strain of asphalt mixture under different stress levels at (**a**) 10 °C, (**b**) 30 °C, and (**c**) 50 °C.

**Table 5.** The parameters of damage evolution model.

| Testing Temperature/°C | $p$ | $q$ | $R^2$ |
|---|---|---|---|
| 10 | 0.173 | 0.964 | 0.992 |
| 30 | 0.055 | 0.722 | 0.995 |
| 50 | 0.021 | 0.687 | 0.995 |

The nonlinear damage constitutive model proposed in this study has the advantages of fewer parameters and a clear physical meaning of parameters relative to the creep model proposed by Zeng et al. [16]. Relative to the model proposed by Zhang et al. [19], there is no need to assume that there is critical stress. Additionally, relative to other damage models [16,17,19], this study decouples the damage strain from the creep strain and establishes a unified damage evolution model for different stress levels, whose parameters are only related to material properties and temperature, and are independent of the magnitude of the applied stress.

## 5. Conclusions

A fractional viscoelastic creep damage model was proposed to characterize the creep process of asphalt mixtures at different stress levels and temperatures based on the fact that the asphalt mixtures exhibited different deformation mechanisms throughout the entire creep process at different stress levels. The following conclusions were drawn:

(1) The compressive creep tests showed that the creep processes of asphalt mixtures at three different temperatures exhibited a consolidation effect at low stress levels and a tertiary stage at high stress levels. Based on the compressive creep tests, this study considered that the creep hardening and creep damage deterioration mechanisms were at work throughout the creep process. A nonlinear fractional creep damage model was proposed by combining the two models. The results of the compression creep tests were in good agreement with the proposed creep damage model, which confirmed the effectiveness of the model in describing the three-stage creep process of the asphalt mixture at different stress levels and temperatures.

(2) The parameters of the proposed model were explicit in a physical sense, and relationships between the parameters and the applied stresses were established. The creep consolidation stress range could be obtained through the relationship between $K_1$ and the applied stresses.

(3) The damage value was negligible in the primary stage; it increased in an approximately linear fashion in the steady stage and rapidly during the tertiary stage. The damage was initiated in the steady stage and developed in the tertiary stage. The damage value remained at virtually zero throughout the creep process at low stresses; however, it increased rapidly with the increasing stress.

(4) Based on the statistical quantification of the asphalt mixture damage evolution, a unified creep damage evolution model, which represented the relationship between the damage evolution and damage strain, was established. The damage parameters in this damage evolution relationship were related to the material properties and temperature and were independent of the magnitude of the stress levels. The model described well the asphalt mixture damage evolution process, and the damage obeyed the Weibull distribution.

It should be noted that only the relationship between the model parameters and applied stresses was established in the work. In order to generalize the proposed nonlinear damage constitutive model, it is also necessary to provide a way to construct a relationship between the model parameters, temperature, and the applied stress. The establishment of the relationship between the three will be considered in future studies.

**Author Contributions:** The authors confirm the contributions to the paper as follows: Conceptualization: X.G., Q.D.; data curation: Z.Y.; analysis and interpretation of results: Q.Z.; funding acquisition: X.G.; draft manuscript preparation: Q.Z.; review and editing: J.L. All authors reviewed the results and approved the final version of the manuscript.

**Funding:** This research was funded by the National Natural Science Foundation of China (Grant No. 51878162), the National Key Research and Development Program of China (Grant No. 2017YFF0205600).

**Institutional Review Board Statement:** Not applicable.

**Informed Consent Statement:** Not applicable.

**Conflicts of Interest:** The authors declare no conflict of interest.

## References

1. Ahmed, I.; Thom, N.; Bilal Ahmed Zaidi, S.; Carvajal-Munoz, J.S.; Rahman, T.; Dawson, A. Application of a novel linear-viscous approach to predict permanent deformation in simulative inverted pavements. *Constr. Build. Mater.* **2021**, *267*, 120681. [CrossRef]
2. Doll, B.; Ozer, H.; Rivera-Perez, J.J.; Al-Qadi, I.L.; Lambros, J. Investigation of viscoelastic fracture fields in asphalt mixtures using digital image correlation. *Int J. Fract.* **2017**, *205*, 37–56. [CrossRef]
3. Dong, N.; Wang, D.; Zhang, S.; Chen, Z.; Liang, H.; Ni, F.; Yu, J.; Yu, H. Exploring creep and recovery behavior of hot mix asphalt field cores with multi-sequenced repeated load test. *Constr. Build. Mater.* **2021**, *279*, 122435. [CrossRef]

4. Katsuki, D.; Gutierrez, M. Viscoelastic damage model for asphalt concrete. *Acta Geotech.* **2011**, *6*, 231–241. [CrossRef]
5. Nguyen, Q.T.; Di Benedetto, H.; Nguyen, Q.P.; Hoang, T.T.N.; Bui, V.P. Effect of time–temperature, strain level and cyclic loading on the complex Poisson's ratio of asphalt mixtures. *Constr. Build. Mater.* **2021**, *294*, 123564. [CrossRef]
6. Zbiciak, A.; Brzeziński, K.; Michalczyk, R. Constitutive models of pavement asphaltic layers based on mixture compositions. *J. Civ. Eng. Manag.* **2017**, *23*, 378–383. [CrossRef]
7. Ghorbani, B.; Arulrajah, A.; Narsilio, G.; Horpibulsuk, S. Experimental and ANN analysis of temperature effects on the permanent deformation properties of demolition wastes. *Transp. Geotech.* **2020**, *24*, 100365. [CrossRef]
8. Alrashydah, E.A.I.; Abo-Qudais, S.A. Modeling of creep compliance behavior in asphalt mixes using multiple regression and artificial neural networks. *Constr. Build. Mater.* **2018**, *159*, 635–641. [CrossRef]
9. Ghorbani, B.; Arulrajah, A.; Narsilio, G.; Horpibulsuk, S.; Win Bo, M. Thermal and mechanical properties of demolition wastes in geothermal pavements by experimental and machine learning techniques. *Constr. Build. Mater.* **2021**, *280*, 122499. [CrossRef]
10. Bai, F.; Yang, X.; Zeng, G. Stochastic Viscoelastic–Viscoplastic Response of Asphalt Mixture under Uniaxial Compression. *J. Eng. Mech.* **2017**, *143*, 04017049. [CrossRef]
11. Gao, D.; Wang, P.; Li, M.; Luo, W. Modelling of nonlinear viscoelastic creep behaviour of hot-mix asphalt. *Constr. Build. Mater.* **2015**, *95*, 329–336. [CrossRef]
12. Huang, C.; Wang, F.; Gao, T.; Gao, D.; Kachanov, L.M. A New Viscoelastic Mechanics Model for the Creep Behaviour of Fibre Reinforced Asphalt Concrete. *Frat. Integrita Strut.* **2018**, *12*, 108–120. [CrossRef]
13. Lagos-Varas, M.; Movilla-Quesada, D.; Arenas, J.P.; Raposeiras, A.C.; Castro-Fresno, D.; Calzada-Pérez, M.A.; Vega-Zamanillo, A.; Maturana, J. Study of the mechanical behavior of asphalt mixtures using fractional rheology to model their viscoelasticity. *Constr. Build. Mater.* **2019**, *200*, 124–134. [CrossRef]
14. Lagos-Varas, M.; Raposeiras, A.C.; Movilla-Quesada, D.; Arenas, J.P.; Castro-Fresno, D.; Muñoz-Cáceres, O.; Andres-Valeri, V.C. Study of the permanent deformation of binders and asphalt mixtures using rheological models of fractional viscoelasticity. *Constr. Build. Mater.* **2020**, *260*, 120438. [CrossRef]
15. Luo, W.; Li, B.; Zhang, Y.; Yin, B.; Dai, J. A Creep Model of Asphalt Mixture Based on Variable Order Fractional Derivative. *Appl. Sci.* **2020**, *10*, 3862. [CrossRef]
16. Zeng, G.; Yang, X.; Bai, F.; Gao, H. Visco-elastoplastic damage constitutive model for compressed asphalt mastic. *J. Cent. South. Univ.* **2014**, *21*, 4007–4013. [CrossRef]
17. Zhang, J.; Li, Z.; Chu, H.; Lu, J. A viscoelastic damage constitutive model for asphalt mixture under the cyclic loading. *Constr. Build. Mater.* **2019**, *227*, 116631. [CrossRef]
18. Zhang, J.; Zhu, C.; Li, X.; Pei, J.; Chen, J. Characterizing the three-stage rutting behavior of asphalt pavement with semi-rigid base by using UMAT in ABAQUS. *Constr. Build. Mater.* **2017**, *140*, 496–507. [CrossRef]
19. Zhang, Y.; Liu, X.; Yin, B.; Luo, W. A Nonlinear Fractional Viscoelastic-Plastic Creep Model of Asphalt Mixture. *Polymers* **2021**, *13*, 1278. [CrossRef] [PubMed]
20. Luo, W.; Jazouli, S.; Vu-Khanh, T. Modeling of Nonlinear Viscoelastic Creep of Polycarbonate. *e-Polymers* **2007**, *7*, 1. [CrossRef]
21. Darabi, M.K.; Huang, C.-W.; Bazzaz, M.; Masad, E.A.; Little, D.N. Characterization and validation of the nonlinear viscoelastic-viscoplastic with hardening-relaxation constitutive relationship for asphalt mixtures. *Constr. Build. Mater.* **2019**, *216*, 648–660. [CrossRef]
22. Im, S.; You, T.; Ban, H.; Kim, Y.-R. Multiscale testing-analysis of asphaltic materials considering viscoelastic and viscoplastic deformation. *Int. J. Pavement. Eng.* **2017**, *18*, 783–797. [CrossRef]
23. Chang Kuo-Neng, G.; Meegoda Jay, N. Micromechanical Simulation of Hot Mix Asphalt. *J. Eng. Mech.* **1997**, *123*, 495–503. [CrossRef]
24. Cheng, Y.; Li, H.; Li, L.; Zhang, Y.; Wang, H.; Bai, Y. Viscoelastic Properties of Asphalt Mixtures with Different Modifiers at Different Temperatures Based on Static Creep Tests. *Appl. Sci.* **2019**, *9*, 4246. [CrossRef]
25. Xu, X.; Cui, Z. Investigation of a fractional derivative creep model of clay and its numerical implementation. *Comput. Geotech.* **2020**, *119*, 103387. [CrossRef]
26. Gu, L.; Zhang, W.; Ma, T.; Qiu, X.; Xu, J. Numerical Simulation of Viscoelastic Behavior of Asphalt Mixture Using Fractional Constitutive Model. *J. Eng. Mech.* **2021**, *147*, 04021027. [CrossRef]
27. Xu, Y.; Shan, L.; Tian, S. Fractional Derivative Viscoelastic Response Model for Asphalt Binders. *J. Mater. Civil. Eng.* **2019**, *31*, 04019089. [CrossRef]
28. Celauro, C.; Fecarotti, C.; Pirrotta, A.; Collop, A.C. Experimental validation of a fractional model for creep/recovery testing of asphalt mixtures. *Constr. Build. Mater.* **2012**, *36*, 458–466. [CrossRef]
29. Sun, L.; Zhu, H.; Zhu, Y. Two-Stage Viscoelastic-Viscoplastic Damage Constitutive Model of Asphalt Mixtures. *J. Mater. Civil. Eng.* **2013**, *25*, 958–971. [CrossRef]
30. Koeller, R.C. Applications of Fractional Calculus to the Theory of Viscoelasticity. *J. Appl. Mech.* **1984**, *51*, 299–307. [CrossRef]
31. Kachanov, L.M. Rupture Time Under Creep Conditions. *Int J. Fract.* **1999**, *97*, 11–18. [CrossRef]
32. Lemaitre, J. How to use damage mechanics. *Nucl. Eng. Des.* **1984**, *80*, 233–245. [CrossRef]
33. Luo, W.; Liang, S.; Zhang, Y. Fractional Differential Constitutive Model for Dynamic Viscoelasticity of Asphalt Mixture. *China J. Highw. Transp.* **2020**, *33*, 34–43.

34. Al-rub, R.K.A.; You, T.; Masad, E.A.; Little, N. Mesomechanical modeling of the thermo-viscoelastic, thermo-viscoplastic, and thermo-viscodamage response of asphalt concrete. *Int J. Eng. Sci.* **2011**, *3*, 14–33. [CrossRef]
35. Tashman, L.; Masad, E.; Little, D.; Zbib, H. A microstructure-based viscoplastic model for asphalt concrete. *Int J. Plast.* **2005**, *21*, 1659–1685. [CrossRef]

*Article*

# An Improved Mechanistic-Empirical Creep Model for Unsaturated Soft and Stabilized Soils

Xunli Jiang, Zhiyi Huang and Xue Luo *

College of Civil Engineering and Architecture, Zhejiang University, 866 Yuhangtang Road, Hangzhou 310058, China; jxunli@zju.edu.cn (X.J.); hzy@zju.edu.cn (Z.H.)
* Correspondence: xueluo@zju.edu.cn; Tel.: +86-0571-8820-6542

**Abstract:** Soft soils are usually treated to mitigate their engineering problems, such as excessive deformation, and stabilization is one of most popular treatments. Although there are many creep models to characterize the deformation behaviors of soil, there still exist demands for a balance between model accuracy and practical application. Therefore, this paper aims at developing a Mechanistic-Empirical creep model (MEC) for unsaturated soft and stabilized soils. The model considers the stress dependence and incorporates moisture sensitivity using matric suction and shear strength parameters. This formulation is intended to predict the soil creep deformation under arbitrary water content and arbitrary stress conditions. The results show that the MEC model is in good agreement with the experimental data with very high R-squared values. In addition, the model is compared with the other classical creep models for unsaturated soils. While the classical creep models require a different set of parameters when the water content is changed, the MEC model only needs one set of parameters for different stress levels and moisture conditions, which provides significant facilitation for implementation. Finally, a finite element simulation analysis of subgrade soil foundation is performed for different loading levels and moisture conditions. The MEC model is utilized to predict the creep behavior of subgrade soils. Under the same load and moisture level, the deformation of soft soil is largest, followed by lime soil and RHA–lime-stabilized soil, respectively.

**Keywords:** soft soil; stabilized soil; rice husk ash; Mechanistic-Empirical creep model; matric suction

**Citation:** Jiang, X.; Huang, Z.; Luo, X. An Improved Mechanistic-Empirical Creep Model for Unsaturated Soft and Stabilized Soils. *Materials* **2021**, *14*, 4146. https://doi.org/10.3390/ma14154146

Academic Editor: Jacek Tejchman

Received: 15 June 2021
Accepted: 23 July 2021
Published: 26 July 2021

**Publisher's Note:** MDPI stays neutral with regard to jurisdictional claims in published maps and institutional affiliations.

**Copyright:** © 2021 by the authors. Licensee MDPI, Basel, Switzerland. This article is an open access article distributed under the terms and conditions of the Creative Commons Attribution (CC BY) license (https://creativecommons.org/licenses/by/4.0/).

## 1. Introduction

Soft soils are usually regarded to be problematic due to their poor engineering properties, for instance, high water content, low undrained shear strength, poor permeability and remarkable rheological properties [1]. With the increased demand of urbanization, more and more foundations of buildings and infrastructure have been built on soft soil areas [2]. As a result, a series of practical problems have arisen because of insufficient strength and/or excessive deformation of soft foundations and subgrades [3]. In order to mitigate the problems, the following three methods are generally used to treat soft soils [4]: (i) consolidation of soft soil by surcharge preloading; (ii) foundation reinforcement by compaction piles; and (iii) strengthening of soil by chemical stabilization.

These treatment methods are applicable to different engineering situations. The surcharge preloading method has lower cost but a longer consolidation period, which can be used when the construction period permits. The compaction pile method is usually applicable to sandy soil, loose soil and miscellaneous fill foundation with gravel, brick and rubble. The chemically stabilized method has a wider range of applications because the curing period is relatively short and the effect of stabilization is better controlled. Moreover, compared with the other two methods, the chemically stabilized method is simpler and more economical, so it is one of most popular treatments and has been widely used in engineering practice.

Chemical stabilization is a method of adding a certain amount of curing agent to the soft soil to change the surface properties and connections between the particles through physical and chemical reactions, which improve the soil engineering properties [5]. The curing agents for chemical stabilization of soft soils can be mainly divided into inorganic curing agents and organic curing agents. Among them, inorganic curing agents like Portland cement, fly ash and lime are the most popular and effectively enhance the stability and shear strength of soils [6–10]. In recent years, with the emphasis on resource conservation and environmental protection, more and more industrial and agricultural wastes have been used for soil stabilization, such as wood industry ash, coconut shell ash, rice husk ash (RHA), etc. [11–17].

The engineering properties of soft soils are effectively improved by chemical stabilization, which meets the standards of engineering applications. At present, strength and water stability are the two engineering properties of stabilized soils that have been given the most attention, while less attention has been paid to deformation performance [18]. This is because creep behaviors of stabilized soils are often considered negligible. However, there have been a large number of engineering practices showing that soft soils after chemical stabilization still have creep deformation [19]. Such deformation is not only affected by the loading level, but also has connections with time and humidity. Therefore, accurate prediction of creep characteristics of soft soils and stabilized soils is of great significance to engineering applications.

A great number of models have been built to characterize the soils' creep behavior, including empirical semi-empirical (ESE) models, rheological theory-based models, and elastic–viscoplastic (EVP) models [20]. The empirical semi-empirical models are relatively simple and rely on statistical analysis [21]. Researchers have summarized empirical semi-empirical models based on laboratory or field tests of different soils under different test conditions. The representative ones include the Taylor model, Singh–Mitchell model and Mesri model, etc. [22–24]. Since then, many improved creep models have been proposed on the basis of the Singh–Mitchell model and Mesri model. For example, Lin et al. [25] proposed a new creep equation considering the over-consolidation factor of clay. Lu et al. [26] utilized a power function to express the stress–strain relationship and a hyperbola equation to express the strain–time relationship.

The models based on rheological theory often refer to component models such as the Maxwell model, Burgers model and Xiyuan rheological model [27–29]. Among them, the Burgers model is more often used to characterize the soft soil deformation behavior. For example, Rajesh et al. [30] used the Burgers model to characterize the soft soil subgrade when studying the performance of geosynthetic-reinforced railway track system lain on soft clay. Huang et al. [31] proposed an improved Burgers model to characterize the creep deformation behavior of soil. The establishment of an EVP model depends on the theory of elasticity and viscoplasticity. There are many classical EVP models for soils, such as the Perzyna model [32] and the EVP models proposed by Yin and Graham [33,34].

The three types of constitutive models mentioned above have their own advantages and disadvantages [21]. The physical implication about the empirical semi-empirical model is not very clear, but it is readily applicable in engineering practice; the rheological theory-based models find it difficult to reflect the creep deformation of rock and soil under complicated conditions; the EVP model can reflect the three-dimensional stresses of soils more accurately, but is relatively complex.

What is more, most models are formulated based on saturated soils; less consideration is given to unsaturated characteristics of soft soils. In fact, most soils in the field experience unsaturated conditions. When soft soils become unsaturated, they exhibit different volume, strength and hydraulic properties, and these properties are greatly affected by the degree of saturation. The distinctive volume, strength and hydraulic behaviors of soft soils at unsaturated states make the soils exhibit obvious nonlinear characteristics. These characteristics should be considered consistently and coherently in the process of modeling. In other words, an appropriate creep model should reflect the soil behavior under arbitrary pore water pressure and stress value, that is, the soil behavior under the whole hydraulic path and stress path [35,36].

As discussed, soil is often in an unsaturated state in the field. Therefore, this study aims at developing a Mechanistic-Empirical creep model (MEC) for unsaturated soft and stabilized soils. In this model, it considers the stress dependence and incorporates moisture sensitivity using matric suction and shear strength parameters, which could characterize the long-term creep deformation of these soils under various stress levels and water content conditions by one set of parameters, so as to enable more convenient and accurate predictions of the settlement of the foundation and the subgrade in the soft soil areas.

## 2. Development of Mechanistic-Empirical Creep Model for Unsaturated Soils

### 2.1. Typical Empirical Semi-Empirical Creep Models

The empirical semi-empirical creep models adopt simple mathematical expressions to describe creep properties of soils. Because of its simplicity and convenient parameter acquisition, it is widely used in engineering practice. Mesri and Godlewski [37] introduced a compression index and proposed the following creep model:

$$e = R_c C_c \lg(1 + \frac{t}{t_1}) \tag{1}$$

in which:

$$R_c = C_\alpha / C_c \tag{2}$$

where $e$ is the void ratio; $t_1$ is the time of the creep process at a certain time; $C_\alpha$ is the coefficient of secondary compression; and $C_c$ is the compression index.

In order to better consider the nonlinear creep characteristics, Singh and Mitchell [23] proposed a three-parameter creep model, which is shown in Equation (3):

$$\varepsilon = Be^{\beta \bar{D}_1}(\frac{t}{t_1})^\lambda \tag{3}$$

where $\varepsilon$ is the strain at any moment; $\bar{D}_1 = \frac{(\sigma_1 - \sigma_3)}{(\sigma_1 - \sigma_3)_f}$ is the deviator stress level at $t = t_1$; $\sigma_1 - \sigma_3$ is deviator stress of soil at $t = t_1$; $(\sigma_1 - \sigma_3)_f$ is damage deviator stress; and $B$, $\beta$ and $\lambda$ are model parameters. The Singh–Mitchell model can better show the creep behaviors of soils, but it is only suitable for describing the strain–time relationship in the range of 20% to 80% of the deviator stress level at the failure point. Later, Mesri [24] improved the creep model on the basis of the Singh–Mitchell empirical model, which is shown in Equation (4):

$$\varepsilon = \frac{2}{(E_u/S_u)_1} \frac{\bar{D}_1}{1 - (R_f)_1 \bar{D}_1}(\frac{t}{t_1})^\lambda \tag{4}$$

where $E_u/S_u$ is the undrained modulus to undrained shear strength ratio; $(R_f)_1$ and $\lambda$ are model parameters. When $t = t_1$, Equation (4) can be written as Equation (5):

$$\frac{\varepsilon}{\bar{D}_1} = (\frac{2}{E_u/S_u})_1 + (R_f)_1 \varepsilon \tag{5}$$

so $(\frac{2}{E_u/S_u})_1$ and $(R_f)_1$ can be obtained directly from the $\frac{\varepsilon}{\bar{D}} - \varepsilon$ graph of $t = t_1$ time, and the intercept is $(\frac{2}{E_u/S_u})_1$ and the slope is $(R_f)_1$; $\lambda$ is the mean slope of $\lg \varepsilon - \lg t$ curves under different stress levels. The Mesri creep model can characterize the deformation behavior of soils under any shear stress level, not being limited to the range of 20–80%.

Thereafter, Tseng and Lytton [38] proposed a popular empirical model for the deformation behavior of subgrade soils and granular materials in 1989, which is shown in Equation (6):

$$\varepsilon_p = \varepsilon_0 e^{-(\rho/N)^\beta} \tag{6}$$

where $\varepsilon_p$ is the permanent strain of the granular material; $\varepsilon_0$ is the maximum permanent strain; $\rho$ is the scale factor; $\beta$ is the shape factor; and $N$ is the number of cyclic loads. The Tseng–Lytton model can well describe the deformation behavior of soil materials at a certain stress level, but it cannot consider the variation of the stress level. Thus, Gu et al. [39] improved the model by adding $\sqrt{J_2}$ and $\alpha I_1 + K$ into the Tseng–Lytton model. The stress term $\sqrt{J_2}$ indicates the influence of the deviatoric shear stress on the materials: when $\sqrt{J_2}$ is higher, the deformation will be higher. The stress term $\alpha I_1 + K$ represents the influence of the hydrostatic stress on the materials, and it is closely related to shear strength parameters. Such a model is called a Mechanistic-Empirical (ME) model, which is expressed as:

$$\varepsilon_p = \varepsilon_0 e^{-(\rho/N)^\beta} \left(\sqrt{J_2}\right)^m (\alpha I_1 + K)^n \tag{7}$$

$$\alpha = \frac{2\sin\phi}{\sqrt{3}(3-\sin\phi)} \tag{8}$$

$$K = \frac{c \cdot 6\cos\phi}{\sqrt{3}(3-\sin\phi)} \tag{9}$$

where $J_2$ is second invariant of the deviatoric stress tensor; $I_1$ is first invariant of the stress tensor; $\varepsilon_0$, $\rho$, $\beta$, $m$, and $n$ are model coefficients; and $c$ and $\phi$ are cohesion and friction angle, respectively.

With the alteration of water content, the saturation degree of soil changes, and its basic mechanical properties also change greatly. The fundamental difference between unsaturated soils and saturated soils lies in the addition of a new stress state variable, matric suction. Its mechanical properties are related to this variable. The effect of the matric suction on the soil's properties is essentially a reflection of the impact of water. Therefore, in order to find out the effect of humidity level on the creep deformation of soils, it is necessary to develop a creep model containing the matric suction [40]. In recent years, some researchers have tried to establish an empirical creep model of unsaturated soils by introducing the matric suction to the existing models. For example, on the basis of the Mesri creep model, Lai et al. [41] established a creep model for unsaturated soils as:

$$\varepsilon(t) = \left(\frac{(\sigma_1-\sigma_3)_f}{E_u}\right) \frac{\bar{D}_1}{1-(R_f)_1 \bar{D}_1} \left(\frac{t}{t_1}\right)^\lambda \tag{10}$$

in which

$$R_f = a(h_m/P_a) + b \tag{11}$$

Combine Equations (10) and (11) to Equation (12):

$$\varepsilon(t) = \left(\frac{(\sigma_1-\sigma_3)_f}{E_u}\right) \frac{\bar{D}_1}{1-(a(h_m/P_a)+b)_1 \bar{D}_1} \left(\frac{t}{t_1}\right)^\lambda \tag{12}$$

where $P_a$ is the atmospheric pressure (101.33 kPa); $h_m$ is the matric suction; and $a$ and $b$ are model parameters that can be obtained from the $R_f$ versus $h_m/P_a$ curve. As shown in Equation (11), Lai et al. [41] only established a simple regression relationship to consider the matric suction, and replaced the term of $(R_f)_1$ in the Mesri model by the matric suction expression. In other words, the improved creep model is inherently still an empirical model, and the fitting parameters $a$ and $b$ are greatly affected by the data samples, so the accuracy of model prediction needs to be improved.

Based on the discussions above, it is seen that in order to establish an empirical semi-empirical model which accurately reflects the creep deformation behavior of unsaturated soils on the basis of classical creep models, the original model should contain terms or parameters that are appropriately related to the matric suction so as to accurately reflect its influence on creep. By comparing the creep models above, it is found that the ME model developed by Gu et al. [39] based on the Tseng–Lytton model is promising. The ME model contains mechanical terms of $\sqrt{J_2}$ and $(\alpha I_1 + K)$ with clear physical meanings, and $(\alpha I_1 + K)$ reflects the effect of the shear strength of the material. Therefore, this paper tries to further develop this ME model by introducing the matric suction to reflect the effect of unsaturated characteristics on the long-term deformation behavior of soils, which is elaborated on next.

*2.2. Formulation of Mechanistic-Empirical Creep Model*

When the ME model was proposed by Gu et al [39]. it only reflected the multi-stress creep deformation of materials under a certain moisture content, but soils in engineering practice are mostly unsaturated and the moisture content changes frequently. Thus, there is a need to establish a creep model for unsaturated soils which can simultaneously respond to various moisture conditions and multi-stress levels. To this end, based on the ME model, this study introduces the matric suction parameter, and establishes the relationship between the water content and the matric suction through the soil–water characteristic curve (SWCC). The improved model, called the MEC model, is formulated as follows:

$$\varepsilon_p(t) = \varepsilon_0 e^{-(\rho/t)^\beta} \left(\frac{\sqrt{J_2}}{p_a}\right)^m \left(\frac{\alpha_1 I_1 + K_1}{p_a}\right)^n (p_a)^{-(m+n)} \tag{13}$$

$$\alpha_1 = \frac{2 \sin \phi_1'}{\sqrt{3}(3 - \sin \phi_1')} \tag{14}$$

$$K_1 = \frac{c_1' \cdot 6 \cos \phi_1'}{\sqrt{3}(3 - \sin \phi_1')} \tag{15}$$

where $c_1'$ and $\phi_1'$ are the cohesion and internal friction angle for unsaturated soil, respectively.

The relationship between the matric suction and shear strength is established as Equation (16) [42]:

$$\tau_f = c\prime + (\sigma_n + \theta f h_m) \tan \phi' \tag{16}$$

where $c'$ and $\phi'$ are the cohesion and internal friction angle of soil in a saturated state, respectively; $\theta$ is the volumetric water content; $f$ is the saturation factor; and $h_m$ is the matric suction.

Equation (17) is used to calculate the saturation factor $f$:

$$\begin{cases} f = \frac{1}{\theta} & pF < 2 \\ f = 1 + \frac{S-85}{15}(\frac{1}{\theta} - 1) & 2 \leq pF \leq 3.5 \\ f = 1 & pF > 3.5 \end{cases} \tag{17}$$

The matric suction can be expressed in the form of pressure (Pa), water head (cm), or *pF*. The *pF* is a unit introduced from soil science, which represents the logarithmic value of pore water potential energy in centimeter head. In Equation (17), the *pF* is the number of matric suction expressed in *pF* form. The conversion relationships are as follows: $pF = \log 10(h_m \text{ in } cmH_2O); 1020 \, cmH_2O = 10^5 Pa$.

For unsaturated soils, the apparent cohesion that affects the shear strength of soils includes two terms: one is the conventional cohesion, which represents the shear force produced by the physicochemical interaction between particles such as van der Waals; the second part represents the shear strength produced by capillary action, that is, capillary cohesion, which is mainly related to matric suction in soil [43]. According to the research of Fredlund et al. [44], the influence of matric suction on soil shear deformation is mainly manifested in the suction stress, that is, the effect of capillary cohesion on soil. Fredlund et al. [44] proposed an extended Mohr–Coulomb criterion to represent the shear strength characteristics of unsaturated soils, and found that in the three-dimensional space of stress state variables $(\sigma - u_a)$, $(u_a - u_w)$ and shear stress $\tau$, the failure envelope is a planar surface. If the failure plane is projected on the shear stress and net normal stress plane, the expanded Mohr–Coulomb criterion is shown in Figure 1. According to this criterion and Equation (16), the total cohesion of unsaturated soil can be calculated, and is as follows:

$$c_1' = c' + \theta f h_m \tan \phi' \tag{18}$$

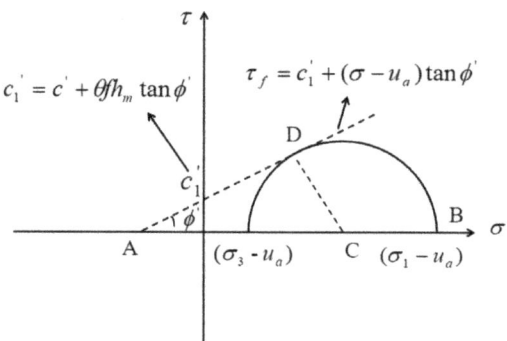

**Figure 1.** Mohr stress circle expression of failure envelope in two-dimensional plane of net normal stress and shear stress of unsaturated soil.

Substitute Equation (18) into Equations (14) and (15), which yields:

$$K_1 = \frac{c' \cdot 6 \cos \phi_1'}{\sqrt{3}(3 - \sin \phi_1')} + \frac{6\theta f h_m \cos \phi_1' \tan \phi'}{\sqrt{3}(3 - \sin \phi_1')} = K_n + K_m \tag{19}$$

Therefore, the improved creep model of unsaturated soils is obtained as follows:

$$\varepsilon_p(t) = \varepsilon_0 e^{-(\rho/t)^\beta} \left(\frac{\sqrt{J_2}}{p_a}\right)^m \left(\frac{\alpha_1 I_1 + K_n + K_m}{p_a}\right)^n (p_a)^{-(m+n)} \tag{20}$$

$$\alpha_1 = \frac{2 \sin \phi_1'}{\sqrt{3}(3 - \sin \phi_1')} \tag{21}$$

$$K_n = \frac{c' \cdot 6 \cos \phi_1'}{\sqrt{3}(3 - \sin \phi_1')} \tag{22}$$

$$K_m = \frac{6\theta f h_m \cos \phi_1' \tan \phi'}{\sqrt{3}(3 - \sin \phi_1')} \tag{23}$$

In this model, $J_2$ and $I_1$ reflect the effect of the stress on creep behavior of soils. The term of $\alpha_1 I_1 + K_n + K_m$ represents the hardening influence of the hydrostatic stress on the soils, which is closely related to these factors, like the material cohesion, internal friction angle, matric suction and moisture status. Among them, $K_m$ is a newly introduced item, which is used to further consider the effect of humidity and matric suction on soil deformation. The three moisture-related terms of $\theta$, $f$ and $h_m$ are used to compute $K_m$.

Through the establishment of the MEC model in Equation (20), it can be used to accurately describe the creep behavior of unsaturated soft soils and unsaturated stabilized soils. Three steps are involved in the process of model application:

(1) Determine the SWCCs of soft and stabilized soils.
(2) Determine the relevant shear strength parameters of soft and stabilized soils.

(3) Determine the coefficients $\varepsilon_0$, $\rho$, $\beta$, $m$ and $n$ from the creep tests at different stress levels and moisture conditions.

## 3. Materials and Laboratory Tests

### 3.1. Materials

The test soil is silt clay. Basic physical properties of the soils are list in Table 1. In this study, the soil was remolded after being filtered through a 4.75 mm sieve.

Table 1. Basic physical properties of soil samples.

| Soil Type | Liquid Limit, $W_L$/% | Plastic Limit, $W_P$/% | Plastic Index, $I_P$ | Optimum Moisture Content/% | Maximum Dry Density/(kg/m$^3$) |
|---|---|---|---|---|---|
| Silt clay | 38 | 19 | 19 | 18 | 1798 |

Lime and RHA are used as the stabilization materials in this study. Table 2 shows their chemical compositions. RHA is black powder, and Figure 2 shows the particle size distribution. The microstructure of RHA was studied by scanning electron microscopy (SEM). The magnification was 2000 times and 5000 times, as shown in Figures 3 and 4, respectively. From these figures, it can be seen that RHA contains many fine spherical particles, which are silica.

Table 2. Chemical composition of stabilized materials.

| Materials | $SiO_2$ | CaO | $Al_2O_3$ | MgO | Others |
|---|---|---|---|---|---|
| Lime | - | 86.2 | - | 0.68 | - |
| RHA | 88.09 | 0.98 | 1.25 | 0.34 | - |

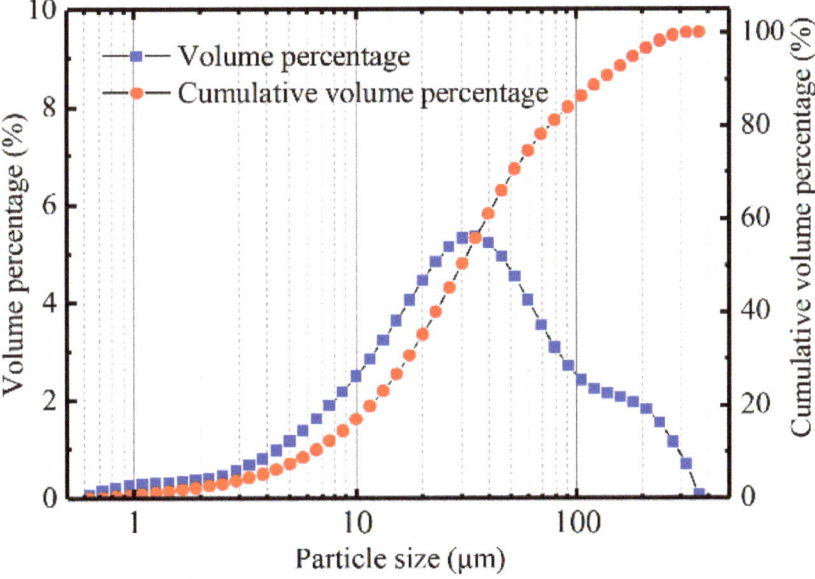

Figure 2. Particle size distribution of RHA.

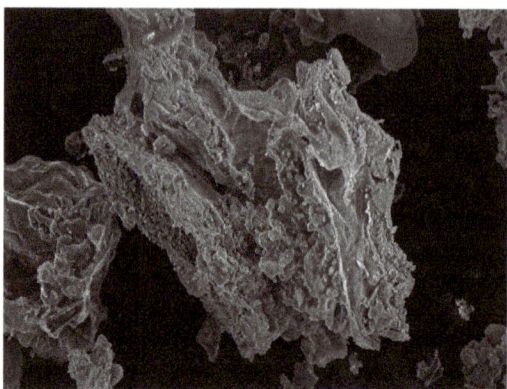

**Figure 3.** SEM diagram of RHA at 2000 times.

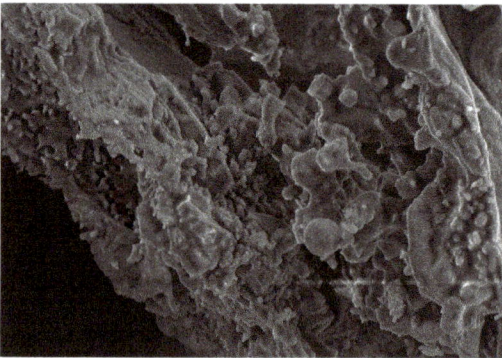

**Figure 4.** SEM diagram of RHA at 5000 times.

*3.2. Test Design*

Three types of soil, soft soil (silt clay), lime soil and rice husk ash–lime (RHA–lime) composite stabilized soil, are prepared for testing. The lime content in the lime soil is 5%; the content of lime and RHA in RHA–lime soil is 5% and 4%, respectively. The laboratory tests include a SWCC test, shear strength test, unconfined compressive strength (UCS) test and creep test. The purpose of the SWCC test and shear strength test is to confirm the model parameters $\alpha_1$, $K_n$ and $K_m$, and the UCS test is to determine the stress levels used in the creep test.

*3.3. Test Methods*

3.3.1. Preparation of Specimens

The specimens are remolded at 95% degree of compaction under the optimum moisture contents. The optimum moisture contents were obtained using the light compaction test in the specification of "Test Methods of Soils for Highway Engineering" [45]. The value of the optimum moisture of soft soil, lime soil and RHA–lime soil are 18%, 18.5% and 19%, respectively. The curing age of stabilized specimens is 28 days. The dimension of the specimen for each test is: 70 mm in diameter and 40 mm high for the SWCC test; 61.8 mm in diameter and 20 mm high for the direct shear test; and 70 mm in diameter and 140 mm high for the UCS test and creep test.

### 3.3.2. Shear Strength Test

Direct shear test [45] is used to determine the shear strength parameters. Three water contents (15%, 18% and 21%) are selected, and four specimens are tested for each water content. The shear strength test is carried out at 50 kPa, 100 kPa, 150 kPa and 200 kPa stress levels.

### 3.3.3. Soil–Water Characteristic Curve Test

The test is carried out in accordance with ASTM D5298 [46]. The filter paper and test pieces are put into the sealed tank together and stored at the constant temperature and humidity curing box for 7 days. Finally, the filter paper is weighed with a high-precision balance, and the matric suction of the sample is calculated according to the calibration curve of the filter paper. The calibration equation is according to the matric suction curve equation measured by ASTM:

$$\begin{cases} \log h_m = -0.0673 w_{fp} + 4.945 & w_{fp} < 47\% \\ \log h_m = -0.0229 w_{fp} + 2.909 & w_{fp} \geq 47\% \end{cases} \quad (24)$$

where $h_m$ is the matric suction, unit kPa, and $w_{fp}$ is the moisture content when the filter paper is balanced, unit %.

### 3.3.4. Unconfined Compressive Strength Test and Creep Test

The universal testing machine was used to carry out the UCS test and creep test. The loading rate of the UCS test is kept at 1 mm/min. The creep test adopts the form of a single-stage test, the loading time for each stage is 15,000 s. For each kind of soil, three moisture contents are set, which are 15%, 18% and 21%, respectively. Under the same water content of the same soil, five different stress levels are chosen. The results of the UCS test are used to determine the magnitude of the stress level. The five stress levels are approximately 20%, 40%, 60%, 80% and 95% of the ultimate compressive strength. In the test, an extensometer is used to monitor the deformation of materials. The measurement range is 60 mm in the middle of the specimen (140 mm high).

## 4. Results of Laboratory Soil Tests

### 4.1. SWCC Test Results

After obtaining the matric suction of the three kinds of soils under different water conditions, the Fredlund and Xing model [47] is used to determine the SWCC of each specimen. The Frendlund–Xing model is:

$$S = C(h_m) \times \left[ \frac{1}{\left\{ \ln \left[ \exp(1) + \left(\frac{h_m}{a_f}\right)^{b_f} \right] \right\}^{c_f}} \right] \quad (25)$$

where $C(h_m)$ is the correction factor, defined as

$$C(h_m) = \left[ 1 - \frac{\ln(1 + \frac{h_m}{h_r})}{\ln(1 + \frac{10^6}{h_r})} \right] \quad (26)$$

where $S$ is the degree of saturation; $h_r$, $a_f$, $b_f$ and $c_f$ are model coefficients. Figure 5 shows the fitting results, and Table 3 [48] shows the values of the model coefficients.

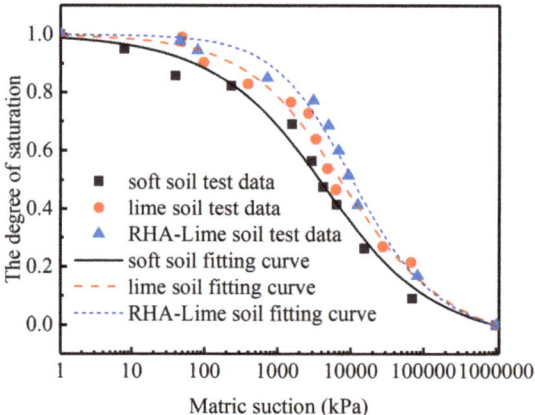

**Figure 5.** SWCC of the selected three kinds of soils.

**Table 3.** Results of the SWCC model coefficients.

| Materials | $h_r$ | $a_f$ | $b_f$ | $c_f$ |
|---|---|---|---|---|
| Soft soil | 3000 | 2000 | 0.509 | 1.581 |
| Lime soil | 3000 | 3059 | 0.589 | 1.192 |
| RHA–lime soil | 3000 | 7980 | 0.787 | 1.241 |

### 4.2. Shear Strength Test Results

Figure 6 shows the relationship between the shear strength and vertical pressure of each soil type. The corresponding shear strength parameters $c'$ and $\phi'$ are under a saturated state, and $\phi_1'$ of unsaturated soils under different water contents can be obtained from the test result.

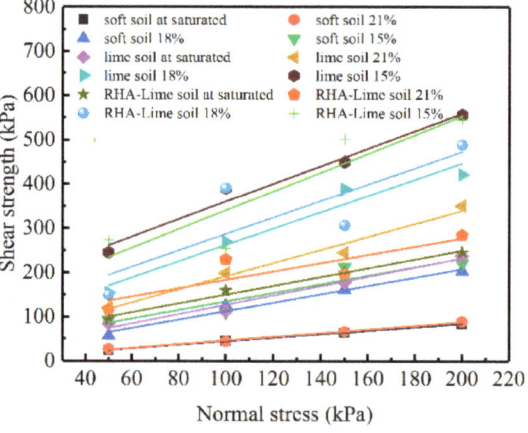

**Figure 6.** Relationship between shear strength and vertical pressure.

Based on the results of the SWCC and shear strength test, the matrix suction $h_m$ and shear strength parameters $c'$, $\phi'$ and $\phi_1'$ can be obtained. The saturation factor $f$ can then be obtained according to Equation (17). Finally, the model parameters $\alpha_1$, $K_n$ and $K_m$ of each soil under different water contents can be calculated by substituting the relevant parameters into Equations (21)–(23). The results are presented in Table 4.

Table 4. Values of the parameters in the MEC model.

| Materials | $w$ (%) | $\theta$ (%) | $f$ | $h_m$ (kPa) | $c'$ (kPa) | $\phi'$ (°) | $\phi'_1$ (°) | $\alpha_1$ | $K_n$ | $K_m$ |
|---|---|---|---|---|---|---|---|---|---|---|
| Soft soil | 21 | 36.4 | 1 | 124 | 3.885 | 21.25 | 22.88 | 0.172 | 4.75 | 20.92 |
|  | 18 | 31.2 | 1 | 993 | 3.885 | 21.25 | 43.54 | 0.344 | 4.22 | 130.86 |
|  | 15 | 26.0 | 1 | 2439 | 3.885 | 21.25 | 44.11 | 0.349 | 4.20 | 266.14 |
| Lime soil | 21 | 36.7 | 2.11 | 71 | 52.105 | 46.54 | 47.00 | 0.372 | 54.28 | 60.39 |
|  | 18 | 31.5 | 1 | 723 | 52.105 | 46.54 | 55.90 | 0.440 | 46.62 | 214.85 |
|  | 15 | 26.2 | 1 | 2276 | 52.105 | 46.54 | 61.54 | 0.479 | 40.59 | 489.82 |
| RHA–lime soil | 21 | 36.0 | 1 | 504 | 62.57 | 44.98 | 53.21 | 0.420 | 59.03 | 171.04 |
|  | 18 | 30.9 | 1 | 2214 | 62.57 | 44.98 | 61.73 | 0.480 | 48.48 | 529.32 |
|  | 15 | 25.8 | 1 | 5995 | 62.57 | 44.98 | 64.80 | 0.498 | 44.10 | 1088.41 |

### 4.3. Unconfined Compressive Strength Test Results

Before the creep test, the UCS test should be carried out in order to determine the loading magnitude at each level of the creep test. Figure 7 shows the UCS test curves and the change of the UCS with the water content. Comparisons among the three kinds of soils show that the strength of the RHA–lime soil is highest while that of the soft soil is smallest under the same water content. The effect of the degree of water on UCS of the three kinds of soils is also different. Among them, the soft soils are most affected by the moisture level. The lime soils and RHA–lime soils are less affected and their strength decreases slightly. It can be seen from Figure 8 that the failure patterns of soft and stabilized soils are different. For the soft soil specimen, it presents a lateral swelling failure pattern, while for lime soil and RHA–lime soil, it presents a conical failure pattern. The different failure patterns also make the stress–strain curves of soft soil and stabilized soil different, which can be seen in Figure 7a. With the addition of curing agent, the rising straight line becomes more apparent, and the slope becomes steeper. Meanwhile, the descending section of the stress–strain curve becomes steeper, and the strain becomes smaller during failure. These indicate that the brittleness of soft soil increases after stabilization. The reason for these phenomena is that the admixtures such as lime and RHA to the soil will produce chemical reactions and produce certain cementitious minerals, which will make the soil particles more closely connected and enhance the water stability of the soil.

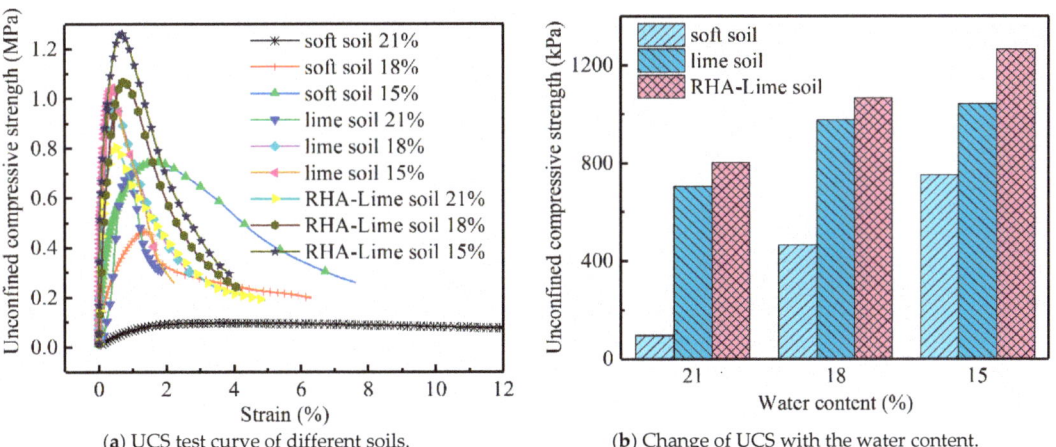

Figure 7. Unconfined compressive strength test results.

(a) The failure pattern of soft soil.  (b) The failure pattern of stabilized soils.

**Figure 8.** The failure patterns of soils.

### 5. Modeling of Creep Test Results of Unsaturated Soils

The creep tests provide a complete dataset for soft soils, lime soils and RHA–lime soils at three moisture conditions and four to five stress levels. To appraise the accuracy and reliability of the proposed MEC model, this model is applied to all of the testing cases. In addition, two existing creep models are selected for comparative analysis. The two models are the Mesri creep model (denoted as M model) and the improved unsaturated soil creep model proposed on the basis of the Mesri creep model (denoted as IM model). These two empirical semi-empirical creep models are described in Section 2.1 above.

*5.1. Determination of Creep Model Parameters*

The MEC model introduces the stress level term and moisture level term, which can predict the creep deformation behavior well under various water contents and various stress conditions. The MEC model is used to fit the soft soils, lime soils and RHA–lime soils under the conditions of multiple water contents and various stress levels. The fitting process is as follows: according to the results of the SWCC and shear strength tests shown in Table 4 above, the test data points are subjected to multivariate nonlinear fitting analysis using Equation (20). In the fitting analysis, all the test data points of the same soil sample are put together to obtain a set of fitting parameters. For example, for a soft soil sample, there are 15 sets of test data, which are five sets of test data under 21% water content at five stress levels D, five sets of data under 18% water content and five sets under 15% water content. Then the 15 sets of test data are fitted by the MEC model.

Figure 9 shows the fitting results, where diagrams a, b and c correspond to soft soil, lime-stabilized soil and RHA–lime-stabilized soil, respectively. The creep test data are shown in the black dots in Figure 9. The MEC model is fitted to the measured data using the software Origin. The fitting curves are shown in the red curves in Figure 9, and the values of the model coefficients are given in Table 5. By comparing the experimental data points and fitting curves in the figure, it can be found that the fitting curves are consistent with the experimental data. The goodness of fitting is demonstrated in Figure 9 with the R-squared values and Root Mean Square Error (RMSE). From the figures, it can be seen that the fitting accuracy of the three soils are very high, with the R-squared values of 0.9739, 0.9986 and 0.9823, respectively.

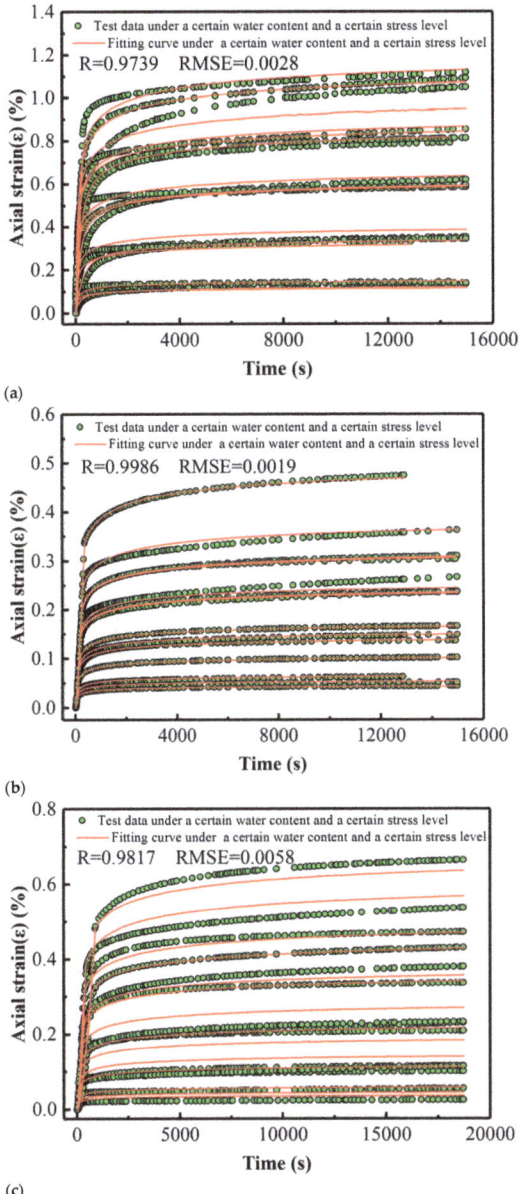

**Figure 9.** Fitting results of the MEC model under multi-moisture levels. (**a**) Soft soil (including 15 sets of test data, $D$ = 0.214, 0.427, 0.646, 0.834 and 0.945 when the water content is 21%, $D$ = 0.195, 0.391, 0.587, 0.754 and 0.950 when the water content is 18%, and $D$ = 0.172, 0.345, 0.552, 0.759 and 0.966 when the water content is 15%). (**b**) Lime soil (including 14 sets of test data, $D$ = 0.184, 0.368, 0.551, 0.735 and 0.919 when the water content is 21%, $D$ = 0.160, 0.346, 0.532 and 0.665 when the water content is 18%, and $D$ = 0.184, 0.368, 0.551, 0.735, 0.919 when the water content is 15%). (**c**) RHA–lime soil (including 14 sets of test data, $D$ = 0.184, 0.368, 0.551, 0.735 and 0.919 when the water content is 21%, $D$ = 0.195, 0.391, 0.587, 0.754 and 0.950 when the water content is 18%, and $D$ = 0.172, 0.345, 0.552, 0.759 and 0.966 when the water content is 15%).

Table 5. Values of the parameters of different creep models.

| | MEC Model Parameters | | | | | M Model Parameters | | | IM Model Parameters | | | |
|---|---|---|---|---|---|---|---|---|---|---|---|---|
| | $\varepsilon_0$ | $\rho$ | $\beta$ | $m$ | $n$ | $(\frac{2}{E_u/S_u})_1$ | $(R_f)_1$ | $\lambda$ | $(\frac{2}{E_u/S_u})_1$ | $\lambda$ | $a$ | $b$ |
| Soft soil 21% | | | | | | 0.523 | 0.519 | 0.084 | 0.523 | 0.084 | | |
| Soft soil 18% | 5.948 | 0.012 | 0.318 | 1.706 | −1.327 | 0.648 | 0.435 | 0.056 | 0.276 | 0.056 | −0.010 | 0.540 |
| Soft soil 15% | | | | | | 0.751 | 0.309 | 0.066 | 0.751 | 0.066 | | |
| Lime soil 21% | | | | | | 0.169 | 0.576 | 0.061 | 0.169 | 0.061 | | |
| Lime soil 18% | 19.047 | 0.012 | 0.340 | 1.225 | −0.021 | 0.289 | 0.509 | 0.052 | 0.289 | 0.052 | −0.008 | 0.572 |
| Lime soil 15% | | | | | | 0.326 | 0.391 | 0.079 | 0.326 | 0.079 | | |
| RHA–lime soil 21% | | | | | | 0.158 | 0.868 | 0.051 | 0.158 | 0.051 | | |
| RHA–lime soil 18% | 58.449 | 0.013 | 0.311 | 1.674 | −0.253 | 0.239 | 0.768 | 0.073 | 0.239 | 0.073 | −0.009 | 0.938 |
| RHA–lime soil 15% | | | | | | 0.478 | 0.374 | 0.083 | 0.478 | 0.083 | | |

The parameter-solving processes for the M model and IM model are different from that of MEC model. Its parameters cannot be directly fitted; instead, they need to be solved according to the specific solution method of parameters. The process of solving the parameters of the M model is as follows: take $t_1 = 1$ h and draw the $\frac{\varepsilon}{D} - \varepsilon$ diagram when $t_1 = 1$ h, as shown in Figure 10, based on which the values $\left(\frac{2}{E_u/S_u}\right)_1$ and $(R_f)_1$ are obtained. Specifically, according to Equation (5), it is known that $\left(\frac{2}{E_u/S_u}\right)_1$ and $(R_f)_1$ are the intercept and slope of the $\frac{\varepsilon}{D} - \varepsilon$ curve respectively. Therefore, according to Figure 10, the $\left(\frac{2}{E_u/S_u}\right)_1$ and $(R_f)_1$ values of soft soil under the condition of 18% water content can be calculated. The parameters about $\left(\frac{2}{E_u/S_u}\right)_1$ and $(R_f)_1$ of other soils under different moisture contents are obtained in the same way as above. Then plot the $\lg\varepsilon - \lg t$ diagram, as shown in Figure 11, based on which the value of $\lambda$ is gained. $\lambda$ is the mean slope of $\lg\varepsilon - \lg t$ curves under different stress levels. The process to obtain the parameters for the IM model is similar. The results of $\left(\frac{2}{E_u/S_u}\right)_1$ and $\lambda$ are the same as those in the M model, but the difference is that the parameter $(R_f)_1$ will be replaced by a and b. According to Equation (11), the relationship between $(R_f)_1$ and $h_m/P_a$ can be obtained, from which a and b are the slope and intercept of the curve, respectively. Figure 12 shows the $R_f - h_m/P_a$ curve of soft soil, from which parameters a and b in the IM model of soft soil can be obtained. The method for obtaining a and b parameters of other types of soil is the same as above. Finally, the parameters of these two models are given in Table 5.

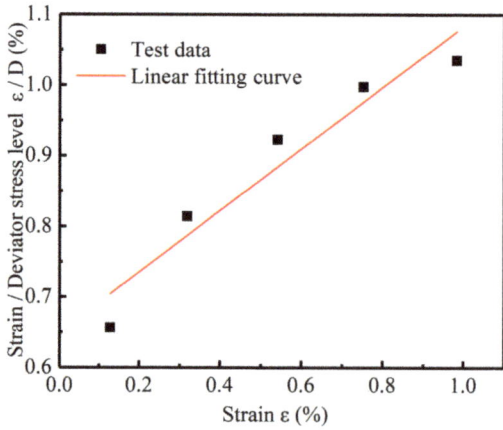

Figure 10. The $\frac{\varepsilon}{D} - \varepsilon$ diagram when $t_1 = 1$ h of soft soil at 18% water content.

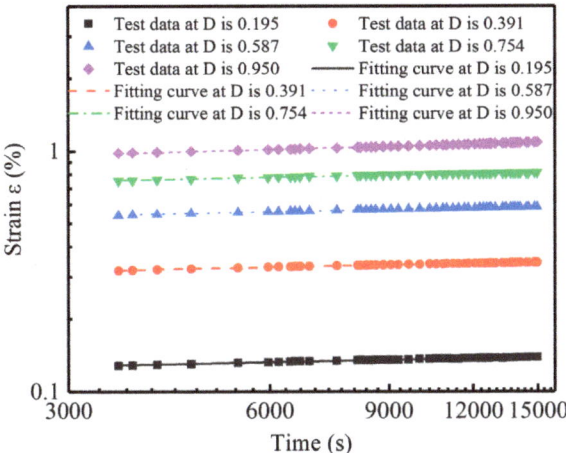

**Figure 11.** The lg $\varepsilon$ – lg $t$ curve of the soft soil at 18% water content.

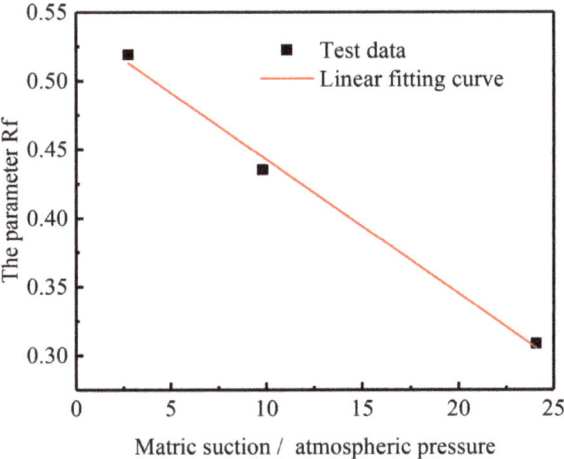

**Figure 12.** The $R_f - \frac{h_m}{P_a}$ curve of the soft soil.

*5.2. Comparison of Different Creep Models*

To compare the fitting effects of different creep models, the prediction results of the three models and the test data are plotted in the same graph, with some of the results illustrated in Figure 13. Based on the figures, the prediction results of the three models are in good agreement with the test data, signifying that the MEC model has as good prediction accuracy as the other two classical models. Moreover, it should be emphasized that the most prominent difference between the MEC model and the other two models lies in the ability to predict at multiple moisture conditions by one set of parameters. This conclusion can be directly obtained from Table 5. From Table 5, it can be found that for the same type of soil, the MEC model only has one set of model parameters, while the M and IM models need different sets of parameters at different moisture levels. In other words, once the model parameters are determined, the MEC model could predict the creep deformation behaviors of a soil at any water content. However, for the M and IM models, the process of model determination should be repeated for each water content. For instance, to predict the deformation of soft soil at 10% water content, it is necessary to undertake indoor creep tests and repeat the previous analysis to determine the model parameters at 10% water content when using the M or IM model. However, for the MEC model, the deformation of soft soil with 10% water content can be directly predicted by the MEC model parameters of soft soil in Table 5.

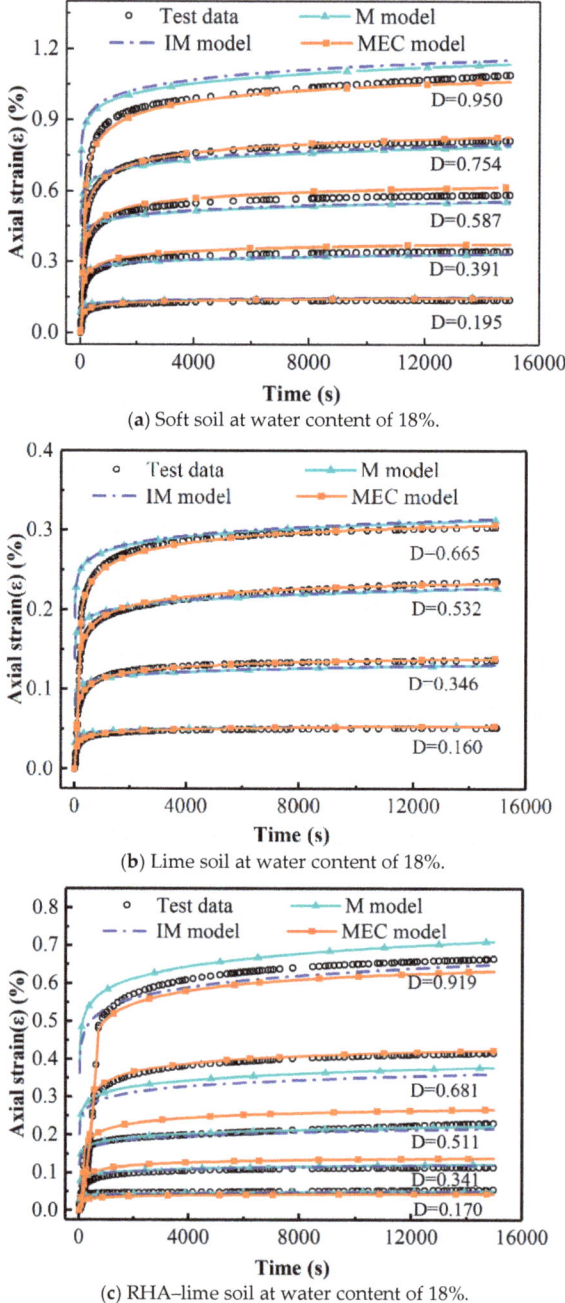

(a) Soft soil at water content of 18%.

(b) Lime soil at water content of 18%.

(c) RHA–lime soil at water content of 18%.

**Figure 13.** Fitting results of different creep models.

In the process of engineering practice, geotechnical structures such as foundation or subgrade are deformed under the condition of changing stress and moisture, and there is a coupling effect between these two factors. Therefore, in order to aim to predict the creep behavior of soils accurately in engineering practice, the stress and moisture levels need to be considered at the same time. That is to say, an appropriate mode should predict the creep deformation behavior under different water contents and different stress conditions. It is obvious that the MEC model is more convenient and suitable for this purpose. That is to say, compared to the other two classical models, the MEC model not only has good prediction accuracy, but also has greater convenience and practicability. It can be easily used to predict the deformation of unsaturated soil under any stress level and water content condition.

## 6. Model Implementation for Predicting Subgrade Deformation

To analyze the deformation of the three types of soil under various water content conditions, in the finite element analysis, the MEC model is implemented to compute the creep deformation of a typical flexible pavement structure through a simplifying method [39,49]. In this way, the curing effect of the two kinds of solidification materials can be visualized, which provides a certain reference for the material selection of engineering in a soft soil area. The specific process and analysis results are as follows.

The pavement structure model is shown in Figure 14. The viscoelastic model is used for the surface layer and the elastic model is employed for the other layers. Among them, the resilient modulus of the subgrade is calculated according to the approximate formula (27) between UCS and resilient modulus of the subgrade [50].

$$M_r(\text{MPa}) = 0.124 q_u(\text{kPa}) + 68.8 \qquad (27)$$

where $M_r$ is resilient modulus of subgrade; $q_u$ is UCS.

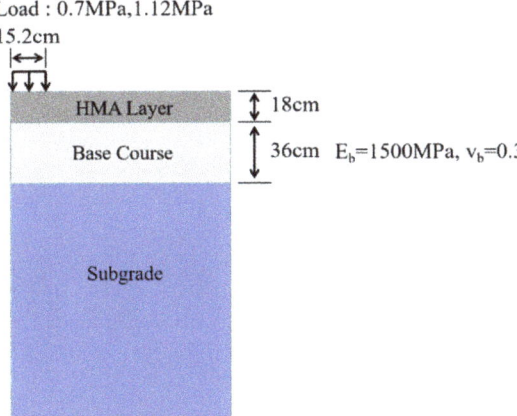

**Figure 14.** Illustration of pavement structure model and corresponding parameters.

In the FE analysis, two common traffic loading levels are selected as 0.7 MPa and 1.12 MPa [51] to compute the stress distributions of the subgrade. Then, the MEC model is applied to calculate the vertical compressive creep strain at corresponding locations for one year. Figure 15 presents the distributions of the creep strains of soft and stabilized soils at different moisture levels. Then the total creep deformation was computed by the multi-layered incremental approach, as shown in Equation (28).

$$\delta_s(t) = \int_0^h \varepsilon_0 e^{-(\rho/t)^\beta} \left(\frac{\sqrt{J_2}}{p_a}\right)^m \left(\frac{\alpha I_1 + K + K_m}{p_a}\right)^n (p_a)^{-(m+n)} dz \qquad (28)$$

where $\delta_s$ is the creep deformations of subgrade; $t$ is the creep time; and $h$ is depth of the effective deforming zone of the subgrade.

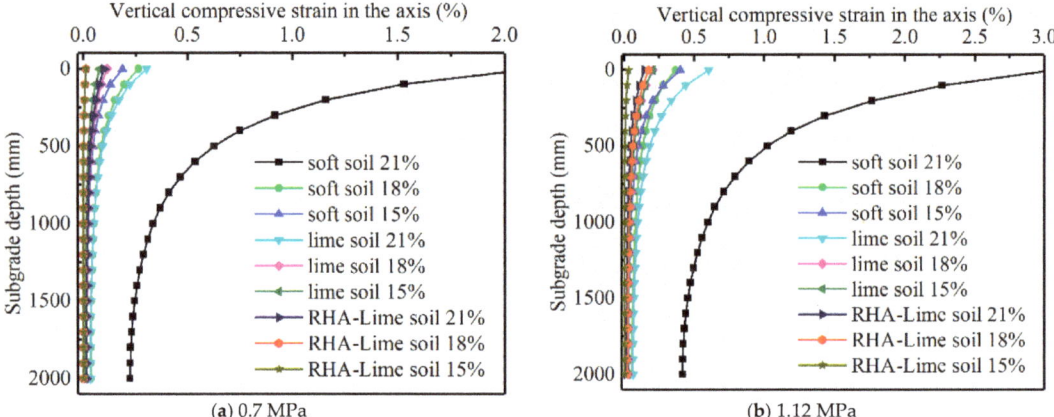

**Figure 15.** Distribution of vertical compressive creep strain in the subgrade.

The depth of the effective deforming zone of a soil foundation is set as the location where the stress is 1/10 of the gravity stress [52]. The creep deformations of the three types of soils under different stress conditions with different water contents are obtained. Figure 16 present the results.

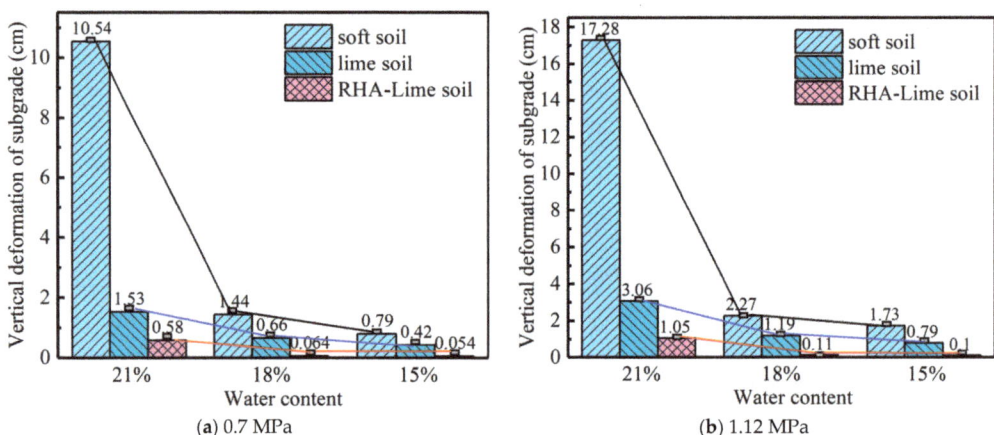

**Figure 16.** The creep deformation of the subgrade in the vertical direction.

From Figure 16, it can be seen that under the same load and moisture level, the deformation is from small to large, in the order of RHA–lime-stabilized soil, lime soil and soft soil. For the same kind of soil, water content has a great influence on its deformation, and the deformation in the wet state is much larger than that in the dry state. The deformation of soft soils is most affected by the moisture level. Through the analysis results of two stress levels, it can be seen that the deformation of the subgrade under a heavy load is obviously larger than that under standard vehicle load. In addition, the deformation of the soft soils is much larger than that of stabilized soil when the same load is applied. Among the three types of soils, the RHA–lime composite soils deform least. In other words, the creep deformation of the soft soil subgrade can be effectively reduced by chemical stabilization, and the effect of RHA–lime composite stabilization is better than that of only adding lime. The reason for these phenomena is that the addition of the curing agent changes the soil structure, which make the deformation behaviors of the soil change.

## 7. Conclusions

This study proposed an improved creep model and validated for soft and stabilized soils, which can characterize the deformation behavior of unsaturated soils very well. The main findings of this work are as follows:

(1) The MEC model takes into account the stress dependence based on mechanical principles, and incorporates moisture sensitivity using matric suction and shear strength parameters. This formulation is intended to characterize the creep deformation behavior of unsaturated soils under arbitrary water content and arbitrary stress condition.

(2) The deformation of unsaturated soils was analyzed by the MEC model under various stress and moisture conditions. The results show that the predicted results of the MEC model are consistent with the experimental data with very high R-squared values.

(3) Compared with the classical unsaturated soil creep models, the MEC model only needs one set of parameters for different stress levels and moisture conditions, while the classical models (like the Mesri and improved Mesri models) require a different set of parameters when the water content is changed. In addition, the MEC model agrees with the experimental data better for stabilized soils, and provides better accuracy in predicting creep deformations at high stress levels.

(4) In the FE analysis, the MEC model is implemented to analyze the creep behavior of subgrade soils. Loading level and moisture have a great effect on the deformation of soil foundation, especially soft soils. Under heavy loading and a wet state, the deformation of soft soil increases rapidly.

(5) After stabilization, the deformation of the soil foundation is obviously reduced. Under the same load and moisture level, the deformation of soft soil is largest, followed by lime soil and RHA–lime-stabilized soil, respectively.

**Author Contributions:** Conceptualization, X.J. and X.L.; methodology, X.J.; software, X.J.; validation, Z.H. and X.L.; formal analysis, X.J.; investigation, X.J.; resources, X.L.; data curation, X.J.; writing—original draft preparation, X.J.; writing—review and editing, X.L.; visualization, Z.H.; supervision, Z.H.; project administration, X.L.; funding acquisition, X.L. All authors have read and agreed to the published version of the manuscript.

**Funding:** This research was funded by the National Key R&D Program of China (No. 2019YFE0117600) and the Zhejiang Provincial Natural Science Foundation of China (No. LZ21E080002).

**Institutional Review Board Statement:** Not applicable.

**Informed Consent Statement:** Not applicable.

**Data Availability Statement:** Not applicable.

**Acknowledgments:** This research was supported by the National Key R&D Program of China (No. 2019YFE0117600) and the Zhejiang Provincial Natural Science Foundation of China (No. LZ21E080002).

**Conflicts of Interest:** The authors declare no conflict of interest.

## References

1. Feng, S.J.; Lu, S.F.; Shi, Z.M.; Shui, W.H. Densification of loosely deposited soft soils using the combined consolidation method. *Eng. Geol.* **2014**, *181*, 169–179. [CrossRef]
2. Cai, Y.; Chen, Y.; Cao, Z.; Ren, C. A combined method to predict the long-term settlements of roads on soft soil under cyclic traffic loadings. *Acta Geotech.* **2018**, *13*, 1215–1226. [CrossRef]
3. Araújo, G.L.; Palmeira, E.M.; Macêdo, Í.L. Comparisons between predicted and observed behaviour of a geosynthetic reinforced abutment on soft soil. *Eng. Geol.* **2012**, *147*, 101–113. [CrossRef]
4. Arulrajah, A.; Bo, M.W.; Leong, M.; Disfani, M.M. Piezometer measurements of prefabricated vertical drain improvement of soft soils under land reclamation fill. *Eng. Geol.* **2013**, *162*, 33–42. [CrossRef]

5. Consoli, N.C.; Winter, D.; Rilho, A.S.; Festugato, L.; dos Santos Teixeira, B. A testing procedure for predicting strength in artificially cemented soft soils. *Eng. Geol.* **2015**, *195*, 327–334. [CrossRef]
6. Consoli, N.C.; Rosa, D.A.; Cruz, R.C.; Rosa, A.D. Water content, porosity and cement content as parameters controlling strength of artificially cemented silty soil. *Eng. Geol.* **2011**, *122*, 328–333. [CrossRef]
7. Guidobaldi, G.; Cambi, C.; Cecconi, M.; Deneele, D.; Paris, M.; Russo, G.; Vitale, E. Multi-scale analysis of the mechanical improvement induced by lime addition on a pyroclastic soil. *Eng. Geol.* **2017**, *221*, 193–201. [CrossRef]
8. Yilmaz, Y. Compaction and strength characteristics of fly ash and fiber amended clayey soil. *Eng. Geol.* **2015**, *188*, 168–177. [CrossRef]
9. Sargent, P.; Hughes, P.N.; Rouainia, M.; White, M.L. The use of alkali activated waste binders in enhancing the mechanical properties and durability of soft alluvial soils. *Eng. Geol.* **2013**, *152*, 96–108. [CrossRef]
10. Yi, Y.; Zheng, X.; Liu, S.; Al-Tabbaa, A. Comparison of reactive magnesia- and carbide slag-activated ground granulated blastfurnace slag and portland cement for stabilisation of a natural soil. *Appl. Clay Sci.* **2015**, *111*, 21–26. [CrossRef]
11. Zhang, T.; Liu, S.; Cai, G.; Puppala, A.J. Experimental investigation of thermal and mechanical properties of lignin treated silt. *Eng. Geol.* **2015**, *196*, 1–11. [CrossRef]
12. Fatehi, H.; Abtahi, S.M.; Hashemolhosseini, H.; Hejazi, S.M. A novel study on using protein based biopolymers in soil strengthening. *Constr. Build. Mater.* **2018**, *167*, 813–821. [CrossRef]
13. Kharade, A.S. Waste product 'bagasse ash' from sugar industry can be used as stabilizing material for expansive soils. *Int. J. Res. Eng. Technol.* **2014**, *3*, 506–512.
14. Ilie, N.M.; Crcu, A.P.; Nagy, A.C.; Ciubotaru, V.C.; Kisfaludi-Bak, Z. Comparative study on soil stabilization with polyethylene waste materials and binders. *Procedia Eng.* **2017**, *181*, 444–451. [CrossRef]
15. Rahgozar, M.A.; Saberian, M.; Li, J. Soil stabilization with non-conventional eco-friendly agricultural waste materials: An experimental study. *Transp. Geotech.* **2018**, *14*, 52–60. [CrossRef]
16. Liu, S.Y.; Zhang, T.; Cai, G.Q. Research on Technology and Engineering Application of Silt Subgrade Solidified by Lignin-based Industrial By-product. *China J. Highw. Transp.* **2018**, *31*, 1–11.
17. Gu, F.; Moraes, R.; Chen, C.; Yin, F.; Watson, D.; Taylor, A. Effects of Additional Antistrip Additives on Durability and Moisture Susceptibility of Granite-Based Open-Graded Friction Course. *J. Mater. Civ. Eng.* **2021**, *33*, 04021245. [CrossRef]
18. Oliveira, P.V.; Correia, A.A.; Garcia, M.R. Effect of stress level and binder composition on secondary compression of an artificially stabilized soil. *J. Geotech. Geoenviron. Eng.* **2013**, *139*, 810–820. [CrossRef]
19. Oliveira, P.V.; Correia, A.A.; Garcia, M.R. Effect of organic matter content and curing conditions on the creep behavior of an artificially stabilized soil. *J. Mater. Civ. Eng.* **2014**, *24*, 868–875. [CrossRef]
20. Sivasithamparam, N.; Karstunen, M.; Bonnier, P. Modelling creep behaviour of anisotropic soft soils. *Comput. Geotech.* **2015**, *69*, 46–57. [CrossRef]
21. Augustesen, A.; Liingaard, M.; Lade, P.V. Evaluation of time-dependent behavior of soils. *Int. J. Geomech.* **2004**, *4*, 137–156. [CrossRef]
22. Taylor, D.W. *Research on Consolidation of Clays*; Massachusetts Institute of Technology: Cambridge, MA, USA, 1942.
23. Singh, A.; Mitchell, J.K. General stress-strain-time function for soils. *ASCE Soil Mech. Found. Div. J.* **1968**, *94*, 21–46. [CrossRef]
24. Mesri, G.; Febrescordero, E.; Shields, D.R.; Castro, A. Shear stress-strain-time behaviour of clays. *Géotechnique* **1981**, *32*, 407–411. [CrossRef]
25. Lin, H.D.; Wang, C.C. Stress-strain-time function of clay. *J. Geotech. Geoenviron. Eng.* **1998**, *124*, 289–296. [CrossRef]
26. Lu, P.; Zeng, J.; Sheng, Q. Creep tests on soft clay and its empirical models. *Rock Soil Mech.* **2008**, *29*, 1041.
27. Wang, S.; Qi, J.; Yin, Z.; Zhang, J.; Wei, M. A simple rheological element based creep model for frozen soils. *Cold Reg. Sci. Technol.* **2014**, *106*, 47–54. [CrossRef]
28. Ma, W.B.; Rao, Q.H.; Peng, L.; Guo, S.C.; Kang, F. Shear creep parameters of simulative soil for deep-sea sediment. *J. Cent. South Univ.* **2014**, *21*, 4682–4689. [CrossRef]
29. Zhao, M.H.; Xiao, Y.; Chen, C.F. Laboratory experiment of the rheological property of soft clay and the improved xiyuan model. *J. Hunan Univ.* **2004**, *31*, 48–52.
30. Rajesh, S.; Choudhary, K.; Chandra, S. A generalized model for geosynthetic reinforced railway tracks resting on soft clays. *Int. J. Numer. Anal. Methods Geomech.* **2015**, *39*, 310–326. [CrossRef]
31. Huang, W.; Liu, D.Y.; Zhao, B.Y.; Feng, Y.B.; Xia, Y.C. Study on the rheological properties and constitutive model of shenzhen mucky soft soil. *J. Eng. Sci. Technol. Rev.* **2014**, *7*, 55–61.
32. Perzyna, P. The constitutive equations for work-hardening and rate sensitive plastic materials. *Proc. Vib. Probl.* **1963**, *4*, 281–290.
33. Yin, J.H.; Graham, J. Viscous-elastic-plastic modelling of one-dimensional time-dependent behaviour. *Can. Geotech. J* **1989**, *26*, 199–209. [CrossRef]
34. Yin, J.H.; Graham, J. Elastic viscoplastic modelling of the time-dependent stress-strain behaviour of soils. *Can. Geotech. J.* **2011**, *36*, 736–745. [CrossRef]
35. Sheng, D. Review of fundamental principles in modelling unsaturated soil behaviour. *Comput. Geotech.* **2011**, *38*, 757–776. [CrossRef]
36. Gens, A.; Sánchez, M.; Sheng, D. On constitutive modelling of unsaturated soils. *Acta Geotech.* **2006**, *1*, 137. [CrossRef]
37. Mesri, G. Time and stress-compressibility interrelationship. *J. Geotech. Eng. Div.* **1977**, *103*, 417–430. [CrossRef]

38. Tseng, K.H.; Lytton, R.L. *Prediction of Permanent Deformation in Flexible Pavement Materials*; ASTM International: West Conshohocken, PA, USA, 1989.
39. Gu, F.; Zhang, Y.; Droddy, C.V.; Rong, L.; Lytton, R.L. Development of a new mechanistic empirical rutting model for unbound granular material. *J. Mater. Civ. Eng.* **2016**, *28*, 04016051. [CrossRef]
40. Lai, X.L.; Wang, S.M.; Qin, H.B.; Liu, X.F. Unsaturated creep model of the sliding zone soils of qianjiangping landslide in three gorges and its empirical models. In Proceedings of the International Symposium on Unsaturated Soil Mechanics & Deep Geological Nuclear Waste Disposal, Shanghai, China, 24 August 2009.
41. Lai, X.L.; Wang, S.M.; Ye, W.M.; Cui, Y.J. Experimental investigation on the creep behavior of an unsaturated clay. *Can. Geotech. J.* **2014**, *51*, 621–628. [CrossRef]
42. Lytton, R.L. Foundations and pavements on unsaturated soils. In Proceedings of the International Conference on Unsaturated Soils, Unsat 95, Paris, France, 6–8 September 1996.
43. Lu, N.; Likos, W.J. *Unsaturated Soil Mechanics*; Wiley: Hoboken, NJ, USA, 2004.
44. Fredlund, D.G.; Morgenstern, N.R.; Widger, R.A. The shear strength of unsaturated soils. *Can. Geotech. J.* **1978**, *15*, 313–321. [CrossRef]
45. Ministry of Transport of the People's Republic of China. *The Methods of Soils for Highway Engineering*; JTG E40-2007; China Communications Press: Beijing, China, 2007; pp. 130–138, 194–217. (In Chinese)
46. ASTM D5298-92. Standard test method for measurement of soil potential(suction) using filter paper. In *Annual Book of ASTM Standards*; American Society of Testing and Materials: Philadelphia, PA, USA, 1992; pp. 156–161.
47. Fredlund, D.G.; Xing, A. Equations for the soil-water characteristic curve. *Can. Geotech. J.* **1994**, *31*, 521–532. [CrossRef]
48. Jiang, X.; Huang, Z.; Ma, F.; Luo, X. Analysis of strength development and soil–water characteristics of rice husk ash–lime stabilized soft soil. *Materials* **2019**, *12*, 3873. [CrossRef] [PubMed]
49. Omairey, E.L.; Gu, F.; Zhang, Y. An equation-based multiphysics modelling framework for oxidative ageing of asphalt pavements. *J. Clean. Prod.* **2020**, *280*, 124401. [CrossRef]
50. American Association of State Highway and Transportation Officials. *Mechanistic-Empirical Pavement Design Guide: A Manual of Practice*; American Association of State Highway and Transportation Officials (AASHTO): Washington, DC, USA, 2008; p. 1000.
51. Hu, X. Measurements of the Distribution of the Vehicle Tire and the Response Stress in the Asphalt Pavement. Ph.D. Thesis, Department of the Road and Airport Engineering, Tongji University, Shanghai, China, 2003. (In Chinese).
52. Zhang, L. *Urban Road Engineering*; People's Communications Publishing House: Beijing, China, 2008. (In Chinese)

MDPI
St. Alban-Anlage 66
4052 Basel
Switzerland
www.mdpi.com

*Materials* Editorial Office
E-mail: materials@mdpi.com
www.mdpi.com/journal/materials

Disclaimer/Publisher's Note: The statements, opinions and data contained in all publications are solely those of the individual author(s) and contributor(s) and not of MDPI and/or the editor(s). MDPI and/or the editor(s) disclaim responsibility for any injury to people or property resulting from any ideas, methods, instructions or products referred to in the content.